Adipose Stem Cells and Regenerative Medicine

Yves-Gérard Illouz • Aris Sterodimas
Editors

Adipose Stem Cells and Regenerative Medicine

Springer

Editors
Prof. Yves-Gérard Illouz
Plastic Surgery
Department, Saint Louis Hospital
Avenue Claude-Vellefaux
Paris
France
yillouz2000@hotmail.com

Prof. Aris Sterodimas
Plastic and Reconstructive Surgery
Department, IASO General Hospital
264 Mesogeion Avenue
Athens
Greece
aris@sterodimas.com

ISBN 978-3-642-20011-3 e-ISBN 978-3-642-20012-0
DOI 10.1007/978-3-642-20012-0
Springer Heidelberg Dordrecht London New York

Library of Congress Control Number: 2011929933

Cover design: eStudioCalamar, Figueres/Berlin

Printed on acid-free paper

Springer is part of Springer Science+Business Media (www.springer.com)

This book is dedicated to my parents,
Mr Konstantinos Sterodimas and Mrs Aikaterini Eleftheriou.

My thanks to my wife Beatriz and my daughter Kate, whose patience and support were endless in order to complete this book project.

Aris Sterodimas

Because I think that adipose stem cells will be the most efficient therapy in the future, this book is dedicated to all the people, scientists or doctors working in that field.

Yves-Gérard Illouz

Foreword

It has been said that man is entering a new era of discoveries, the truly first revolution since the discovery and harnessing of fire. In a word: biotechnology. This vast and growing field of knowledge presents to us, on a daily basis, inventions that only a few years ago would have seemed to be taken out of a story by Jules Verne. The correlated sciences that surround biotechnology stretch from engineering to computers, from biology to nanotechnology. As physicians, we are challenged with new concepts and terminology that already are part of our lexicon. It is in this spirit that my fellow plastic surgeons, Drs. Illouz and Sterodimas, present to the medical reader the book *Adipose Stem Cells and Regenerative Medicine*.

Dr. Illouz is recognized as a pioneer in liposuction, and it is natural that, with his curious mind, he follows his interest in adipose tissues, immersing into the ever-expanding discipline of cell culture and derivation. The promise of perpetuating and renovating whole tissues is leading us into a new era of medical practice, called regenerative medicine. Together with Dr. Sterodimas, a younger colleague, recently graduated from our school, Dr. Illouz stimulates our intellect, opening new vistas to an unknown future.

The authors are to be congratulated for this pioneering publication.

Ivo Pitanguy

Preface

In the last decades, investments in basic research have yielded extensive knowledge about the many and complex processes involved in the development of an organism. Since human pluripotent stem cells were first isolated, research on stem cells has received much public attention, both because of its extraordinary promise and because of relevant legal and ethical issues. Regenerative medicine is classified into cell therapy that does not require a scaffold and tissue engineering that requires a scaffold and bioactive substances such as growth factors; though both need adult stem cells. Bioengineers, life scientists, and physicians across all specialties are synergistically coupling expertise in areas such as cell culture technology, tissue transfer, cell differentiation, angiogenesis, computer modeling, and polymer chemistry to use adipose tissue as a base for regenerative medicine.

On November 2008, during a meeting of Yves-Gérard Illouz and Aris Sterodimas in Rio de Janeiro, Brazil, the idea to write a book focusing on adipose stem cells and their role in regenerative medicine was born. This book is the most up-to-date text on regenerative medicine based on adipose stem cells. There are 24 chapters by international experts with the newest techniques explained in detail. Bioengineers, life scientists, and physicians from Brazil, Canada, France, Germany, Greece, Indonesia, Israel, Italy, Korea, Japan, Switzerland, Turkey, United Arabic Emirates, UK, and USA joined forces in order to ensure the reader is provided information both about the basic biology of adipose stem cells, and their therapeutic potential. This book contains chapters focused on the applications of adipose stem cells on specific fields, like cardiology, orthopedic surgery, neurology, urology, otolaryngology, plastic and reconstructive surgery, organ transplantation, and dentistry. Also included are chapters on adipose stem cells as therapeutic delivery tools for gene therapy, in the field of pharmacology and obesity. The science represented in this book focuses exclusively on scientific publications, paying extreme attention to the safety and ethics considerations for developing adipose stem cell-based therapies.

The compilation of this book on the latest advances in the field of adipose stem cell research required the participation of many individuals. We wish to sincerely thank all the authors for contributing to this book. We would like to acknowledge, in particular, the contribution of Mrs. Mahalakshmi and the Springer team for the meticulous type-editing of the chapter in this book and Mrs. Ellen Blasig from Springer-Verlag for her continuous support since we embarked on this project in January 2010.

December 2010

Yves-Gérard Illouz
Aris Sterodimas

Contents

1 **Cellular and Molecular Aspects of Adipose Tissue** 1
Tahsin Murad Aktan, Selcuk Duman, and Bulent Cihantimur

2 **Isolation of Stem Cells from Human Adipose Tissue: Technique, Problems, and Pearls** 13
Ithamar Nogueira Stocchero and Guilherme Flosi Stocchero

3 **Principles of Liposuction** 19
Juarez M. Avelar

4 **Clinical Applications of Liposuction** 31
Melvin A. Shiffman

5 **Principles of Autologous Fat Transplantation** 45
Luiz S. Toledo

6 **Clinical Applications of Autologous Fat Transplantation** 57
Luiz Haroldo Pereira, Beatriz Nicaretta, and Yves-Gérard Illouz

7 **Adipose Stem Cells: From Liposuction to Adipose Tissue Engineering** 67
Kotaro Yoshimura, Hitomi Eto, Harunosuke Kato, Kentaro Doi, and Hirotaka Suga

8 **Storage of Adipose Stem Cells** 83
Giorgio Bronz and Gianni Soldati

9 **Engineered Scaffolds and Matrices: Tailored Biomaterials for Adipose Stem Cell Engineering** 93
Karen J.L. Burg, Nichole Myers Cavin, and Katherine Neser

10 **Fundamentals and Principles of Biomolecules in Adipose Stem Cell Engineering** 103
Michael Magarakis, Sachin M. Shridharani, Navin K. Singh, and Richard J. Redett

11 **Gene Therapy Used for Adipose Stem Cell Engineering Applications** 111
Ulrich R. Goessler

12 Stem Cells from Human Adipose Tissue: A New Tool for Pharmacological Studies and for Clinical Applications 121
Claude A. Dechesne and Christian Dani

13 Role of Adipose Stem Cells Therapy in Obesity 133
Yves-Gérard Illouz, Aris Sterodimas, and Alexandra Condé Green

14 Stem Cell Based Cardioregeneration and Adipose Tissue 141
Atta Behfar and Andre Terzic

15 Adipose Stem Cell Engineering: Characterization and Current Application in Neurological Tissue Repair 155
Guangfan Chi and Youngsook Son

16 Adipose Stem Cell Engineering: Clinical Applications in Plastic and Reconstructive Surgery 165
Aris Sterodimas

17 Orthopedic Use of Adipose-Derived Stem Cells 181
Josh E. Schroeder, Shaul Beyth, and Meir (Iri) Liebergall

18 Adipose-Derived Stem Cells: Characterization and Application in Urology 193
Ching-Shwun Lin and Tom F. Lue

19 Adipose Stem Cell Engineering: Characterization and Current Application in Otolaryngology 209
Suzanne N. King and Susan Thibeault

20 Adipose Stem Cell Technologies for Tissue Regeneration in Dentistry 221
Andrea Cochis and Lia Rimondini

21 Adipose Stem Cells, Tissue Engineering, and Solid Organ Transplantation and Regeneration 229
Benoit Labbé, Valérie Trottier, Maryse Proulx, Caroline Vincent, and Julie Fradette

22 Industrial Approaches to Adipose Stem Cells Engineering 245
Filippo Boriani, Robert Warr, and Raimondo Ascione

23 Future Research in Adipose Stem Cell Engineering 257
Jeanne Adiwinata Pawitan

24 Conclusions and Future Directions 273
Yves-Gérard Illouz and Aris Sterodimas

Index 277

Cellular and Molecular Aspects of Adipose Tissue

1

Tahsin Murad Aktan, Selcuk Duman, and Bulent Cihantimur

Contents

1.1 **Introduction** 1

1.2 **Cell Biology of Adipose Tissue** 3
1.2.1 WAT Adipocyte Properties................... 3
1.2.2 BAT Adipocyte Properties 3
1.2.3 Extracellular Matrix (ECM).................. 4

1.3 **Secretions of Adipose Tissue Cells** 5

1.4 **Stem Cell Concept in Adipose Tissue**......... 8

1.5 **Adipose Stem Cells and Tissue Engineering** ... 9

1.6 **Conclusion**............................... 10

References.................................. 10

T.M. Aktan (✉) • S. Duman
Histology Embryology Department, Meram Medical Faculty, Selcuk University, Konya, Turkey
e-mail: muradaktan@gmail.com; dumanselcuk@gmail.com

B. Cihantimur
Aesthetic Plastic Surgery Division, Aesthetic Plastic Surgery Branch Center, Estetik Plastik Cerrahi Dal Merkezi, Dikkaldırım Cad. No: 25 Çekirge, Bursa, Turkey
e-mail: tct@tctmedical.net

1.1 Introduction

Adipose tissue is a biological entity that awaits extensive investigation; for a long time, its function was believed to be limited to storage of excess energy from food intake. Adipose tissue secretes molecules that directly interact with the brain, and it also has immunological functions. This tissue shapes the body contours of individuals, which has a sociological impact in terms of determining sexual attractiveness. The literature contains several papers on adipose tissue, but the majority of experimental work was performed on animals. Thus, the results of these studies may not be applicable to the human tissue. Adipose tissue is also increasingly being recognised in cell transfer–based therapies as offering the widest spectrum of applications for cell-based therapies for treatments while being the easiest tissue to access. Adipose tissue exists as two types: brown adipose tissue (BAT) and white adipose tissue (WAT). These two tissues have different functions in the body but are both named adipose tissue because of their intracellular triglyceride deposits [16]. BAT is involved in heat generation mechanisms of the body. On the other hand, WAT has several functions in the body: it provides thermal insulation; affects the image and appearance of the body, which is especially important in females to determine their sexual attractiveness; allows for energy storage; serves as a shock absorber on the basis of its anatomical location; has endocrine functions; fills body spaces; and helps lubricate neighbouring muscles to allow ease in movement. BAT and WAT usually exist in a mixed manner without a strict boundary between them [16, 38, 66]. In non-obese males and females, adipose tissue weighs

Y.-G. Illouz and A. Sterodimas (eds.), *Adipose Stem Cells and Regenerative Medicine*,
DOI: 10.1007/978-3-642-20012-0_1, © Springer-Verlag Berlin Heidelberg 2011

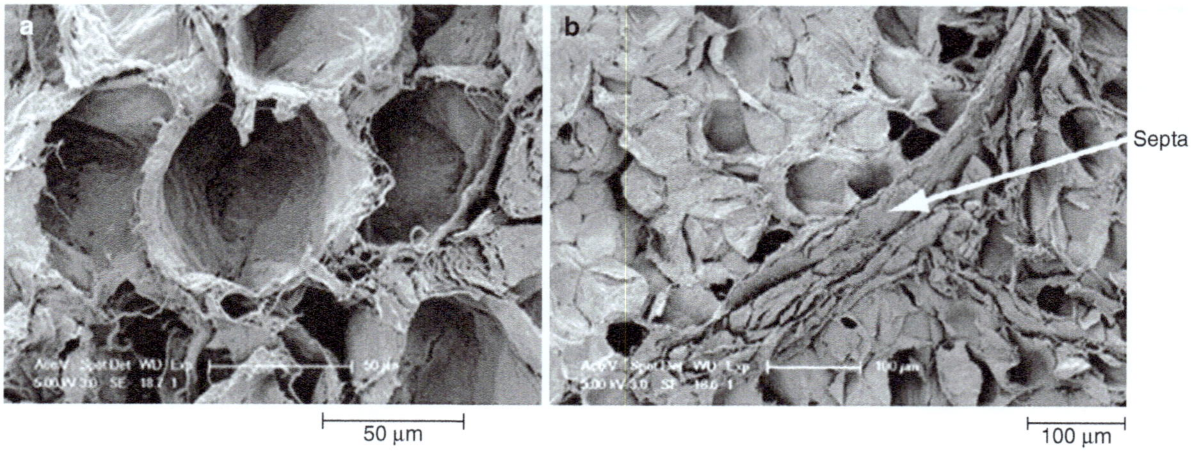

Fig. 1.1 (**a**) The collagen component of WAT is abundant in collagen type IV surrounding each adipocyte (adipocytes are emptied). (**b**) The connective tissue septum extends into the tissue and divides the tissue into several smaller lobes (Scanning electron micrograph figures are courtesy of editors of *International Journal of Solids and Structures* [17])

around 15% and 22% of total body mass, respectively. The anatomical distribution of adipose tissue also differs between male and females.

Adipose tissue is widely distributed all over the body. The amount of adipose tissue within an organism may increase or decrease according to the calorie balance between the body's energy usage and nutritional intake. When there is excess energy intake, adipose tissue can develop around small vessels in undifferentiated cells. At the primary stages of adipogenesis, cells do not contain lipid droplets; then, a lipid proliferation phase takes place, and as this new tissue island develops, lipid begins to accumulate in cells and pre-adipocytes differentiate into adipocytes. If an adipocyte loses its lipid mass (e.g., in cases of fasting and weight loss), the cell acquires an irregular morphology with abundant mitochondria distributed in the entire cytoplasm. These lipid-depleted adipocytes (post-adipocytes) are surrounded by a dense collagen matrix which can cause fibrosis of the adipose tissue [60]. Adipose tissue lobules are surrounded by a connective tissue composed capsule. Collagen compartmentalisation goes until each adipocyte is individually surrounded by a collagen scaffold (Fig. 1.1a). Connective tissue septa extend (Fig. 1.1b) into the tissue and subdivide tissue into several smaller lobes. Adipose tissue has two kinds of nerve fibres – innervations from sympathetic and sensory fibres to the adipose tissue. There is strong evidence that sympathetic innervations suppress the increase in adipocyte cell numbers, which suggests that at least some obesity syndromes can be attributed to sympathetic insufficiency [33, 89]. At least in rat and mouse, it has been proven that parasympathetic innervation does not exist in adipose tissue [27]. For at least one anatomical site – the epididymis – adipose tissue has a direct neuronal pathway from the brain (i.e., through para-ventricular nucleus of the hypothalamus [PVH] neurons) [76]. Because of its direct connection to the liver, this pathway is believed to be involved in controlling the metabolic processes of an organism. PVH has many neuronal circuits within other parts of brain, and when the possible effects of these connections are considered, unexpected biological interactions can arise. A direct neuronal pathway between the central nervous system and BAT is seen in the interscapular region in hamsters. Neurons at the medial pre-optic area, PVH, ventro-medial hypothalamic nucleus, and supra-chiasmatic and lateral hypothalamic nuclei are connected to interscapular BAT by a route passing to the spinal cord, brainstem, midbrain, and forebrain [5].

Adipose tissue exhibit different properties according to their anatomical localisation. The bulk of the tissue is roughly divided into two main localisations – subcutaneous and omentum. This bulk mass easily expands or shrinks depending on the nourishment obtained. Gender difference also affects the adipose tissue mass and distribution. On the other hand, in certain anatomical locations such as eyeball surrounding, palms, and soles of the feet, the adipose tissue content is not involved in the metabolic processes of the organism, and the size of the tissue does not change even in conditions of severe fasting [38, 66]. The smallest adipocytes, which are localised in the mesenteric region, lower insulin sensitivity. Adipose tissue in the mesenteric region is the most innervated site of the body and is the site with the highest blood circulation [33].Blood supply to adipose tissue differs according to the body mass index (BMI) of the individual; the cardiac output to adipose tissue is 3–7% in a lean person and 15–30% in a severely obese person. This amount of increase in cardiac output may have detrimental haemodynamic effects and can result in the development of cardiomegaly and congestive heart failure [33]. The macrophage percentage in the tissue also depends on the BMI: as obesity becomes more severe, the number of macrophages also increases, which results in fat droplet phagocytosis of dead adipocytes [16].

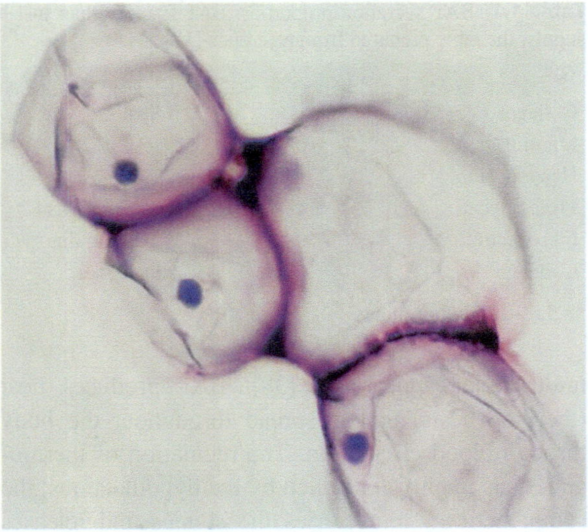

Fig. 1.2 Adipocytes without their lipid component have a typical signet-ring shape. In routine tissue processing, xylene and alcohol are used to dissolve lipids. Haematoxylin and eosin staining; bar represents 25 μm

1.2 Cell Biology of Adipose Tissue

1.2.1 WAT Adipocyte Properties

WAT adipocytes contain one large lipid droplet in each cell. This droplet does not have a well-defined limiting unit, but has a half membrane–like structure (monolayer membrane) between the intracellular lipid component and the cytoplasm. The cells are round or polygonal, ranging between 25 and 200 μm in size. Although these cells contain many organelles, it is difficult to recognise them because the large lipid droplet pushes the organelles, including the nucleus, towards the thin cytoplasm under the plasmalemma. During routine histological processing, lipid becomes dissolved, leaving an empty space that can be seen as a typical signet-ring shape under a light microscope (Fig. 1.2). Abundant pinocytotic vesicles exist near the plasma membrane [16, 38, 66]. Histochemical techniques such as Sudan III or Scarlet Red staining are used to reveal adipocytes. WAT has a very rich vascular supplement; each single adipocyte is especially in contact with a vessel. The resident cell population of adipose tissue comprises mature adipocytes, pre-adipocytes, post-adipocytes, mesenchymal stem cells, endothelial cells, pericytes, mast cells, macrophages, fibroblasts, circulating blood cells, reticulocytes, and nervous system elements. Half of the cell population is made up of mature adipocytes [9, 38, 66].

1.2.2 BAT Adipocyte Properties

BAT is classified as adipose tissue because its cells contain triglyceride deposits in the cytoplasm. The brown colour of this tissue is due to the high cytochrome oxidase content of its mitochondria [12, 38], which are abundantly found in the cytoplasm. The anatomical distribution of BAT differs from that of WAT. During the first 10 years of human life, BAT can be found in almost all sites where WAT is located (e.g., in the interscapular area and anterior abdominal wall); however, as the years advance, peripherally localised BAT disappears while more deeply localised BAT, especially around the kidneys, supra-renals, aorta, neck, and mediastinum, usually persists until the 8th decade of life [34]. A recent study (done on 3,604 patients) based on 18F-FDG-PET/CT examinations showed that the mass and activity of BAT are more steadily maintained by women than men. The male body loses BAT mass and activity as age progresses. This difference is suggested to be due to hormonal differences; that is, testosterone inhibits UCP1 mRNA expression in a dose-dependent manner [62]. Brown adipocytes may be polygonal or ellipsoid in shape, with a diameter ranging between 15 and 50 μm [16]. Many lipid droplets are widely distributed in the cytoplasm and differ in size, which give the adipocytes a

Table 1.1 BAT secretion properties. BAT is more than just heat generation tissue, it has secretion capacity with its rich blood supply, the table prepared from reference [12]

Extracellular matrix components	Autocrine	Paracrine	Endocrine
Collagen IV	Adipsin	NGF	Fatty acids
Laminin	FGF-2	VEGF-A	Leptin
Heparan sulphate	IGF-I	VEGF-B	Adiponectin
Proteoglycan	Prostaglandins E2	VEGF-C	T*
Fibronectin	Prostaglandins Fα	Nitric oxide	
Collagen IV	Adenosine	Angiotensinogen	

*3,5,3'-triiodothyronines

multi-ocular appearance [38]. BAT produces heat (thermogenesis) that is spread throughout the body through blood circulation. The regulation of thermogenesis is mainly controlled by the hypothalamus; the sympathetic system carries the signals and releases nor-epinephrine, which induces fatty acid metabolism in the mitochondria of brown adipocytes [12]. Heat production is accomplished by the action of a special protein in the inner membrane of the mitochondrion, uncoupling protein 1 (UCP1). In the literature, the same protein is also called 'thermogenin'. UCP1 is strictly limited to BAT adipocytes; thus, the presence of this protein in cells can positively identify a tissue as BAT [93]. UCP1 is activated in the presence of free fatty acids (which are released from triacylglycerols through the effect of nor-epinephrine-activated β adrenergic receptors) and suppresses ATP generation while fat is being oxidised; in this manner, the excess energy is released as heat. The mechanism of heat generation is still being investigated. UCP1 is a triglyceride carrier rather than a proton carrier; free protons are pumped out of the mitochondria into the cytoplasm where they combine with fatty acids, making the fatty acids soluble. These soluble fatty acids return to mitochondria and protons are not used for ATP production (this is for the uncoupling concept); instead, heat is generated from the release of protons. The fatty acid thus becomes insoluble again and is carried back to the cytoplasm by UCP1 [12]. The residual products of oxidised fat seem to be the suppressors of UCP1 activity [57]. BAT has a dense nerve supply from the sympathetic nervous system. When nor-epinephrine is secreted from nerves, the interaction of the stimulated receptors of BAT determines the fate of heat generation. If β3-receptors are stimulated, the heat generation pathway is activated; however, if α2-receptors are stimulated, then heat generation is inhibited. A mechanism to regulate balance seems to exist but is still awaiting further research [11, 12]. To completely understand the thermoregulation nerve circuit, it may be useful to remember that nerve signalisation begins from cutaneous (environment-sensitive) and body core thermoreceptors, especially fibres conveying information from the cutaneous site to the area localised at the rostral pole of the hypothalamus, which is the pre-brain part of thermoregulation circuit established by BAT [53]. Brown adipocytes also have secretion capacity besides their heat generation function. Some BAT secretions are basement membrane components, autocrines, paracrines, and endocrine molecules, as listed in Table 1.1.

1.2.3 Extracellular Matrix (ECM)

Adipose tissue harbours extracellular matrix (ECM) components; typically, collagen, reticular fibres, elastin, nerve fibres, vascular, stroma, and the lymphatic system all exist. The ECM of the tissue is extremely important for the survival of mature adipocytes, which bear large lipid droplets. These lipid droplets can add a greater weight load than the living part of the cell; thus, the mechanical support of the ECM is required to prevent cell disruption [47]. Moreover, the ECM architecture seems to deflect the force to other parts of the tissue, thus further diminishing the disruption effect [17]. The ECM of the tissue can be divided into two parts: the basement membrane (Fig. 1.1a) and the region surrounding the adipocytes. Each adipocyte has its own basement membrane composed of a network of collagen type IV, laminin, heparan sulphate proteoglycan, perlecan, and entactin [58, 60]. The main component of the ECM between adipocytes is collagen type VI [35, 47]; a brief focus on this protein can help better understand adipose tissue. Collagen type VI has interesting properties; this protein binds to the basement membrane type IV collagen and interacts with proteoglycans and fibronectin. Collagen

VI maintains the structural integrity of the ECM. The 3-dimensional (3D) form of this collagen is composed of tetramers that are highly branched, flexible basket-like fibres that anchor different structures such as blood vessels, nerves, and other collagens with or without basement membranes in the connective tissue. The spatial form of this collagen also differs according to tissue localisation [19, 39]. This collagen can be secreted from fibroblasts, muscle cells [19], macrophages [71], and adipocytes [3, 69], as well as from tumour cells [21]. In certain cells, collagen VI enhances migration and invasion (macrophage-like cells) [21, 48], and aids in cell survival [21, 35]; however, these effects also allow tumour cells to proliferate in the tissues [21]. This protein enriches the cytokines, ligands, and growth factors around it and can itself contribute to signalisation of the surrounding cells. To understand the importance of collagen VI, muscle weakness under conditions of specific collagen VI deficiency [41] can be considered. Understanding the properties of collagen VI provides insights about the function of the ECM in adipose tissue.

Measurements of the toughness (JC) of adipose tissue, as done on porcine models, revealed the value as being 4.1 (1.2) kJ m^{-2}. As a comparison, the dermis of porcine tissue was found to be approximately four times tougher (around 17 kJ m^{-2}). Tissue toughness gives an idea about the viscoelastic properties of the tissue, which, at least for adipose tissue, can be essential information for tissue engineering processes such as tissue manipulation, tissue transplantation, or aspiration vacuum adjustments. The toughness of adipose tissue is mainly contributed by collagen type IV and its 3D micro-architecture [17].

1.3 Secretions of Adipose Tissue Cells

Besides its fat storage function, adipose tissue also functions as an endocrine organ. This tissue contains adipocytes, pre-adipocytes, fibroblasts, stromal vascular cells, and mast cells, which all secrete biological molecules. These cells may have similar or different secretions between them; that is, one secretion can be from a single cell type while another secretion can be from 3 to 4 different cells. However, because cells such as macrophages and mast cells are constantly migrating, the cell number constantly varies, thus complicating the evaluation of the secretion property of adipose tissue. In WAT, dynamic cell traffic exists and it is affected by several parameters [16]. Adipose tissue secretions can be classified according to ECM components, fatty acids, paracrines, and hormones (Table 1.2). It is a dynamic endocrine organ that releases some newly recognised but well-known hormones such as leptin, adiponectin, resistin, and cytokines including TNFα [1, 74]. Understanding the properties of some of these secretions will give an idea about the scope of the effects of WAT on the whole body. Leptin is secreted from adipocytes and directly affects the brain, especially the ventro-medial hypothalamus (VMH) region. Leptin was discovered in 1949, through The Jackson Laboratory's work on gene mapping on obese mice. After recognising the agent, long-term and extensive work was carried out in several laboratories, and in 1994 the peptide with 167 amino acids was identified through the expression of a 4.5-kb transcript. Finally, the name 'leptin' was coined in 1995, from the Greek word *leptos* which means thin. This protein is now widely recognised, and research suggests that the effect of leptin on the brain is not limited to the VMH region [24]. Leptin is mainly removed from circulation through renal tubules by metabolic degradation followed by glomerular filtration [51]. Its half-life is around 30 min; it is commonly secreted during midnight hours [92]. Leptin is a circulating protein; by targeting cells in the arcuate nucleus of the hypothalamus it affects our feeding behaviour [1]. When BMI increases, leptin secretion also increases and the physiological response is to lower food intake. When there is tissue damage at the VMH or a resistance mechanism in the brain, eating behaviour will not be suppressed and the body will continue to gain excess weight. Leptin resistance can be at the receptor level or the blood–brain barrier level and can be affected by circulating triglycerides. When there is leptin resistance, blood levels increase and the protein can have clinical effects, for example, increased sympathetic activity, increased angiogenic activity, risk of arterial thrombosis due to the platelet leptin receptors, and hyper-leptinaemia, which can harm the cardiovascular system by inducing hypertension and atherosclerosis [6, 8, 74].

Besides its effects on feeding behaviour, leptin also influences other parts of the brain which have regulation properties on thyroid secretion, sympathetic nervous system [81], and GnRH neurons [92]. Leptin itself acts as a neuro-protective molecule for some neurons, influences the excitability of hippocampal neurons, and has neuro-trophic and anti-apoptotic

Table 1.2 WAT as a source of paracrine, endocrine and autocrine source. Some molecules have been identified on experimental animal models or under cell culture conditions

Extracellular matrix components	Autocrine	Paracrine	Endocrine
Collagen I	Glycerol	NEFA	NEFA
Collagen III	NEFA	Monoglyceride	Leptin
Collagen IV	Monoglyceride	Eicosanoids	Resistin
Collagen VI	Eicosanoids	Oleoyl-estrone	Omentin
Laminin	Oleoyl-estrone	Prostaglandin E2, I2	Estrogen and Oleoyl-estrone
Heparan sulphate Proteoglycan	Prostaglandin E2, I2	Acylation-stimulating protein	Androgens
Perlecan	Lipoprotein lipase	Fasting-induced adipose factor	Cortisol, cotisone
Entactin	Acylation-stimulating protein	Cholesteryl ester transfer protein	Adiponectin
	Fasting-induced adipose factor	Tumor necrosis factor α	Visfatin
	Cholesteryl ester transfer protein	Interleukin-6	Vaspin
	Tumor necrosis factor α	IL-1β	Fasting-induced adipose factor
	Interleukin-6	IL-8	Retinol binding protein
	Complement components	IL-10	Interleukin-6
	Angiotensinogen	IL-18	IL-1β
	VEGF	IL-17D	IL-8
	Metalloproteinase inhibitors	Transforming growth factor-β	IL-10
	Matrix metalloproteinases	Monocyte chemoattractant protein 1	IL-18
	Secreted protein acidic and rich in Cysteine	Macrophage migration inhibiting factor	IL-17D
	Collagens	Complement components	Transforming growth factor-β
		Haptoglobin	Monocyte chemoattractant protein 1
		Serum amyloid A3	Macrophage migration inhibiting factor
		Plasminogen activator inhibitor-1	Complement components
		Angiotensinogen	Haptoglobin
		Pigment epithelial-derived factor	Serum amyloid A3
		Adrenomedullin	Plasminogen activator inhibitor-1
		VEGF	Pigment epithelial-derived factor
		Apelin	Adrenomedullin
		NGF	VEGF
		Fibroblast growth factors	Apelin
		Metalloproteinase inhibitors	NGF
		Matrix metalloproteinases	Gelsolin[a]
		Secreted protein acidic and rich in cysteine	Heparin-binding epidermal growth
		Collagens	Insulin-like growth factor II
			Colligin YKL-40

NEFA non-esterified fatty acid, *VEGF* vascular endothelial growth factor, *NGF* nerve growth factor
[a]Gelsolin is both used intracellularly and secreted extracellularly, the parameters are collected from; [25, 36, 46, 64, 65, 67, 84]

properties [81]. These properties make leptin a potential molecule for use in studying brain tissue pathologies [86]. Leptin is also involved in the reproduction mechanisms of humans: it triggers puberty, and on the basis of the BMI and fasting status of the experimental models, it is assumed to be involved in the functions of the hypothalamic–gonadotropic axis. Among different animal species, there are important differences in leptin interactions, especially at the reproductive level [92]. Leptin is mainly secreted from adipocytes, but gastrointestinal tissue, skeletal muscle, placenta [1], foetal cartilage, and the pituitary gland [92] also secrete this protein in small amounts.

In industrialised countries, the percentage of the elderly population is increasing, which also increases the incidence of health problems such as dementia and Alzheimer disease (AD). These two diseases incur large social and financial costs to communities [91]. Interestingly, leptin has a regulatory effect on the neuron uptake of amyloid-beta (Aβ) from circulation; there is strong evidence that leptin inhibits glial cholesterol synthesis and uptake, which has an effect on Aβ production and uptake in neurons [23]. In AD, the main pathological sign is deposition of neuro-fibrillary tangles in neurons; leptin and insulin decreases hyperphosphorylation of tau (a major component of tangles) in a dose-dependent manner [29]. A prospective study on 785 persons, with an 8.3-year follow up, reported that in subjects with high circulating levels of leptin, reduced incidence dementia and AD was observed. The same study reported that individuals with high circulating levels of leptin had a larger brain parenchyma and smaller ventricle volumes than healthy persons [42]. Thus, leptin seems to have potential as a therapeutic agent for AD and dementia. Additionally, other properties of leptin include its role during foetal life, interaction with testicular cells [82], effects on endometrial receptivity [2], ability to show seasonal variation, different blood level–dependent effects depending on ethnicity [43], regulatory effects on the immune system, and so on.

Omentin is a newly discovered protein suggested to regulate insulin action [52, 88]. It is secreted from stromal vascular cells, but has not been found to be secreted from adipocytes [88]. This adipokine (size 34 kDa) is produced mainly by visceral adipose tissue and in smaller amounts by subcutaneous adipose tissue [70, 88]; it has also been identified in human epicardial fat [20]. Circulating levels of omentin decreases as the organism's obesity increases [52]. Omentin causes arterial vessels to vasodilate through endothelial signalling as well as through a non-endothelial pathway. This adipokine is believed to contribute to the development of hypertension in obese people. No specific receptor for omentin has yet been identified [87].

Resistin is a dimeric protein (108 amino acids) thought to be involved in the insulin resistance seen in obese people; its circulating levels increase with obesity [37]. This concept was developed on the basis of animal experimentations [63], but the mechanism in humans is still unclear. Resistin is mainly secreted from circulating monocytes [68] and in lower amounts from adipocytes [37].

Non-esterified fatty acids are not only energy sources for organisms but also function as signalling molecules and substrates for liver synthesis of lipoproteins [4].

Adiponectin is one of the main secretion products of adipocytes. This multifunctional hormone is composed of 244 polypeptides; the 3D form has parts that resemble collagenous helices while other parts have similarities with TNF-α topology [72]. Blood adiponectin levels decrease during obesity and increase as weight decreases. It has diverse effects on the body; it stimulates the oxidation of fatty acids, suppresses gluconeogenesis in the liver, exerts anti-atherogenic effects by inhibiting monocyte adhesion to endothelial cells, directs macrophages to turn to foam cells and directs smooth muscle migration in vessels [1], enhances human trophoblastic cells for a successful invasion of the uterus by modulating tissue matrix metalloproteinases (MMP) and Tissue Inhibitor of Metalloproteinase (TIMP) balance [7], and may help in the biochemical evaluation of patients with non-traumatic osteonecrosis of femoral head [73] and polycystic ovary syndrome [30]. It has two identified receptors (AdipoR1 and 2), and these receptors are located in the brain regions where the sympathetic system is activated. When intracerebroventricular adiponectin injection was done on mice, the animals showed increased thermogenesis [1]. The mechanism of adiponectin clearance is still under research, and there is a doubt that the kidney has a role in this process [79]. Adiponectin is also known by different names such as GBP-28, apM1, AdipoQ, and Acrp30 [45]. It has shown potential as a candidate molecule for obesity treatment.

Mast cells in the adipose tissue contribute to the secretion function of the tissue. They secrete similar

Table 1.3 Adipocyte lipid composition. List of some lipid composition of adipocytes studied in the mesenteric region

Myristic	Myristoleic	Margaric
Margaroleic	Stearic	Oleic
Linoleic	α-linolenic	Palmitic
Palmitoleic	Eicosapentaenoic	Docosahexaenoic
Pentadecanoic	Pentadecanoic	Arachidonic
Palmitic	Palmitoleic	

substances as other cells in the tissue, for example, VEGF, HGF, MMP, FGF, TGF-β, TNF-α, NGF, and IL-1 [13]. It is also interesting that a deficiency in mast cells increases the risk of adipose tissue hyper-lipidaemia [32]. Lipids in adipocyte tissue are generally classified as triacylglycerols, but a detailed analysis further distinguishes them (Table 1.3). The lipid composition may differ according to the anatomical site of the tissue, and health status and dietary habits of individuals. Many biochemical isoforms of these molecules also exist [85].

1.4 Stem Cell Concept in Adipose Tissue

The basic properties of stem cells are exemplified by their high-proliferation, self-renewal, and differentiation capacities. Mammalian stem cells can be derived from embryonic or adult tissues, and their properties vary according to origin. The main difference lies in the differentiation potential – stem cells from adult tissues have low plasticity while those from embryonic stem cells can give rise to a wider spectrum of cells/tissues. WAT stem cells have the highest plasticity potential among all adult tissues [22]. A cell population with stem cell properties has been identified in WAT. After collagenase digestion of WAT, a group of different cells was obtained, and this cell population is named stromal vascular fraction (SVF) cells [28]. Approximately 2% of the SVF cell population showed stem cell properties with multi-lineage differentiation capacity. It is estimated that 2% of WAT cell population contains cells with multilineage differentiation properties (to compare with bone marrow this is only 0.002%) [28, 83]. Adipose-derived stromal cells (ADSC) differentiate into a wide spectrum of cell lineages. For current and potential clinical applications, the cells can be targeted to differentiate into a variety of cells, for example, chondrogenic, neurogenic, osteogenic, myogenic, cardiogenic, vascular, endocrine, hepatic, or haematopoietic. ADSC has a high potential for cartilage [31, 44, 90] and skeletal/joint repair for treatment of osteoarthritis [59, 44, 69], while ADSC transplantation can be used as a substitute for liver transplantation [80]. A wide range of pathologies due to ischemia can be cured by utilising the angiogenic capacity of ADSC. Studies on acute renal ischemia, limb ischemia, and vascular stroke have been done, which gave promising results [26, 69]. The list of potential applications may still grow as experience advances and manipulation techniques are developed.

When SVF is cultured, a group of cells adhered to the plastic surface of the culture dish; these cells showed stem cell properties. In the literature, different groups have given different names to identify this same cell population, such as adipose-derived stem/stromal cells (ASC), adipose-derived adult stem (ADAS) cells, adipose-derived adult stromal cells/ADSCs, adipose stromal cells (ASC), adipose mesenchymal stem cells (AdMSC), lipoblast, pericyte, pre-adipocyte, and processed lipoaspirate (PLA) cells [10, 22]. Currently, ADSC seems to be gaining wider acceptance. It is estimated that 10% of WAT cells are ADSCs [14]. When multilineage differentiation is aimed for a WAT-derived cell population, cell cultivation must be properly carried out. After or during cultivation, special stimulants are used for differentiation; for example, insulin, dexamethasone, and indomethacin are used to obtain adipogenic cells whereas dexamethasone, ascorbate, and glycerophosphate are added to obtain osteogenic cells [28].

ADSC transplantation has been shown to contribute to tissue healing, especially due to cell secretions that have angiogenic functions. The most significant ADSC secretions that cause angiogenesis are VEGF [75], HGF, SDF-1, and TGF-β [55]. The optimal number of cells to be transplanted for efficient angiogenesis is still under debate [55].The results of ADSC transplantation are affected by several uncontrollable parameters. Because the source of cells is autogenic tissues, the age, sex, and life history cannot be changed. Source localisation (subcutaneous or visceral), selection of surgical operation techniques, and culture media can enhance or limit the proliferation and plasticity potential of ADSCs [40, 69]. If the patient's body profile is suitable a surgeon can collect adipose tissue from the

Table 1.4 ADSC markers. With the help of ADSC markers cell separation and research studies can be made in a more detailed style, while some markers are constant some others are reported occasionally

Stabile expressed surface markers	Non-stabile expressed surface markers	Other cell markers
CD9 CD10 CD13 CD29 CD44 CD49d CD49e CD51	CD 34	Stro-1
	CD 45	A-smooth muscle actin
CD54 CD55	CD 56	Collagen type I
CD59 CD71	CD 61	Collagen type II
CD73 CD90 CD105 CD117 CD146 CD166	CD 104	HLA-ABC
	CD 106	Osteopontin
		Osteonectin
		Vimentin

lower abdomen, which contains a higher percentage of ADSC than other sites. Higher fibrous connective tissue containing WATs contains higher ADSC content [40]. Other factors that can influence results are the duration and/or velocity of WAT centrifugation during lipoaspirate manipulation, and optimisation of these parameters are still under discussion [18]. The results are also dependent on the handling or culturing medium, and even slight differences in medium composition can cause variations in gene expression; for example, in a study that compared two different media, 441 gene expression differences were observed [69].

Some in vivo and in vitro techniques depend on cellular markers for identification of the progressions; the use of these markers may help understand the mechanisms of in vivo tissue transplantation treatments. ADSC has several cellular markers, as shown in Table 1.4. If ADSC transplantation to a new body site is not successful, a necrosis and granulation tissue formation will occur, next this the absorption of granulation tissue will take place. Moreover, for a successful implantation, tissue remodelling must occur. Both the processes of implantation and tissue remodelling are mainly regulated by the secretory functions of ADSC. In cell transplantation or tissue engineering applications, not only autocrine, paracrine, and endocrine secretory products but also enzymes and their inhibitors are used. If there is a balance between the activity of matrix metalloproteinases (MMP) and their inhibitors (TIMPs), then regenerative processes like vascularisation and turnover of connective tissue elements (e.g., collagen) can occur [61]. Although current experimental research is insightful, the mechanisms of transplantation in human tissues need to be investigated further through in vivo experiments. Currently, the use of autologous additives such as platelet-rich plasma (PRP) and LED light stimulation are promising approaches in enhancing the applications of ADSCs [56, 59].

In some cases, adipose tissue itself may need to be manipulated for cosmetic procedures, mastectomy, trauma management, or abnormalities. In such procedures, lipoaspirates from one site are usually transferred as a graft to the site that needs treatment; however, a large volume reduction (around 50%) in total adipose tissue is a major complication that may result in the need for a second operation. ADSC enrichment approaches (e.g., cell-assisted lipotransfer) [77] and tissue engineering techniques are being developed to resolve this issue [49].

1.5 Adipose Stem Cells and Tissue Engineering

Another emerging concept in stem cell application is tissue engineering. Here, clinical treatment is aimed; a tissue-like product composed of a combination of cells (scaffold), which can be transferred to a living body as a tissue component or, if possible, as a whole organ, is generated in a controlled microenvironment [28]. For a biomaterial to be used in tissue engineering, it must have an ideal pore size to permit cell migration; it must be loadable with signalling molecules and growth factors; it should have low immunity, if any; it should be suitable for cell cultivation; it must have mechanical strength and uniformity; and it must be made up of materials suitable for patient comfort [50]. Tissue engineering applications have great potential for adipose tissue restoration, congenital defect correction, or regenerative treatments [28].

Tissue engineering techniques require cell scaffolds which are compatible with the cell population that will be transferred and which will not be rejected by the recipient tissue. Natural biomaterials such as collagen have long been considered to be the most favourable material for tissue engineering, while extensive studies have been done on collagen micro-beads, gelatin sponges, and de-cellularised adipose or placenta matrix [15, 83]. Recent studies suggest the use of silk-derived materials (e.g., polyglycolic or polylactic acid) for their greater strength, slower biodegradation, and lower immunity. Silk-derived materials have been successfully used in ligament, cartilage, and bone biotechnological applications, and these materials are compatible with ADSCs [49].

Although adipose tissue has wide applications in cell or tissue engineering, the potential for undesired results also exists. For example, if vascularisation of implanted tissue is not sufficient, the transplanted volume will decrease; thus, factors affecting angiogenesis must be carefully considered [28]. Another complication is that ADSC secretions can suppress local immunity, which can promote tumour cell proliferation and migration. Therefore, cancer evaluation must be performed before any treatment and follow-up after surgery may be important; for example, after breast surgery, long-term mammography may help screen for possible complications [14]. Another interesting complication is based on the capacity of ADSCs to differentiate into epithelial cells. In breastfeeding women who had undergone breast augmentation using ADSC, the transplanted cells can differentiate into epithelial cells that can be excreted in milk; when breastfeeding ceases, these cells undergo apoptosis and the breast may show a significant volume reduction [54].

1.6 Conclusion

In summary, adipose tissue has a wealth of potential applications; however, significant further research is still needed to understand in detail its physiological, histological, and cell transplantation properties. Moreover, manipulation techniques for tissue engineering need to be optimised [78].

References

1. Ahima RS (2005) Central actions of adipocyte hormones. Trends Endocrinol Metab 16(7):307–313
2. Alfer J, Müller-Schöttle F, Classen-Linke I et al (2000) The endometrium as a novel target for leptin: differences in fertility and subfertility. Mol Hum Reprod 6(7):595–601
3. Alvarez-Llamas G, Szalowska E, de Vries MP et al (2007) Characterization of the human visceral adipose tissue secretome. Mol Cell Proteomics 6:589–600
4. Arner P (2005) Human fat cell lipolysis: biochemistry, regulation and clinical role. Best Pract Res Clin Endocrinol Metab 19:471–482
5. Bamshad M, Song CK, Bartness TJ (1999) CNS origins of the sympathetic nervous system outflow to brown adipose tissue. Am J Physiol Regul Integr Comp Physiol 276:1569–1578
6. Beltowski J (2006) Leptin and atherosclerosis. Atherosclerosis 189:47–60
7. Benaitreau D, Santos ED, Leneveu MC et al (2010) Effects of adiponectin on human trophoblast invasion. J Endocrinol 207:45–53
8. Brydon L (2011) Adiposity, leptin and stress reactivity in humans. Biol Psychol 86(2):114–120
9. Bukowiecki L, Collet AJ, Follea N et al (1982) Brown adipose tissue hyperplasia: a fundamental mechanism of adaptation to cold and hyperphagia. Am J Physiol 242: E353–E359
10. Bunnell BA, Flaat M, Gagliardi C et al (2008) Adipose-derived stem cells: isolation, expansion and differentiation. Methods 45(2):115–120
11. Cannon B, Nedergaard J (2004) Brown adipose tissue: function and physiological significance. Physiol Rev 84: 278–359
12. Cannon B, Houstek J, Nedergaard J (1998) Brown adipose tissue. More than an effector of thermogenesis? Ann NY Acad Sci 856:171–187
13. Chaldakov GN, Tonchev AB, Tuncel N et al (2007) Adipose tissue and mast cells, adipokines as yin-yang modulators of inflammation. In: Fantuzzi G, Mazzone T (eds) Adipose tissue and adipokines in health and disease. Humana Press, New Jersey
14. Chan CW, McCulley SJ, Macmillan RD (2008) Autologous fat transfer – a review of the literature with a focus on breast cancer surgery. J Plast Reconstr Aesthet Surg 61(12):1438–1448
15. Choi JS, Yang HJ, Kim BS et al (2009) Human extracellular matrix (ECM) powders for injectable cell delivery and adipose tissue engineering. J Control Release 139(1):2–7
16. Cinti S (2007) The adipose organ. In: Fantuzzi G, Mazzone T (eds) Adipose tissue and adipokines in health and disease. Humana Press, New Jersey
17. Comley K, Fleck NA (2010) A micromechanical model for the young's modulus of adipose tissue. Int J Solids Struct 47:2982–2990
18. Condé-Green A, de Amorim NF, Pitanguy I (2010) Influence of decantation, washing and centrifugation on adipocyte and mesenchymal stem cell content of aspirated adipose tissue: a comparative study. J Plast Reconstr Aesthet Surg 63(8): 1375–1381
19. Engvall E, Hessle H, Klier G (1986) Molecular assembly, secretion, and matrix deposition of type VI collagen. J Cell Biol 102:703–710
20. Fain JN, Sacks HS, Buehrer B et al (2008) Identification of omentin-1 mRNA in human epicardial adipose tissue: comparison to omentin-1 in subcutaneous, internal mammary artery periadventitial and visceral abdominal depots. Int J Obes (Lond) 32:810–815
21. Fang X, Burg MA, Barritt D et al (1999) Cytoskeletal reorganization induced by engagement of the NG2 proteoglycan leads to cell spreading and migration. Mol Biol Cell 10:3373–3387
22. Fang B, Luo S, Song Y et al (2009) Hemangioblastic characteristics of human adipose tissue-derived adult stem cells in vivo. Arch Med Res 40(4):311–317

23. Fewlass DC, Noboa K, Pi-Sunyer FX (2004) Obesity-related leptin regulates Alzheimer's Ab. FASEB J 18:1870–1878
24. Flier JS, Maratos-Flier E (2010) Lasker lauds leptin. Cell 143:9–12
25. Fonseca-Alaniz MH, Takada J, Alonso-Vale MI et al (2007) Adipose tissue as an endocrine organ: from theory to practice. J Pediatr (Rio J) 83(5 Suppl):S192–S203
26. Gimble JM, Guilak F, Bunnell BA (2010) Clinical and preclinical translation of cell-based therapies using adipose tissue-derived cells. Stem Cell Res Ther 1(2):19
27. Giordano A, Song CK, Bowers RR et al (2010) White adipose tissue lacks significant vagal innervation and immunohistochemical evidence of parasympathetic innervation. Am J Physiol Regul Integr Comp Physiol 291:R1243–R1255
28. Gomillion CT, Burg KJL (2006) Stem cells and adipose tissue engineering. Biomaterials 27:6052–6063
29. Greco SJ, Sarkar S, Johnston JM et al (2008) Leptin reduces Alzheimer's disease-related tau phosphorylation in neuronal cells. Biochem Biophys Res Commun 376(3):536–541
30. Groth SW (2010) Adiponectin and polycystic ovary syndrome. Biol Res Nurs 12(1):62–72
31. Han Y, Wei Y, Wang S et al (2010) Cartilage regeneration using adipose-derived stem cells and the controlled-released hybrid microspheres. Joint Bone Spine 77(1):27–31
32. Hatanaka K, Tanishita H, Ishibashi-Ueda H et al (1986) Hyperlipidemia in mast cell-deficient W/WV mice. Biochim Biophys Acta 878(3):440–445
33. Hausman DB, DiGirolamo M, Bartness TJ et al (2001) The biology of white adipocyte proliferation. Obes Rev 2:239–254
34. Heaton JM (1972) The distribution of brown adipose tissue in the human. J Anat 112(1):35–39
35. Iyengar P, Espina V, Williams TW et al (2005) Adipocyte-derived collagen VI affects early mammary tumor progression in vivo, demonstrating a critical interaction in the tumor/stroma microenvironment. J Clin Invest 115:1163–1176
36. Iwata T, Kuwajima M, Sukeno A et al (2009) YKL-40 secreted from adipose tissue inhibits degradation of type I collagen. Biochem Biophys Res Commun 388:511–516
37. Janke J, Engeli S, Gorzelniak K et al (2002) Resistin gene expression in human adipocytes is not related to insulin resistance. Obes Res 10:1–5
38. Johnson PR, Greenwood MRC (1988) The adipose tissue. In: Weiss L (ed) Cell and tissue biology: a textbook of histology. Urban and Schwarzenberg, Baltimore
39. Keene DR, Engvall E, Glanville RW (1988) Ultrastructure of type VI collagen in human skin and cartilage suggests an anchoring function for this filamentous network. J Cell Biol 107:1995–2006
40. Kishi K, Imanishi N, Ohara H et al (2010) Distribution of adipose-derived stem cells in adipose tissues from human cadavers. J Plast Reconstr Aesthet Surg 63(10):1717–1722
41. Lampe AK, Bushby KMD (2005) Collagen VI related muscle disorders. J Med Genet 42:673–685
42. Lieb W, Beiser AS, Vasan RS et al (2009) Association of plasma leptin levels with incident Alzheimer disease and MRI measures of brain aging. JAMA 302(23):2565–2572
43. Lilja M, Rolandsson O, Shaw JE et al (2010) Higher leptin levels in Asian Indians than Creoles and Europids: a potential explanation for increased metabolic risk. Int J Obes (Lond) 34(5):878–885
44. Locke M, Windsor J, Dunbar PR (2009) Human adipose-derived stem cells: isolation, characterization and applications in surgery. ANZ J Surg 79:235–244
45. Maeda K, Okubo K, Shimomura I et al (1996) cDNA cloning and expression of a novel adipose specific collagen-like factor, apM1 (adipose most abundant gene transcript 1). Biochem Biophys Res Commun 221(2):286–289
46. Maeda K, Okubo K, Shimomura I et al (1997) Analysis of an expression profile of genes in the human adipose tissue. Gene 190:227–235
47. Mariman EC, Wang P (2010) Adipocyte extracellular matrix composition, dynamics and role in obesity. Cell Mol Life Sci 67:1277–1292
48. Matsumoto H, Kumon Y, Watanabe H et al (2008) Accumulation of macrophage-like cells expressing NG2 proteoglycan and Iba1 in ischemic core of rat brain after transient middle cerebral artery occlusion. J Cereb Blood Flow Metab 28(1):149–163
49. Mauney JR, Nguyen T, Gillen K et al (2007) Engineering adipose-like tissue in vitro and in vivo utilizing human bone marrow and adipose-derived mesenchymal stem cells with silk fibroin 3D scaffolds. Biomaterials 28(35):5280–5290
50. Merceron C, Vinatier C, Clouet J et al (2008) Adipose-derived mesenchymal stem cells and biomaterials for cartilage tissue engineering. Joint Bone Spine 75(6):672–674
51. Meyer C, Robson D, Rackovsky N et al (1997) Role of the kidney in human leptin metabolism. Am J Physiol 273E:903–907
52. Moreno-Navarrete JM, Catalán V, Ortega F et al (2010) Circulating omentin concentration increases after weight loss. Nutr Metab 7:27–33
53. Morrison SF, Nakamura K, Madden CJ (2008) Central control of thermogenesis in mammals. Exp Physiol 93(7):773–797
54. Morroni M, Giordano A, Zingaretti MC et al (2004) Reversible transdifferentiation of secretory epithelial cells into adipocytes in the mammary gland. Proc Natl Acad Sci USA 101(48):16801–16806
55. Murohara T (2009) Autologous adipose tissue as a new source of progenitor cells for therapeutic angiogenesis. J Cardiol 53(2):155–163
56. Mvula B, Mathope T, Moore T et al (2008) The effect of low level laser irradiation on adult human adipose-derived stem cells. Lasers Med Sci 23(3):277–282
57. Nicholls DG, Bernson VS, Heaton GM (1978) The identification of the component in the inner membrane of brown adipose tissue mitochondria responsible for regulating energy dissipation. Experientia Suppl 32:89–93
58. Niemela SM, Miettinen S, Konttinen Y et al (2007) Fat tissue: views on reconstruction and exploitation. J Craniofac Surg 18(2):325–335
59. Niemeyer P, Fechner K, Milz S et al (2010) Comparison of mesenchymal stem cells from bone marrow and adipose tissue for bone regeneration in a critical size defect of the sheep tibia and the influence of platelet-rich plasma. Biomaterials 31(13):3572–3579
60. Orci L, Cook WS, Ravazzola M et al (2004) Rapid transformation of white adipocytes into fat-oxidizing machines. PNAS 101(7):2058–2063
61. Peroni D, Scambi I, Pasini A et al (2008) Stem molecular signature of adipose-derived stromal cells. Exp Cell Res 314(3):603–615

62. Pfannenberg C, Werner MK, Ripkens S et al (2010) Impact of age on the relationships of brown adipose tissue with sex and adiposity in humans. Diabetes 59(7): 1789–1793
63. Rajala MW, Obici S, Scherer PE et al (2003) Adipose-derived resistin and gut-derived resistin-like molecule-beta selectively impair insulin action on glucose production. J Clin Investig 111:225–230
64. Roelofsen H, Dijkstra M, Weening D et al (2009) Comparison of isotope-labeled amino acid incorporation rates (CILAIR) provides a quantitative method to study tissue secretomes. Mol Cell Proteomics 8:316–324
65. Ronti T, Lupattelli G, Mannarino E (2006) The endocrine function of adipose tissue: an update. Clin Endocrinol 64:355–365
66. Ross MH, Reith EJ, Romrell LJ (1989) Adipose tissue. In: Kist K (ed) Histology: a text and atlas. Williams and Wilkins, Baltimore
67. Ru M, Sahin N, Johannsen M (1999) Soluble collagen VI drives serum-starved fibroblasts through S phase and prevents apoptosis via down-regulation of Bax. J Biol Chem 274(48):34361–34368
68. Savage DB, Sewter CP, Klenk ES et al (2001) Resistin/Fizz3 expression in relation to obesity and peroxisome proliferator-activated receptor action in humans. Diabetes 50:2199–2202
69. Schäffler A, Büchler C (2007) Concise review: adipose tissue-derived stromal cells-basic and clinical implications for novel cell-based therapies. Stem Cells 25(4):818–827
70. Schäffler A, Neumeier M, Herfarth H et al (2005) Genomic structure of human omentin, a new adipocytokine expressed in omental adipose tissue. Biochim Biophys Acta 1732:96–102
71. Schnoor M, Cullen P, Lorkowski J et al (2008) Production of type VI collagen by human macrophages: a new dimension in macrophage functional heterogeneity. J Immunol 180: 5707–5719
72. Shapiro L, Scherer PE (1998) The crystal structure of a complement-1q family protein suggests an evolutionary link to tumor necrosis factor. Curr Biol 8(6):335–338
73. Shuai B, Shen LS, Yang YP et al (2010) Low plasma adiponectin as a potential biomarker for osteonecrosis of the femoral head. J Rheumatol 37(10):2151–2155
74. Sierra-Honigmann MR, Nath AK, Murakami C et al (1998) Biological action of leptin as an angiogenic factor. Science 281:1683–1686
75. Song YH, Gehmert S, Sadat S et al (2007) VEGF is critical for spontaneous differentiation of stem cells into cardiomyocytes. Biochem Biophys Res Commun 354(4):999–1003
76. Stanleya S, Pintoa S, Segala J et al (2010) Identification of neuronal subpopulations that project from hypothalamus to both liver and adipose tissue polysynaptically. PNAS 107(15):7024–7029
77. Sterodimas A, de Faria J, Nicaretta B et al (2010) Cell-assisted lipotransfer. Aesthet Surg J 30(1):78–82
78. Sterodimas A, de Faria J, Nicaretta B (2010) Tissue engineering with adipose-derived stem cells (ADSCs): current and future applications. J Plast Reconstr Aesthet Surg 63(11):1886–1892
79. Taherimahmoudi M, Ahmadi H, Mehrsai A et al (2010) Plasma adiponectin concentration and insulin resistance: role of successful kidney transplantation. Transplant Proc 42(3):797–800
80. Taléns-Visconti R, Bonora A, Jover R et al (2007) Human mesenchymal stem cells from adipose tissue: differentiation into hepatic lineage. Toxicol In Vitro 21(2):324–329
81. Tang BL (2008) Leptin as a neuroprotective agent. Biochem Biophys Res Commun 368:181–185
82. Tena-Sempere M, Barreiro ML (2002) Leptin in male reproduction: the testis paradigm. Mol Cell Endocrinol 188:9–13
83. Vallée M, Côté JF, Fradette J (2009) Adipose-tissue engineering: taking advantage of the properties of human adipose-derived stem/stromal cells. Pathol Biol (Paris) 57(4):309–317
84. Wang P, Mariman E, Renes J (2008) The secretory function of adipocytes in the physiology of white adipose tissue. J Cell Physiol 216:3–13
85. Westcott E, Windsor A, Mattacks C et al (2005) Fatty acid compositions of lipids in mesenteric adipose tissue and lymphoid cells in patients with and without Crohn's disease and their therapeutic implications. Inflamm Bowel Dis 11(9):820–827
86. Wójcik-Gładysz A, Wankowska M, Misztal T et al (2009) Effect of intracerebroventricular infusion of leptin on the secretory activity of the GnRH/LH axis in fasted prepubertal lambs. Anim Reprod Sci 114:370–383
87. Yamawaki H, Tsubaki N, Mukohda M et al (2010) Omentin, a novel adipokine, induces vasodilation in rat isolated blood vessels. Biochem Biophys Res Commun 393:668–672
88. Yang RZ, Lee MJ, Hu H et al (2006) Identification of omentin as a novel depot-specific adipokine in human adipose tissue: possible role in modulating insulin action. Am J Physiol Endocrinol Metab 290:1253–1261
89. Youngstrom TG, Bartness TJ (1998) White adipose tissue sympathetic nervous system denervation increases fat pad mass and fat cell number. Am J Physiol Regul Integr Comp Physiol 275:1488–1493
90. Zhang H, Li L, Leng P (2009) Uninduced adipose-derived stem cells repair the defect of full-thickness hyaline cartilage. Chin J Traumatol 12(2):92–97
91. Zhu CW, Scarmeas N, Torgan R (2006) Longitudinal study of effects of patient characteristics on direct costs in Alzheimer disease. Neurology 67:998–1005
92. Zieba DA, Amstalden M, Williams GL (2005) Regulatory roles of leptin in reproduction and metabolism: a comparative review. Domest Anim Endocrinol 29:166–185
93. Zingaretti MC, Crosta F, Vitali A et al (2009) The presence of UCP1 demonstrates that metabolically active adipose tissue in the neck of adult humans truly represents brown adipose tissue. FASEB J 23:3113–3120

Isolation of Stem Cells from Human Adipose Tissue: Technique, Problems, and Pearls

2

Ithamar Nogueira Stocchero and Guilherme Flosi Stocchero

Contents

2.1 Introduction 13
2.2 Basic Concepts of Stem Cells 14
2.3 Sources of Stem Cells 15
2.4 Stem Cell Culture 15
2.5 Animal Models 16
2.6 Pitfalls 17
2.7 Pearls 17
2.8 Conclusion 17
References 18

2.1 Introduction

If you are looking for volume, take the leaves. If you are looking for leaves, flowers and fruits, take the stem and the branches. With Stem Cells it is the same.

Plastic surgery is essentially the search for form. And form means volume (or its absence). When the volume is insufficient, it is necessary to look for alternatives to increase it, either by the inclusion of a material in the deficient area, or by the transposition of tissue from donor sites of the body; when the volume is excessive, the removal of a part of it is a viable option. The use of such techniques may either last or be abandoned depending on the tolerance of an inclusion by the human body and/or the cost–benefit relation that the surgical methods may offer.

Such mentality is currently applied in plastic surgery, as in many other fields of Medicine: removal of excessive tissue and replacement of absent tissue. Restoration is a rare option. Some time ago, scientists began to seek the real possibility of regenerating the human body in its deficient sites, based on a concept in which the cure is contained in the body itself. The essence of such treatments would fall within the simplicity of putting back what is missing, using the patient's own body as the source.

Certainly, the idea of partially or integrally restoring a missing organ dates back to ancient times. After all, it was observed that, after a gecko had lost its tail to escape a predator, leaving the still-shaking lost organ behind, the lizard would surprisingly grow back a new tail in a few days after the episode, being identical in functionality, texture, color, and vascular and nerve supply, bearing

I.N. Stocchero (✉) • G.F. Stocchero
Centro Médico Viver Melhor, Rua Abílio Soares, 1337, 04005–005 São Paulo, SP, Brazil
e-mail: dr.ithamar@vivermelhor.com.br; guilherme@vivermelhor.com.br

Y.-G. Illouz and A. Sterodimas (eds.), *Adipose Stem Cells and Regenerative Medicine*,
DOI: 10.1007/978-3-642-20012-0_2, © Springer-Verlag Berlin Heidelberg 2011

self-limited growth and without mutations. Why would it not be the same with humans? Millenia have passed, and the search for a magical treatment that would permit the restoration of tissues, organs, and their functions is still continuing.

The miraculous expression that seems to encompass this almost-divine gift is *stem cell*. Ultimately, what are living creatures, if not the perfect engineering originated from the union of only two cells? From a tiny code resulting from the mixture of two sources of DNA (ovum + spermatozoon = zygote), there will be a unique individual, bearing its own characteristics, and with a body full of rich details, in the end complete.

2.2 Basic Concepts of Stem Cells

This chapter is not intended to be a thorough source of learning about stem cells, but indeed a guide for the plastic surgeon to build a notion of this new edge of human knowledge, thus instigating the curiosity of those who identify with scientific research, and also offering the more pragmatic ones a panorama that permits them to deal with these new challenges that have entered plastic surgery, therefore enriching it and consolidating it as a state-of-the-art medical specialty.

The ability of two gametes to originate the whole universe of more than 200 different types of tissues present in the human body gives wings to wondering how far the power of human intervention in Nature may get, as long as man is capable of coordinating the pathways to be followed by those cells, which carry the potential to develop everything within a species, hence justifying their name – *totipotent cells*. And they will maintain this faculty during their first divisions. The early blastocyst, when still made up of eight cells, is the stage that allows a relatively easy harvest of embryonic cells, since they still have loose bonds between them, being almost free before compaction occurs. Those cells that are in the inner cell mass (around 4 or 5 days after fertilization), *if harvested and cultivated in vitro*, will originate the *totipotent embryonic stem cells*. The external layer of the blastocyst (trophoblast or trophectoderm), in its turn, will originate the placenta. With the progression of mitoses, cells continue differentiating and, between the fifth and eighth weeks, there will be *embryonic germinal cells*, or *pluripotent embryonic stem cells*. This coincides with the end of the embryonic stage; from that phase onward, the concept is considered to be a *fetus* (ninth week until birth).

Totipotent stem cells, as mentioned above, originate all fetal tissues (placental cells and body cells); *pluripotent stem cells* may originate *all body tissues*: ectoderm (neurons, skin), mesoderm (muscles, cartilage, bones, blood, fat), and endoderm (endocrine glands). Stem cells will keep undergoing mitoses and will become more and more specific, therefore losing their ability to differentiate.

Adult stem cells can be defined as those that are extracted after the embryonic stage. Those can be divided into *mesenchymal stem cells* (MSCs) (which will originate osteoblasts, chondroblasts, myoblasts, adipocytes), *hematopoietic stem cells* (HSCs) (which will originate blood cells), and *tissue-specific stem cells* (also known as *precursor cells* or *unipotent cells*, they will originate specific tissues, from which they can be extracted). *Adult stem cells* can be called *somatic stem cells* as well; besides the aforementioned unipotent cells, some may be *oligopotent* or *multipotent*, depending on how many types of cells may be originated. Such cells might be present in *all human tissues*.

Stem cells show the following features:

- They are undifferentiated (i.e., not specialized).
- They may multiply for long periods of time without differentiating, which can lead to a great population of similar cells.
- They are able to perform *asymmetric divisions*, which means that a part of their descendant cells will be identical to them; the other part will carry the characteristics of specialized cells.

Like every structure in living creatures, stem cells also undergo an aging process (*senescence*). After a certain number of mitoses, stem cells may simply halt the process of cell division or, as they multiply, they lose a small part of the distal portion of the DNA (*telomere*). When the DNA reaches a size too small to allow for the continuation of mitoses, the multiplication of cells is stopped. Another point to be considered in cell transplantation is *apoptosis* (programmed cell death): will an adult stem cell, once transplanted to another organism, follow its original path, or will it be able to live longer in its new host? This question emphasizes the following issue: for example, if tissues obtained from an adult with a hypothetical lifespan of thirty more years are transplanted to an 8-year-old child, will this define an "expiration date" of the transplant, therefore limiting the life expectancy of the host to 38 years?

Table 2.1 Stem-cell yield according to donor sources

Source	Umbilical cord	Bone marrow	Liposuctioned fat	Sliced fat
Yield	200–20,000 CFU/mL	100–1,000 UFC/mL	3,600–10,700 UFC/mL	28,000 UFC/g

2.3 Sources of Stem Cells

Embryonic stem cells are usually obtained from the eight-cell blastula stage of mammal embryos. Such cells may have their original nucleus replaced by that of an adult cell, thus originating the process called *cloning*. This biomolecular engineering maneuver will create a new embryo, carrying the characteristics of the adult whose nucleus was transplanted.

Due to ethical, religious, and legal controversies, we will not present a deeper analysis of the methods used to obtain human embryonic stem cells. Another negative aspect regarding that type of stem cell is that there are studies on murines indicating that these cells display a strong tendency of developing tumors (*teratomas*) after a few divisions, owing to their ability of generating multiple tissues. The current lack of capacity to control their growth would be a predisposing factor for these conditions.

In adult humans, *adult stem cells* can offer a well-defined, relatively abundant extraction when harvested from bone marrow, peripheral blood, or fat (the three most frequent sources).

Bone marrow has been considered the usual source of *ASTs* (adult stem cells), although their harvest is a painful procedure, with significant morbidity and low yield [5]; it offers, according to several authors, *from 100 to 1,000 colony-forming units (CFUs)* per milliliter of extracted material. It has its greatest use in hematologic diseases, with well-established results in hematopoietic replacement. The basis for the treatment consists in storing bone marrow cells (either from the patient themselves or from a compatible donor), after selection and exclusion of tumor cells, sterilizing the cells and the sick bone marrow of the patient, and then reinfusing the hematopoietic stem cells (*HSCs*), which will adequately repopulate the bone marrow, without originating teratomas.

In order to harvest stem cells from peripheral blood, it is necessary to administrate stimulants (*hematopoietic growth factors*), with the purpose of releasing stem cells into the peripheral blood, which is easily collected. After collection, the blood is filtered to concentrate the greatest number possible of HSCs to be reinfused. There are controversies regarding the best treatment option between direct harvest of bone marrow stem cells or their obtention from peripheral blood. Diseases in more advanced stages seem to better respond to peripheral blood stem cell transplant.

Umbilical cord is a source of stem cells that has been used for more than 2 decades after the first successful transplant in France. It has been known that both placental blood contained in the cord and Wharton's jelly (*HUCM – human umbilical cord matrix*) are stem cell rich [5]. The storage of umbilical cord blood in specific blood banks has become frequent worldwide. A range of 10,000 samples can be enough to serve an entire population.

Fat obtained from humans is a rich source of *mesenchymal stem cells* (*MSCs*, also known as *ADSCs – adipose-derived stem cells*), which, due to their *plasticity* (ability to generate several types of tissues), show an extremely promising future as the first option for obtaining ASTs [3]. The low-morbidity extraction (through liposuction) and the high yield (*5,000 CFUs per gram of extracted material*) make human fat a potential weapon for stem cell therapy and engineering [7, 14, 15]. The storage of ADSC in tissue banks will permit every person to have a technical supply of their own stem cells for future use (Table 2.1).

2.4 Stem Cell Culture

Stem cells have well-standardized in vitro culture methods [9]. The obtention of stem cells from fat will be described, because it is comprehensive and concerns plastic surgeons. *ASTs* exhibit the feature of adhering to glass or plastic. That allows one to confirm their successful culture only after that process occurs. Once human fat is obtained, either through liposuction or lipectomy, it is placed in sterile, ice-filled plastic bags and sent to the laboratory. The extracted tissue is aseptically homogenized and then taken to a Biological Exposure Chamber, where the tissue is repeatedly washed with PBS (Phosphate Buffered Saline) until all visible blood and excessive fluids are eliminated, in order to leave fat as clean as possible. The washed fat is

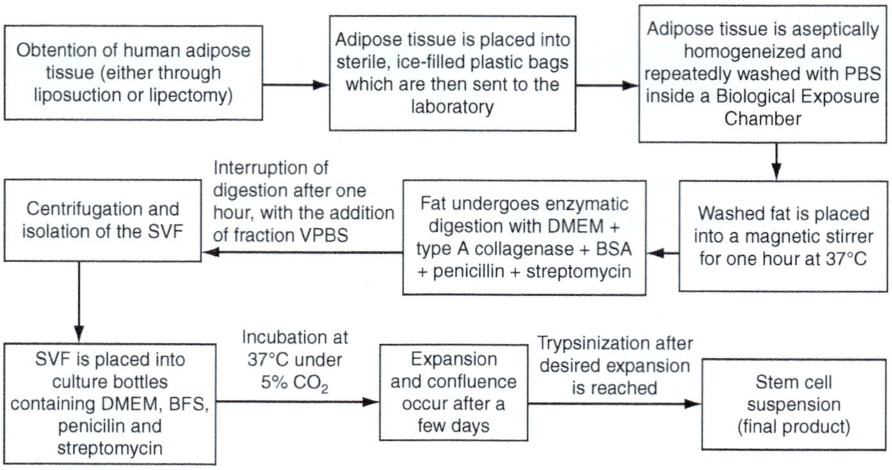

Fig. 2.1 Diagrammatic representation of stem cell culture

Table 2.2 ADSC yield according to donor sites at the moment of harvest, according to author's projection

Donor site	Submental	Arm	Pre-axillay	Gynecomasta	Abdomen	Flank	Riding breeches	Inner tigh	Knee	Sliced abdominal fat
Yield	3,600 UFC/mL	3,800 UFC/mL	7,100 UFC/mL	5,700 UFC/mL	5,500 UFC/mL	5,300 UFC/mL	5,300 UFC/mL	4,300 UFC/mL	7,100 UFC/mL	28,000 UFC/mL

placed in a magnetic stirrer for 1 h at 37°C. Afterwards, the fat undergoes enzymatic digestion, with the addition of a solution prepared with DMEM (Dulbecco's Modified Eagle Medium), type A collagenase, Fraction V BSA (Bovine Serum Albumin), penicillin, and streptomycin. After 1 h, the digestion is interrupted with the addition of BFS (Bovine Fetal Serum). The resultant material is centrifuged, and the floating fraction is discarded. The portion sitting at the bottom of the centrifuged tube (called *pellet*), which is the one containing the so-called stromal vascular fraction (SVF), is transferred to culture bottles filled with DMEM, BFS, penicillin, and streptomycin. The cells are taken to an incubator at 37°C, under 5% CO_2, where they will expand. The liquid content of the bottles is changed daily, and the cells are washed with PBS, resulting in their adhesion to the plastic bottom of the bottles. After washing, the culture medium is replaced. Within a few days, cells will proliferate (*expansion*) and adhere to one another (*confluence*). At this point, stem cells will have become *undifferentiated mesenchymal cells* (UMCs). When cells reach the desired expansion, they are treated with trypsin-EDTA (*trypsinization*), in order to rupture the intercellular junctions. This procedure results in a *stem cell suspension*, for clinical use or culture transfer (Fig. 2.1).

Based on studies conducted at the Hospital Santa Catarina and at the Laboratory of Medical Investigation of the Discipline of Human Structural Topography of the Department of Surgery at the University of São Paulo Medical School, Stocchero projected a Stem Cell Growth Curve based on fat obtained through liposuction of different topographic areas [10], coming to the conclusion that, on the day of collection, there are an average of 5.3×10^3 UMCs per gram of extracted material, and that such UMCs multiply at an average rate of *18.5%* per day. In another study conducted at the Division of Clinical and Toxicologic Analyses at the University of São Paulo School of Pharmaceutical Sciences [1] was concluded that fat donor sites located in the human torso could offer a *23.6%* higher yield when compared to limb donor sites (Table 2.2).

2.5 Animal Models

Most laboratory studies are carried out using several murine species, owing to their easy obtention, confinement, and follow up. Examples well-known to the greater public have involved successful experiments on cattle and sheep. An excellent animal model of easy care was described as a source of ADSCs at the Second

Laboratory of Medical Investigation of the Department of Surgery at the University of São Paulo Medical School – the New Zealand white rabbit [12].

2.6 Pitfalls

Stem cells do not carry a GPS navigation device!

Effective use of ASCs in clinical practice still faces some obstacles that have been traversed quite rapidly. One of the issues to be solved regards the *homing* (the *target*) of treatments. That means the ability to conduct ASCs to their adequate destination. Even though such cells seem to display a certain tropism toward their expected sites of action, it is fundamental that this is an absolutely precise process. It seems perfectly acceptable that HSCs seek the bone marrow as their main target due to their own nature. But when we want MSCs to differentiate into a more specific type of tissue (e.g., cardiomyoblasts), it is essential that the site of interest be reached.

Nowadays, among the greatest challenges, there is the search for an adequate *scaffold* to create three-dimensional structures that can contain stem cells in adequate size and shape, allow angiogenesis and innervation and, above all, be transplantable from the laboratory to humans.

Difficulties increase when the subject in question is the need to replace an organ (which is three-dimensional) starting from two or three imbricate germ layers. And in case of a limb that must be replaced (a situation that is currently not possible to reproduce in an animal model), the solution for this issue possibly relies on the implantation of a germinal bud in the remaining stump to grow back the missing part.

2.7 Pearls

Stem cells suffer from personality disorders! They do not know exactly who they are or what they will be!

Cell differentiation will occur by means of adequate cell differentiation inductors (*triggers*), which are well-known biomolecules capable of determining what tissues MSCs will originate. By linking the development of MSCs to growth factors, one would only have to wait for the desired amount of cells and then halt their proliferation. This would be valid for any tissue-deficient site. In case of organs constituted by few differentiated tissues (e.g., breast), it will be possible to place an MSC *carrier* [4] in the desired place and, with the use of a trigger plus the interaction between neighboring cells, there may be an induction to cell differentiation, through microRNA (*μRNA*), which is a messenger displayed by the surrounding tissue cells. Therefore, plasticity depends on these factors. It is necessary to reach the target. And one can only be sure that the relocated stem cells have indeed differentiated into the desired cells if one uses *cell markers*, which will replicate along with the already transdifferentiated descendent cells [2, 6].

There are situations when stem cells *fuse* with the cells of the target tissue, without the occurrence of true differentiation. Although intended results may apparently be achieved in this manner, it was not owing to stem cell differentiation, but to their fusion to other types of cells. The stem cell-drenched carrier may be the best solution, since it will permit the development of stem cells directed in the site of implantation, using both vascularization and innervation available at the receptor site itself.

2.8 Conclusion

Throughout this chapter, whose intention is no more than that of giving plastic surgeons an idea about the names, practices, and tactics involved in this field of medicine which is still not a part of the daily work of most who practice plastic surgery, we cited some expectations for the future use of these little wonders that are stem cells, in their multiple intriguing features.

It is important to emphasize that the adipocyte in itself will not originate anything. When we discuss fat, we assume that there is a stroma supporting such tissue. MSCs are indeed located within blood vessels and the pericytes surrounding them, where stem cells are responsible for the turnover of fat cells, whose replacement is currently known to take from 2 to 10 years, depending on the body region where they are located. Stem cells work to promote angiogenesis when necessary (e.g., when fat mass increases), to restore the vascular endothelium, etc.

The fact that such stroma is the great source of MSCs can be confirmed by observing that sliced fat tissues obtained through lipectomy show a much higher stem cell yield when compared to liposuctioned fat samples, owing to the increased amount of supporting stroma present in the former.

Extreme care must be taken when referring to the use of stem cells in a treatment. What has been used in wound in healing and breast reconstruction is the *stromal vascular fraction*, which contains stem cells, sometimes enriched with the addition of a cell concentrate that can be obtained even inside the operating room, although it is not a pure stem cell composition [8, 11, 13].

Based on what is currently known about MSCs, it is possible to imagine that, within a relatively short period of time, there will be uses for stem cells such as

- Excellent-quality skin, cultivated from the patient's own cells, to cover burnt areas, or to replace unaesthetic scars and sequelae of tumor resection.
- "Custom-built" replacement bones and cartilage for accident victims and patients who have undergone large resections or suffered from degenerative diseases, or production of new limbs, in case of agenesis.
- Nerve repair after trauma or tumor excision.
- Insertion of collagen in resorption areas of the face, leading to facial rejuvenation through the laxity of folds and rhytids.
- Replacement of allogeneic prostheses using the patient's own tissues. Surely, it will be possible to induce growth in hypoplastic breasts, or to offer new ones to women who have lost them.

Although this seems to be a divine power, man's dedication to Science is making it closer to all of us. From small cells we will get great results (Y.-G. Illouz, 2003, personal communication).

References

1. Almeida KA, Campa A, Alonso-Vale MIC et al (2008) Stromal vascular fraction from fat tissue: obtaining stem cells and their yield according to the topography of the donor areas: previous note. Cir Plást Iberolatinoam 34(1):71–77
2. Barry FP, Murphy JM (2004) Mesenchymal stem cells: clinical applications and biological characterization. Int J Biochem Cell Biol 36:568–584
3. Gimble JM, Guilak F (2003) Adipose-derived adult stem cells: isolation, characterization and differentiation potential. Cytotherapy 5:362–369
4. Iacovella T, Bentini R, Filippin FB et al (2009) Proliferation of adipose-derived mesenchymal stem cells in polylactic acid-derived biomembranes. Poster session presented at the 24th Annual Conference of the Federation of Brazilian Societies for Experimental Biology (FeSBE), Águas de Lindóia, Brazil, 19–22 Aug 2009
5. Lu LL, Liu YJ, Yang SG et al (2006) Isolation and characterization of human umbilical cord mesenchymal stem cells with hematopoiesis-supportive function and other potentials. Haematologica 91:1017–1026
6. Mitchell JB, McIntosh K, Zvonic S et al (2006) Immunophenotype of human adipose-derived cells: temporal changes in stromal-associated and stem cell-associated markers. Stem Cells 24:376–385
7. Ogawa R, Mizuno H, Watanabe H et al (2004) Osteogenic and chondrogenic differentiation by adipose-derived stem cells harvested from GFP transgenic mice. Biochem Biophys Res Commun 313:871–877
8. Rigotti G, Marchi A, Galiè M et al (2007) Clinical treatment of radiotherapy tissue damage by lipoaspirate transplant: a healing process mediated by adipose-derived adult stem cells. Plast Reconstr Surg 119(5):1409–1422
9. Rodbell M (1964) Metabolism of isolated fat cells. Effects of hormones on glucose metabolism and lipolysis. J Biol Chem 239:357–380
10. Stocchero IN, Stocchero GF, Stocchero G et al (2006) Liposuctionable fat: a hypothetic model. Plast Reconstr Surg 117(1):337–338
11. Stocchero IN, Stocchero GHF, Stocchero GF et al (2009) Lamellar fatgrafting: a promising treatment with stromal vascular fraction in recurrent vulvovaginitis. Cir Plást Iberolatinoam 35(3):179–186
12. Torres FC, Rodrigues CJ, Stocchero IN et al (2007) Stem cells from the fat tissue of rabbits: an easy-to-find experimental source. Aesth Plast Surg 31(5):574–578
13. Yoshimura K (2008) Clinical use of adipose-derived stem cells for breast volume enhancement. ISAPS News 2(3):6
14. Zuk PA, Zhu M, Mizuno H et al (2001) Multilineage cells from human adipose tissue: implications for cell-based therapies. Tissue Eng 7:211–228
15. Zuk PA, Zhu M, Ashjian P et al (2002) Human adipose tissue is a source of multipotent stem cells. Mol Biol Cell 13:4279–4295

Principles of Liposuction

3

Juarez M. Avelar

Contents

3.1 **Introduction** 19

3.2 **Surgical Principles** 20
3.2.1 Tunnelization 20
3.2.2 Cannula 22
3.2.3 Forward and Backward Movements 22
3.2.4 Deep Plane Movements of the Cannula 24
3.2.5 Local Infiltration 24
3.2.6 Skin and Panniculus Retraction 24
3.2.7 Derived Procedures Stemming from Liposuction 26

3.3 **Combined Procedures of Liposuction with Conventional Panniculus Resection** 26
3.3.1 Abdominoplasty Combined with Liposuction Without Panniculus Undermining: Lipoabdominoplasty 27

3.4 **Flankplasty and Torsoplasty** 27

3.5 **Aesthetic Plastic Surgery of the Axilla** 27

3.6 **Medial Thigh Lift** 28

3.7 **Conclusion** 28

References 28

J.M. Avelar
Professor Assistent of Plastic Surgery, Unit of Medical Shcool of Marilia, São Paulo, SP, Brazil
e-mail: juarezavelar@bol.com.br

3.1 Introduction

Liposuction is a superb contribution by Yves-Gérard Illouz for the treatment of localized fat deposits and nowadays is one of the most frequently performed surgical procedures all over the world. His first presentation was done in 1977 and the first publication in 1980 [19]. Until then correction of those deformities was a great challenge, performed only through techniques based in panniculus undermining. The results of all the previous techniques performed by outstanding plastic surgeons used to leave long scars and very often the final results were ungraceful. The final scars often were irregular due to excessive resection or even due to the damaged *fascia superficialis.* The trochanteric regions are the most important areas concerning localized adiposity which bring severe alterations of the silhouette especially in women.

All principles of liposuction were very well advocated, presented, and published by Illouz:

- He created the cannula, which is the fundamental instrument for liposuction that creates tunnels inside the adipose tissue avoiding any undermining or cavities.
- He developed a machine that promotes negative pressure in order to aspirate the fat.
- He demonstrated that the method is indicated only for treatment of localized fat in all regions of the human body.
- He described that the cannula should work in deep tissue planes.
- He emphasized the use of local infiltration to avoid bleeding.
- He taught his methodology worldwide with lectures, surgical demonstrations, and scientific publications.

Y.-G. Illouz and A. Sterodimas (eds.), *Adipose Stem Cells and Regenerative Medicine*,
DOI: 10.1007/978-3-642-20012-0_3, © Springer-Verlag Berlin Heidelberg 2011

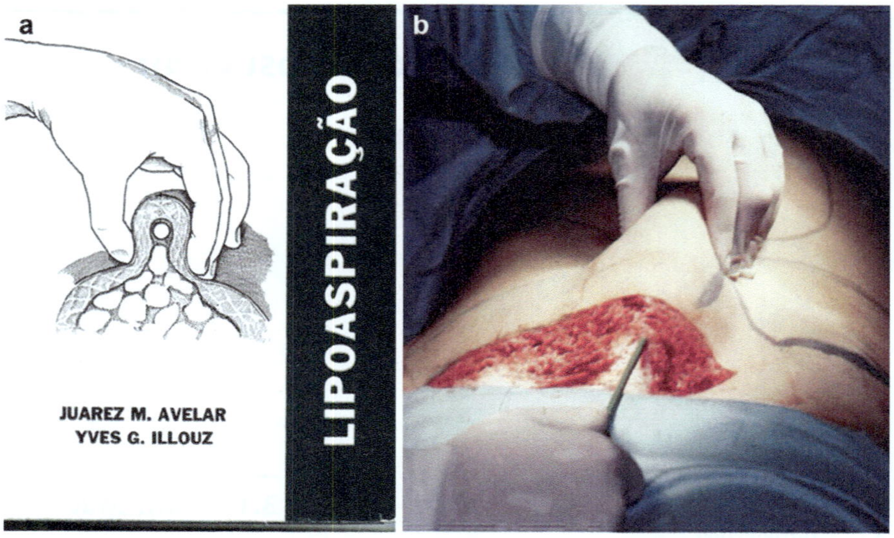

Fig. 3.1 Pinching of the panniculus can control the position and depth of the cannula in order to perform liposuction on the abdominal wall. (**a**) The cover of our book published in Brazil in 1986 with its illustration showing how to pinch the abdominal panniculus for liposuction. (**b**) Photo during operation showing the surgeon's left hand pinching the panniculus and with his right hand holding the cannula. The tip of the cannula passes between the surgeon's fingers

3.2 Surgical Principles

Didactically I will present in this chapter the basic principles of liposuction that have been already described by Illouz and several other authors in scientific journals and in books as well. They are:
Tunnelization

- Creation of the cannula
- Forward and backward movements of the cannula
- Deep plane movements of the cannula
- Local infiltration
- Skin and panniculus retraction
- Derived procedures stemming from liposuction

3.2.1 Tunnelization

The main surgical principle of the liposuction technique is to create tunnels on subcutaneous layer in order to remove fat tissue and to preserve all vascular network, arterial and venous, lymphatic and nerve endings. Therefore, the volume of fat tissue is reduced without damaging the structures between the muscular level and the cutaneous surface (Figs. 3.1 and 3.2).

Since I was taught how to perform liposuction by Illouz in 1981, I introduced some procedures which have been very useful. Pretunneling is recommended when the surgeon inserts the cannula with the purpose of reducing the thickness of the panniculus, which gives the surgeon adequate judgment of the volume that needs to be aspirated. The surgeon works more comfortably with less physical effort. The effect of the negative pressure of the machine associated with the surgeons' effort to insert the cannula make the surgeons' movement more difficult as I have already described several years ago [4, 5]. Pretunneling is an excellent step helping to keep an adequate thickness of the remaining panniculus on the areolar layer, which is located between the *fascia superficialis* and the cutaneous level (Fig. 3.3). In my publications [6, 7], the anatomy of the panniculus is described, based on dissections on cadaver, describing the lamellar layer, the fascia superficialis, and the areolar layer (Figs. 3.2 and 3.3). Illouz [23] and Hetter [18] have also described the anatomy of the panniculus. The surgeon needs to fold the panniculus with one hand and with the other hand must introduce the cannula and by gentle movements should perform the liposuction (Fig. 3.1).

The thickness of the remaining panniculus will be useful for comparison when the other side will be aspirated [4]. To evaluate the regularity and thickness of the panniculus the Illouz' "holling test" maneuver is very useful [20].

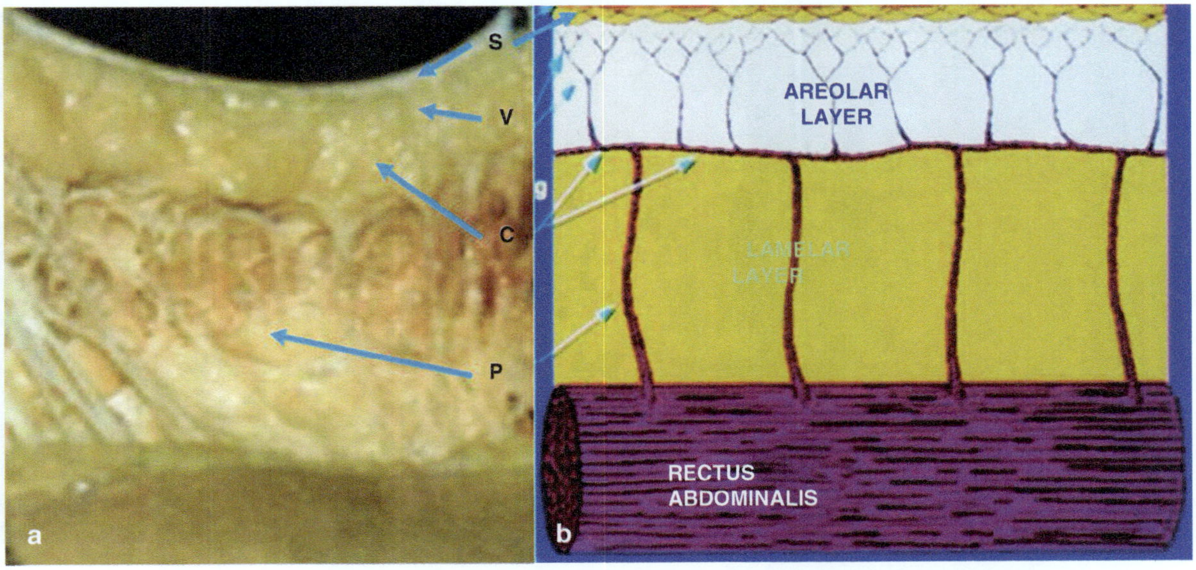

Fig. 3.2 Blood supply of the abdominal panniculus. (**a**) Photo of a segment of an abdominal panniculus of a cadaver after liposuction performed on the lamelar layer. One can see the areolar layer that is well preserved with its normal fat layer. (**b**) Drawing shows the relationship of the vascular network with areolar and lamelar layers. The perforating vessels (*arrows P*) go from the muscle to the *fascia superficialis* where the communicating vessels (*arrows C*) make a "bridge" in the fascia superficialis, in order to provide blood supply (*arrows V*) to the skin (*S*) and subdermal layer

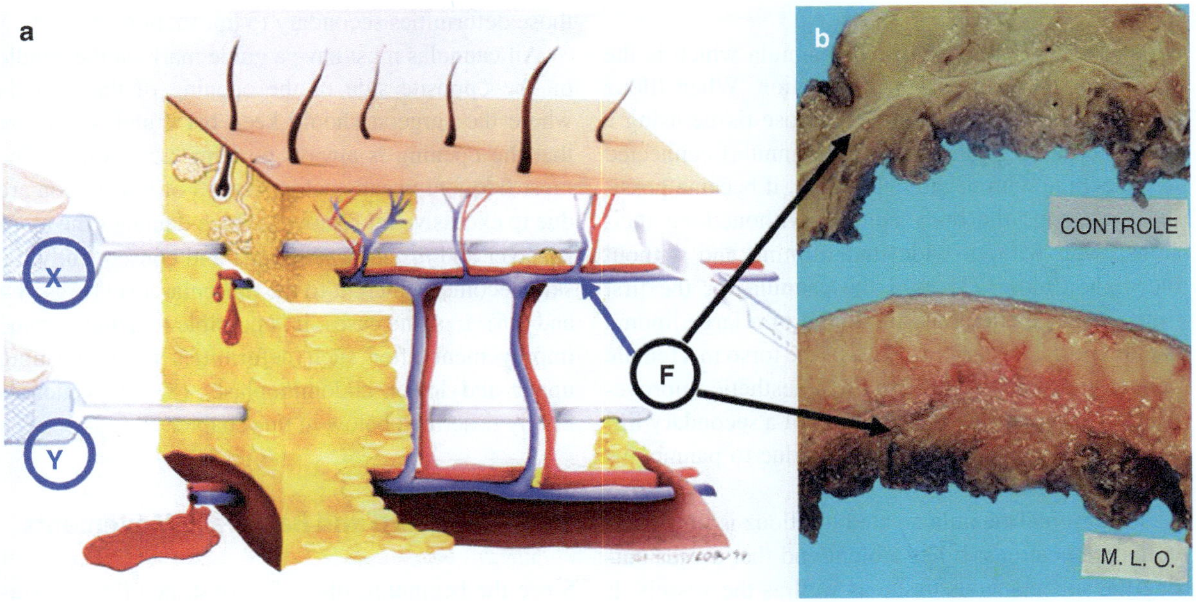

Fig. 3.3 Anatomy of abdominal panniculus. (**a**) Drawing shows the relationship of the vascular network with the areolar and lamelar layers. The perforating vessels go from the muscle to the *fascia superficialis* where the communicating vessels make a "bridge" in order to provide blood supply to the skin and subdermal layer. The cannula *X* is on wrong level, too superficial on the lamelar layer. The cannula *Y* is placed on correct level on the areolar layer. (**b**) Photo of the two segments of the abdominal panniculus. *On top* – a segment of a cadaver without localized fat deposits shows that the *fascia superficialis* is thin and the areolar layer presents normal and regular thickness. *On bottom* – a segment of the abdomen, 6 months after liposuction properly performed on lamelar layer shows that the *fascia superficialis* is preserved as well as the areolar layer. The *arrows* (*F*) indicate the *fascia superficialis* on the drawing, also on the segment of the abdomen of a cadaver and on segment of the panniculus of the abdomen after liposuction (Photo was kindly given by Prof. Callia and Dr. Batuira)

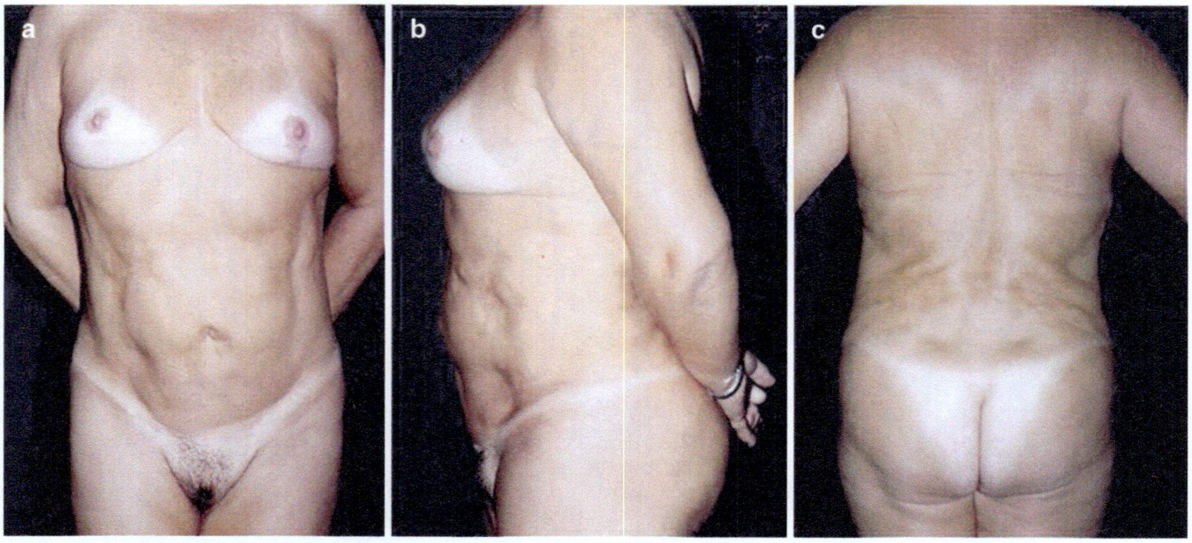

Fig. 3.4 Unfavorable result after liposuction on the abdominal wall and flanks. (**a**) *Front view*, irregularities can be seen on the surface with unaesthetic depressions. (**b**) *Lateral view* show asymmetry with deep depressions. (**c**) *Posterior view*, severe asymmetry can be seen with unaesthetic depressions on each side of the torso. One can see that it was done too much liposuction with severe damage of the abdominal panniculus on the areolar layer and fascia superficialis

3.2.2 Cannula

Illouz created and developed a cannula which is the key factor for a successful liposuction. When Illouz had the genial idea to remove adipose tissue using a very simple surgical instrument (cannula) connected to a machine with negative pressure, it became possible to reshape the body contour without long incisions, even without wide undermining and without bleeding. Illouz [23] used the cannula for the first time, when he attempted treatment of a large lipoma located on the posterior surface of the torso in a female patient with good result and great aesthetic improvement. Afterwards he treated a case of a secondary lipodistrophy on trochanteric regions due to panniculus over correction.

The original cannula created by Illouz is illustrated in his publications; it has a blunt end that avoids cutting the subcutaneous tissue as well as the vessels. It has one suction hole (1–1.5 cm of diameter) close to the tip. After he created the first cannula, many plastic surgeons developed a series of liposuction cannulas. When liposuction is performed with that suction hole pointing against the subdermal layer, it may be considered as an incorrect maneuver or even a technical mistake. The opening of the cannula may damage the subdermal layer causing irregularities and unfavorable retractions of the skin. Vilain [30] has published about those deformities secondary to liposuction.

All cannulas must have a guide mark on the handle on the opposite side of the opening of the cannula where the surgeon should keep his thumb to ensure that the opening is always pointing downwards. The most frequent unsatisfactory results of liposuction are due to excessive fat removal, which damages the areolar layer and *fascia superficialis* and consequently the skin becomes adherent to the muscular level (Figs. 3.4 and 3.5). In some cases, it is possible to achieve some improvement after such deformities by performing upper and lower abdominoplasty [8, 12] which is briefly described below in this chapter.

3.2.3 Forward and Backward Movements

Since the beginning Illouz demonstrated that the surgeon should hold the cannula with one hand and pinch the panniculus with the other one in order to introduce the instrument with forward and backward movements [22]. This concept is so important that in 1986 [13] the cover of our book included an image demonstrating this maneuver (Fig. 3.3). Unfortunately, surgeons who did not practice this maneuver had severe complications post-operatively (Fig. 3.4). When the surgeon

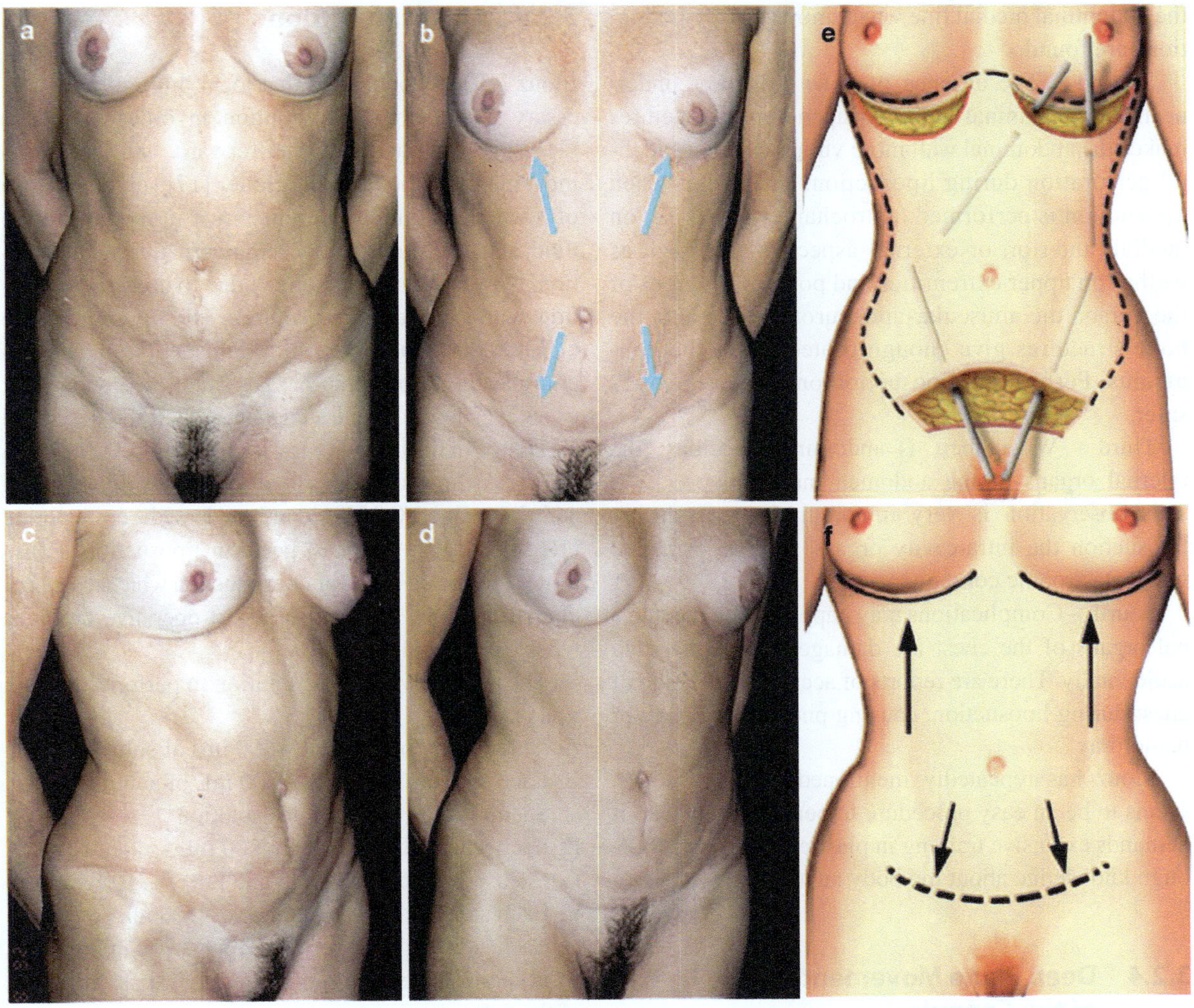

Fig. 3.5 Unfavorable result after liposuction on the abdominal wall. On photos (**a**) and (**c**) one can see unaesthetic result after liposuction on the abdominal wall. (**b**, **d**) Photos after abdominoplasty was performed. The upper abdominal panniculus was pulled upwards and the lower abdomen pulled downward after tunnelization with cannula. (**e**) Drawings of the lower and upper abdominoplasty method without panniculus undermining or dissection. With (**f**) The *arrows* show the direction of the traction upward and downward with the final scars on suprapubic area and submammary folds

performs the movements of the cannula pinching the panniculus with one hand and introducing the cannula with the other one, it is possible to evaluate how deep the instrument inside the abdominal panniculus is.

The most dangerous region of the human body to perform liposuction is the abdominal wall. The cannula can perforate the muscular-aponeurotic structures when the forward and backward movements are not done correctly.

The abdominal wall presents some peculiar anatomical characteristics:

First – Female patients may have anatomical alterations after pregnancies, presenting cutaneous laxity, striae, and decreased elasticity. Also, repeated weight loss may damage gradually the *rectus abdominalis* as well as the aponeurosis. These structures become very thin and as a result it is quite dangerous to perform liposuction on these cases. It does not mean that those patients cannot undergo liposuction, but the surgeon must be cautious before planning surgery. I recommend my patients to do a pre-operative ultrasound and tomography of the abdominal wall in order to evaluate the anatomy of the muscular-aponeurotic structures. Even if those preoperative examinations do not present anatomical disturbances, it is advisable to perform liposuction with special care especially on the abdominal medial line where mostly of the herniations are found.

Second – The absence of bone structures underneath the abdominal muscular-aponeurotic structures makes the abdominal wall more vulnerable to accidental perforation during liposuction. Therefore, when liposuction is performed on trochanteric regions, on medial, anterior, or external aspect of the thigh as well as on upper extremities and posterior regions of the torso, the muscular-aponeurotic wall and the bone structures give enough protection to the panniculus. Performing liposuction on these regions is safer.

Third – When there is abdominal flaccidity, the internal organs of the abdomen may prolapse [27]. This observation is very important because during liposuction the intra-cavity organs may be damaged with catastrophic consequences.

Fourth – Complications may happen when the internal organs of the chest are damaged by the cannula accidentally. There are reports of accidental damage of chest during liposuction, causing pneumotorax and/or hemotorax.

Illouz has repeatedly mentioned that liposuction seems to be an easy procedure to perform, however, it demands extensive training in plastic surgery and thorough knowledge about the body anatomy.

3.2.4 Deep Plane Movements of the Cannula

The cannula should be kept deep, close to the muscular level and should maintain a significant fat layer underneath the skin [6]. Some bad results after liposuction can be seen in Figs. 3.4 and 3.5. The amount of remaining fat is more important than the amount of aspirated fat [4, 5].

I personally make meticulous demarcation of each region to be liposuctioned, 1 day before the operation. In this way, the patient can see her/his body in the front of a mirror in order to confirm all the demarcations.

Anatomical studies have been performed on female and male cadavers of different ages. In my publications [6, 7], the peculiar organization of the vascular network as well as its relationship with the *fascia superficialis* has been described. The anatomy of the abdominal wall panniculus after liposuction on a cadaver can be seen in Figs. 4.2 and 4.3.

3.2.5 Local Infiltration

Illouz [19] has published that local infiltration is necessary using hypotonic solution in order to reduce bleeding and facilitate the forward and backward movements of the cannula. Hetter [17] studied the use of low concentration of epinephrine in the infiltration fluid and the variation of hematocrit during liposuction. Nowadays, it is recommended to do local infiltration with normal saline solution adding epinephrine (1:1,000,000) and performing liposuction under general anesthesia or epidural. The solution I personally use is even less concentrated than studied and described by Hetter in his book [18], which is 1:435,000. When I perform liposuction in small areas, it is possible to add lidocaine (0.25%) combined with epinephrine (1:1,000,000) in order to perform liposuction. Echymoses may be observed after liposuction which disappear after 8–12 days. I do not use post-operative drainage when I perform liposuction.

The average volume of solution to perform liposuction is:

A. Anterior abdomen wall – 300 mL of solution
B. Each side of the flanks – 200 mL of solution
C. Sacral region – 200 mL of solution
D. Each trochanteric region – 200 mL
E. Each side of medial thigh and knees – 100 mL
F. Each side of the face and neck – 100 mL
G. Each side of the upper extremity – 200 mL

It is advisable to not infiltrate more than 1,000 mL of the solution on the patient.

It is possible to perform liposuction without any infiltration, which is called the "dry" technique as described by Fournier [16]. I do not recommend it because the risk of hemorrhage during surgery and hematoma and seroma formation after surgery is quite high.

Blood transfusion used to be done quite often after liposuction as has been described by Uebel [28]. Nowadays, it is not necessary to be performed as long as local infiltration is injected.

3.2.6 Skin and Panniculus Retraction

In young patients, isolated liposuction gives a very good result on the face and neck due to skin retraction after removal of the excess fat [6, 7]. When patients present accumulated adiposity without excess and

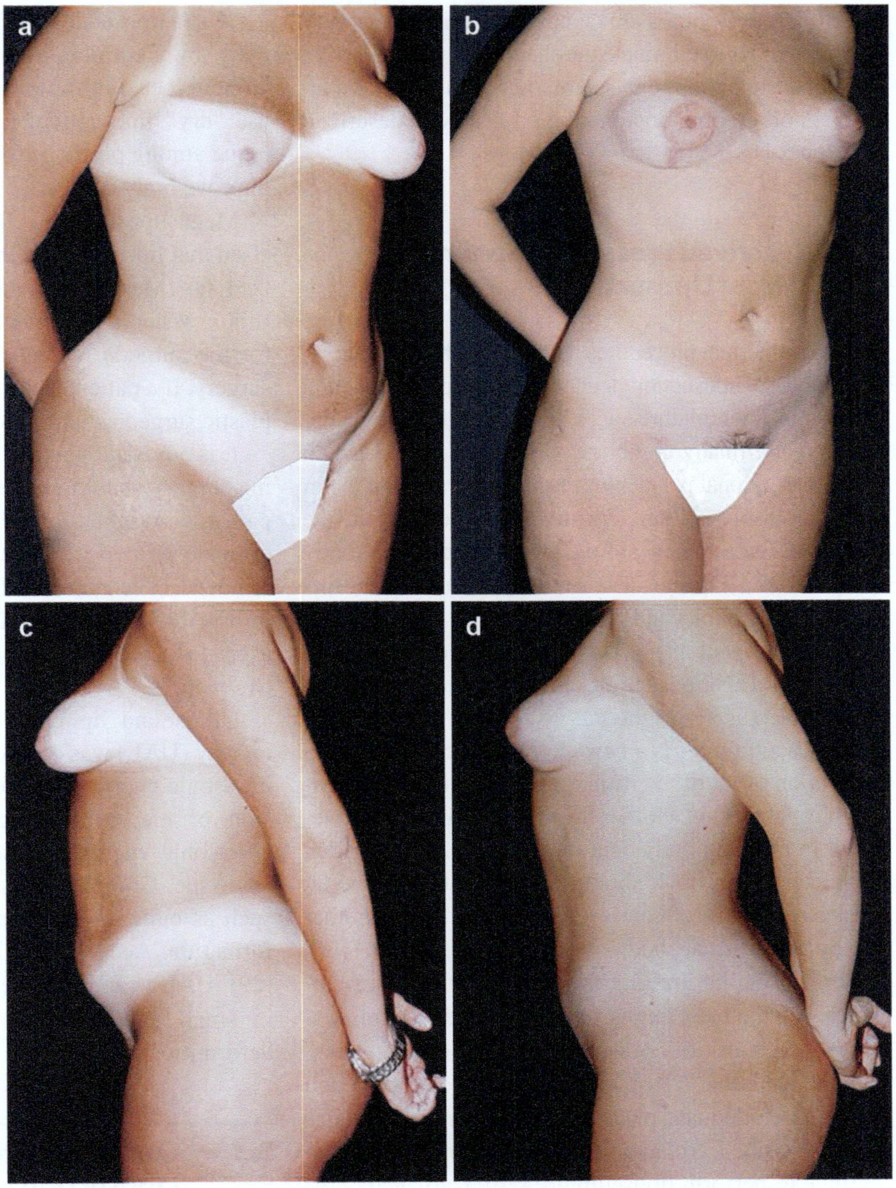

Fig 3.6 Favorable result after liposuction on the abdominal wall, flanks, and buttocks. Photos (**a**, **c**) Pre-operative views of 20-year-old girl showing abundant accumulated fat on abdomen, flanks, buttocks, and trochanteric regions. Photos (**b**, **d**) the final result 1 year after liposuction

flaccidity of the skin, liposuction may be performed without cutaneous resection (Fig. 3.6). However, if a patient presents accumulated adiposity combined with skin flaccidity and less elasticity, liposuction should be performed combined with skin resection (full abdominoplasty or lower abdominoplasty or even lower and upper abdominoplasty) [5, 9, 12, 13].

The pinching test recommended by Illouz is an important step pre-operatively to evaluate the amount of fat tissue accumulated in the panniculus. The surgeon must examine the entire body contour with the patient in a standing position as well as in sitting position.

I have in my examination room two large mirrors placed vertically one in front of the other and another two mirrors located on top of the vertical ones making a 45° angle with the ceiling. When I am examining a patient I have him/her stand in front of both vertical mirrors so he/she can see his/her entire body. I feel that this simple arrangement of the mirrors enables the patient to identify some irregularities on the body that may have not been observed before. When pre- and post-operative photos are taken, those mirrors are very useful in order to eliminate the shadow on patients' back side and the patient can see his/her natural body contour. Photographic documentation has been described by Hetter [18].

3.2.7 Derived Procedures Stemming from Liposuction

Several important procedures have been published following Illouzs' liposuction technique. Lipoplasty, suction lipectomy, liposculpture, mega liposuction, ultrasound lipoplasty, and finally combined procedures of liposuction with conventional panniculus resection are some of them.

Lipoplasty – When liposuction is performed, the surgeon creates a new silhouette on the patients' body [21]. Illouz [23] and also Hetter [18] have used this term in their books and publications.

Suction lipectomy – This term has been introduced by Illouz. The liposuction cannula is a blunt instrument which aspirates the fat.

Liposculpture – is a new expression combining the aspiration of fat tissue from one region and transplanting it as fat graft in another in order to achieve improvement in face and body contour. Illouz [22] has written a chapter in our book about the reutilization of the fat tissue aspirated from one region and grafted to another in order to treat body irregularities. Toledo [25] described the fat grafting for the correction of irregularities on the face, thighs, and buttocks due to unsatisfactory results of previous liposuction. Toledos' cannula with the "V" tip to treat the retracted scars and to prepare adequate place to embed the fat graft has been described. Toledo has also developed a special syringe to perform liposculpture [26]. Carpaneda [14] has presented outstanding results after fat injection. Fournier [15] has presented a wide variety of fat-graft techniques for body contouring improvement.

I am convinced that fat grafting is a very useful procedure to improve body contour. Fat injection combined with face lifting may give excellent results [14]. For buttocks augmentation, I perform liposuction of the supra-iliac regions as a donor area and then inject fat in the intramuscular plane.

I have never though performed fat grating for breast augmentation. It may cause concerns on mammary pathology during palpation as well as during a mammogram.

"Dry" liposuction – This expression was used initially by Fournier [16] who performed liposuction without local infiltration. Although I personally do not use "dry" liposuction, it is a well-popularized technique among plastic surgeons and it is another way to improve body contour.

Mega liposuction – Illouz repeatedly has emphasized that liposuction is indicated for removal of localized fat. Mega liposuction is not recommended by Illouz. When a patient is overweight, it is mandatory for the surgeon to recommend losing weight before surgery. If a patient is not motivated to lose weight, the plastic surgeon should not indicate liposuction.

It is advisable that the volume of fat aspirated during surgery should not be more than 4% or 5% of the patient's weight. For that reason the surgeon must evaluate properly the amount of fat that needs to be removed. The nurse in the operating theatre should inform the surgeon about the volume that is being aspirated from each region according to the pre-operative plan.

Ultrasound lipoplasty – Ultrasound-assisted lipoplasty (UAL) was developed by Zocchi [31, 32]. The ultrasonic energy is based on the principle of conversion of electrical energy to a mechanical wave. Ultrasonic energy produces selective destruction of fat tissue. Complications though have been reported such as burns of the skin and post-operative seroma formation. According to Perez [24] and Van Tetering [29], UAL is an effective method of body contouring in small and moderate cases of accumulated fat deformities.

3.3 Combined Procedures of Liposuction with Conventional Panniculus Resection

The use of liposuction combined with abdominoplasty [5] has been well documented. In that publication, it is described that panniculus resection of the abdominal wall is done after the liposuction procedure. Obviously, in order to perform resection of the excess panniculus a wide undermining needs to done according to the conventional methods.

In my practice, I used to have complications that encouraged me to study the anatomy of the panniculus again as well as to analyze all aspects looking for an adequate method that could lessen the risk of intraoperative and postoperative complications. After performing research, I concluded that during traditional abdominoplasty, there is significant trauma on the vascular network of the abdominal panniculus. Based on this research [6, 7], I developed a new technique of abdominoplasty [8, 12] based on the principle that the perforating vessels of the abdominal wall do not need to be cut and consequently the arterial, venous, and lymphatic networks are preserved. The main principle is to maintain the normal blood supply to the panniculus, avoiding major complications during and after surgery.

3.3.1 Abdominoplasty Combined with Liposuction Without Panniculus Undermining: Lipoabdominoplasty

I will describe this procedure in more detail since it was the beginning of my personal method to perform the combined approach without panniculus undermining.

Two areas must be well demarcated before surgery:

A. The infraumbilical area to be resected;
B. The area to be liposucted.

When lipoabdominoplasty is performed, our preference is epidural anesthesia combined with intravenous sedation. Local infiltration with saline solution with epinephrine (1:1,000,000) is done.

3.3.1.1 Liposuction Procedure

The first step of the operation is to perform liposuction on:

- The infraumbilical area, where skin resection will be done, liposuction is performed on the full thickness of the panniculus. All perforating vessels (arteries, veins, and nerves), connective tissue and *fascia superficialis* are preserved.
- On the supraumbilical area, where the skin will not be resected, deep liposuction is done which means on lamellar layer, below the *fascia superficialis*. Therefore, at the end, the thickness of the remaining panniculus will be formed only by the areolar layer.

In my previous publications [8, 12], I emphasized that the areolar layer must be preserved in order to achieve regular thickness of the remaining panniculus, giving a harmonious result and smooth balance to the body contour.

3.3.1.2 Full-Thickness Skin Resection

When it is necessary, reinforcement of the musculo-aponeurotic system of the abdominal wall is done by plication. The plication may be performed on midline and also obliquely on each side of the umbilicus. In our anatomical dissections on cadavers, we found that the perforating vessels are preserved and go from the muscles to the panniculus, from the *rectus abdominalis* on each side to the remaining panniculus.

3.3.1.3 Pulling the Remaining Abdominal Panniculus Flap

The next step of the operation is to pull downward the remaining abdominal flap. Demarcation of the umbilicus' new position by the use of my personal surgical instrument is then done [3].

3.3.1.4 Transposition of the Umbilicus and Suture of the Surgical Wound

After demarcation of the new umbilical area, I employ my personal approach [1, 2] in order to create a natural umbilicus. A circle is drawn and three incisions are marked in order to create three small cutaneous flaps on the abdominal flap. Afterward the umbilicus is pulled outward and it is sutured in place. I do not use drains.

3.4 Flankplasty and Torsoplasty

Torsoplasty and flankplasty [11] are very useful procedures for the body contour although the resulting scars are visible. Patients after massive weight loss present with excess skin that folds over itself giving an unaesthetic appearance.

Liposuction procedure is performed all over the marked areas but in two different levels. In the area where cutaneous resection will be done liposuction is performed on full thickness of the panniculus. However, on the regions where there are localized fat deposits, liposuction is performed only on the deep level below the *fascia superficialis,* preserving the areolar layer.

3.5 Aesthetic Plastic Surgery of the Axilla

The axilla is a difficult region to perform aesthetic surgery due to its complex anatomy. Patients presenting with hyperhydrosis and with flaccidity of the skin are good candidates [9]. The operation may be performed under local anesthesia combined with intravenous sedation and the demarcation is done 1 day prior to the operation.

A subdermal fat "island" flap needs to be created in order to give a natural support to the future axillary cavity. The first step of the operation is full-thickness skin resection of the elliptical area already drawn on the center of the axilla from the anterior to the posterior border of the axillary region. In order to create the subdermal fat "island" flap two incisions are done on each border of the flap. The *fascia superficialis* is the deep limit of those incisions in order to avoid any damage to the axillary structures. The next step of the operation is a subcutaneous tunnelization on each border of cutaneous incisions using a 2 mm cannula. Liposuction may be performed when the patient presents localized fat deposits on the inner side of the arm and on the lateral side of the chest wall.

The anterior border of the subdermal fat "island" flap is sutured to the major pectoralis muscle and the posterior one is sutured to the latissimus dorsi muscle. In order to perform the cutaneous lifting of the lateral aspect of the chest and the inner side of the arm, the cutaneous flap is pulled upwards and sutured to the dermal fat "island" flap.

The number of sweat glands is reduced during operation, and the skin flaccidity of the axilla is improved since the cutaneous excess is removed without undermining.

3.6 Medial Thigh Lift

The medial thigh may present localized fat deposits, cutaneous flaccidity which may cause many problems to patients, and the aesthetic correction of such deformities is not a routine procedure. The operation has been described as a technical evolution and refinements of old techniques that used to be performed with wide undermining of the medial and anterior aspects of the thigh [10].

The demarcation prior the operation is a mandatory procedure since the patient needs to understand where the final scars will be placed. The operation is done under local anesthesia combined with intravenous sedation. Local infiltration is done in two different levels: intradermally on the inguinal fold and deep infiltration below the localized fat deposits.

Liposuction if needed is performed in two levels: full-thickness liposuction of the panniculus where the excess skin will be excised and deep liposuction is done, below the *fascia superficialis* in order to remove localized fat deposits. After liposuction is performed, the excess of the skin is removed on the inguinal sulcus without damage of the vascular network underneath. The final scars lie on the inguinal fold.

3.7 Conclusion

Liposuction is an outstanding technique created, described, and popularized by Yves-Gérard Illouz. Liposuction brings many advantages for the treatment of localized fat deposits and a very high rate of satisfaction to patients. Liposuction is used for the treatment of localized fat in all regions of the human body and it is not a method for treatment of obesity.

References

1. Avelar JM (1976) Umbilicoplastia - uma técnica sem cicatriz externa, An do XIII Cong Bras de Cir Plast Porto Alegre, pp 81–82
2. Avelar JM (1978) Abdominoplasty; methodization of a technique without external umbilical scar. Aesthet Plast Surg 2:141
3. Avelar JM (1983) Abdominoplasty: technical refinement and analysis of 130 cases in a 8-year follow-up. Aesthet Plast Surg 7:205
4. Avelar JM (1985) Fat-suction versus abdominoplasty. Aesthet Plast Surg 9:265–276
5. Avelar JM (1985) Combined liposuction with traditional surgery in abdomen lipodistrophy. XXIV. Instructional Course of Aesthetic Plastic Surgery of the ISAPS, Madrid
6. Avelar JM (1986) Anatomia cirúrgica e distribuição do tecido celular subcutâneo no organismo humano. In: Avelar JM, Illouz YG (eds) Lipoaspiração. Hipócrates, Sao Paulo, pp 45–57
7. Avelar JM (1989) Regional distribution and behavior of the subcutaneous tissue concerning selection and indication for liposuction. Aesthet Plast Surg 13:155–165
8. Avelar JM (1999) A new technique for abdominoplasty – closed vascular system of subdermal flap folded over itself combined to liposuction. Rev Bras Cir Cardiovasc 88/89(1/6):3–20
9. Avelar JM (1999) Aesthetic plastic surgery of the axilla. Rev Bras Cir Cardiovasc 88/89(1/6):41–54
10. Avelar JM (1999) Aesthetic plastic surgery in the inner side of the thigh. Rev Bras Cir Cardiovasc 88/89(1/6): 57–67
11. Avelar JM (1999) Flankplasty and torsoplasty – a new surgical approach. Rev Bras Cir Cardiovasc 88/89(1/6): 21–35

12. Avelar JM (2002) Abdominoplasty without panniculus undermining and resection: analysis and 3 year follow up of 97 consecutive cases. Aesthet Surg J 22:16–25
13. Avelar J, Illouz YG (1986) Lipoaspiração. Hypócrates, São Paulo
14. Carpaneda CA, Ribeiro MT (1994) Percentage of graft viability versus injected volume in adipose autotransplants. Aesthet Plast Surg 18:17–19
15. Fournier P (1989) Liposculpture – ma technique. Arnette, Paris
16. Fournier PE, Otteni FM (1983) Lipodissection in body sculpturing: the dry procedure. Plast Reconstr Surg 72(5):598
17. Hetter GP (1984) The effect of low dose epinephrine on the hematocrit drop following lipolysis. Aesthet Plast Surg 8(1):19
18. Hetter GP (1990) Lipoplasty – the theory and practice of blunt suction lipectomy. Little Brown and Company, Boston
19. Illouz YG (1980) Une nouvelle technique por lês lipodystrophies localizes. Rev Chir Esthet Lang Franç 6:19
20. Illouz YG (1983) Body contouring by lipolysis: a 5 year experience with over 3000 cases. Plast Reconstr Surg 72(5):591
21. Illouz YG (1985) Surgical remodeling of the silhouette by aspiration lipolysis or selective lipectomy. Aesthet Plast Surg 9(1):7
22. Illouz YG (1986) Reutilização do tecido adiposo lipoaspirado. In: Avelar JM, Illouz YG (eds) Lipoaspiração. Hipócrates, Sao Paulo, pp 117–122
23. Illouz YG, Villers YT (1989) Body sculpturing by liposuction. Churchill Livingstone, New York
24. Perez JA (1999) Ultrasonic-assisted lipoplasty endoscopic view in vivo. American Society of Plastic Surgeons Congress, New Orleans
25. Toledo LS (1991) Syringe liposculpture. A two year experience. Aesthet Plast Surg 15:321–326
26. Toledo LS (1993) Total liposculpture. In: Gasparotti M, Lewis CM, Toledo LS (eds) Superficial liposculpture. Springer, New York, p 44
27. Toledo LS (2001) Refinamentos em cirurgia estética da face e do corpo. Revinter, Rio de Janeiro
28. Uebel CO (1986) Autotransfusão e hemodiluição aplicadas à lipoaspiração. In: Avelar JM, Illouz YG (eds) Lipoaspiração. Hipócrates, Sao Paulo, pp 123–126
29. Van Tetering JPB, Thio EA (2001) Ultrasound-assisted lipoplasty: personal experience with the contour genesis system with 71 consecutive patients. Cir Plast Iberolatinamer 27:119
30. Vilain R (1986) Prevention and treatment of waves after suction lipectomy. Ann Plast Surg 17(3):194–205
31. Zocchi ML (1992) Ultrasonic liposclupturing. Aesthet Plast Surg 16:287
32. Zocchi ML (1995) Ultrasonic-assisted lipectomy. Plast Reconstr Surg 11:197–221

Clinical Applications of Liposuction

4

Melvin A. Shiffman

Contents

4.1 **Introduction** 31

4.2 **Modifications in Instruments and Techniques for Liposuction** 31
4.2.1 Orthostatic Liposculpture 31
4.2.2 External Ultrasound 31
4.2.3 Low-Level Laser-Assisted Liposuction 32
4.2.4 Powered Liposuction 33
4.2.5 Current Instruments 33

4.3 **Liposuction for Cosmetic Purposes** 35
4.3.1 Breast Reduction 36
4.3.2 Gynecomastia 36
4.3.3 Cellulite 36
4.3.4 Apocrine Gland Disorders 37
4.3.5 Hematoma 37
4.3.6 Involuted Hemangiomas 37
4.3.7 Lipoma and Lipomatosis 37
4.3.8 Liposuction-Assisted Nerve-Sparing Hysterectomys 38
4.3.9 Lymphedema 38
4.3.10 Obesity 38

4.4 **Reconstructive Surgery** 39
4.4.1 Silicone Removal 39
4.4.2 Steroid-Induced Lipodystrophy 39

4.5 **Non-cosmetic Applications** 39

4.6 **Conclusions** 39

References 39

M.A. Shiffman
Chair, Department of Surgery, Newport Specialty Hospital, 17501 Chatham Drive, Tustin, CA 92780, USA
e-mail: shiffmanmdjd@yahoo.com

4.1 Introduction

Since the introduction of liposuction for cosmetic purposes, the procedure has evolved to include, not just the body, but the face and extremities [47, 82, 83]. Klein described local tumescent anesthesia that was the prelude for safer liposuction and increased the surgeon's ability to remove more adipose tissue [92–94]. Fournier reported the use of syringe liposuction that led to less vacuum pressure resulting in less bleeding and preservation of fat for fat transfer [51–53]. Ostad described the safe use of 55 mg/kg of lidocaine in liposuction [123].

4.2 Modifications in Instruments and Techniques for Liposuction

4.2.1 Orthostatic Liposculpture

Orthostatic liposculpture was originated by Fischer in order to perform liposuction in the same standing position as the markings are made [46]. A table was developed that would bring the patient to a standing position for liposuction and could be reversed if the patient became dizzy or faint.

4.2.2 External Ultrasound

External ultrasound may be used immediately before liposuction to facilitate the procedure or to speed healing on follow-up visits during the postoperative period.

Y.-G. Illouz and A. Sterodimas (eds.), *Adipose Stem Cells and Regenerative Medicine*,
DOI: 10.1007/978-3-642-20012-0_4, © Springer-Verlag Berlin Heidelberg 2011

The application of ultrasonic energy to the adipose tissue effectively liquefies the fat, releasing a combination of triglycerides, normal interstitial fluid, and the infused tumescent solution. These components form an emulsion, which can be removed using vacuum suction. Because of the predilection of the ultrasonic waves for low-density tissue such as fat, there is felt to be a selective targeting of the fat cells without affecting the intervening connective tissue and neurovascular structures. The depth of penetration is inversely proportional to the frequency used. It is felt that ultrasonic energy affects the adipose tissue via several mechanisms: thermally, micromechanically, and through the phenomenon of cavitation. Internal ultrasonic liposuction primarily utilizes the principles of cavitation. The exact mechanism by which external ultrasound affects fatty tissues is not currently clear; however, it is felt to be a micromechanical effect [168].

Because of the problems associated with internal ultrasound, the author has worked to develop the concept of external ultrasound [29]. External ultrasonic energy may be used preoperatively to produce a more favorable result without the side effects and complications associated with internal ultrasound. External ultrasound may also be applied postoperatively to reduce swelling and shorten the recovery course.

Any surgeon utilizing ultrasonic energy for any purpose should be knowledgeable about its usage and side effects as well as possible complications. Also, any ancillary personnel should be adequately trained and experienced in the use of ultrasonic devices.

In 1989, Gasperoni presented subdermal superficial liposuction [57, 58]. The technique consists of suctioning the superficial subdermal fat through small gauge cannulas (1.8–2.0 mm diameter) and then proceeding with the same thin cannulas to aspirate the deeper fat as well. The procedure is begun with a thin cannula and gradually increased in gauge. The advantages of using this technique includes suction of the subdermal fat layer making it possible to obtain effective skin retraction, the treatment of all the layers of fat is made in an even fashion that good results are predictable, it is possible to treat patients with slight adiposities as well as areas with small thickness of the fat layers such as the ankles, and the procedure is similar for patients with large adiposities and those with small ones.

When the subdermal fat increases its volume, the vertical septae of the subdermal layer are stretched pulling the dermis thus giving the skin the so-called "cellulite" aspect. Other skin dimples and hollows may be due to considerable irregularities of a firm deep fat or may be iatrogenic. In these cases and in regions where the fat is fibrous and hard, external ultrasound (EU) is indicated to soften the fat crushing it with the mechanical impact of its waves [29, 73, 97, 146]. EU may be applied not only to the superficial fat but also to the deep one. This layer should be treated whenever the deep fat is fibrous and hard. When external ultrasound is used, care should be taken to infiltrate immediately under the dermis to facilitate the propagation of EU waves in the superficial layer of fat. When the deep fat must be effectively reached by the EU waves, it must be infiltrated conspicuously to allow a successful cavitation induced by the EU. A thin layer of ultrasonic gel is spread on the skin of the areas to be treated with ultrasound. The ultrasound is then delivered through a 3 MHz probe to treat the superficial fat and with a 1 MHz probe to treat the deeper layers. A 2 MHz probe may be additionally used when we must be sure that all layers are treated.

4.2.3 Low-Level Laser-Assisted Liposuction

Laser is defined as Light Amplification by Stimulated Emission of Radiation. Neira described the use of the low-level laser to assist in liposuction [45, 112, 114–117]. The laser used for liposuction is an external beam cold laser, electric diode with 635 nm wavelength that irradiates adipose tissue at 1.2, 2.4, and 3.6 J/cm^2 at 2, 4, and 6 min exposure in each area.

Low-level laser energy has an impact on the adipose cell consisting in opening a transitory pore in the cell membrane which permits the fat content to go from inside to outside the cell. The cells interstice and capillaries remain intact. Partial disruption of the adipose cell has observed. The irradiated cells were recultured and showed that they recover the normal anatomy and were alive [113].

4.2.3.1 Nd: YAG (Neodimiun: YAG)

The Nd:YAG laser is a solid laser formed by a Granate Aluminum Ytrium crystal (the YAG) contaminated with an unusual soil (the Neodimiun) that emits an infrared band in 1064 nm [62]. Histological studies were performed 30 min after laserlipolysis from a piece of tissue resected in a dermolipectomy

[136, 139]. It was possible to observe areas with necrobiotic adipose tissue with accumulation of lipophagic macrophages cells forming granulomatous lipophagic nodules. Twenty-five days after laserlipolysis it was possible to observe adipose tissue with breakage of the cell walls surrounded by histiocytes with foamy cytoplasm. There were also areas with scar fibrosis, and the nerve threads were intact. The use of the laser causes a destruction of the fat cell specifically protecting the nerves, while in a tumescent liposuction the fat cell is evacuated intact. The remaining tissue is immediately phagocytosed by the macrophages and the immune system while fibrosis covers and retracts the empty spaces.

4.2.4 Powered Liposuction

Development of Powered Liposuction Technology

The concept of using mechanical instrumentation with liposuction technology is as old as liposuction itself. Arpad and Giorgio Fischer introduced the concept of liposuction in combination with instruments they developed that they called the cellusuctiotome [45]. The Fischers' early instruments contained blades with moving internal components designed to cut fat when it was aspirated into the cannula. Later, blunt cannulas were developed with side ports and other designs that aspirated fat with little blood loss.

Gross used an existing cannula that had an exposed internal blade driven by a motorized handpiece, which was modified for use in fat removal [67]. His "liposhaving" procedure was an open technique in which the fat harvesting unit was used for neck liposuction using submental incisions. Fat cells could thoroughly be removed from the platysma to allow for an even and complete fat extraction. This was revived in 2000 by Schaefer using an endoscope rather than direct vision [138].

Coleman developed an oscillating blade within a cannula that facilitates removal of fat removal [27].

The oscillating cutting cannulas demonstrated decreased work on the part of the liposuction surgeon. This led to the development of a number of reciprocating cannula systems. The instruments contained a motor, driven either electrically or by air, which moved the tip of the liposuction cannula forward and backward. These designs have been found to decrease the work of performing liposuction on the part of the surgeon and increase the rate of fat removal.

4.2.5 Current Instruments

Flynn clinically assessed available instruments and an independent engineering firm measured each instrument [49]. Laboratory measurements such as the degree of torque, amount of heat produced, size and weight, amount of torque force, and degree of vibration were among the measurements taken by the independent engineering firm. A concise practical description of each instrument was featured. Stroke force was variable with instruments having a range of 9.5–30 lb. The noise of the units varied between 60 and 87 dB. Units produced variable heat with surface temperature measurements ranging from 77 to 102°F. Build quality and reliability varied from instrument to instrument. The air-driven devices were clumsy and loud.

4.2.5.1 Power Assisted

Coleman evaluated the efficacy of powered liposuction. A variety of electrical and air-driven instruments were used [28]. All cannulas had 3 mm outside diameter. The amount of fat extracted was measured using a mucous specimen trap, widely used by respiratory therapists, in series between the cannula aspiration hose and the aspirator. They documented that there was increased efficiency in fat removal (Fig. 4.1).

4.2.5.2 Ultrasonic-Assisted Liposuction

Scuderi and Zocchi pioneered the application of ultrasonic vibration to fat emulsification and removal [140, 163–165, 167, 168]. The hope and objective of this effort was to create both technology and associated techniques that consistently produced safer and more effective means of aesthetic body contouring when compared to liposuction. The benefits of tissue selectivity were expected to produce a method of lipoplasty that was more "fat specific" than the existing and well-known suction cannula. This technology and technique were named UAL for Ultrasound-Assisted Lipoplasty. The first-generation UAL device was produced by the SMEI Company of Italy and utilized smooth, solid probes at a frequency of 20 kHz. The solid probes had a stepped design with diameters at the tip as small as 3.0 mm (small probe) and diameters at the base as large as 6.0 mm (large probe). The basic technique involved good surgical practice and two fundamental rules: the essential use of a wet environment produced by infiltration of sufficient wetting solution and a constantly moving the probe to prevent thermal injury [167].

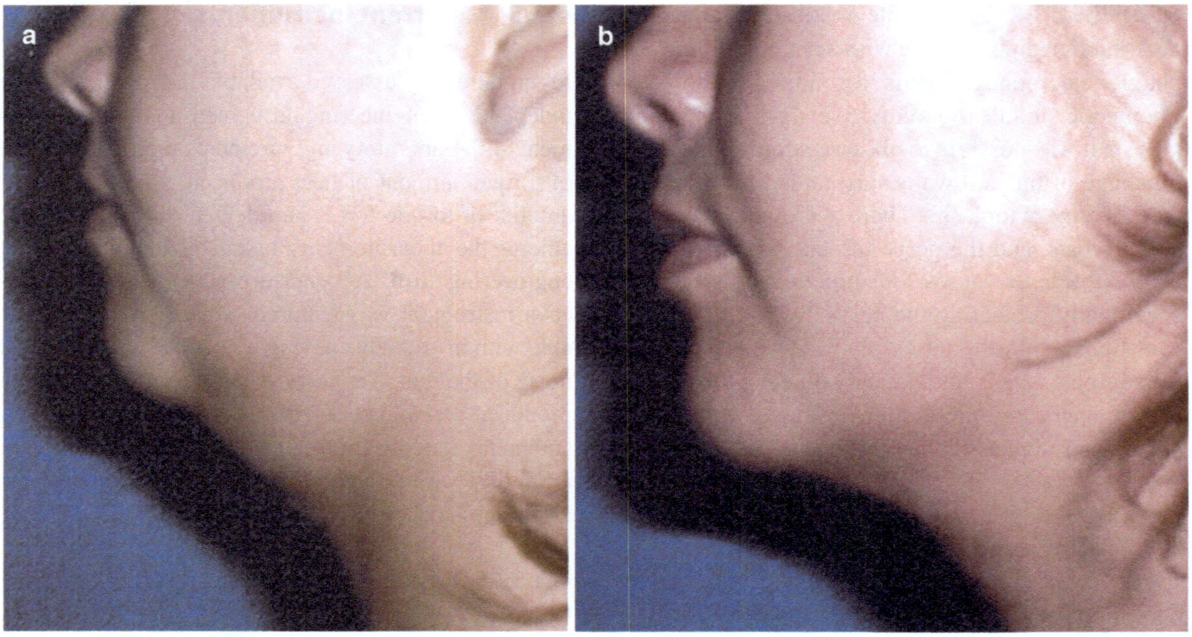

Fig. 4.1 Power-assisted liposuction of neck. (a) Preoperative. (b) Postoperative

Fodor published his experience on 100 patients using a contra-lateral study model [50]. His conclusions comparing SAL to UAL found no significant differences between SAL and UAL and failed to prove the claimed benefits attributed to UAL.

The UAL technique and instruments continued to evolve. Originally, application times were long, significant complications and were reported, and safety was questioned [12, 59, 66, 97, 128, 156]. As application times were reduced the complication rate declined. The concept of "loss of resistance" became widely known as a realistic surgical endpoint. Rapid probe movement was introduced as another means to safely control the energy presented by the second-generation machines [152]. Results ranged from safe and effective use of UAL to high complication rates and questionable safety. A number of surgeons continued to use the ultrasonic instrumentation safely and effectively [61, 95, 106, 132]. Their evolving technique allowed them to get effective results without the complications noted at the introduction of the technology (Fig. 4.2).

4.2.5.3 VASER (Vibration Amplification of Sound Energy at Resonance)

VASER-Assisted Lipoplasty (VAL) is a third-generation ultrasound-assisted liposuction. The VASER® system is highly selectivity for fatty tissue resulting in decreased overall damage to the vessels, nerves, structural tissues, and lymph tissue. VASER-assisted lipoplasty uses ultrasonic frequency vibrations to emulsify the fatty component of the tissue matrix but in a fundamentally different manner than earlier versions of ultrasonic instrumentation for lipoplasty. The VASER system delivers significantly less power to the tissues while simultaneously increasing fragmentation/emulsion efficiency compared to UAL devices and eliminates the simultaneous aspiration feature of UAL devices [24, 89].

4.2.5.4 Water Jet-Assisted Liposuction

Taufig devised a method of water jet-assisted liposuction that allows a controlled and selective removal of fat tissue within the epifascial/subcutaneous area via usage of a cannula system [150]. The technique uses the energy of the pressurized fluid using a specialized cannula in which an infusion tube and nozzle are integrated as well as a suction unit. An infinitely variable force pump dispenses the fluid in a controlled manner via a nozzle at the top of the cannula system. The cannula is attached to a common and well-proven suction device for liposuction.

The water jet technique uses an isotonic sodium chloride solution with an additive of adrenaline in the ratio of 1 mL–3 L of sodium chloride solution that is

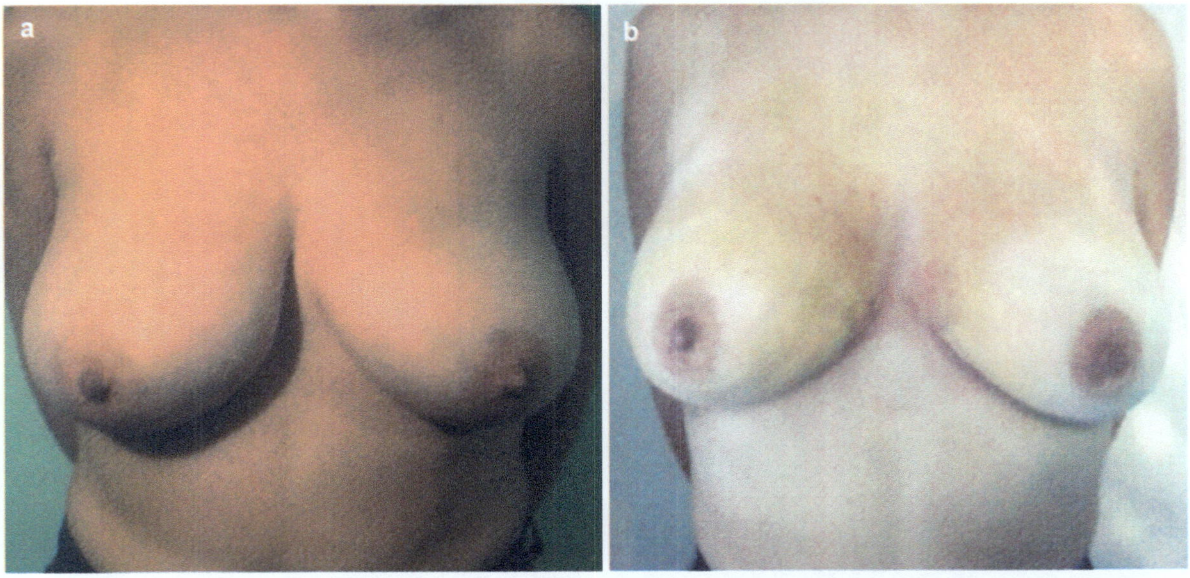

Fig. 4.2 Ultrasound-assisted breast lift. (**a**) Preoperative. (**b**) Postoperative

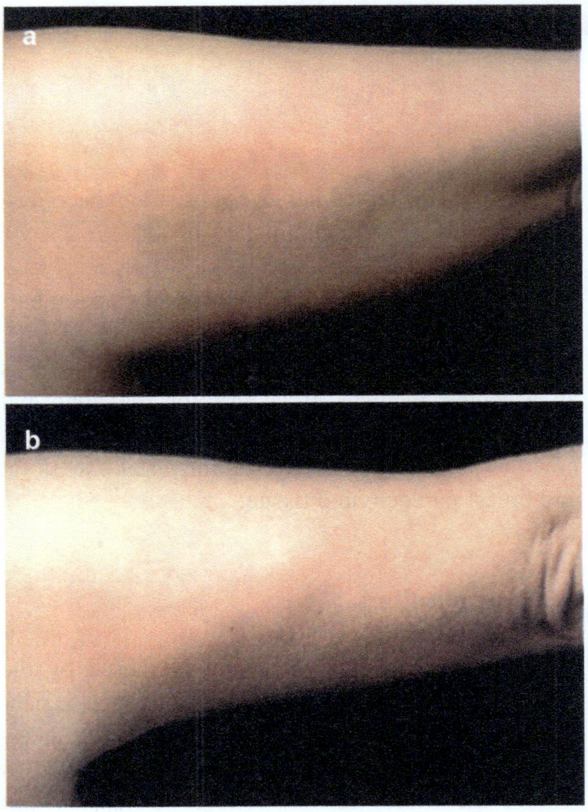

Fig. 4.3 Percussion-assisted liposuction of arms. (**a**) Preoperative. (**b**) Postoperative. *Arrow* shows triceps muscle

suctioned off almost at the same time as the dissolved fat particles [151]. Therefore, no side effects are caused by the solution. This avoids a separate fragmentation step as with the tumescent technique.

4.2.5.5 External Percussion Massage-Assisted Liposuction

Shiffman and Mirrafati (Fig. 4.3) reported the use of a double-headed percussion massage instrument that is used externally on areas that have been infiltrated with tumescent anesthesia fluid prior to liposuction causes emulsification of the fat and makes removal easier [144]. The instrument (selling from $30 to $300) is less expensive than all other instruments and is simple to use.

4.3 Liposuction for Cosmetic Purposes

Liposuction is used for cosmetic purposes on the face and neck, chest, back (Fig. 4.4), and abdomen, upper and lower extremities, and buttocks and breasts [84, 87, 110, 118, 135]. Fat obtained from liposuction has been used for fat grafting for the face and hemifacial atrophy [23, 51]. Gasparotti and Toledo described superficial liposculpture to smooth out areas that are not removed with deep liposuction and also results in skin contraction [56, 155].

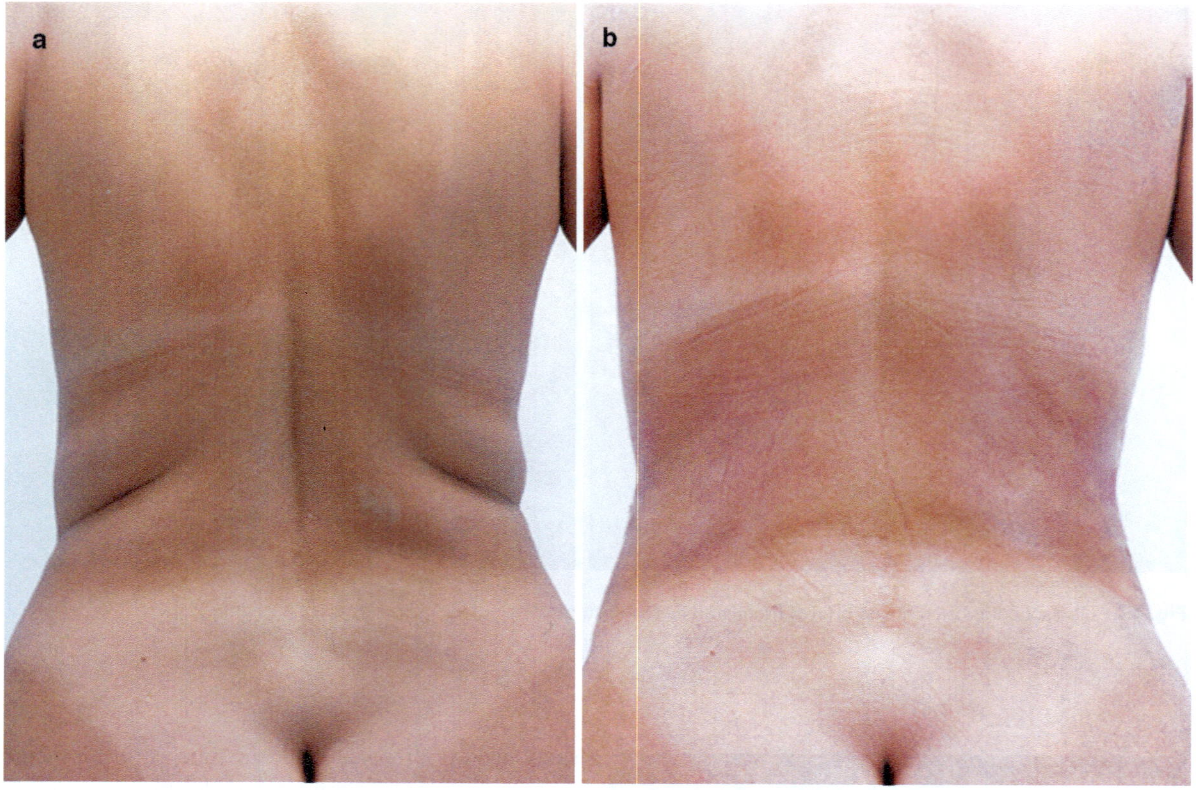

Fig. 4.4 Liposuction of back. (**a**) Preoperative. (**b**) Postoperative

4.3.1 Breast Reduction

The problems of macromastia and gigantomastia actually have significant medical symptoms (neck and upper back pain, grooving of the shoulders from the bra straps, inframammary fold irritation and dermatitis) that are treatable with breast reduction. The utilization of liposuction to reduce breast volume and, at times in conjunction with breast lift to relieve ptosis, can achieve resolution of the symptoms [2, 7, 17, 30, 63, 64, 85, 98, 105, 131].

4.3.2 Gynecomastia

Gynecomastia may appear in the adolescent male as part of the physiologic and hormonal changes taking place, in the elderly male because of hormonal changes, from a variety of drug therapies in males that are associated with stimulation of the breast tissue, and in a male with a breast tumor that may very well be malignant. True gynecomastia consists of excessive breast tissue and fat but pseudogynecomastia is the excessive accumulation of fat [11, 25, 99, 133, 158, 166].

Breast enlargement in the male is often an embarrassment to the patient because of the large breasts. Surgical removal of the breast for gynecomastia is an accepted medical procedure for the abnormal enlargement of the breast in males. Liposuction is now the preferred method for removal of excess fat and breast parenchymal tissue.

4.3.3 Cellulite

This condition involves indentations caused by the increased accumulation of fat with restricted expansion by the fibrous attachments of the skin to the underlying fascia [100]. Liposuction has been utilized to remove the excess fat and relieve the tension that causes the indentations because of the fascial attachments and also by transecting some of the fascial attachments [96].

4.3.4 Apocrine Gland Disorders

The apocrine or sweat glands may be involved with of variety of problems such as excessive sweating that can become foul-smelling or infected. Liposuction may be the only minimal incision solution to the problem [3, 6, 22, 65, 76, 102, 103, 122, 124, 126, 127, 141, 149, 157].

(a) Bromhidrosis: axillary (apocrine) sweat, which has become foul-smelling as a result of its bacterial decomposition
(b) Hyperhidrosis (polyhidrosis): excessive perspiration
(c) Osmidrosis: Same as bromhidrosis
(d) Emotional hyperhidrosis: an autosomal dominant disorder of the eccrine sweat glands, most often of the palms, soles, and axillae, in which emotional stimuli (anxiety) and sometimes mental or sensory stimuli elicit volar or axillary sweating
(e) Fox–Fordyce disease: a chronic, usually pruritic disease chiefly seen in women, characterized by the development of small follicular popular eruptions of apocrine gland-bearing areas, especially the axillae and pubes, and caused by obstruction and rupture of the intraepidermal portion of the ducts of the affected apocrine glands, resulting in alteration of the regional epidermis, apocrine secretory tubule, and adjacent dermis

There have been no reports concerning the treatment of hidradenitis suppurativa with liposuction. Hidradenitis suppurativa is a chronic suppurative and cicatricial disease of the apocrine gland-bearing areas, chiefly the axilla, usually in young women, and anogenital region, usually in men. The disorder is caused by poral occlusion with secondary bacterial infection of apocrine sweat glands. It is characterized by the development of tender red abscesses that enlarge and eventually break through the skin resulting in purulent and seropurulent drainage. Healing occurs with fibrosis and recurrences lead to sinus tract formation and progressive scarring. This disorder would have to be treated by liposuction to remove the apocrine glands in its resting phase when there is no apparent infection. The earlier in the disease that treatment is instituted, the less likely infection will be stirred up. The author is presently observing a patient with early hidradenitis suppurativa since the infections have responded well to antibiotics each time of recurrent symptoms and the patient is reticent about surgery. Field described the use of axillary liposuction for osmidrosis and hyperhidrosis with an aggressive approach that has minor scarring but removes more glandular tissue [42].

4.3.5 Hematoma

Hematomas can be liposuctioned through small incisions rather than opening the total wound [5, 34, 107, 121]. This less invasive method reduces the morbidity associated with postoperative hematomas. The aspiration of seromas probably could be performed with the liposuction cannula but a simple needle and syringe usually suffices. However, in very large seromas, the use of liposuction with machine would be easier than aspirating 60 cc at a time with a syringe.

4.3.6 Involuted Hemangiomas

Liposuction of hemangiomas should be performed when the hemangioma is involuted otherwise there may be extensive bleeding. There have been two reports of hemangiomas having been removed with liposuction [13, 48].

4.3.7 Lipoma and Lipomatosis

A lipoma is a benign, soft, rubbery, encapsulated tumor of adipose tissue, usually composed of mature fat cells generally occurring in the subcutaeous tissues of the trunk, nuchae, or forearms but may occur intramuscular, intermuscular, intraarticular, intraspinal, intradural, epidural. These can become an annoyance to the patient because of size and because of the cosmetic appearance [4, 14, 21, 26, 31, 33, 36, 38, 40, 68, 70, 78, 81, 86, 88, 104, 109, 134, 137, 159, 161, 166].

(a) Lipomatosis dolorosa: Lipomatosis in which lipomas are tender or painful
(b) Lipomatosis gigantea: Adipose deposits form large masses nodular circumscribed lipomatosis: formation of multiple circumscribed or encapsulated lipomas that may be symmetrically distributed (symmetrical lipomatosis) or haphazardly
(c) Dercum's disease (adiposis dolorosa, Anders syndrome, adiposalgia, adipositas tuberosa simplex, fibrolipomatosis dolorosa, neurolipomatosis, lipalgia, lipomatosis doloroas): a disease accompanied by painful localized fatty swellings and by various nerve lesions. Usually seen in women and may cause death from pulmonary complications.
(d) Madelung's disease, asymmetric adenolipomatosis, Launois–Bensaude syndrome, Buschke's II syndrome: Onset is between 35 and 40 years of

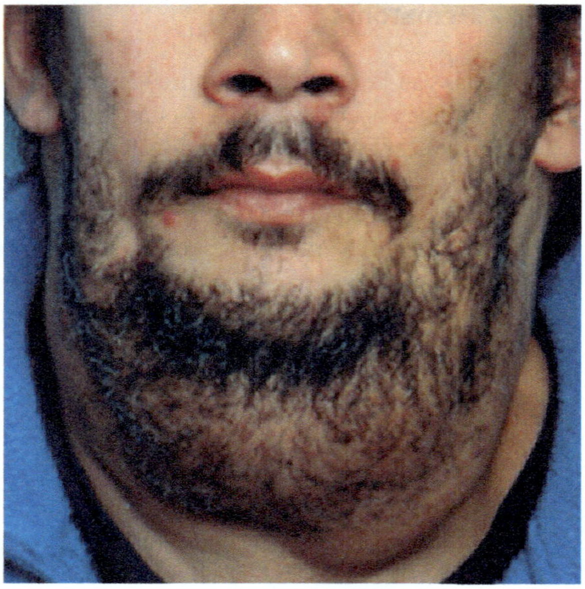

Fig. 4.5 Madelung's neck

age, more prevalent in males, with diffuse tumefaction in the posterior part of the neck [16]. This is followed by symmetric accumulation of masses in the submandibular region and other lipomas on the chest and the rest of the body. Asthenia and apathy are usually present. Compression of peripheral nerves results in pain. Dyspnea, cough, cyanosis, and exophthalmus may develop.

(e) Madelung's deformity or Madelung's neck (annulare colli): Haphazard accumulation of lipomas around the neck (Fig. 4.5).

(f) Bannayan syndrome (Bannayan–Zonana syndrome, microcephaly–hamartomas syndrome): A familial syndrome characterized by symmetrical microcephaly, mild neurological dysfunction, postnatal retardation, mesodermal hamartomas, discrete lipomas, and hemangiomas [8, 108].

(g) Proteus syndrome: A sporadic disorder that causes postnatal overgrowth of multiple tissues that include skin, subcutaneous tissue, connective tissue (including bone), the central nervous system, and viscera [15]. Progressive skeletal deformities occur with invasive lipomas, and benign and malignant tumors.

(h) Angiolipomas may also be removed with liposuction. Liposuction is a method to remove the tumor with minimal incisions [90].

4.3.8 Liposuction-Assisted Nerve-Sparing Hysterectomy

Nerve-sparing hysterectomy can be performed more easily with the use of liposuction to remove excess fat and better exposing the surrounding structures [75, 79]. There may be other surgical procedures that can be made easier through better exposure from removal of fatty tissue accumulation.

4.3.9 Lymphedema

The treatment of persistent obstructive lymphedema can be aided with liposuction, especially with the limited incisions utilized [18–20, 119, 120, 160]. There can be uniform removal of the lymphedematous tissue with liposuction without the need for major surgery to aid in the discomfort of a large extremity. Long-term results need to be reported.

4.3.10 Obesity

Obesity is defined as an increase in body weight beyond the limitation of skeletal and physical requirements [35, 37, 39, 54, 55, 80, 101, 125, 145, 154]. Endogenous obesity is excess weight due to metabolic (endocrine) abnormalities or genetic defects that affect the synthesis of enzymes involved in intermediate metabolism. Exogenous obesity is obesity due to overeating.

The treatment of endogenous obesity requires resolution of any endocrine problem but also may include liposuction for improving contour and reducing total fat deposits. Exogenous obesity should be treated with diet and exercise but if this is not successful, liposuction may be performed. This can sometimes stimulate the patient to continue with weight loss regimens.

Giese has shown that the cardiovascular profile can be improved with large-volume liposuction [60]. Large-volume liposuction has an impact on serum lipids [77]. Perez noted that large-volume and small-volume liposuction have been clinically shown to improve insulin sensitivity in obese patients thus reducing their risk of developing type 2 diabetes [129]. Evidence is presented to support the hypothesis that liposuction disrupts the pathway that brings about insulin insensitivity in the obese patient. It is proposed that using liposuction in the overall treatment of obesity could lead to an improvement in insulin sensitivity and thus greatly improve the quality of life of the obese patient.

4.4 Reconstructive Surgery

Liposuction has been used to aid in reconstructive surgery, especially in debulking flaps without injuring the blood supply [26, 40, 43, 69, 148]. The author has used liposuction to correct dog ears following reconstructive procedures and there are probably many other problems in reconstructive surgery that can be corrected with the use of liposuction.

4.4.1 Silicone Removal

Silicone is almost impossible to remove from the tissues without removing some normal tissue even with a siliconomas [26, 162]. The resulting defect may be cosmetically unacceptable. Attempting to remove silicone from a mammary prosthesis pocket is very difficult because the silicone is as sticky as gum and clings to gloves, skin, and pads. If even small drops of silicone accidentally drop to the floor, there is extreme danger of slipping by persons in the operating room. Silicone can be extracted from tissues with less deformity and from the implant pocket with the use of liposuction.

4.4.2 Steroid-Induced Lipodystrophy

A better cosmetic appearance in patients with steroid induced lipodystrophy can be achieved with liposuction of the excessive areas of fat [9, 41, 71, 72, 74, 111, 130]. The underlying endocrine problem also needs to be addressed at the same time.

4.5 Non-cosmetic Applications

There have been single reports of the use of liposuction to treat certain problems. Denneny described the obliteration of the frontal sinus with liposuction [32]. Shenoy used liposuction to aid in correcting a buried penis [142]. Sonenshein and Lepoudre treated a critically ill obese patient with massive fat accumulation in the neck with the use of liposuction, removing 225 mL of fat, to allow visualization of the tracheal stoma in order to insert a tracheostomy tube [147]. Ad-El reported the use of liposuction to relieve chronic facial swelling following multiple bee stings [1]. Apesos and Chami used liposuction in the treatment of congenital body asymmetry and fat necrosis [4]. Illouz reported the use of liposuction to treat scar deformity [86]. Field showed that liposculpturing can be used to improve submental scar revision by removing submental and submandibular adipose tissue followed by cannula dissection of the submental skin flap [44]. Babovic reported the use of liposuction in debulking plexiform neurofibromas and Thomas showed that the tumescent technique can aid in the resection of neurofibromas [10, 143, 153]. Shiffman described the use of liposuction to remove infected Vicryl sutures in the breast that did not respond completely to resections of breast and fat tissue. There were recurrent infections and granulomas over 4 years from the infected Vicryl sutures. The final result with liposuction was clearing of the recurring infections and allowed reconstruction of the breasts.

As physicians become aware of the multiple uses of liposuction outside the cosmetic surgery field, further disorders will be found that can be amenable to liposuction.

4.6 Conclusions

Liposuction is a procedure that has yet to reach its limitations. Started as a limited incision cosmetic operation, liposuction has progressed to uses that were not even imagined by its founders and many of the early surgeons utilizing the procedure. The future of liposuction in surgery needs physicians who will find innovative uses in areas outside the cosmetic surgery field.

References

1. Ad-El DD (2002) Chronic facial edema caused by multiple bee stings: effective treatment with liposuction. Plast Reconstr Surg 110(4):1192–1193
2. Aiache AE (1984) Lipolysis of the female breast. In: Hetter GP (ed) Lipoplasty: the theory and practice. Little, Brown and Co, Boston, pp 227–231

3. Altchek E (2003) Hyperhidrosis. Plast Reconstr Surg 11(2):943
4. Apesos J, Chami R (1991) Functional applications of suction-assisted lipectomy: a new treatment for old disorders. Aesthet Plast Surg 15(1):73–79
5. Ascari-Raccagni A, Baldari U (2000) Liposuction for the treatment of large hematomas of the leg. Dermatol Surg 26(3):263–265
6. Atkins JL, Butler EM (2002) Hyperhidrosis: a review of current management. Plast Reconstr Surg 110(1): 222–228
7. Baker TM (2000) Suction mammaplasty: the use of suction lipectomy to reduce large breasts. Plast Reconstr Surg 106(1):227
8. Bannayan GA (1971) Lipomatosis, angiomatosis, and macrocephalia. Arch Pathol 92:1–5
9. Barak A, Har-Shai Y, Ullmann Y, Hirshowitz B (1996) Insulin-induced lipohypertrophy treated by liposuction. Ann Plast Surg 37(4):415–417
10. Barbovic S, Bite U, Karnes PS, Babovic-Vuksanovic D (2003) Liposuction: a less invasive surgical method of debulking plexiform neurofibromas. Dermatol Surg 29: 785–787
11. Bauer T, Gruber S, Todoroff B (2001) Periareolar approach in pronounced gynecomastia with focus-plasty and liposuction. Chirurg 72(4):433–436
12. Baxter RA (1999) Histologic effects of ultrasound-assisted lipoplasty. Aesthet Surg J 19:109–114
13. Berenguer B, de Salamanca JE, Gonzalez B, Rodriguez P, Zambrano A, Perez Higueras A (2003) Large involuted facial hemangioma treated with syringe liposuction. Plast Reconstr Surg 111(1):314–318
14. Berntorp E, Berntorp K, Brorson H, Frick K (1998) Liposuction in Dercum's disease: impact on haemostatic factors associated with cardiovascular disease and insulin sensitivity. J Intern Med 243(3):197–201
15. Biesecker LG (2001) The multifaceted challenges of Proteus syndrome. J Am Med Assoc 285(17):2240–2243
16. Biou C, Illouz G, Langman JC, Mokdad R, Beydoun B (1984) Place of lipolysis in the surgical treatment of Lanois-Bensaude syndrome. Rev Stomatol Chir Maxillofac 85(6): 497–500
17. Brauman D (1994) Reduction mammaplasty by suction alone. Plast Reconstr Surg 94(7):1095–1096
18. Brorson H (2000) Liposuction gives complete reduction of chronic large arm lymphedema after breast cancer. Acta Oncol 39(3):407–420
19. Brorson H, Svensson H (1998) Liposuction with controlled compression therapy reduces arm lymphedema more effectively than controlled compression therapy alone. Plast Reconstr Surg 102(4):1058–1067
20. Brorson H, Svensson H, Norrgren K, Thorsson O (1998) Liposuction reduces arm lymphedema without significantly altering the already impaired lymph transport. Lymphology 31(4):156–172
21. Carlin MC, Ratz JL (1988) Multiple symmetric lipomatosis: treatment with liposuction. J Am Acad Dermatol 18: 359–362
22. Chae KM, Marschall MA, Marschall SF (2002) Axillary Fox-Fordyce disease treated with liposuction-assisted curettage. Arch Dermatol 138(4):452–454
23. Chajchir A, Benzaquen I (1986) Liposuction fat grafts in face wrinkles and hemifacial atrophy. Aesthet Plast Surg 10(2):115–117
24. Cimino WW (2001) Ultrasonic surgery: power quantification and efficiency optimization. Aesthet Surg J 21(3):233–240
25. Cohen IK (1987) Gynecomastia: suction lipectomy as a contemporary solution. Plast Reconstr Surg 80:386
26. Coleman WP (1988) Noncosmetic applications of liposuction. J Dermatol Surg Oncol 14:1085–1090
27. Coleman WP III (2000) Powered liposuction. Dermatol Surg 26(4):315–318
28. Coleman WP III, Katz B, Bruck M, Narins R, Lawrence N, Flynn TC, Coleman WP, Coleman KM (2001) The efficacy of powered liposuction. Dermatol Surg 27(8):735–738
29. Cook WR Jr (1997) Utilizing external ultrasonic energy to improve the results of tumescent liposculpture. Dermatol Surg 23(12):1207–1211
30. Courtiss EH (1993) Reduction mammaplasty by liposuction alone. Plast Reconstr Surg 92(7):1276–1284
31. Darsonval V, Duly T, Munin O, Houet JF (1990) Le tratment chirugical de la maladie Lanoise-Bensaude. Interet de la lipoaspiration. Ann Chir Plast Esthét 35:128–133
32. Denneny JC III (1986) Frontal sinus obliteration using liposuction. Otolaryngol Head Neck Surg 95(1):15–19
33. Dolsky RL, Asken S, Nguyen A (1986) Surgical removal of lipoma by lipo-suction surgery. Am J Cosmet Surg 3(3):27–34
34. Dowden RV, Bergfeld JA, Lucas AR (1990) Aspiration of hematomas with liposuction apparatus. A technical note. J Bone Joint Surg Am 72(10):1534–1535
35. Eed A (1999) Mega-liposuction: analysis of 1520 patients. Aesthet Plast Surg 23(1):16–22
36. Ersek RA (2000) Removal of lipomas by liposuction. Plast Reconstr Surg 105(2):807
37. Ersek RA, Philips C, Schade K (1991) Obesity can be treated by suction lipoplasty when combined with other procedures. Aesthet Plast Surg 15(1):67–71
38. Faga A, Valdatta LA, Thione A, Buoro M (2001) Ultrasound assisted liposuction for the palliative treatment of Madelung's disease: a case report. Aesthet Plast Surg 25(3): 181–183
39. Ferreira JA (1997) Lipoplasty in the treatment of obesity. Am J Cosmet Surg 14(3):251–255
40. Field LM (1988) Liposuction surgery (suction-assisted lipectomy) for symmetrical lipomatosis. J Am Acad Dermatol 18(6):1370
41. Field LM (1988) Successful treatment of lipohypertrophic insulin lipodystrophy with liposuction surgery. J Am Acad Dermatol 19(3):570
42. Field LM (2003) Tumescent axillary liposuction and curettage with axillary scarring: not an important factor. Dermatol Surg 29:317
43. Field LM, Skouge J, Anhalt TS, Recht B, Okimoto J (1988) Blunt liposuction cannula dissection with and without suction-assisted lipectomy in reconstructive surgery. J Dermatol Surg Oncol 14(10):1116–1122
44. Field LM, Ostertag J, Krekels G, Sneets N, Neumann H (2003) Submental scar revision via cervicomental liposculpturing and submental scar resection. Dermatol Surg 29(1):114–115

45. Fischer G (1991) History of my procedure, the harpstring technique and the sterile fat safety box. In: Fournier PF (ed) Liposculpture: the syringe technique. Arnette, Paris, pp 9–21
46. Fischer G (2006) Orthostatic liposculpture. In: Shiffman MA, Di Giuseppe A (eds) Liposuction: principles and practice. Springer, Berlin, pp 217–221
47. Fischer A, Fischer G (1977) Revised technique for cellulitis fat reduction in riding breeches deformity. Bull Int Acad Cosm Surg 2(4):40–43
48. Fisher MD, Bridges M, Lin KY (1999) The use of ultrasound-assisted liposuction in the treatment of an involuted hemangioma. J Craniofac Surg 10(6):500–502
49. Flynn TC (2002) Powered liposuction: an evaluation of currently available instrumentation. Dermatol Surg 28(5): 376–382
50. Fodor PB, Watson J (1998) Personal experience with ultrasound-assisted lipoplasty: a pilot study comparing ultrasound-assisted lipoplasty with traditional lipoplasty. Plast Reconstr Surg 101(4):1103–1116
51. Fournier PF (1987) Body sculpturing. Samuel Rolf International, California
52. Fournier PF (1988) Why the syringe and not the suction machine? J Dermatol Surg Oncol 14(10):1062–1071
53. Fournier PF (1988) Who should do syringe liposculpturing? J Dermatol Surg Oncol 14(10):1055–1056
54. Fournier PF (1993) Is reduction liposculpturing justified? Am J Cosmet Surg 10(4):259–262
55. Fournier PF (1997) Therapeutic megalipoextraction or megaliposculpture: indications, technique, complications, and results. Am J Cosmet Surg 14(3):297–309
56. Gasparotti M (1990) Superficial liposculpture for face and body. In: Toledo LS (ed) Annals of the II international symposium: recent advances in plastic surgery – RAPS/90, Sao Paulo, Mar 1990, p 441
57. Gasperoni C, Salgarello M, Emiliozzi P, Gargani G (1989) Subdermal liposuction. In: Abstract of the 10th Congress of the International Society of Aesthetic Plastic Surgery, Zurich 11–14 Sept 1989, p 95
58. Gasperoni C, Salgarello M, Emiliozzi P, Gargani G (1990) Subdermal liposuction. Aesthet Plast Surg 14(2):137–142
59. Gerson RM (1997) Avoiding end hits in ultrasound-assisted lipoplasty. Aesthet Surg J 17:331–332
60. Giese SY, Bulan EJ, Commons GW, Spear SL, Yanovski JA (2001) Improvements in cardiovascular risk profile with large-volume liposuction: a pilot study. Plast Reconstr Surg 108(2):510–519
61. Gilliand MD, Commons GW, Halperin B (1999) Safety issues in ultrasound-assisted large volume lipoplasty. Clin Plast Surg 26(2):317–335
62. Goldman A, Schavelzon DE, Blugerman GS (2002) Laserlipolysis: liposuction using Nd-Yag laser. Rev Soc Bras Cir Plast 17(1):17–26
63. Gray LN (1998) Liposuction breast reduction. Aesthet Plast Surg 22(3):159–162
64. Gray LN (2001) Update on experience with liposuction breast reduction. Plast Reconstr Surg 108(4):1006–1010
65. Grazer FM (1992) Special reconstructive and therapeutic procedures. In: Grazer FM (ed) Atlas of suction assisted lipectomy in body contouring. Churchill Livingstone, New York, pp 319–329
66. Grolleau JL, Rouge D, Chavoin JP, Costagliola M (1997) Severe cutaneous necrosis after ultrasound lipolysis. Medicolegal aspects and review. Ann Chir Plast Esthét 42(1):31–36
67. Gross CW, Becker DG, Lindsey WH, Park SS, Marshall DD (1995) The soft-tissue shaving procedure for removal of adipose tissue. A new, less traumatic approach than liposuction. Arch Otolaryngol Head Neck Surg 121(10): 1117–1120
68. Halachmi S, Moskovitz B, Calderon N, Nativ O (1996) The use of an ultrasonic assisted lipectomy device for the treatment of obstructive pelvic lipomatosis. Urology 48(1): 128–130
69. Hallock GG (1986) Liposuction for debulking free flaps. J Reconstr Microsurg 2:235–239
70. Hallock GG (1987) Suction extraction of lipomas. Ann Plast Surg 18(6):517–519
71. Hardy KJ, Gill GV, Bryson JR (1993) Severe insulin-induced lipohypertrophy successfully treated by liposuction. Diabetes Care 16(6):929–930
72. Hauner H, Olbrisch RR (1994) The treatment of type-1 diabetics with insulin-induced lipohypertrophy by liposuction. Dtsch Med Wochenschr 119(12):414–417
73. Havoonjian HH, Luftman DB, Menaker GM, Moy RL (1997) External ultrasonic tumescent liposuction. A preliminary study. Dermatol Surg 23(12):1201–1206
74. Hetter GP (1984) Treatment of insulin induced fat hypertrophy. In: Hetter GP (ed) Lipoplasty: the theory and practice of blunt suction lipectomy. Little, Brown and Co, Boston, p 323
75. Hockel M, Konerding MA, Heussel CP (1998) Liposuction-assisted nerve-sparing extended radical hysterectomy: oncologic rationale, surgical anatomy, and feasibility study. Am J Obstet Gynecol 178(5):971–976
76. Hong JP, Shin HW, Yoo S-C, Chang H, Park SH, Koh KS, Hur JY, Lee TJ (2004) Ultrasoun-assisted lipoplasty treatment for axillary bromhidrosis: clinical experience of 375 cases. Plast Reconstr Surg 113(4):1264–1269
77. Hong YG, Kim HT, Seo SW, Chang CH, Rhee EJ, Lee WY (2006) Impact of large-volume liposuction on serum lipids in orientals: a pilot study. Aesthet Plast Surg 30(3): 327–332
78. Horl C, Biemer E (1992) Benigne symmetrische lipomatose. Lipektomie und liposuction in der behandlung des morbus Madelung. Handchir Mikrochir Plast Chir 24:93–96
79. Horn LC, Fischer U, Hockel M (2001) Occult tumor cells in surgical specimens from cases of early cervical cancer treated by liposuction-assisted nerve-sparing radical hysterectomy. Int J Gynecol Cancer 11(2):159–163
80. Hunstad JP (1996) Body contouring in the obese patient. Clin Plast Surg 23(4):647–670
81. Ilhan H, Tokar B (2002) Liposuction of a pediatric giant superficial lipoma. J Pediatr Surg 37(5):796–798
82. Illouz YG (1983) Body contouring by lipolysis: a 5-year experience with over 3000 cases. Plast Reconstr Surg 72(5): 591–597
83. Illouz YG (1984) Illouz's technique of body contouring by lipolysis. Clin Plast Surg 11(3):409–417
84. Illouz YG (1985) Liposuction: the Franco-American experience. Medical Aesthetics, Inc, Beverly Hills

85. Illouz YG (1985) New applications of liposuction. In: Illouz YG (ed) Liposuction: the Franco-American experience. Medical Aesthetics, Inc, Beverly Hills, CA, pp 365–414
86. Illouz YG (1985) Principles of liposuction. In: Illouz YG (ed) Liposuction: the Franco-American experience. Medical Aesthetics, Inc, Beverly Hills, pp 21–31
87. Illouz YG, Devillers YT (1989) Body sculpturing by lipoplasty. Edinburgh, Churchill Livingstone
88. Illouz YG, de Villers YT (1989) Extensions of the technique. In: Illouz YG, de Villers YT (eds) Body sculpturing by lipoplasty. Churchill Livingstone, Edinburgh, pp 367–382
89. Jewell ML, Fodor PB, de Souza Pinto EB, Al Shammari MA (2002) Clinical application of VASER-assisted Lipoplasty: a pilot clinical study. Aesthet Surg J 22(2):131–146
90. Kaneko T, Tokushige H, Kimura N, Moriya S, Toda K (1994) The treatment of multiple angiolipomas by liposuction surgery. J Dermatol Surg Oncol 20(10):690–692
91. Kinney BM (1999) Body contouring with external ultrasound. Plast Reconstr Surg 103(2):728–729
92. Klein JA (1986) The tumescent technique for liposuction surgery. Presented at the second world congress of liposuction surgery, American Academy of Cosmetic Surgery, Philadelphia, June 1986
93. Klein JA (1987) The tumescent technique for liposuction surgery. Am J Cosmet Surg 4:263–267
94. Klein JA (1989) Tumescent technique for regional anesthesia permits lidocaine doses of 35 mg/kg for liposuction. Presented at the 1989 annual scientific meeting of the American Academy of Cosmetic Surgery, Los Angeles, 15 Jan 1989
95. Kloehn RA (1996) Liposuction with "sonic sculpture": six years' experience with more than 600 patients. Aesthet Surg J 16(2):123–128
96. Konstantinow A (2001) Best method against cellulite. Liposuction and freezing. MMW Fortschr Med 143(4):8
97. Lack EB (1998) Safety of ultrasonic-assisted liposuction (UAL) using a non-water-cooled ultrasonic cannula. A report of six cases of disproportionate fat deposits treated with UAL. Dermatol Surg 24(8):871–874
98. Lejour M (1994) Vertical mammaplasty and liposuction of the breast. Plast Reconstr Surg 94(1):100–114
99. Lewis CM (1985) :Lipoplasty: treatment for gynecomastia. Aesthet Plast Surg 9(4):287–292
100. Lieberman C (1999) Surgical treatment of cellulite and its results. Am J Cosmet Surg 16(4):299–303
101. Lieberman C, Cohen J (1997) Why abdominoplasty when you have liposuction? Am J Cosmet Surg 14(3):257–261
102. Lillis PJ, Coleman WP III (1990) Liposuction for treatment of axillary hyperhidrosis. Dermatol Clin 8(3): 479–482
103. Mangus DJ (2003) Hyperhidrosis. Plast Reconstr Surg 11(2):943
104. Martinez-Escribano JA, Gonzalez R, Quecedo E, Febrer I (1999) Efficacy of lipectomy and liposuction in the treatment of multiple symmetric lipomatosis. Int J Dermatol 938(7):551–554
105. Matarasso A (2000) Suction mammaplasty: the use of suction lipectomy to reduce large breasts. Plast Reconstr Surg 105(7):2604–2607
106. Maxwell GP (1999) Use of hollow cannula technology in ultrasound-assisted lipoplasty. Clin Plast Surg 26(2): 255–260
107. McEvan C, Jackson I, Stice R (1987) The application of liposuction for removal of hematomas and fat necrosis. Ann Plast Surg 19:480–481
108. Miles JH, Zonana J, Mcfarlane J, Aleck KA, Bawle E (1984) Macrocephaly with hamartomas: Bannayan-Zonana syndrome. Am J Med Genet 19:225–234
109. Mole B (2000) Assisted liposuction of lipomas. Ann Chir Plast Esthét 45(5):522–525
110. Morgan WR, Berkowitz F (1984) Body sculpture: a guide to the permanent removal of fat. Premier Publishing Co, Anaheim
111. Narins RS (1989) Liposuction for a buffalo hump caused by Cushing's disease. J Am Acad Dermatol 52(2):307
112. Neira R, Solarte E, Reyes M, Arroyave J, Montoya A (2001) Laser irradiation effect in adipocyte dilutions. Colombian National Congress Physics and Optic. Rev Colombiana Fisica 33(2):325–328
113. Neira R, Arroyave J, Solarte E, Isaza C, Gutierrez O, Gutierrez MI (2001) In vitro culture of adipose cell after irradiating them with a low level laser device. Presented at the Bolivian plastic surgery meeting, Lima, Peru, 6–9 Oct 2001
114. Neira R, Solarte E, Isaza C, Creole W, Rebolledo A, Arrogate J, Ramirez H (2002) In vitro effects of 635 nm low intensity diode laser irradiation on the fat distribution of one adipose cell "ICO 19 Florence, Italy (2002) SPIE (Int Soc Optical Engineering) Proceeding, 4829(2):961–962
115. Neira R, Rebolledo AF, Solarte E, Reyes MA (2002) Diffraction and dispersion of coherent light in adipose tissue samples. Colombian national congress physics and optic. Rev Colombiana Fisica 33(1):191–195
116. Neira R, Rebolledo AF, Solarte E, Reyes MA, Montoya JA, Ortiz C, Arroyave J (2002) Coherent light dispersion in adipose tissue samples. Colombian national congress physics and optic. Rev Colombiana Fisica 34(1):210–213
117. Neira R, Arroyave J, Ramirez H, Ortiz CL, Solarte E, Sequeda F (2002) Fat liquefaction: effect of low-level laser energy on adipose tissue. Plast Reconstr Surg 110(3):912–922
118. Newman J, Dolsky RL, Mai ST (1984) Submental liposuction extraction with hard chin augmentation. Arch Otolaryngol 110(7):454–457
119. O'Brien BM, Khazanchi RK, Kumar PA, Dvir E, Pederson WC (1989) Liposuction in the treatment of lymphoedema; a preliminary report. Br J Plast Surg 42(5):530–533
120. O'Brien B, Mellow CG, Khazanchi MC, Dvir E, Kumar V, Pederson WC (1990) Long term results after microlymphaticovenous anastomoses for the treatment of obstructive lymphedema. Arch Otolaryngol Head Neck Surg 85(4):562–572
121. Oliver DW, Inglefield CJ (2002) Liposuction of haematoma. Br J Plast Surg 55(3):269–279

122. Ong WC, Lim TC, Lim J, Leow M, Lee SJ (2003) Suction-curettage: treatment for axillary hyperhidrosis and hidradenitis. Plast Reconstr Surg 11(2):958–959
123. Ostad A, Kageyama N, Moy RL (1996) Tumescent anesthesia with a lidocaine dose of 55 mg/kg is safe for liposuction. Dermatol Surg 22(11):921–927
124. Ou LF, Yan RS, Chen IC, Tang YW (1998) Treatment of axillary bromhidrosis with superficial liposuction. Plast Reconstr Surg 102(5):1479–1485
125. Palmieri B, Bosio P, Palmieri G, Gozzi G (1997) Ultrasound lipolysis and suction lipectomy for treatment of obesity. Am J Cosmet Surg 14(3):289–296
126. Park DH (1999) Treatment of axillary bromhidrosis with superficial liposuction. Plast Reconstr Surg 104(5): 1580–1581
127. Payne CM, Doe PT (1998) Liposuction for axillary hyperhidrosis. Clin Exp Dermatol 23(1):9–10
128. Perez JA (1999) Treatment of dysesthesias secondary to ultrasonic lipoplasty. Plast Reconstr Surg 103(5):1534
129. Perez RA (2007) Liposuction and diabetes type 2 development risk reduction in the obese patient. Med Hypotheses 68(2):393–396
130. Ponce-de-Leon S, Iglesias M, Cellabos J, Ostrosky-Zeichner L (1999) Liposuction for protease-inhibitor-associated lipodystrophy. Lancet 353(9160):1244
131. Price MF, Massey B, Rumbolo PM, Paletta CE (2001) Liposuction as an adjunct procedure in reduction mammaplasty. Ann Plast Surg 47(2):115–118
132. Rohrich RJ, Beran SJ, Kenkel JM, Adams WP Jr, DiSpaltro F (1998) Extending the role of liposuction in body contouring with ultrasound-assisted liposuction. Plast Reconstr Surg 101(4):1090–1102; discussion 1117–1119
133. Rosenberg GJ (1987) Gynecomastia: suction lipectomy as a contemporary solution. Plast Reconstr Surg 80: 379–385
134. Rubenstein R, Roenigk H, Garden JM, Goldberg NS, Pinski JB (1985) Liposuction for lipomas. J Dermatol Surg Oncol 11(11):1070–1074
135. Samdal F (1991) The female breast and reduction liposculpturing. In: Fournier PF (ed) Liposculpture: the syringe technique. Arnette, Paris, pp 233–235
136. Sandhofer M, Douwens KE, Sandhofer-Novak R, Blugerman GS (2002) Laserlipolyse und liposkulptur. Äesthet Chir 3:20–26
137. Sasaki GH (1996) Endoscopically assisted suctioning of lipomas. In: Sasaki GH (ed) Endoscopic, aesthetic, and reconstructive surgery. Lippincott-Raven, Philadelphia
138. Schaefer BT (2000) Endoscopic liposhaving for neck recontouring. Arch Facial Plast Surg 2(4):264–268
139. Schavelzon DE (2002) Laserlipolysis for the treatment of localized adiposity. World Congress on Liposuction Surgery, Westminster, Colorado, 4–6 Oct 2002
140. Scuderi N, Devita R, D'Andrea F, Vonella M (1987) Nuove prospettive nella liposuzione la lipoemulsificazone. Giorn Chir Plast Riconstr Estetica 2(1):33–39
141. Shenag SM, Spira M (1987) Treatment of bilateral axillary hyperhidrosis by liposuction assisted lipoplasty technique. Ann Plast Surg 19:548–551
142. Shenoy MU, Srinivasan J, Rance CH (2000) Buried penis: surgical correction using liposuction and realignment of skin. BJU Int 86(4):527–530
143. Shiffman MA (2006) Non-cosmetic applications of liposuction. In: Shiffman MA, Di Giuseppe A (eds) Liposuction: principles and practice. Springer, Berlin, pp 408–414
144. Shiffman MA, Mirrafati S (2006) External percussion massage-assisted liposuction. In: Shiffman MA, Di Giuseppe A (eds) Liposuction: principles and practice. Springer, Berlin, pp 309–311
145. Sidor V (1997) Megalipotherapy: problems and results. Am J Cosmet Surg 14(3):241–249
146. Silberg BN (1998) The technique of external ultrasound-assisted lipoplasty. Plast Reconstr Surg 101(2):552
147. Sonenshein H, Lepoudre C (1985) Suction assisted lipectomy – a functional use in the neck. Am J Cosmet Surg 2:42–44
148. Stallings J (1984) Defatting of flaps by lipolysis. In: Hetter GP (ed) Lipoplasty: the theory and practice of blunt liposuction lipectomy. Little, Brown and Co, Boston, pp 309–320
149. Swinehart JM (2000) Treatment of axillary hyperhidrosis: combination of the starch-iodine test with the tumescent liposuction technique. Dermatol Surg 26(4): 392–396
150. Taufig AZ (2000) Hydro-jet-liposuction: a new method for liposuction. Presented at Vereinigung der Deutschen Plastischen Chirurgen meeting, Cologne, Sept 2000
151. Taufig AZ (2006) Water-jet-assisted liposuction. In: Shiffman MA, Di Giuseppe A (eds) Liposuction: principles and practice. Springer, Berlin, pp 326–330
152. Tebbetts JB (1999) Rapid probe movement ultrasound-assisted lipoplasty. Aesthet Surg J 19(1):17–23
153. Thomas J (2001) Adjunctive tumescent technique in massive resections. Aesthet Plast Surg 25:343–346
154. Tobin AH (1987) Large-volume lipo-suction: planned staged treatment in the obese patient. Am J Cosmet Surg 4(1):61–66
155. Toledo LS (1990) Superficial syringe liposculpture. In: Toledo LS (ed) Annals of the II international symposium: recent advances in plastic surgery – RAPS/90, Sao Paulo, Mar 1990, p 446
156. Troilius C (1999) Ultrasound-assisted lipoplasty: is it really safe? Aesthet Plast Surg 23(5):307–311
157. Tsai RY, Lin JY (2001) Experience of tumescent liposuction in the treatment of osmidrosis. Dermatol Surg 27(5): 446–448
158. Voigt M, Walgenbach KJ, Andree C, Bannasch H, Looden Z, Stark GB (2001) Minimally invasive surgical therapy of gynecomastia: liposuction and exeresis technique. Chirurg 72(10):1190–1195
159. Wilhelmi BJ, Blackwell SJ, Mancoll JS, Phillips LG (1999) Another indication for liposuction: small facial lipomas. Plast Reconstr Surg 103(7):1864–1867
160. Winslow RB (1984) Treatment of congenital lymphedema of the lower extremity. In: Hetter GP (ed) Lipoplasty: the theory and practice of blunt suction lipectomy. Little, Brown and Co, Boston, pp 326–329

161. Yoho R (2000) Benign symmetrical lipomatosis treated with tumescent liposuction. Int J Cosmet Surg Aesthet Derm 2(2):141–143
162. Zandi I (1985) Use of suction to treat soft tissue injected with liquid silicone. Plast Reconstr Surg 76(2):307–309
163. Zocchi ML (1988) Metodo di trattamento del tessuto adiposo con energia ultrasonica. Congresso dell Societa Italiana di Medicina Estetica, Roma, Apr 1988
164. Zocchi ML (1989) New prospectives in liposcultpuring: the ultrasonic energy. In: Abstract of the 10th ISAPS congress, Zurich, Sept 1989
165. Zocchi ML (1993) Clinical aspects of ultrasonic liposculpture. Perspect Plast Surg 7:153–174
166. Zocchi ML (1995) Ultrasonic-assisted lipectomy. Adv Plast Reconstr Surg 11:197–221
167. Zocchi ML (1996) Ultrasonic assisted lipoplasty. Clin Plast Surg 23(4):575–598
168. Zocchi ML (1999) Basic physics for ultrasound-assisted lipoplasty. Clin Plast Surg 26(2):209–220

Principles of Autologous Fat Transplantation

5

Luiz S. Toledo

Contents

5.1 **Introduction** 45

5.2 **Technique** 46

5.3 **Liposculpture** 46
5.3.1 Syringe Liposculpture 46

5.4 **Fat Grafting** 47
5.4.1 Fat Grafting of Flanks, Buttocks, Thighs, Knees, and Abdomen 47
5.4.2 Fat Grafting Survival 48

5.5 **Cellulite** 49

5.6 **Conclusion** 52

References 54

L.S. Toledo
Mercato Family Clinic,
P.O. Box 213522, Dubai, UAE
e-mail: toledodubai@gmail.com

5.1 Introduction

The first communication about fat transplantation was presented by Neuber [18] in 1893, followed by reports by Czerny [6], Lexer [15], and Rehn [22]. In 1911, Bruning [2] was the first to inject autologous fat into the subcutaneous tissue for the purpose of soft-tissue augmentation.

In 1950, Peer [19] showed that the survival rate for fat grafts could be as high as 50%. In his paper, Peer performed a biopsy 13 years after the procedure of fat grafting to treat a facial depression, showing normal fatty tissue enclosed in a definite connective-tissue capsule.

In the 1980s, liposuction improved the technique. In 1985, Illouz [13] and Fournier [7] developed an easy approach to fat transfer by syringe harvesting, called "microlipoinjection" by Fournier. This process became possible after the advent of liposuction. Fat grafting became the simplest and easiest method among the different techniques used to correct the soft tissues for facial contour and, if needed, to replace soft tissue in body areas.

We started using fat grafting for the face and body in 1985, showed our 18-month experience in 218 patients at the ISAPS Congress in New York City in 1987, and the article was published in 1988 [17]. Fat was aspirated with a liposuction machine and collected in a sterile vial. After decanting, it was transferred to 60-cc catheter tip syringes and injected in different parts of the body using a blunt cannula with the same caliber as the cannula previously used for aspiration.

Still in the 1980s, Loeb [16], Bircoll [1], Chajchir [5], Gonzalez [11], and Pereira [20] were among the

Y.-G. Illouz and A. Sterodimas (eds.), *Adipose Stem Cells and Regenerative Medicine*,
DOI: 10.1007/978-3-642-20012-0_5, © Springer-Verlag Berlin Heidelberg 2011

surgeons who helped establish the parameters for the safe use of fat grafting. In the 1990s, Lewis [14], Carraway [4], Guerrerosantos [12], and Pereira [21] showed long-term results with the use of fat grafting for body contour. In 1994, Carpaneda [3] showed that the percentage of absorption of the fat graft was inversely proportional to the thickness of the graft.

5.2 Technique

In 1988, the only accepted procedure for volume augmentation was the insertion of silicone implants. We showed fat grafting could have major advantages with fewer complications. We used fat volumes ranging from 5 to 450 cc. The majority of the patients were grafted in more than one area and the average age of the patients ranged from 24 to 39 years. Fat was injected where needed. In Brazil the preferred area for augmentation and reshaping was the buttocks (54 patients – 29% of the cases); followed by trochanteric depressions (43 patients – 20.6%); breast augmentation (21 patients – 10.9%); scar depressions (20 patients – 9.6%); thigh, calf, and ankle augmentation (18 patients – 8.6%); small wrinkles; depressions of the face and Romberg's disease (15 patients – 7.2%); the nasolabial fold (13 patients – 6.2%); liposuction sequaelae (10 patients – 4.8%); and fingers and hands (2 patients – 0.9%).

The preferential donor site (for small areas) was the medial part of the knee, then the dorsal region, the lateral thighs, and the abdominal region. We noticed an absorption rate ranging from 20% to 50%. Absorption was more pronounced in the early postoperative months and stabilized by the third month. Consequently, when possible, we injected 35% more fat than necessary.

- Our particular experiences in the first 2 years lead us to believe that we had to
- Avoid high negative pressure during suction
- Use adequate instruments
- Avoid injecting "collections" or "lakes"
- Inject in "rod" or "thread" shapes
- Hypercorrect (35%), if possible
- Rest the grafted site for a week
- Prescribe routine antibiotics
- Perform light massage of the grafted area
- Use molding adhesive tape (1–2 weeks)

We followed up on the 208 cases with a small complication rate: local inflammation signs (4 cases), 1.9%; injection out of the marked receptor area (3 cases), 1.4%; infection (1 case), 0.4%. These patients received anti-inflammatory and corticosteroid treatment and the problems were solved satisfactorily.

Although the method did not show perfect results, fat grafting had many advantages:

The procedure could be repeated several times.

- There were no immunity phenomena.
- Fat was easily obtained.
- Artificial materials (silicone, collagen) could be substituted.
- It is an autograft.
- Results were highly satisfactory.
- The cost factor was very reasonable.

Our conclusion was that injected fat grafting undoubtedly constituted a major step in repositioning the loss of soft tissue, which was very difficult to correct before. It was also an important element for treating certain small iatrogenic liposuction defects.

5.3 Liposculpture

In 1989 [23], we published the improvement of the technique by using syringe liposculpture to treat superficial irregularities, depressions, "cellulite," and liposuction sequaelae to varying degrees depending on the problem, with high patient satisfaction. From 1986, we started using a syringe for facial liposculpture in the sub-mental and nasolabial regions, extracting small quantities of fat from the knee for facial injection. From 1986 to 1988, we performed fat grafting for body sculpturing using fat collected in a sterilized vial by the liposuction machine. Although at the time our results seemed satisfactory, the method of obtaining and injecting fat was still complicated and time consuming. In February 1988, we stopped using the liposuction aspirator; all our body contour surgeries were now performed using only the syringe [8] (Fig. 5.1).

5.3.1 Syringe Liposculpture

In 1989, we introduced a new technique – superficial syringe liposculpture [24], to treat patients with flaccid skin, superficial irregularities or depressions, "cellulite," and liposuction sequaelae.

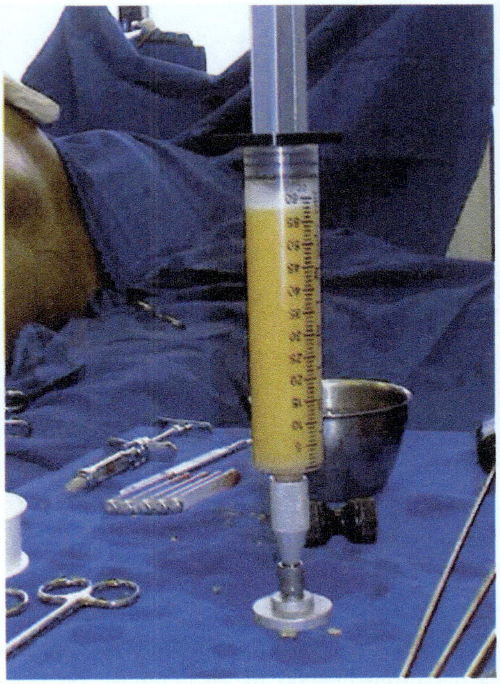

Fig. 5.1 For fat injection (and suction) of the body we use Toomey-tip 60-cc syringes and the stands

The technique combined syringe liposculpture [9], superficial liposuction [10], and our method of treating skin irregularities by breaking the fibrous adherences and injecting fat superficially [25].

5.4 Fat Grafting

In 1991, we published our 2-year experience [26] with the technique, combining syringe liposuction and fat grafting to reshape the face and the body by removing localized fat deposits and reinjecting this fat where needed. When we did not reinject, we called the technique reduction liposculpture. Suction of the face was usually done to improve the submental and the jowl areas. This fat could be reinjected. Fat could also be harvested from the medial knee or any other localized fat deposit. The syringes were centrifuged for 1 min at about 2,000 rpm, the unwanted solution was ejected, and the fat was transferred to the new area. We treated the glabella, malar region, nasolabial folds, lips, and the dorsal part of senile hands. We did not over correct on the face. The injection was repeated in the second and sixth month postoperative, when necessary. The same disposable syringe that harvests the fat was connected to a gun to allow precise injection of very small amounts.

For the body, we used 60-cc syringes, connected to 3, 4, 5, or 6-mm gauge cannula with mercedes, cobra, or pyramid tips and 20–35 cm long. We also used a special 3-mm-gauge V-tip cannula dissector, 20–35 cm long, adapted to the 60-cc syringe, for treating skin irregularities, cellulite depressions, or liposuction sequaelae, dissecting, suctioning, or injecting fat where necessary.

Massaging with the fingers helped break the remaining adhesions. We wash the fat cells very gently with Ringer's solution, transferring them from one syringe to another. Fat have no contact with air. The syringes are put on a stand to decant.

5.4.1 Fat Grafting of Flanks, Buttocks, Thighs, Knees, and Abdomen

Total liposculpture meant suction of the arms, dorsal region, flanks, buttocks, thighs, knees, abdomen, and, when needed, fat injection into the buttocks, trochanteric region, inner thighs, and legs. We usually removed 2–4 L of fat and immediately reinjected from 500 to 1,500 cc. This was done in one procedure, but in some cases we needed to treat certain areas more than once, depending on the defect.

In some cases, we have injected up to 500 cc of fat on each side of the buttocks, in the muscle, on the muscle, into the fatty tissue, and subcutaneously, when needed. In the inner thigh, we have injected from 100 to 300 cc and in the calf from 50 to 150 cc. It has always been a problem to remove the right amount of fat from a patient with flaccid skin. It is generally believed that very deep suction should always be done to avoid skin irregularities. We disagreed and in September 1988 we started treating superficial irregularities.

In 1996, we reported 8-year results with syringe liposculpture [27], for the treatment of localized fat deposits, to remodel the body and the face using disposable syringes and fine tip cannulas. We started working on a combination of syringe liposculpture and superficial liposuction to treat the superficial layer of fat, resulting in "superficial liposculpture." Syringe liposculpture of the face can be offered as an alternative to rhytidoplasty and a complement to other techniques to improve the appearance with minimal morbidity.

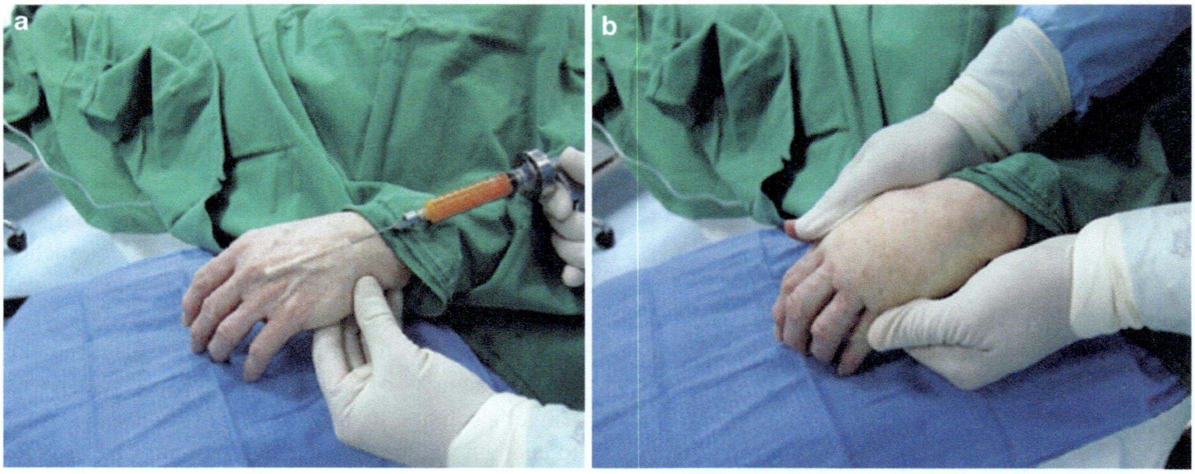

Fig. 5.2 Fat injection for rejuvenation of the hands (**a**) and fingers (**b**)

In 2006, we guest edited an issue of Clinics in Plastic Surgery and published a 20-year revision on fat grafting [28], showing the results obtained by plastic surgeons internationally. In our hands, liposculpture consists of removing adipose tissue with a cannula and syringe and cleaning the aspirated material with Ringer's lactate when necessary. All external contact is carefully avoided, preventing contamination. The fat was harvested in an atraumatic and sterile manner. Fat could be centrifuged at 1,500 rpm for 1 min or decanted for 10 min, then injected with a blunt-tipped cannula. Facial and body volume restoration with the fat autograft technique offered the potential for symmetric, long-term results.

Fat grafting was performed in multiple tunnels in the deep and superficial planes. Fat absorption was estimated by clinical evaluation and measurements to be between 20% and 50% of volume. Fat threads to augment the face and the body should not be thicker than 3-mm to allow for neo-vascularization. In the hands we do not need to inject threads, but a 2–3-mm thick layer of fat in the sub-cutaneous area (Fig. 5.2). A low rate of complications (less than 3%) supported the opinion that this method was an efficient and safe procedure to correct or enhance contour deformities. The grafts will grow if the patient gains weight and lose circumference when the patient reduces weight.

As we had stated in our first publication in 1988, the controversy surrounding the longevity of correction in autologous fat grafts was related to the adipocyte survival. The long-term fat survival rates differs widely depending on the harvesting method, means of reinjection, injection site, and evaluation methods.

5.4.2 Fat Grafting Survival

Survival of aspirated fat cell grafts depends mainly on the anatomic site, the mobility and vascularity of the recipient tissue, and underlying causes and diseases, but also on harvesting and reinjection methods. The preferred area for fat injection in the body continues being the buttocks. We have created the "Brazilian Buttock Technique," which consists of aspiration of fat from the flanks, abdomen, and thighs with injection into the buttocks and trochanter areas (Figs. 5.3 and 5.4).

Fat injection of the face became an option for younger patients wanting to rejuvenate without having to undergo a rhytidoplasty (Fig. 5.5). Fat is aspirated from the sub-mental area and, if needed, from other areas of the body to achieve the necessary amount for grafting in the malar area, nasolabial folds. Fat grafting of the nasojugal area has revolutionized blepharoplasty. We remove less fat from the lower eyelids, during the blepharoplasty procedure, and inject fat in the nasojugal and peri-orbital areas (Fig. 5.6).

Special care should be taken when patients gain weight, because the grafted areas can grow in the same proportion as the donor area. We have seen

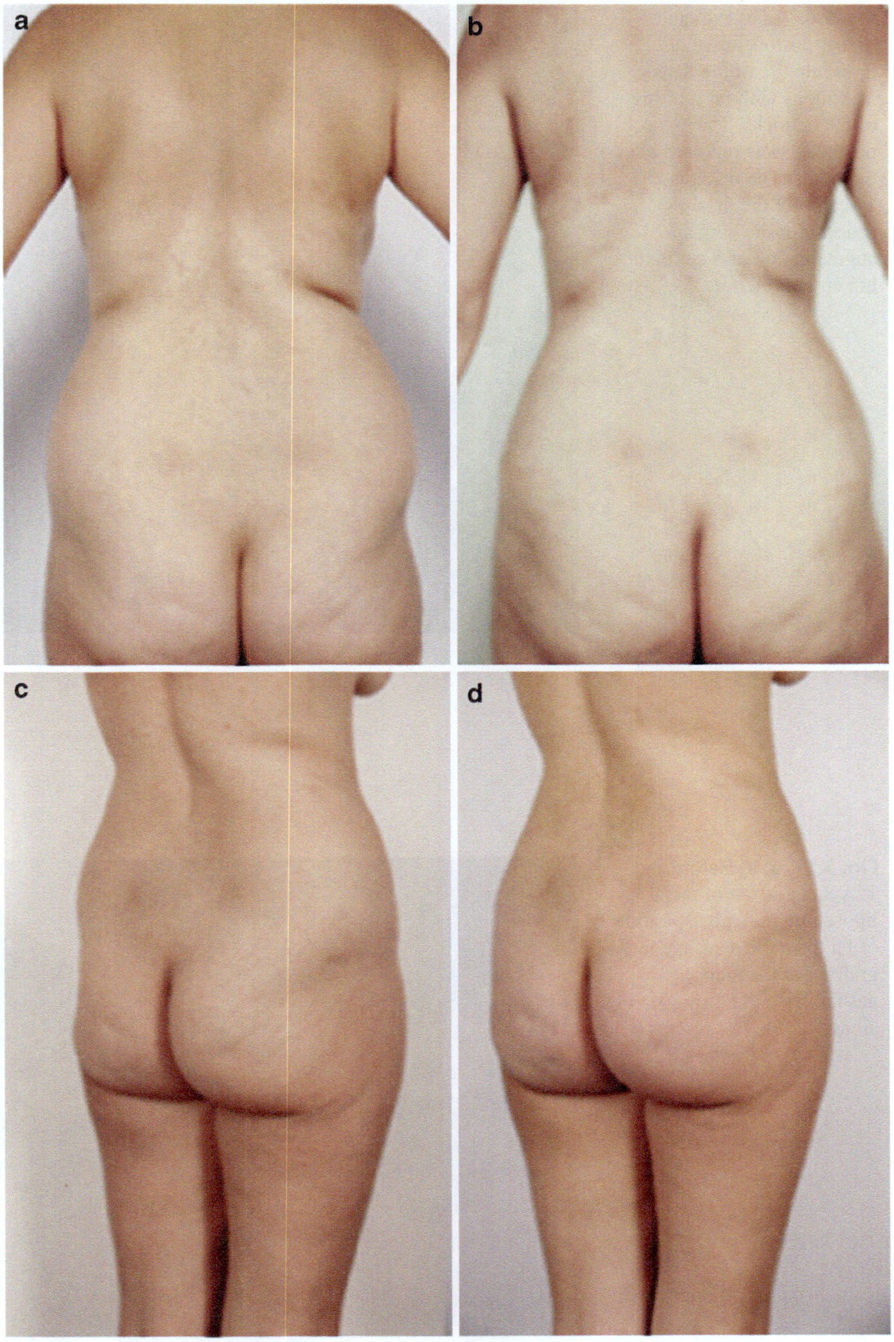

Fig. 5.3 (**a**, **c**) Preoperative picture of a 24-year-old patient with lipodystrophy of the flanks, buttocks, and thighs. (**b**, **d**) One year postoperative after liposculpture of flanks and thighs with fat grafting of the buttocks (500-cc each side)

several cases where we have to aspirate fat from the grafted area and cases where we have to remove grafted fat during a rhytidoplasty. There might be a memory from the donor area, as there is in transplanted hair (Fig. 5.7).

5.5 Cellulite

Cellulite is considered to be a special type of fat with fibrous adherences that pull the skin down like the buttons of a mattress. Deep liposuction alone would not

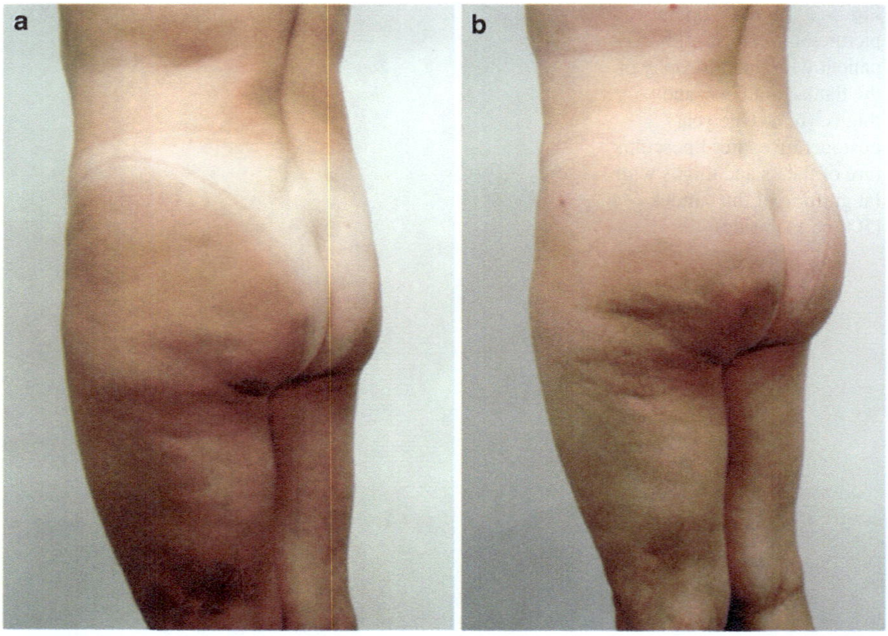

Fig. 5.4 (**a**, **b**) Pre- and 1-year postoperative of liposculpture of the flanks and thighs with fat grafting of the buttocks. The procedure was repeated after 6 months. The result is 1 year after the first procedure, showing an augmentation of 8 cm in the hip circumference

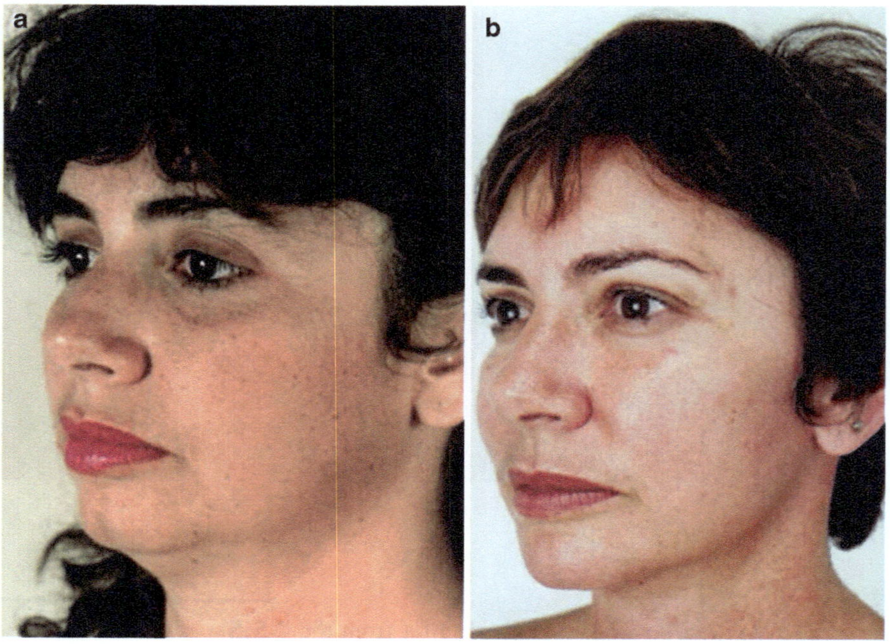

Fig. 5.5 (**a**, **b**) Before and 12 years post-op of facial liposculpture, with aspiration of fat from the neck and fat grafting of the malar, nasolabial, and nasojugal areas

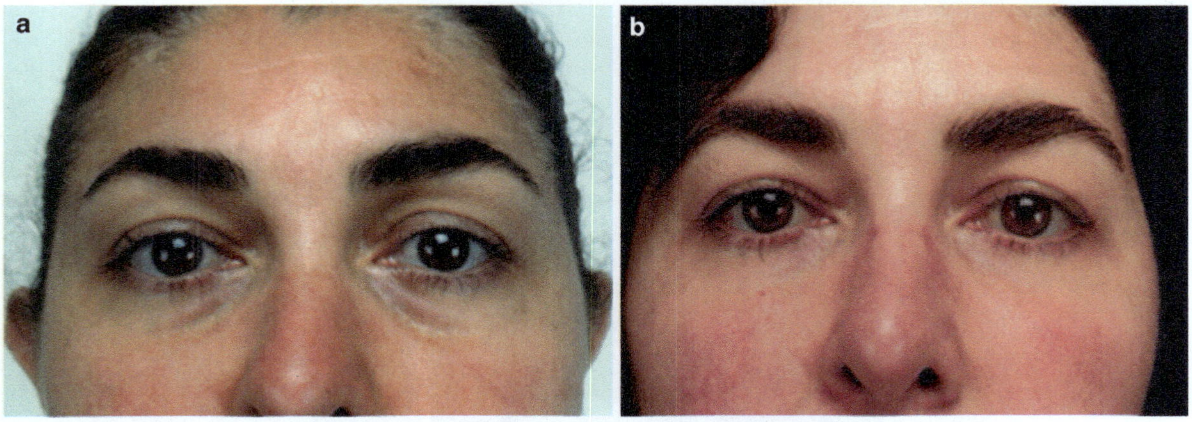

Fig. 5.6 (**a**, **b**) Pre- and 6-month postoperative of fat grafting of the nasojugal areas. No blepharoplasty performed in this case

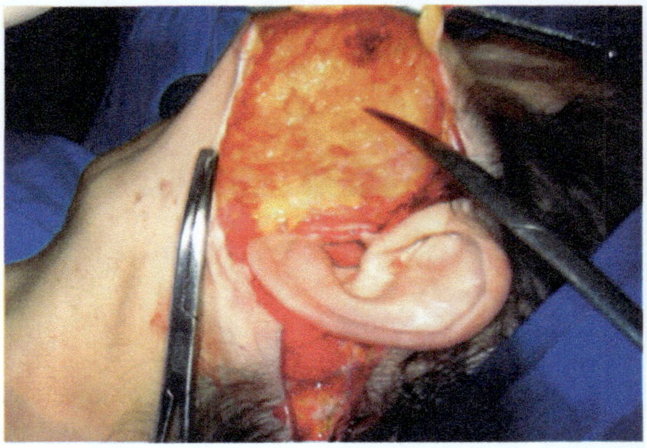

Fig. 5.7 This patient had been submitted to three sessions of fat injection in the malar 6 years before. As she gained weight with age her cheeks became too big. During a rhytidoplasty, we could see and remove the excess fat that had been injected

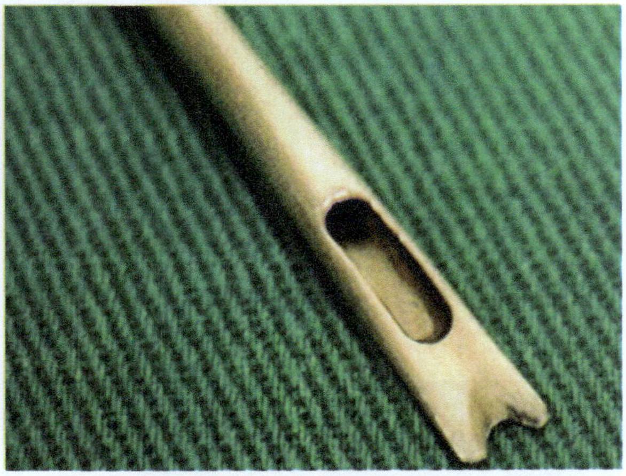

Fig. 5.8 The Toledo V-tip cannula. The tips of the V are blunt and the inside is sharp. It helps with dissection of the fibrous attachments of cellulite, scars, or liposuction sequaelae

solve this problem, and it could sometimes make it worse. Superficial liposuction plus severing the fibrous adherences, combined with superficial fat injection, improved skin irregularities. We could then offer visible improvement of skin irregularities with the first procedure, which could be repeated if necessary. The procedure took from 2 to 2.5 h.

First the contour of the body was improved by suctioning from the deeper layers of fat and then, if necessary, we treated the superficial irregularities. Superficial syringe liposculpture could treat liposuction sequaelae, removing fat where (and if) too much was left and injecting it into depressions. These iatrogenic depressions were freed from the deep layer of fat with the 3-mm-gauge V-tip cannula dissector (Fig. 5.8) and then the fat was injected superficially. This dissector does not cut laterally, only on inward movements. We do not dissect more than we need to in order to free the depression.

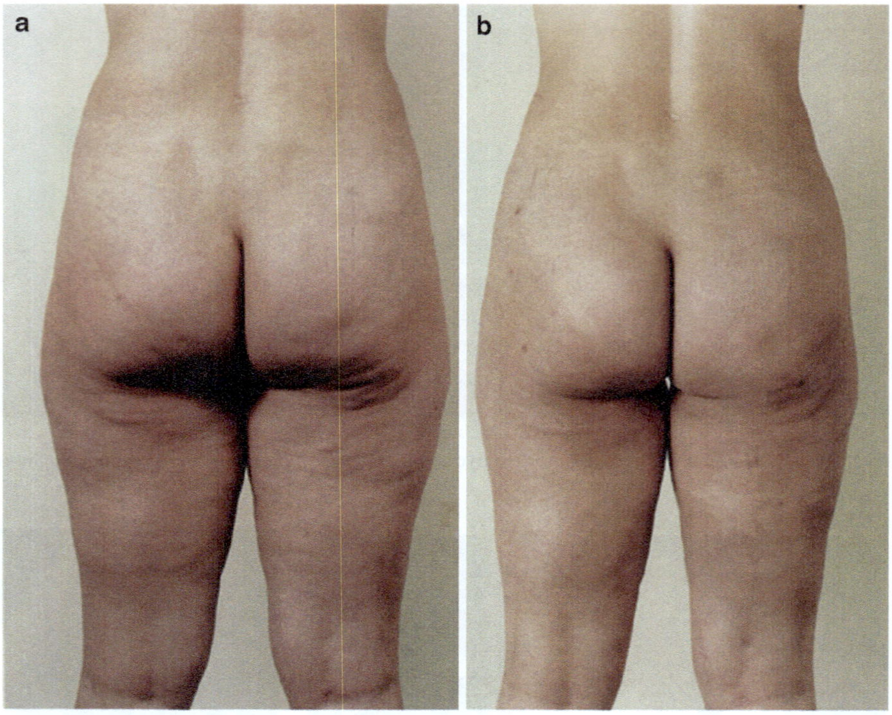

Fig. 5.9 (a, b) Pre- and postoperative result of superficial liposculpture

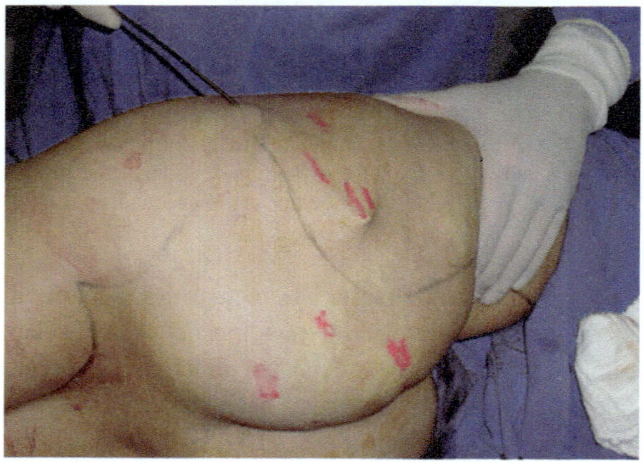

Fig. 5.10 Using the Toledo V-tip cannula to release the "cellulite" depressions that were marked in red with the patient standing. The cannula does not dissect laterally. After the depressions are released, fat is injected to fill in the space and avoid reattachment

This V-tip device allows us to do suction or injection. We over injected by 30% and the patient should be informed of the possibility of repeated treatments.

Cellulite stages 2 and 3 of the Nurnberger and Muller classification could also be treated in the same manner (Figs. 5.9–5.14), with a visible reduction of skin irregularities. Elastic adhesive tape used for 5 days and a girdle for a month to control edema and bruising and to secure the skin in place until it retracts to its normal position.

5.6 Conclusion

Some authorities advocate using newly emerging approaches that apply fat tissue engineering principles. Adipose tissue stem cells are being used to improve

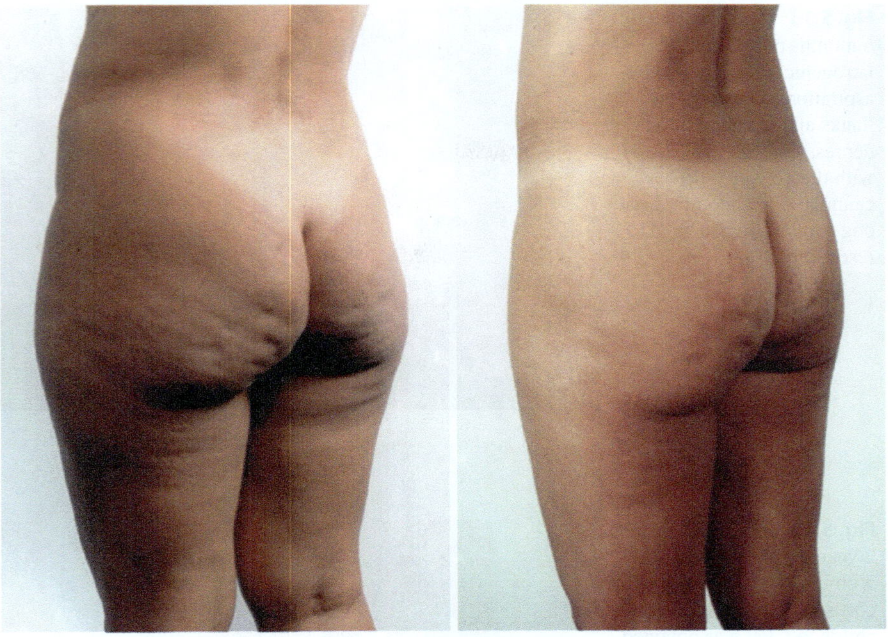

Fig. 5.11 Pre- and 8-year postoperative of superficial liposculpture of flanks, lateral thighs, and buttocks with superficial treatment of the "cellulite" depressions

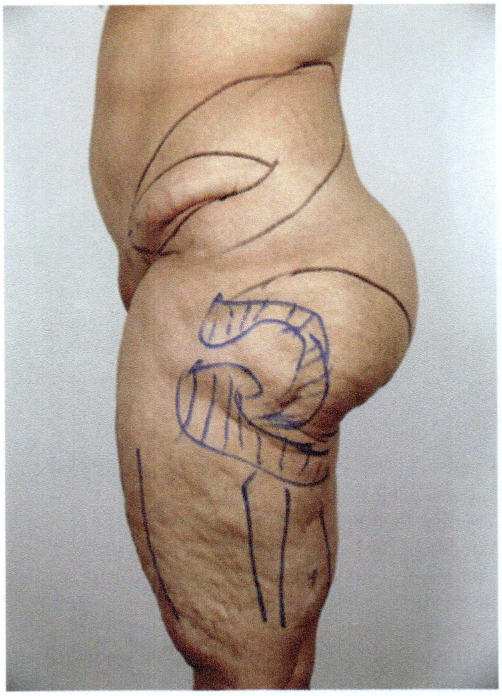

Fig. 5.12 Iatrogenic defect provoked by irregular liposuction of the sub-dermal area in the trochanter and subgluteal areas. The patient had excess fat in the flanks and abdominoplasty dog ears

the results of fat injection. These approaches are sophisticated, complex, and expensive procedures. Fat grafting is commonly performed, and poses little danger to patients. Adipose stem cells were described in the literature only 8 years ago and more research must be done to determine the safety of their use in humans. They can interfere with T-cell proliferation and activation and they can evade an immune response. It seems premature to encourage patients to purchase this technology. Obviously, the media hype on stem cells is widely used by many doctors who now promote that they are doing stem cell injections. To some extent this is true, because if one injects fat, consequently one injects stem cells with it. However, this is not yet done with the level of purity we believe stem cell research will be able to produce in the near future. We have to consider that even using available technology, fat grafting has become in the last 25 years the treatment of choice of many plastic surgeons for facial and body contour soft-tissue augmentation in reconstructive and aesthetic surgery.

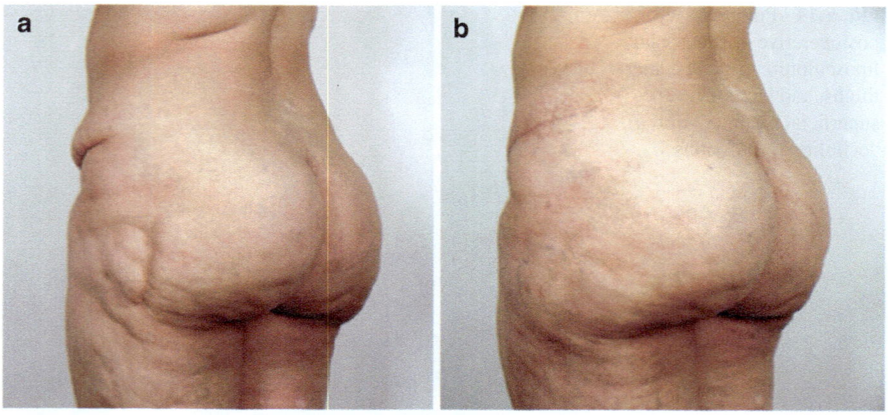

Fig. 5.13 (**a**, **b**) Pre and 6 months post-op of iatrogenic defect treated by aspiration of fat from the flanks and injection into the depressions, after release with the Toledo V-tip cannula. The abdominoplasty dog ear was treated by skin resection

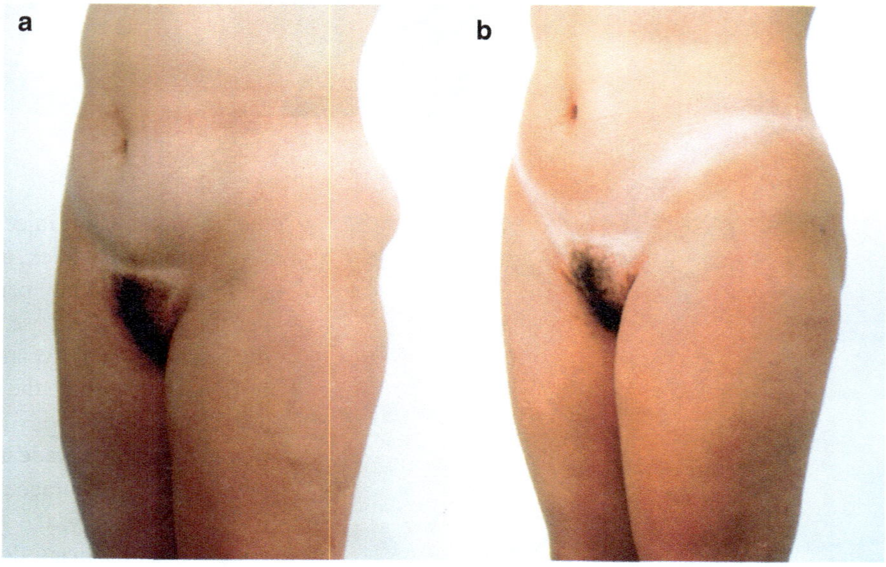

Fig. 5.14 (**a**, **b**) Before and 2 years post-op of a patient with buttock depression after cortisone injections during childhood. Fat was obtained with syringe liposculpture from the abdomen and flanks and injected into the buttocks, after scar tissue release with the Toledo V-tip cannula

References

1. Bircoll M (1987) Autologous fat transplantation. Plast Reconstr Surg 79(3):492–493
2. Bruning P, Broeckaert TJ (1919) Contribution al'étude des greffesadipeuses. Bull Acad R Méd Belg 28:440
3. Carpaneda CA, Ribeiro MT (1994) Percentage of graftviability versus injected volume in adipose autotransplants. Aesthetic Plast Surg 18(1):17–19
4. Carraway JH, Mellow CG (1990) Syringe aspiration and fat concentration: a simple technique for autologous fat injection. Ann Plast Surg 24(3):293–296
5. Chajchir A, Benzaquen I (1986) Liposuction fat grafts in face wrinkles and hemifacial atrophy. Aesthetic Plast Surg 10:115
6. Czerny A (1895) Plastischer Ersatz der Brustdrosedurchein-Lipoma. Chir Kongr Verhandl 216:2
7. Fournier PF (1985) Microlipo-extraction et microlipo-injection. Rev Chir Esthet Lang Franc 10:36–40
8. Fournier P (1988) In: Transactions of the 3rd recent advances in aesthetic plastic surgery symposium, Beverly Hills, February 1988
9. Fournier P (1989) Liposculpture – Ma technique. Arnette, Paris
10. Gasparotti M (1990) Superficial liposuction for flaccid skin patients. In: Toledo LS (ed) Annals of the international symposium recent advances in plastic surgery, Estadão, São Paulo, 28–30 March 1990, p 443
11. Gonzalez R, Spina L (1986) Enxertia de gordura lipoaspirada:detalhes t6cnicos e instrumental. Rev Bras Cir 76:243

12. Guerrerosantos J (1996) Autologous fat grafting for body contouring. Clin Plast Surg 23(4):619–631
13. Illouz YG (1985) De l'utilization de la graisse aspire pour combler les defects cutanés. Rev Chir Esthet Lang Franc 10:13
14. Lewis CM (1992) Correction of deep gluteal depression by autologous fat grafting. Aesthetic Plast Surg 16(3): 247–250
15. Lexer E (1910) Freirefettgewebstranplantation. Dtsch Med Wochenschr 36:46
16. Loeb R (1981) Fat pad sliding and fat grafting for leveling lid depressions. Clin Plast Surg 8(4):757
17. Matsudo PKR, Toledo LS (1988) Experience of injected fat grafting. Aesthetic Plast Surg 12:35–38
18. Neuber GA (1893) Fettransplantation. Verh Dtsch Ges Chir 22:66
19. Peer LA (1950) Loss of weight and volume in human fat grafts. Plast Reconstr Surg 5:217–228
20. Pereira JF, Neves RL (1985) Corre~go de deformidades est-6ticas da face corn gordura lipoaspirada. An XXII Cong Bras Cir Plast, Porto Alegre
21. Pereira LH, Radwanski HN (1996) Fat grafting of the buttocks and lower limbs. Aesthetic Plast Surg 20(5):409–416
22. Rehn E (1912) Die fettransplantation. Arch Klin Chir 98:1
23. Toledo LS (1989) Liposculpture of the face and body. In: Toledo LS (ed) Annals of the international symposium recent advances in plastic surgery, São Paulo, 3–5 March 1989, pp 177–192
24. Toledo LS (1989) Superficial syringe liposculpture for the treatment of "cellulite" and liposuction sequaelae. Presented at the international symposium recent advances in plastic surgery, Buenos Aires, 11 June 1989
25. Toledo LS (1990) Superficial syringe liposculpture. In: Toledo LS (ed) Annals of the international symposium recent advances in plastic surgery, Estadão, São Paulo, 28–30 March 1990, p 446
26. Toledo LS (1991) Syringe liposculpture: a two-year experience. Aesthetic Plast Surg 15(4):321–326
27. Toledo LS (1996) Syringe liposculpture. Clin Plast Surg 23(4):683–693
28. Toledo LS, Mauad R (2006) Fat injection: a 20-year revision. Clin Plast Surg 33:47–53

6 Clinical Applications of Autologous Fat Transplantation

Luiz Haroldo Pereira, Beatriz Nicaretta, and Yves-Gérard Illouz

Contents

6.1 **Introduction** 57

6.2 **Surgical Technique of Fat Grafting** 58

6.3 **Clinical Applications of Autologous Fat Graft** 60
6.3.1 Autologous Fat Graft for Facial Contour Rejuvenation 60
6.3.2 Autologous Fat Graft for Gluteal Augmentation . . 60

6.4 **Autologous Fat Graft as Filler** 60

6.5 **Pitfalls in Autologous Fat Transplantation** 63

6.6 **Stromal Enriched Lipograft (SEL)** 63

6.7 **Conclusion** 63

References 64

L.H. Pereira (✉)
Department of Plastic Surgery, LH Clinic, Rua Xavier da Silveira 45/206, 22061-010 Rio de Janeiro, Brazil
e-mail: haroldo@uninet.com.br

B. Nicaretta
Department of Plastic Surgery, Policlinca Geral, 38 Av Nilo Peçanha, 22020-100 Rio de Janeiro, Brazil
e-mail: bianicareta@hotmail.com

Y.-G. Illouz
Plastic Surgery Department, Saint Louis Hospital, Avenue Claude-Vellefaux, Paris, France
e-mail: yillouz2000@hotmail.com

6.1 Introduction

The German physician Franz Neuber in 1893 used for the first time a small piece of upper arm fat to build up the face of a patient whose cheek had large pit caused by a tubercular inflammation of the bone [24]. With the advent of liposuction in the 1980s, it became possible to aspirate and reinject fat, allowing transplantation of small volumes for soft-tissue augmentation for the correction of contour irregularities [16].

In the last 25 years, several different techniques of lipoinjection have been developed to correct various problems such as buttocks (augmentation and reshaping), trochanteric depressions, breast augmentation, scar depressions, thighs and legs (calf and ankle augmentation), small wrinkles and depressions of the face (Romberg's disease), nasolabial fold, upper outer breast quadrant, liposuction sequel [3, 17, 23, 28, 29, 36].

As in every surgical procedure, the success of autologous fat grafting is dependent upon many factors: the techniques and instruments used to harvest the fat tissue, the fat processing, the volumes of fat implantation, the sites to be implanted, the levels of placement, and even the individual patient. Because of this variability and perhaps because of other factors that are not yet understood, the results of fat grafting with some techniques, in some patients, and in some areas can be unpredictable. A standard procedure has not been adopted by all practitioners. There is no agreement as to the best way of processing the fat to ensure maximal take and viability of the graft [39].

There has been also great disparity in the reported results of fat grafting in terms of survivability and long-term outcomes. Studies evaluating survivability of transplanted fat have reported volume retention of between

Y.-G. Illouz and A. Sterodimas (eds.), *Adipose Stem Cells and Regenerative Medicine*,
DOI: 10.1007/978-3-642-20012-0_6,

20% and 90%. However, most of these data are based on subjective analysis of photographs or anecdotal assessment by the physician's experience [9, 10, 11, 14, 27].

Facial rejuvenation with autologous fat has the advantage of replacing or augmenting tissue with like tissue [32]. Autologous fat transplantation to the face can correct cosmetic defects that are caused by loss of subcutaneous tissue, such as atrophy of the face due to significant weight loss, wrinkles, and facial involution due to aging.

Over the last 30 years, there is a constant interest in breast augmentation by the use of autologous fat transplantation for reconstructive and cosmetic purposes [31]. Illouz presented his technique of autologous fat transplantation to the breast in 1983 and in 1987 Bircoll reported his experience using fat removed by liposuction and transplanted by transcutaneous injection to the breast [2]. A 1987 American Society of Plastic and Reconstructive Surgeons position paper predicted that fat grafting would compromise breast cancer detection and should therefore be prohibited [1]. Up to now, adipose tissue injection to the breast or mammary lipoaugmentation has been stuck by two limiting factors. Firstly, fat injection in and around the breast could result in cyst formation, indurations, and fat necrosis that could be mistaken as cancerous calcifications. Secondly, the degree of reabsorption of the injected adipose tissue is unpredictable. Fat grafting remains shrouded in the stigma of variable results experienced by most plastic surgeons when they first graft fat. In 2005, Spear reported that autologous fat transplantation is a very safe technique that can improve or correct significant contour deformities after breast reconstruction, that otherwise would require more complicated and riskier procedures to improve [34]. There are centers around the world, where autologous fat transfer for the breast reconstruction has become a routine procedure due to its simplicity, safety, and reproducibility [8].

Clinical use of autologous fat grafts for gluteal soft-tissue augmentation has grown in popularity in the plastic surgery community in the past 20 years, despite a perceived drawback of unpredictable results [19, 27].

6.2 Surgical Technique of Fat Grafting

1. Marking of the areas to be liposucted and fat grafted are made while the patient is in standing position (Fig. 6.1).

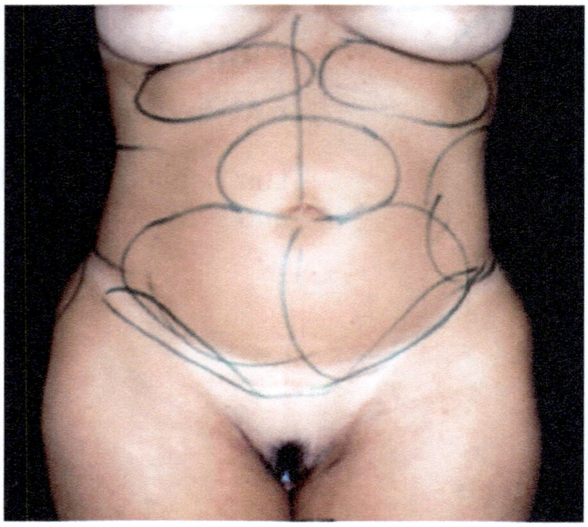

Fig. 6.1 Marking of the areas to be liposucted and fat grafted are made while the patient is in standing position

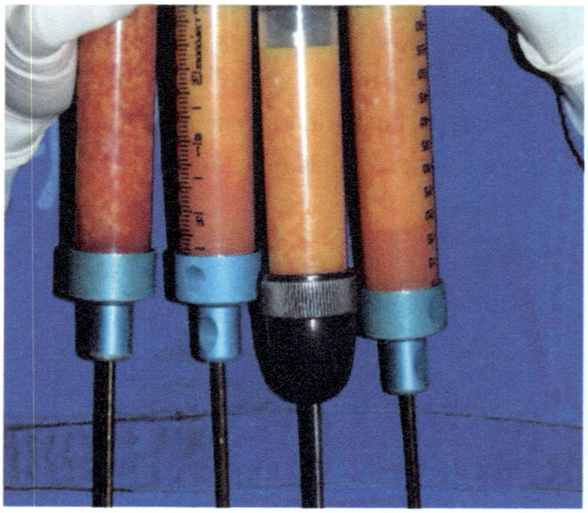

Fig. 6.2 Fat is aspirated using the syringe method

2. Preoperative sedation in the surgical suite is administered. Anesthesia consists of an epidural block and intravenous sedation. The patient is placed in supine position.
3. After the injection of normal saline wetting solution containing 1:500,000 of adrenaline by a small bore cannula and waiting 15 min, a 60-cc syringe attached to a 4 mm blunt cannula is inserted through two small incisions in the abdominal area.
4. Fat is aspirated using the syringe method (Fig. 6.2).

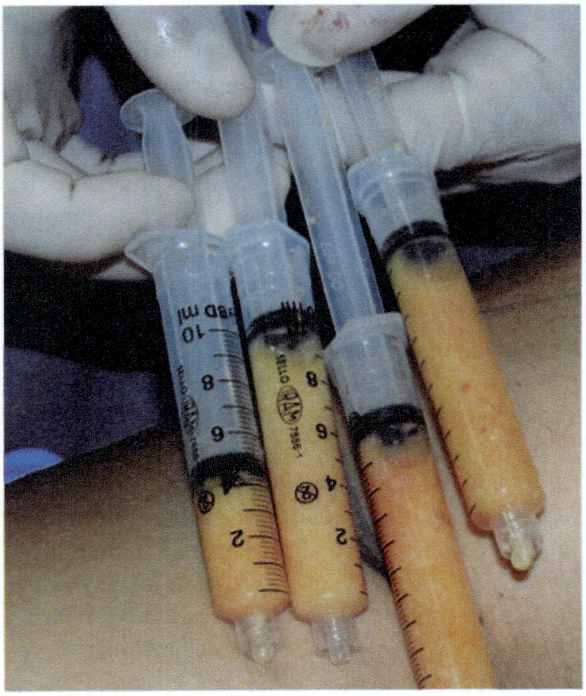

Fig. 6.3 Washed fat by isotonic saline

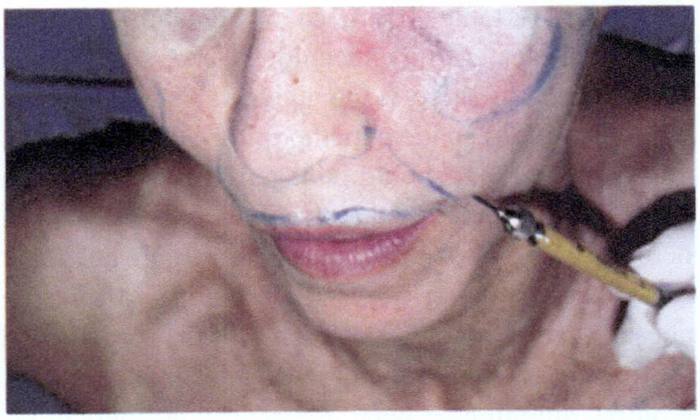

Fig. 6.4 Autologous fat grafting to the face

5. The fatty tissue aspirated is treated in the following manner. With the syringe held vertically with the open end down, the fat and fluid are separated. Isotonic saline is added to the syringe, the fat and saline are separated and the exudate discarded. The procedure is repeated until the fat becomes yellow in color, free of blood and other contaminants (Fig. 6.3).
6. When facial augmentation is scheduled, fat is woven into the deep tissues of the face using a needle and a fine cannula (17 gauge) attached to a 1-mL syringe with multiple passes, injecting only a tiny amount with each pass as the needle is withdrawn, to obtain the most reliable clinical outcome (Fig. 6.4). The entire face is addressed by filling one cosmetic unit at a time.
7. When gluteal augmentation is scheduled, a deep plane to the gluteal muscles is created by the 4 mm cannula. Then other planes are created by the same cannula in different trajectories, always from the deeper aspect to the gluteal surface. The fat is inserted into these tunnels beginning at the deep layer and working up into the intermediate fat compartments. The fat is injected as the cannula is withdrawn. Care should be taken in order to avoid injection of more of fat in the superficial fat compartment. Separate incisions, if necessary can be used in order to treat the whole gluteal region (Fig. 6.5).
8. When breast augmentation by traditional fat grafting is scheduled, the breast is divided in four cosmetic units. Fat is woven into the subcutaneous and intraglandular space of the breast using a 2.5 mm cannula attached to a 10-mL syringe with multiple passes, injecting only a tiny amount with each pass as the cannula is withdrawn, in order to

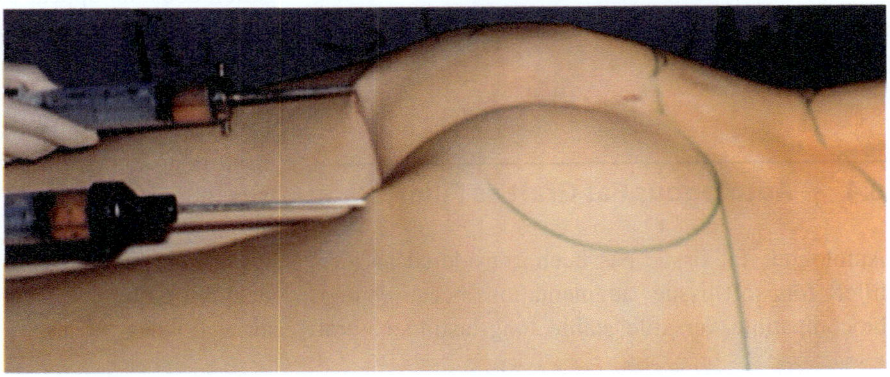

Fig. 6.5 Autologous fat grafting to the gluteal area

obtain the most reliable clinical outcome. The entire breast is addressed by filling one cosmetic unit at a time.

9. The patient remains hospitalized for 24 h. Antibiotics, analgesics, and anti-inflammatory medications are prescribed during the following 7 postoperative days.

6.3 Clinical Applications of Autologous Fat Graft

6.3.1 Autologous Fat Graft for Facial Contour Rejuvenation

This 48-year-old patient who requested facial contour rejuvenation by autologous fat transplantation is shown in Fig. 6.6a and b. She was treated by a total facial fat grafting in the lateral two-thirds of the brow, the nasojugal fold, the malar and buccal fat pads, the nasolabial fold, the lips, and the perioral region by a total of 27 mL. The postoperative result is shown 2 years after the procedure (Fig. 6.6c, d).

6.3.2 Autologous Fat Graft for Gluteal Augmentation

This 21-year-old lady presented complaining of having "no buttocks" which made her "unattractive" (Fig. 6.7a, b). Autologous gluteal lipograft was performed. The following volumes were placed in two sessions: right gluteo intra-muscular 160 mL; left gluteo intra-muscular 150 mL; right gluteo subcutaneous space 50 mL; left gluteo subcutaneous space 60 mL; right subgluteo sulcus 25 mL and left subgluteo sulcus 20 mL. Photos taken 1 year after the procedure (Fig. 6.7c, d).

6.4 Autologous Fat Graft as Filler

Autologous fat tissue has been considered as ideal filler for soft-tissue augmentation because it is biocompatible, versatile, stable, long-lasting, natural-appearing, readily available, abundant, inexpensive, and can be harvested easily and repeatedly, with minimal trauma to the donor sites. Despite unpredictable results, clinical use of autologous fat transfer for soft-tissue augmentation has increased among plastic surgeons. The rising interest in this procedure has paralleled the development and popularity of liposuction for body contouring. While satisfactory results can be achieved in the early postoperative period, long-term results may be disappointing for both the patient and the surgeon. The most common difficulty is the estimation of the resorption rate. Several techniques have been suggested for maximum fat survival including atraumatic fat harvesting technique, centrifugation, graft washing, and addition of growth factors. However, there is no ideal technique described for this purpose. Grafted fat has many attributes of ideal filler, but the results, like those of any procedure, are technique dependent. Quantitative evidence of clinical fat survivability and predictability of volume restoration does not exist, yet reports of patient satisfaction with this procedure are plenty [19]. Fat grafting remains shrouded in the stigma of variable results experienced by most plastic surgeons when they first graft fat. The need of standardization of autologous fat grafting technique needs to be done [16]. Although there is no universal agreement on what constitutes an ideal methodology, there are some proven points that should be taken in to account. No statistical differences in adipocyte viability have been demonstrated among abdominal fat, thigh fat, flank fat, or knee fat donor sites [40]. The donor site can be chosen according to the preference of the surgeon and the patient. Successful, three-dimensional sculpting requires attention to patient preparation, meticulous planning, and fastidious photographic evaluation. Recent reports have shown that mechanical centrifugation does not appear to enhance immediate fat tissue viability before implantation [33]. An important consideration for harvesting and refinement in preparation for grafting is to respect and maintain the tissue architecture of living fat. Any mechanical or chemical insult that damages the fragile tissue architecture of fat will result in eventual necrosis of the injected fat. Recently, it has been reported the preliminary results of en bloc fat grafting that has been shown experimentally to have a greater percentage of adipocyte survival when compared with blunt cannula delivery

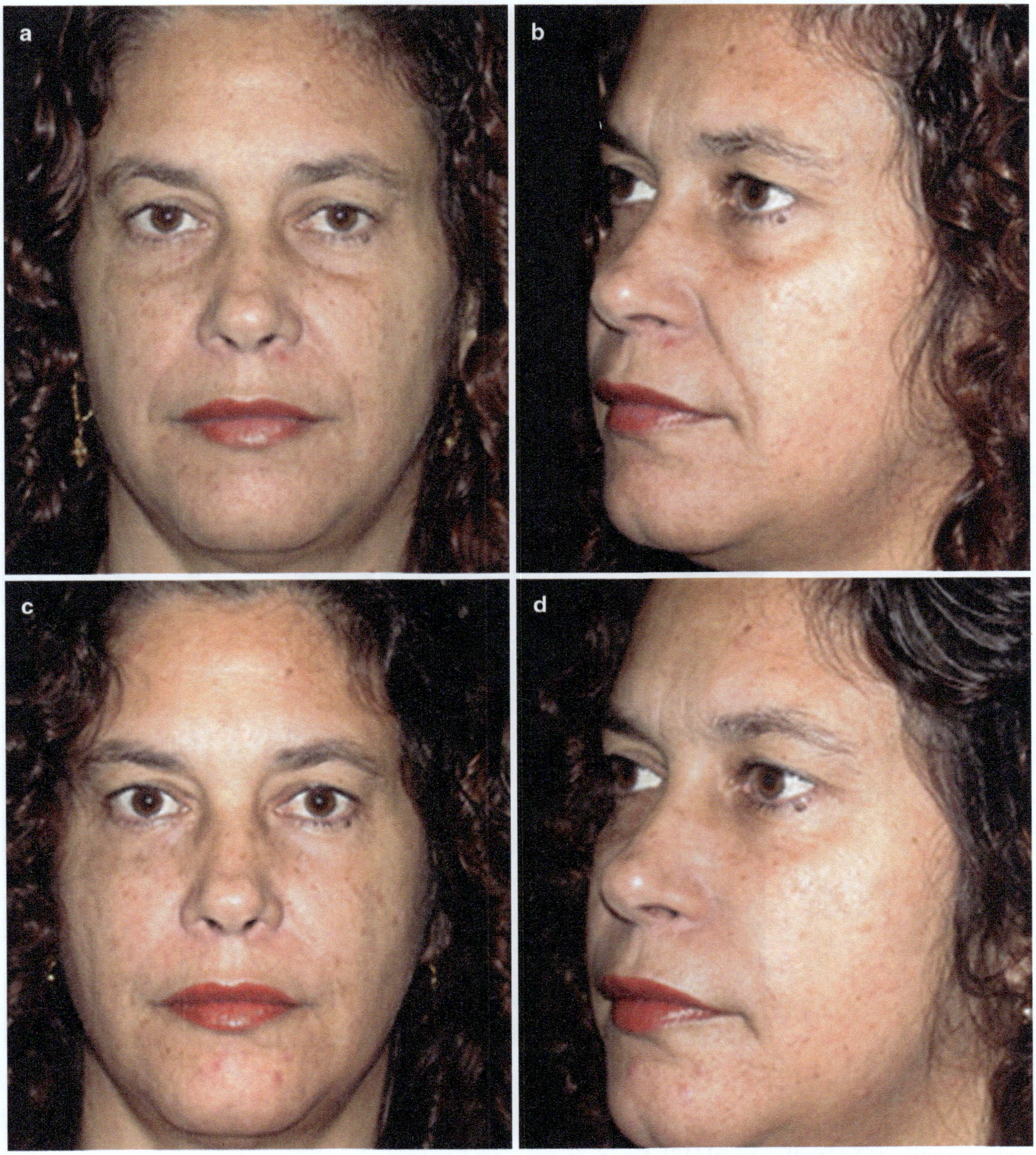

Fig. 6.6 (**a**, **b**) Preoperative photos of a 48-year-old woman requesting facial contour rejuvenation by autologous fat transplantation. (**c**, **d**) Postoperative photos after facial autologous fat transplantation

techniques. En bloc grafting, however, requires an incision for the harvesting and placement of the fat graft with visible scars [13]. Morphometric, as well as histopathologic, analyses published recently, revealed a statistically significant increase of fat graft survival in supramuscular layer (81.95% ± 4.40%) than in subcutaneous (41.62% ± 3.29%) and submuscular layer (37.31% ± 5.77%). This study demonstrates that

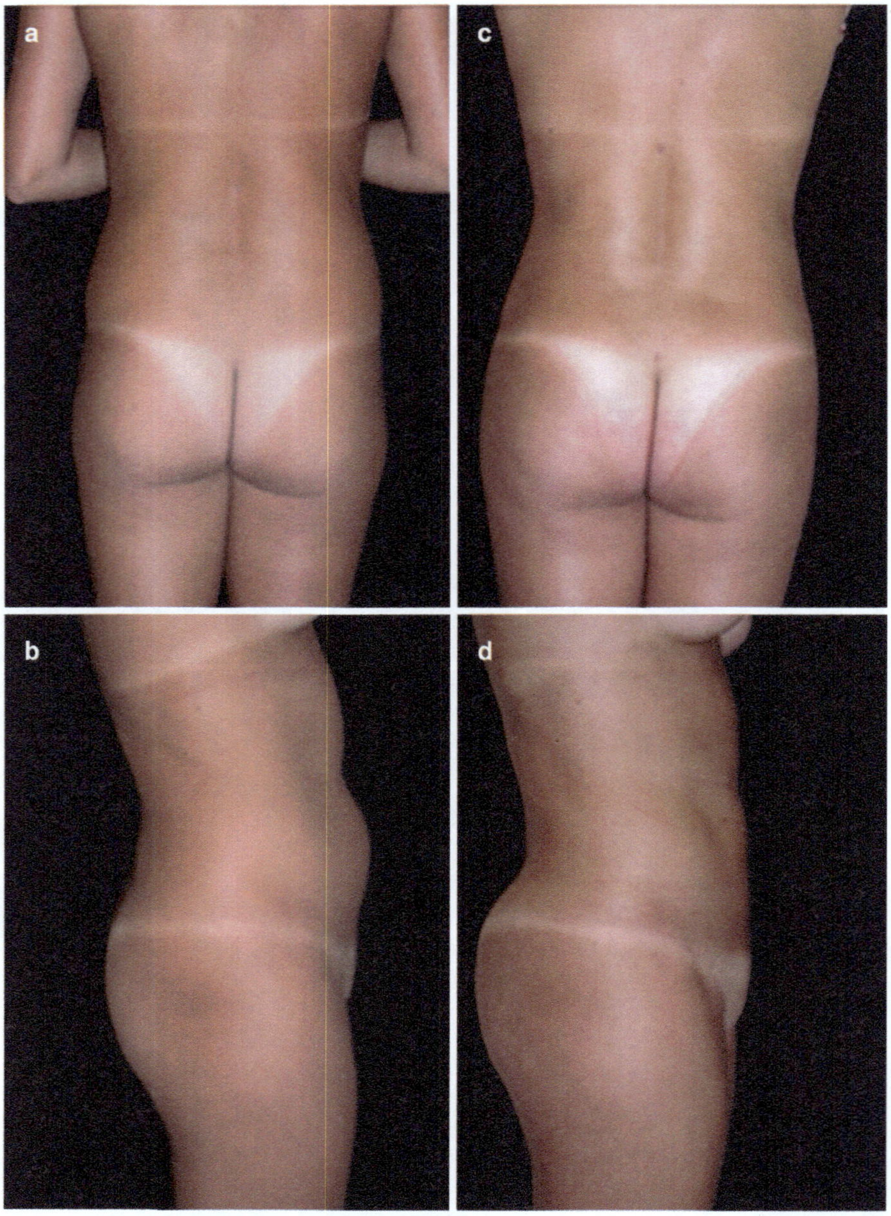

Fig. 6.7 (**a**, **b**) Preoperative photos of a 21-year-old woman requesting gluteal augmentation by autologous fat transplantation. (**c**, **d**) Postoperative photos after gluteal augmentation by autologous fat transplantation

selection of an "appropriate recipient site" should enhance ultimate fat-graft survival [18]. Fat is a living tissue that must be in close proximity to a nutritional and respiratory source to survive. The creation of multiple tunnels ensures adequate blood supply of the grafted fat. Review of the current literature suggests that revascularization may take up to 21 days to reach the center of a microfat graft [22]. The fat grafting injection should be performed in a retrograde mode, in order to avoid intravascular fat injection. Pressure should not be applied in the grafted area. Animal studies have also documented long-term survival of fat grafts up to 5 years, with histologic analysis demonstrating persistent fat survival in biopsy specimens [12]. No studies to date though have delineated whether the volume augmentation is due to tissue fibrosis, adipose survivability and hypertrophy, or stem cell proliferation. Further research is needed to quantify which, if not all, processes contribute to the retained volume.

6.5 Pitfalls in Autologous Fat Transplantation

The most common disadvantages observed with autologous fat grafting in the face, gluteal area, and breast area are edema and echymoses at the donor site for 6–10 days and slight bruising at the injected areas for 3–5 days [7, 17, 25]. The volume of fat injected in secondary procedures is usually considerably less than that used at the original procedure. Touch-up procedures are usually performed with local anesthesia and have less associated downtime. A waiting period of 6 months is generally recommended to allow for initial swelling and resorption to subside. Autologous fat grafting to the breast is not a simple procedure and should be performed by well-trained and skilled surgeons. A recent study confirms that this procedure is being performed incorrectly by untrained and untutored physicians that could result in major complications [15, 26]. An extensive literature review indicated that the major complications observed after lipografting of the breast were mainly related to technical errors and to the wrong anatomic site of harvesting and implantation of the fat [42]. The primary complication of breast lipografting is the formation of liponecrotic cysts which have characteristically benign appearances in sonography, mammography, or MRI [5]. Calcifications in breast parenchyma can be also expected after breast fat injection and according to a recent study lipofilling for breast augmentation should not be performed in patients with a family history of breast cancer [4]. Fat necrosis, cyst formation, and indurations can be seen as in any other surgical manipulation of the breast [6, 21]. A spectrum of mammographic findings such as parenchymal asymmetrical densities, radiolucent cyst, heterogenicity of the subcutaneous tissues, and benign appearing calcifications can be expected after autologous fat transplantation to the breast. Numerous studies support the idea that radiologists have a high level of confidence in differentiating between fat necrosis calcifications after breast surgery and those related to breast cancers [20]. This view can be supported by another study that affirms that the lipomodeling technique does not affect the postoperative follow up of the patients with breast cancer and an imaging controlled biopsy is possible in case of any doubt [30].

6.6 Stromal Enriched Lipograft (SEL)

Regenerative cell-based strategies such as those encompassing the use of stem cells hold tremendous promise for augmentation of the soft-tissue space. Preclinical studies and early clinical series show that adipose-derived stem cells offer the possibility of finally fulfilling the key principle of replacing like with like as an aesthetic filler, without the drawbacks of current technology [37]. In cell-assisted lipotransfer (CAL), autologous adipose-derived stem cells (ADSCs) are used in combination with lipoinjection. A stromal vascular fraction (SVF) containing ADSCs is freshly isolated from half of the aspirated fat and recombined with the other half. This process converts relatively ADSC-poor aspirated fat to ADSC-rich fat [38]. The preliminary results suggest that CAL is effective and safe for soft-tissue augmentation and superior to conventional lipoinjection [35]. Another study has confirmed that the CAL fat can survive better (35% on average) than non-CAL fat, and microvasculature can be detected more prominently in CAL fat, especially in the outer layers of the fat transfer [38, 40, 41].

6.7 Conclusion

The ideal substance for soft-tissue augmentation still eludes physicians, but fat grafting through a blunt cannula seems to be the safest of all of the fillers used; in the hands of an experienced surgeon, it can provide long-lasting, natural-appearing structural changes. In the last 30 years, the results of autologous fat transplantation are satisfying and stable on the condition that a staged treatment consisting of small quantities of adipose tissue fat should be injected in each treatment session. The final expected result should not be the aim of a single procedure, in order to prevent major complications. With experience, the surgeon can predict the amount of volume needed to be grafted in order to produce the desired result. Regenerative cell-based strategies such as those encompassing the use of stem cells hold tremendous promise for augmentation of the soft-tissue space. Preclinical studies and early clinical series show that adipose-derived stem cells offer the possibility of finally fulfilling the key principle of replacing like with like as an aesthetic filler, without the drawbacks of current technology.

References

1. ASPRS Ad-Hoc Committee on New Procedures (1987) Report on autologous fat transplantation. Plast Surg Nurs 7(4):140–141
2. Bircoll M (1987) Cosmetic breast augmentation utilizing autologous fat and liposuction techniques. Plast Reconstr Surg 79(2):267–271
3. Cárdenas-Camarena L, Lacouture AM, Tobar-Losada A (1999) Combined gluteoplasty: liposuction and lipoinjection. Plast Reconstr Surg 104(5):1524–1531, discussion 1532–1533
4. Carvajal J, Patiño JH (2008) Mammographic findings after breast augmentation with autologous fat injection. Aesthet Surg J 28(2):153–162
5. Castelló JR, Barros J, Vázquez R (1999) Giant liponecrotic pseudocyst after breast augmentation by fat injection. Plast Reconstr Surg 103(1):291–293
6. Danikas D, Theodorou SJ, Kokkalis G, Vasiou K, Kyriakopoulou K (2001) Mammographic findings following reduction mammaplasty. Aesthet Plast Surg 25:283
7. Dasiou-Plakida D (2003) Fat injections for facial rejuvenation: 17 years experience in 1720 patients. J Cosmet Dermatol 2(3-4):119–125
8. Delay E, Delpierre J, Sinna R, Chekaroua K (2005) How to improve breast implant reconstructions. Ann Chir Plast Esthét 50(5):582–594
9. Ersek RA (1991) Transplantation of purified autologous fat: a three year follow-up is disappointing. Plast Reconstr Surg 87(2):219–227
10. Fournier PF (2000) Fat grafting: my technique. Dermatol Surg 26(12):1117–1128
11. Fulton JE, Suarez M, Silverton K et al (1998) Small volume fat transfer. Dermatol Surg 24(8):857–865
12. Guerrerosantos J, Gonzalez-Mendoza A, Masmela Y et al (1996) Long-term survival of free fat grafts in muscle: an experimental study in rats. Aesthet Plast Surg 20(5): 403–408
13. Guyuron B, Majzoub RK (2007) Facial augmentation with core fat graft: a preliminary report. Plast Reconstr Surg 120(1):295–302
14. Haroldo Pereira L, Sterodimas A (2008) Aesthetic restoration of axillary contour deformity after lymph node dissection. J Plast Reconstr Aesthet Surg 61(2):231–232, Epub 2007 Nov 19
15. Hyakusoku H, Ogawa R, Ono S, Ishii N, Hirakawa K (2009) Complications after autologous fat injection to the breast. Plast Reconstr Surg 123(1):360–370, discussion 371–372
16. Illouz Y-G (1986) The fat cell "graft": a new technique to fill depressions. Plast Reconstr Surg 78(1):122–123
17. Illouz Y-G, Sterodimas A (2009) Autologous fat transplantation to the breast: a personal technique with 25 years of experience. Aesthet Plast Surg 33(5):706–715
18. Karacaoglu E, Kizilkaya E, Cermik H, Zienowicz R (2005) The role of recipient sites in fat-graft survival: experimental study. Ann Plast Surg 55(1):63–68, discussion 68
19. Kaufman MR, Miller TA, Huang C, Roostaien J, Wasson KL, Ashley RK, Bradley JP (2007) Autologous fat transfer for facial recontouring: is there science behind the art? Plast Reconstr Surg 119(7):2287–2296
20. Kneeshaw PJ, Lowry M, Manton D, Hubbard A, Drew PJ, Turnbull LW (2006) Differentiation of benign from malignant breast disease associated with screening detected micro calcifications using dynamic contrast enhanced MRI. Breast 15(1):29–38
21. Leibman AJ, Styblo TM, Bostwick J (1997) Mammography of the postreconstruction breast. Plast Reconstr Surg 99:698
22. Locke MB, de Chalain TM (2008) Current practice in autologous fat transplantation: suggested clinical guidelines based on a review of recent literature. Ann Plast Surg 60(1): 98–102
23. Matsudo PK, Toledo LS (1988) Experience of injected fat grafting. Aesthet Plast Surg 12(1):35–38
24. Neuber F (1893) Fettransplantation. Chir Kongr Verhandl Dtsch Ges Chir 22:66
25. Nicaretta B, Pereira LH, Sterodimas A, Illouz Y-G (2011) Autologous gluteal lipograft. Aesthetic Plast Surg 35(2): 216–224
26. Parrish JN, Metzinger SE (2010) Autogenous fat grafting and breast augmentation: a review of the literature. Aesthet Surg J 30(4):549–556
27. Pereira LH, Radwanski HN (1996) Fat grafting of the buttocks and lower limbs. Aesthet Plast Surg 20(5):409–416
28. Pereira LH, Sterodimas A (2008) Free fat transplantation for the aesthetic correction of mild pectus excavatum. Aesthet Plast Surg 32(2):393–396
29. Perén PA, Gómez JB, Guerrerosantos J, Salazar CA (2000) Gluteus augmentation with fat grafting. Aesthet Plast Surg 24(6):412–417
30. Pierrefeu-Lagrange AC, Delay E, Guerin N, Chekaroua K, Delaporte T (2005) Radiological evaluation of breasts reconstructed with lipomodeling. Ann Chir Plast Esthét 51(1):18–28
31. Pulagam SR, Poulton T, Mamounas EP (2006) Long-term clinical and radiologic results with autologous fat transplantation for breast augmentation: case reports and review of the literature. Breast J 12(1):63–65
32. Ramirez OM (2001) Full face rejuvenation in three dimensions: a "face-lifting" for the new millennium. Aesthet Plast Surg 25(3):152–164
33. Rohrich RJ, Sorokin ES, Brown SA (2004) In search of improved fat transfer viability: a quantitative analysis of the role of centrifugation and harvest site. Plast Reconstr Surg 13(1):391–395, discussion 396–397
34. Spear SL, Wilson HB, Lockwood MD (2005) Fat injection to correct contour deformities in the reconstructed breast. Plast Reconstr Surg 116(5):1300–1305
35. Sterodimas A, de Faria J, Nicaretta B,Boriani F Autologous fat transplantation versus adipose derived stem cells enriched lipograft: a study. Aesthet Surg J [Epub ahead of print]
36. Sterodimas A, Huanquipaco JC, de Souza Filho S, Bornia FA, Pitanguy I (2008) Autologous fat transplantation for the treatment of Parry-Romberg syndrome. J Plast Reconstr Aesthet Surg 62(11):e424–e426
37. Sterodimas A, de Faria J, Correa WE, Pitanguy I (2009) Tissue engineering in plastic surgery an up-to-date review of the current literature. Ann Plast Surg 62(1):97–103
38. Sterodimas A, de Faria J, Nicaretta B, Papadopoulos O, Papalambros E, Illouz Y-G (2010) Cell-assisted lipotransfer. Aesthet Surg J 30(1):78–81

39. Toledo LS, Mauad R (2006) Fat injection: a 20-year revision. Clin Plast Surg 33(1):47–53, vi
40. Ullmann Y, Shoshani O, Fodor A, Ramon Y, Carmi N, Eldor L, Gilhar A (2005) Searching for the favorable donor site for fat injection: in vivo study using the nude mice model. Dermatol Surg 31(10):1304–1307
41. Yoshimura K, Sato K, Aoi N, Kurita M, Hirohi T, Harii K (2008) Cell-assisted lipotransfer for cosmetic breast augmentation: supportive use of adipose-derived stem/stromal cells. Aesthet Plast Surg 32(1):48–55, discussion 56–57
42. Zocchi ML, Zuliani F (2008) Bicompartmental breast lipostructuring. Aesthet Plast Surg 32(2):313–328

Adipose Stem Cells: From Liposuction to Adipose Tissue Engineering

7

Kotaro Yoshimura, Hitomi Eto, Harunosuke Kato, Kentaro Doi, and Hirotaka Suga

Contents

7.1 Introduction 67

7.2 Cellular Components of Adipose Tissue 68

7.3 Liposuction: Liposuction Aspirates and Stromal Vascular Fraction............. 68

7.4 Aspirated and Intact Fat Tissue 68

7.5 Cellular Components of SVF 69

7.6 Adipose-Derived Stem/Progenitor Cells in Adipose-Tissue Repair/Remodeling........ 71

7.7 Ischemia to Adipose Tissue 72

7.8 Injury to Adipose Tissue 73

7.9 Grafting of Aspirated Adipose Tissue 74

7.10 Supplementation of Adipose Progenitor Cells in Micro Fat Grafting................ 76

7.11 Stem Cells and Signals for Tissue Regeneration........................ 79

7.12 Conclusion 79

References.................................. 80

K. Yoshimura (✉) • H. Eto • H. Kato • K. Doi • H. Suga
Department of Plastic Surgery, University of Tokyo School of Medicine, 7-3-1, Hongo Bunkyo-Ku, 113-8655 Tokyo, Japan
e-mail: kotaro-yoshimura@umin.ac.jp

Abbreviations

ASCs	Adipose stem/progenitor cells
VECs	Vascular endothelial cells
SVF	Stromal vascular faction
PDGF	Platelet-derived growth factor
EGF	Epidermal growth factor
TGF-β	Transforming growth factor-β
b-FGF	Basic fibroblast growth factor
VEGF	Vascular endothelial growth factor
HGF	Hepatocyte growth factor
CAL	Cell-assisted lipotransfer

7.1 Introduction

Adipose tissue has been considered an organ of energy storage, the largest endocrine organ, a soft-tissue filler (augmentation by micro-fat grafting), and a cosmetically unnecessary tissue discarded by liposuction. Adipose tissue physiologically turns over very slowly. Adipocytes have a life span of 10 years [33]; several thousands of adipocytes die and are replaced with new adipocytes every second in our body. Adipose progenitor cells responsible for the tissue turnover are fibroblast-like cells, which have been referred to as stromal vascular cells, adipose stromal cells, or preadipocytes. It was found that the cell population contains not only monopotent progenitor cells but also multipotent mesenchymal cells [51], which are now called adipose (-derived) stem cells (ASCs or ADSCs) or adipose-derived mesenchymal stem cells (AD-MSCs). ASCs are regarded as a potent tool for cell-base therapies, comparable to bone marrow-derived mesenchymal stem cells, because they can be obtained in a large

Y.-G. Illouz and A. Sterodimas (eds.), *Adipose Stem Cells and Regenerative Medicine*,
DOI: 10.1007/978-3-642-20012-0_7, © Springer-Verlag Berlin Heidelberg 2011

amount through a less invasive approach, liposuction. For adipose tissue engineering/regeneration, it is a critical point to understand physiological roles and functions of ASCs; this population, working together with vascular endothelial cells (VECs) and stem/progenitor cells recruited from bone marrow, is a main player for adipose tissue engineering, which means adipogenesis and angiogenesis.

7.2 Cellular Components of Adipose Tissue

Adipose tissue has many cells other than adipocytes, though adipocytes constitute more than 90% of adipose tissue volume. Our rough estimation of cellular components through flow cytometry and two- and three-dimensional histology are as follows: 1 cm^3 adipose tissue contains 1 million adipocytes, 1 million ASCs, 1 million VECs, and 1 million other cells (adipose-resident macrophages and lymphocytes, pericytes, fibroblasts, etc.) [12, 35]. Every single adipocyte has direct contact with a capillary. Adipose tissue has a dense network of capillaries, but they run sparsely with an interval of 100 μm or more between each other because adipocytes are extraordinary large in size such as 70–140 μm. ASCs are thought to localize between adipocytes (co-localize with capillaries), in the vessel walls or in the connective tissue; most of them show perivascular localization [3, 9, 41]. Very recently, it was discovered in mice that adipocyte progenitor cells are present in adipose vasculature [39], but the identity between ASCs and vascular pericytes remains to be verified. Some studies suggest the existence of a cell population localized in the vascular wall (adventitia) which can differentiate into vessels [2, 50], and this also may be related to ASCs.

Very recently, the existence of another multipotent or pluripotent stem cells (multilineage differentiating stress enduring cells: muse cells) in human adipose tissue, labeled by SSEA-3, was suggested [18]; they are not localized in capillary or vessel walls as known adipose progenitors (ASCs), but in the connective tissue near the vessels or between adipocytes very sparsely (not published). This finding suggested that there may be two types of adipose stem/progenitor cells; multipotent "stem" cells (muse cells) located near (but, outside of) larger vessels and adipose "progenitor" cells (corresponding to ASCs) located pericapillary (Fig. 7.1).

7.3 Liposuction: Liposuction Aspirates and Stromal Vascular Fraction

Liposuction is the most frequently performed cosmetic surgery; it was performed approximately 1.6 million times worldwide in 2009 (survey of international society of aesthetic plastic surgery, http://www.isaps.org/stats.php). In the surgical procedure, saline with epinephrine (and local anesthesia) is first infiltrated into the adipose tissue, which is then aspirated manually or with a vacuum machine through a metal cannula with a diameter of 2–4 mm. Substantial amount (100 ml to several liters) of adipose tissue can be safely suctioned with leaving large vascular and neural perforators to the skin left intact. The liposuction aspirates has two portions; the upper portion (yellow color) is fatty portion containing floating shredded adipose tissue, while the infranatant (pink to red color) is fluid portion containing infiltrated saline, peripheral blood, and tissue debris sedimented at the bottom [44]. Through collagenase digestion of the fatty portion and centrifugations, cellular components other than adipocytes are extracted as a cell pellet, which is called stromal vascular fraction (SVF) (Fig. 7.2). Adipocytes are disrupted into oil during the process and discarded as floating tissue and oil after centrifugation.

ASCs can be clinically used without cell expansion if harvested from a large volume of lipoaspirates because a sufficient number of cells can be obtained; 0.1–1 billion nucleate cells can be obtained from 200 ml of aspirated fat tissue and at least 10% of these cells are adipose-derived stromal cells (Fig. 7.2). The use of freshly isolated cells likely leads to higher safety and efficacy in treatments compared to cells expanded by culture; FDA (21 CFR Parts 16, 1,270, and 1,271) regards cells cultured even overnight as more-than-minimally manipulated cells, and cultured ASCs are known to show a distinct phenotype to fresh ASCs [44]. Some ongoing clinical trials employ freshly isolated SVF, rather than purified or cultured ASCs. As the SVF contains other cells such as VECs or macrophages, synergistic effects may be expected.

7.4 Aspirated and Intact Fat Tissue

Aspirated fat tissue (adipose tissue obtained through liposuction; the upper portion of liposuction aspirates), but not excised (intact) fat tissue, can be used as a

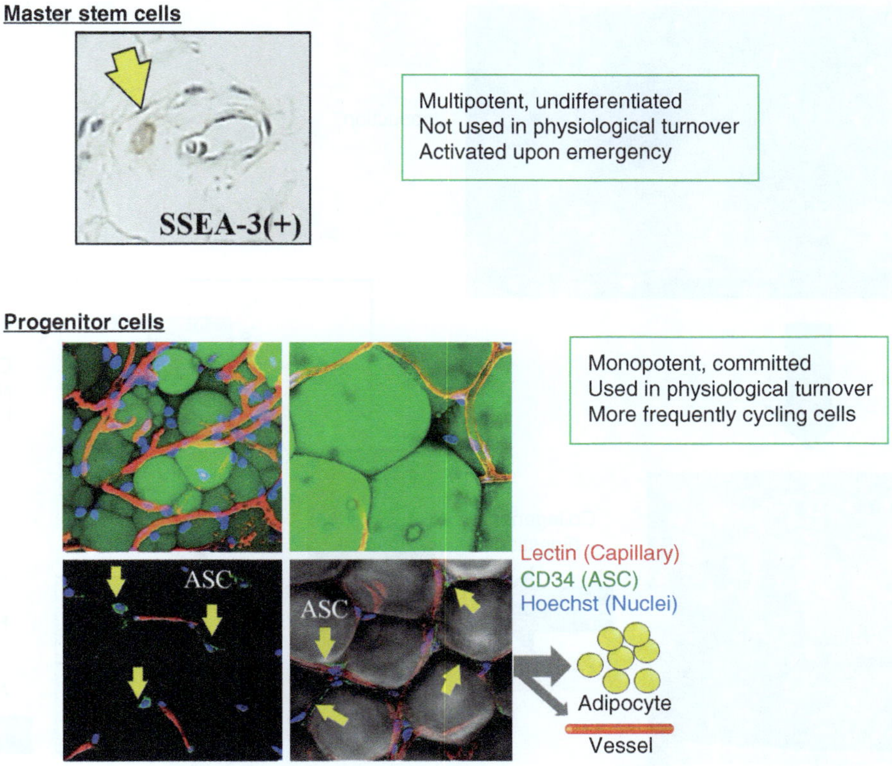

Fig. 7.1 Master stem cells and progenitor cells residing in adipose tissue. It was established that adipose tissue has progenitor cells located perivascularly; they are located in capillaries supporting endothelial cells like pericytes. The identity between the progenitor cells and vascular pericytes remains to be verified. The adipose progenitor cells work not only in physiological turnover but also in incidental remodeling of the tissue such as tissue repair after injury. The progenitor cells may be not only precursor of adipocytes but also of vascular endothelial cells. Recently, another undifferentiated cell population, which may be pluripotent stem cells (named, multilineage differentiating stress enduring cells; muse cells), was suggested to reside in the adipose tissue as well as other tissues. The muse cells may be multipotent stem cells contained in a small number in stromal vascular fraction obtained from adipose tissue

material of adipose-tissue grafting (lipoinjection), because skin incision is necessary to harvest excised fat tissue, which is not acceptable for cosmetic purposes. When aspirated, only fragile parts of adipose tissue are removed with negative pressure through a small suction cannula, while honeycomb-like fibrous structures (connective tissues, vasculatures, and nerves) remain intact in the subcutaneous fatty layer of the donor site [48]. We have found aspirated fat tissue contains only a half number of ASCs compared to intact fat tissue [20] and many adipocytes and capillaries are ruptured and a larger number of dead cells are contained in aspirated fat tissue [12, 20] (Fig. 7.3). In addition, our recent preliminary assay suggested that multipotent stem cells (muse cells) are contained in a much smaller number in aspirated adipose tissue than intact tissue. The relative deficiency of ASCs in aspirated fat tissue may be due to: (1) a substantial portion of ASCs are located around large vessels (within tunica adventitia) and left in the donor tissue and (2) some ASCs are released into the fluid portion of liposuction aspirates [44]. Large-sized vessels are located in the fibrous part of the tissue, present in intact but not aspirated fat tissue. Thus, aspirated fat tissue is regarded as relatively progenitor-poor as compared to intact fat tissue [12, 20].

7.5 Cellular Components of SVF

The SVF contains not only adipose tissue-derived cells but also peripheral blood-derived cells (leukocytes and erythrocytes), because adipose tissue contains some circulating blood and liposuction aspirates contain a substantial amount of hemorrhage [44]. Most of the

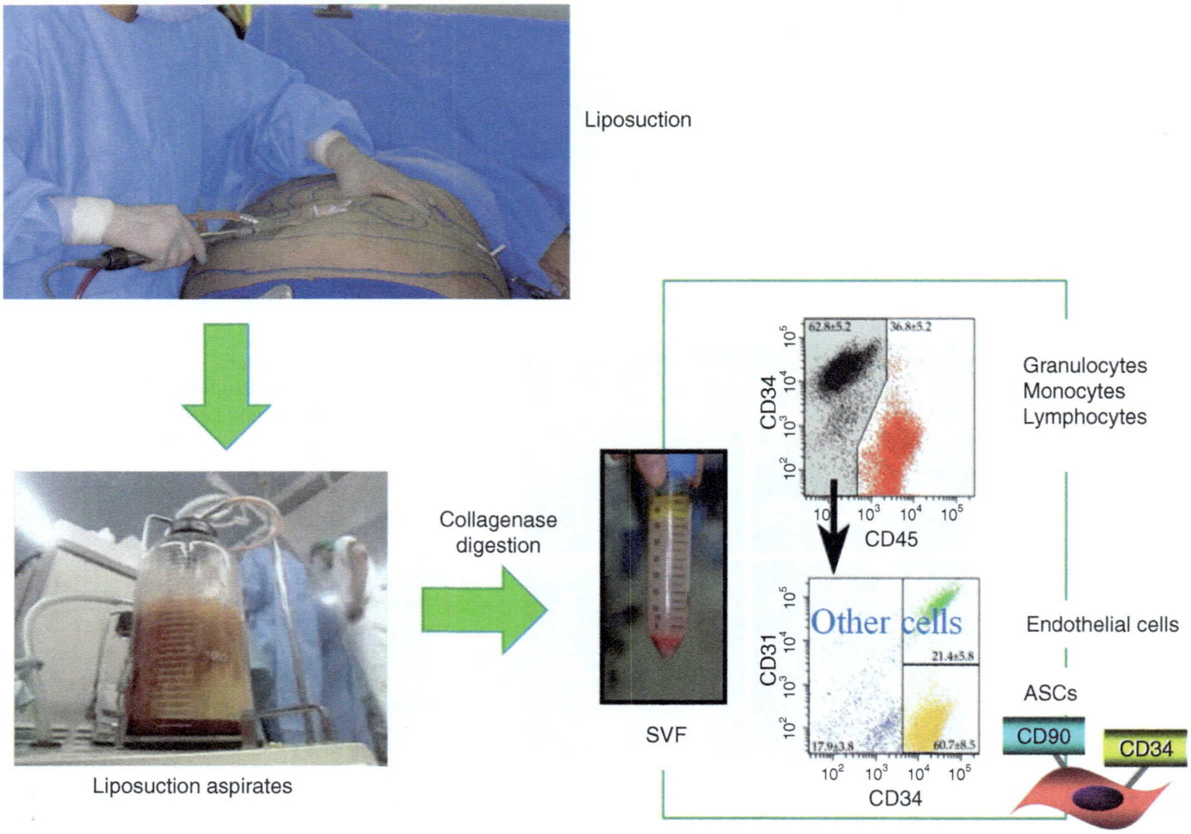

Fig. 7.2 From liposuction to isolation of adipose stem/progenitor cells. Fat tissue can be easily harvested in a large volume (>500 ml) through liposuction. Liposuction aspirates are composed of two parts; floating adipose portion (also called lipoaspirates or aspirated fat tissue) and infranatant fluid portion. Stromal vascular fraction (*SVF*) can be obtained through collagenase digestion of the lipoaspirates. SVF contains many blood-derived cells (CD45+) and adipose-derived cells (CD45−); the latter is further divided into CD31+/CD34+ endothelial cells (*ECs*), CD31−/CD34+ adipose progenitor cells (*ASCs*) and CD31−/CD34− other cells

contaminated erythrocytes can be disrupted with hypotonic-solution processing. The adipose tissue-derived cells are composed of ASCs, VECs, resident macrophages and lymphocytes, and other cells such as vascular pericytes and fibroblasts.

Our study identified freshly isolated ASCs as CD31−CD34+CD45−CD90+CD105−CD146− cells, but they become CD105+ when plated [44]. VECs are identified as CD31+CD34+CD45− cells. Resident macrophages are CD14+CD45+CD206+ and can be discriminated from circulating monocytes, which are CD14+CD45+CD206−. The percentage of peripheral blood-derived cells strongly depends on individual hemorrhage volume [35]. ASCs can be extracted not only from the floating fatty portion but also from the fluid portion of liposuction aspirates; although the infranatant fluid portion contains much fewer ASCs and many more blood-derived cells [44]. In mice, the adipocyte progenitor subpopulation was recently identified as Lin−Sca1+CD24+CD29+CD34+ cells [30].

Regarding resident hematopoietic cells, many remain to be elucidated. We compared CD45+ cells in SVF and those in peripheral blood; most of the adipose-resident hematopoietic cells are macrophages (CD14+/CD206+) or lymphocytes. Recently, some reports suggested that adipose-resident lymphocytes are highly associated with adipose dysfunction and insulin resistance [27]. On the other hand, there are many CD34+ macrophages in human non-obese adipose tissue; they have multipotency including high adipogenic capacity and are strongly suggested to be involved in adipose-tissue remodeling/repair together with ASCs (unpublished data).

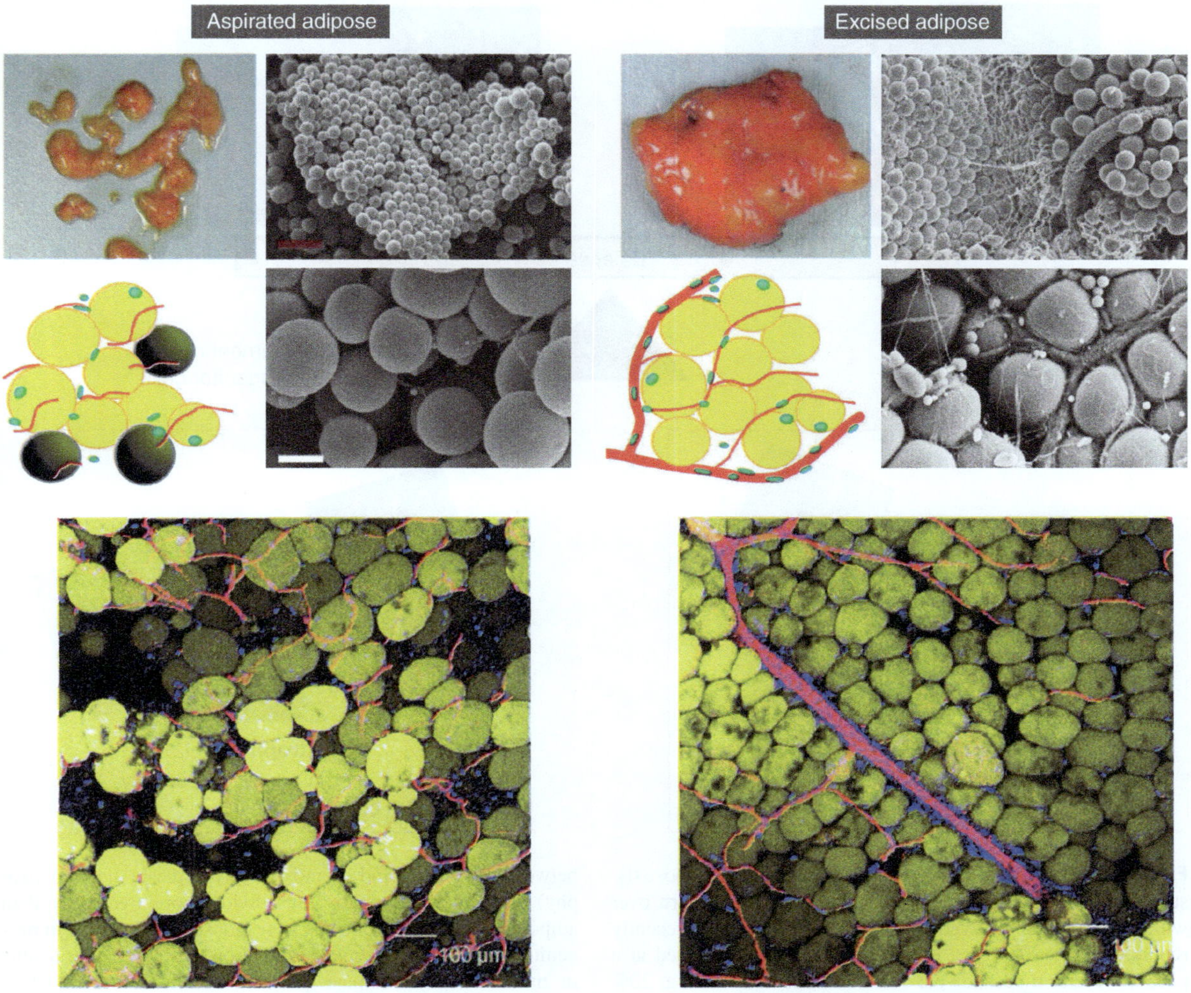

Fig. 7.3 Comparison of human aspirated and excised (intact) fat tissue obtained from a single site of a single patient. Macroscopic views (*left*, *top*), schematic views (*left*, *middle*), electron microscopic (*right*, *top*, and *middle*; red scale bars = 200 μm, white scale bars = 40 μm), and whole mount staining images (*bottom*; scale bars = 100 μm). The basic structure of adipose tissue was preserved in the aspirated fat tissue, but some adipocytes and capillaries were disrupted. Vascular vessels, especially those of large size, were notably less in aspirated fat tissue compared to the excised fat tissue. ASC yield from aspirated fat tissue was considerably less (about a half) than that from excised fat tissue, suggesting that aspirated adipose tissue is relatively progenitor-poor fat tissue

7.6 Adipose-Derived Stem/Progenitor Cells in Adipose-Tissue Repair/Remodeling

ASCs are thought to be the main proliferating cell population in any types of adipose-tissue remodeling, such as developmental growth, hyperplasia in obesity, repair processes after injury [37], or tissue expansion induced by mechanical forces [16]. These remodeling processes are in balance between adipocyte apoptosis/necrosis and adipogenesis managed by ASCs; these degenerative and regenerative changes are always accompanied by capillary remodeling (Fig. 7.4). Adipose-tissue atrophy with age is likely due to a decrease in number of ASCs and consequent impaired physiological turnover, as is commonly seen in other tissues and organs [8, 15].

ASCs have been shown to have angiogenic characteristics, to release angiogenic factors responding to ischemia

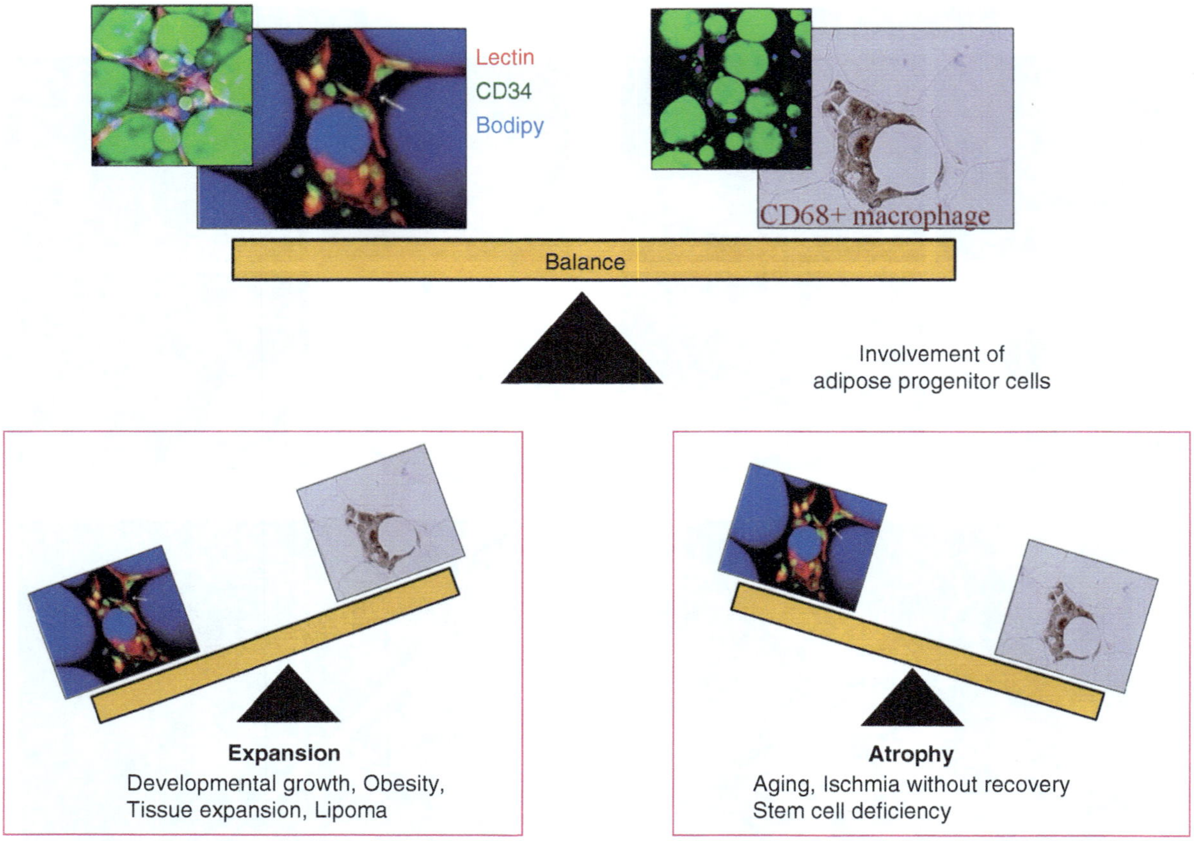

Fig. 7.4 Balance between cell division and death in adipose tissue. Cells and extracellular matrix in adipose tissue turn over with a slow cycle; the average adipocyte life span was recently reported to be ~10 years. Adipocytes are being replaced at a pace of several thousands per second in our body (60 kg, 20% fat). Adipose tissue turnover is on the physiological balance between cell growth and apoptosis. In hyperplasia (not hypertrophy), adipocyte progenitor cells divide more frequently than adipocytes undergo apoptosis. In response to a decrease in progenitor cells such as aging or irradiation, tissue turnover results in tissue atrophy. Adipose progenitor cells are involved in growth, replacement, or repair processes of adipose tissue

[29, 40] or stimulation of growth factors [37], and to experimentally differentiate into VECs [7, 20–22]. Thus, ASCs are now considered to be bipotent progenitor cells for both adipocytes and vascular cells, although the differentiation into VECs was not frequently detected in in vivo studies [17, 26] and a standard in vitro protocol for endothelial differentiation is not yet established. Rather, there is a report showing pericytic differentiation likely occur more frequently [42].

7.7 Ischemia to Adipose Tissue

Adipose tissue has very high (almost highest) partial oxygen tension (pO_2) among organs. The pO_2 of adipose tissue is 40–60 mmHg, between that of arterial blood (≈110 mmHg) and venous blood (≈40 mmHg). The other organs show lower pO_2; for example, the pO_2 of brain, spleen, and thymus were reported to range from 5 to 20 mmHg [5, 6, 11]. The high pO_2 of adipose tissue likely reflects high density of capillaries and low oxygen consumption rate of the tissue. It is known that adipose tissue is more vulnerable to ischemia than the skin, and necrosis of subcutaneous adipose tissue, known as *deep tissue injury*, can seriously affect the vascularity of overlying skin [4].

It is recently suggested that diabetic adipose tissue is relatively ischemic with low-grade chronic inflammation, which may cause adipose endocrine dysfunction, insulin resistance, and metabolic syndrome [28], while lipoma tissue were not ischemic probably due to upregulated angiogenesis [36]. In surgically induced ischemia of non-obese mice model, adipose tissue degenerated and subsequently remodeled with proliferation of ASCs

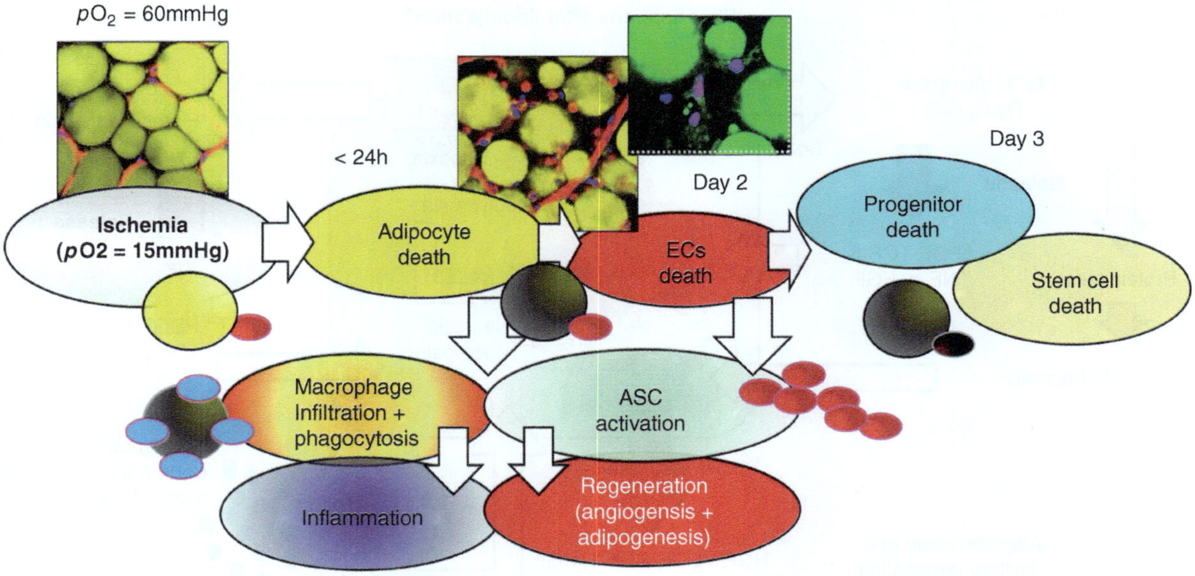

Fig. 7.5 Cellular events in ischemic adipose tissue. Severe ischemia induces adipocyte death. If it prolongs, vascular endothelial cells also start to die. In contrast, adipose stem/progenitor cells can stay alive up to 72 h even under severe conditions; during the 72 h, they are even activated and try to repair the damaged tissue by their regenerative efforts such as proliferation, migration, and differentiation, collaborating with infiltrated cells

and ECs [38]; macrophages infiltrated and were involved in phagocytosis of dead adipocytes.

Among cellular components of adipose tissue, adipocytes die first under severe ischemia such as 15 mmHg of pO_2. When severe ischemia prolongs, ECs and blood-derived cells start to die next. In contrast, ASCs can stay alive up to 3 days even under severe ischemia; rather during the 3 days, they can be activated and contributed to the adaptive repairing process through adipogenesis and angiogenesis [38] (Fig. 7.5).

7.8 Injury to Adipose Tissue

An animal model of ischemia-reperfusion injury (IRI) to adipose tissue can simulate acute injury to adipose tissue; we examined cellular events during the repairing/remodeling process of injured adipose tissue using the mice model [37]. As early as 1 day after IRI, basic fibroblast growth factor (bFGF) released into the injured adipose tissue and degenerative changes such as adipocyte death were observed. Responding to the injury, ASCs proliferated and contributed to the repairing process as a predominant proliferating cell population. Basic FGF stimulated ASC not only to proliferation but also to release hepatocyte growth factor (HGF), a well-known angiogenic growth factor (Fig. 7.6). Through these remodeling processes, the injured adipose tissue almost healed at 2 weeks [37].

In clinical injury to adipose tissue, a number of soluble factors are sequentially released in the injured site. Our study on wound exudates after liposuction surgery showed that contained factors in the wound exudate are different in each phase of adipose wound healing [1]. Basic FGF, PDGF, EGF, and TGFβ are abundant in the early (coagulation) phase (Day 0–1) of wound healing, while VEGF, HGF, IL-8, and MMP-1 gradually increase up to the late (proliferation) phase (Day 5–7). On the other hand, KGF, IL-6, and MMP-8 peak during the inflammatory phase (Day 2–4).

Based on the results above, we prepared a growth factor mixture named adipose injury cocktail (AIC), which contains four major growth factors (bFGF, PDGF, EGF, and TGFβ) in the early phase [13]. AIC was experimentally injected into adipose tissue with surgically induced ischemia and hypoxic adipose tissue in diabetic mice; AIC promoted angiogenesis and improved oxygen tension in both ischemic tissues through activating resident ASCs, suggesting that these factors may be beneficial in therapeutic use for treatment of ischemic diseases.

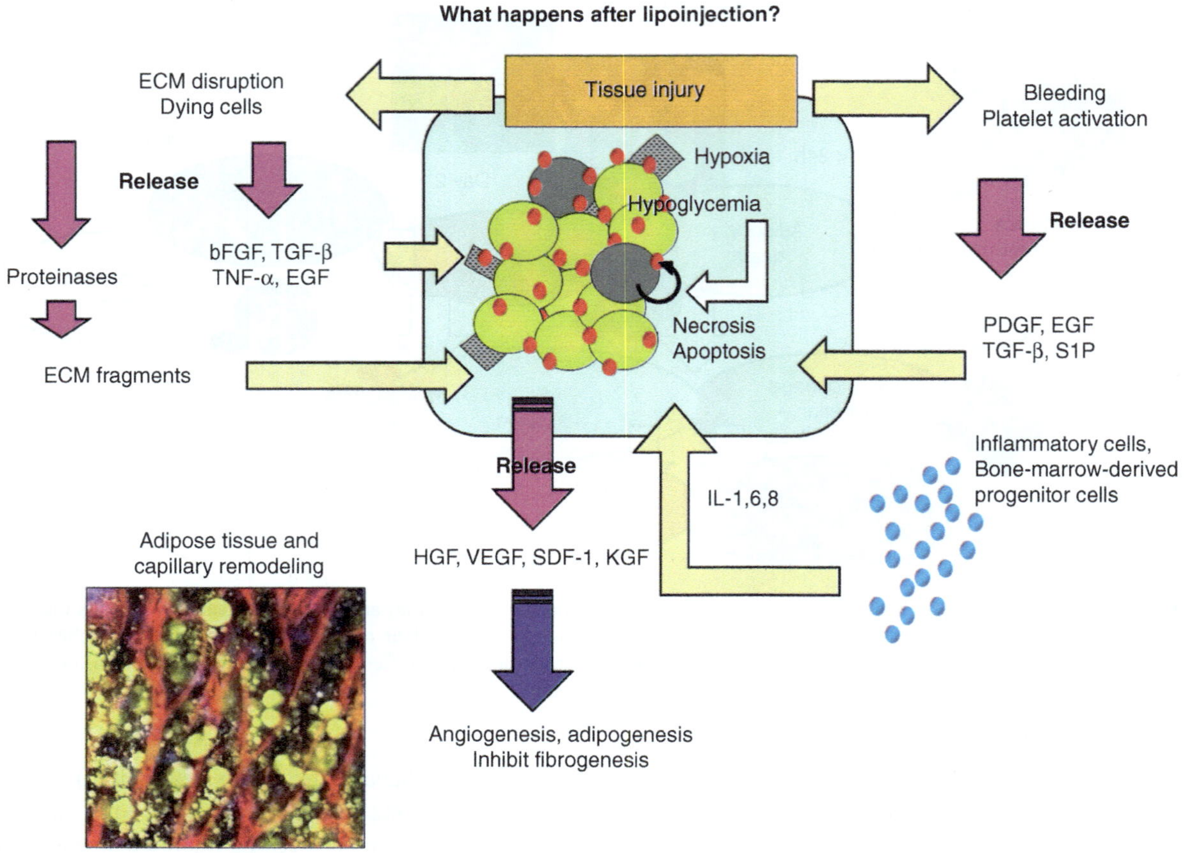

Fig. 7.6 Cellular and molecular events after adipose-tissue grafting. Adipose grafting induces injury in the recipient tissue; bleeding from the host tissue activates platelets, which release various soluble factors activating sleeping stem/progenitor cells. At the same time, transplanted adipose tissue is temporarily placed under severe ischemia; basic fibroblast growth factor (*bFGF*) and other factors are released from injured tissue or dying cells and stimulates adipose-derived progenitor cells to release hepatocyte growth factor (*HGF*), which promotes angiogenesis and inhibits fibrogenesis. Tissue injury further induces inflammatory cell infiltration and release of inflammatory cytokines. Most of differentiated cells in the graft die, but resident stem/progenitor cells are activated or recruited from bone marrow of the host. The dead cells are partly replaced with next-generation cells derived from the stem/progenitor cells after successful vascularization. *ECM* extracellular matrix, *PDGF* platelet derived growth factor, *EGF* epidermal growth factor, *TGF-β* transforming growth factor-β, *SDF-1* stromal derived factor-1, *S1P* sphingosine-1-phosphate, *IL* interleukin

7.9 Grafting of Aspirated Adipose Tissue

Soft-tissue augmentation is performed by grafting autologous tissues or artificial materials to correct inborn or acquired tissue defects; for example, tissue reconstruction after tumor surgery such as mastectomy, or for purely cosmetic purposes such as breast augmentation or facial rejuvenation. Non-vascularized autologous fat grafting (lipoinjection) is a promising option for soft-tissue augmentation because there is no associated incisional scarring or complications derived from foreign materials. Although many innovative efforts and techniques to refine autologous lipoinjection have been reported, problems such as unpredictability and a low rate of graft survival due to partial necrosis remain.

It has not been well documented how adipose grafts survive after non-vascularized transplantation (lipoinjection) (Fig. 7.6). With lipoinjection, the recipient tissue is injured and thus bleeding occurs. The grafted non-vascularized adipose tissue is placed under ischemia (hypoxia with low nutrition) and is temporarily nourished only by diffusion from the surrounding host tissue for a few days until direct capillary supply is

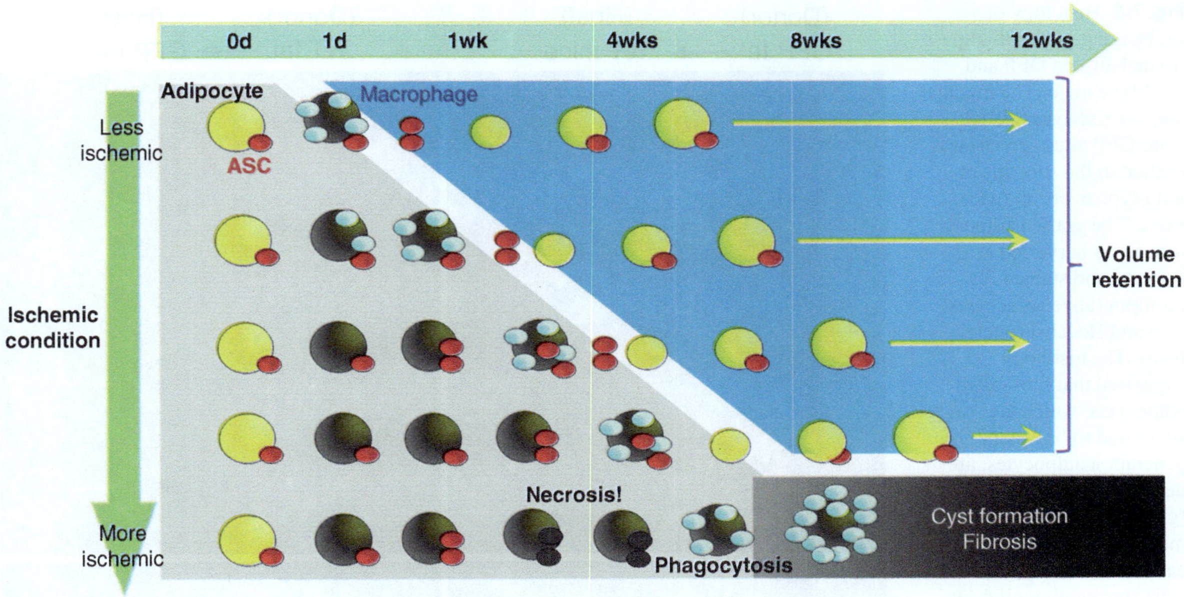

Fig. 7.7 Summarized schema of adipocyte fate after non-vascularized adipose grafting. Animal experiments of non-vascularized adipose-tissue grafting indicated that almost all adipocytes (except for those in the most superficial layers) die within 2–7 days after adipose-tissue grafting. Some of the dead adipocytes are replaced with new adipocytes of next generation, while others are not; in the central area of the graft, even adipose stem/progenitor cells die and the replacement of new adipocytes fails, resulting in scar tissue formation or cyst formation with frequent calcifications. The final volume retention after adipose grafting is determined by the rate of successful replacement of adipocytes

formed (Fig. 7.3) [48]. In response to injury, tissue-bound bFGF, TNF-α and/or transforming growth factor (TGF)-β are released from the injured host tissue and dying donor tissue [24]. Recently, it was also demonstrated that epidermal growth factor (EGF) is released from apoptotic endothelial cells and activates anti-apoptotic response in mesenchymal stem cells [32]. In response to local bleeding from the host tissue, platelet-derived growth factor (PDGF), EGF, and TGF-β are released from activated platelets [1]. Inflammatory cells such as monocytes and lymphocytes are infiltrated and inflammatory cytokines such as interleukins are secreted. During the repairing process, adipocytes, known to be very sensitive to hypoxia, are subject to die within 24 h and subsequently endothelial cells start to die when the ischemia prolongs; while ASCs are likely as resistant to ischemia as bone marrow-derived mesenchymal stem cells [38], which can be functional up to 72 h under ischemia [25].

Our recent studies showed that almost all adipocytes located deeper than 500 μm from the tissue surface likely die within a short period and some of them are replaced with next generation adipocytes [38] (Fig. 7.7). Thus, it was suggested that the rate of this replacement with next-generation adipocytes determines the final volume retention of fat grafting. From the surface of grafted fat tissue, adipocyte remodeling started and progressed slowly toward the center. As described before, unlike other differentiated cells, adipose stem/progenitor cells can stay alive up to 72 h even under severe conditions such as that after non-vascularized grafting (ischemia) and rather start to proliferate, migrate and differentiate in order to try to repair the dying fat tissue. In the case that stem/progenitor cells die due to prolonged severe conditions, the area is not properly replaced with new adipocytes and rather filled with fibrosis and calcification [10].

Our exchanging fat graft experiment using GFP mice and wild-type mice revealed that most of second generation adipocytes and vascular structures seen at 4 weeks were derived from the donor animal, while other cells such as endothelial cells, resident hematopoietic cells, and oil-cyst surrounding cells (macrophages) are derived from the host animal (Fig. 7.8). Even at 4 weeks after fat grafting, many adipocytes remained dead especially in the central areas, but surprisingly,

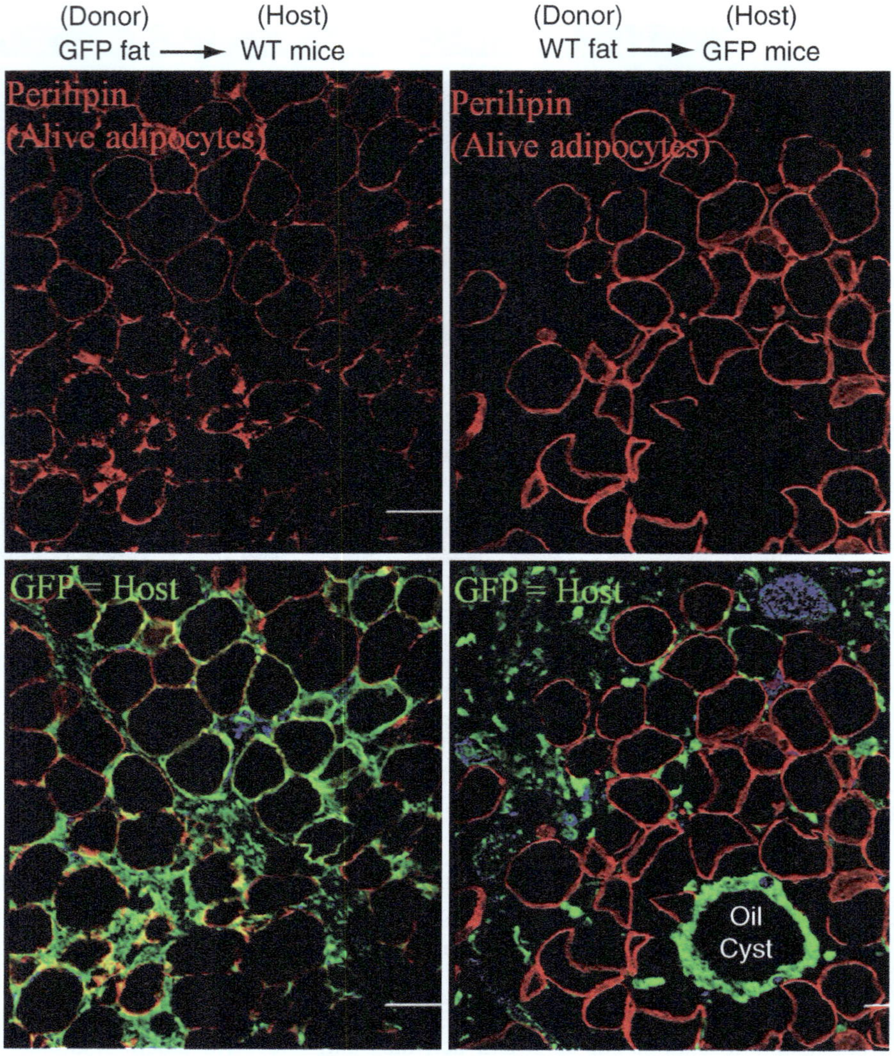

Fig. 7.8 Histology of exchanging grafts of adipose tissue between GFP and wild-type mice. The inguinal adipose tissue harvested from GFP mice (*green*) was grafted to the subcutis of wild-type mice, and vice versa. The grafted adipose tissue was harvested at 4 weeks and stained for perilipin (alive adipocytes: *red*) and Hoechst (nuclei: *blue*). The histology suggested that most alive adipocytes, which are suggested to be second-generation adipocytes, are derived from the donor (graft). On the other hand, most of other cells such as capillaries and infiltrated cells surrounding oil cysts are derived from the recipient (host)

the entire volume of grafted fat was sustained because dead adipocytes kept their volume as lipid droplets and not scavenged by inflammatory cells within a short time (Fig. 7.9). This phenomenon has not been detected before, because grafted fat did not even temporally reduce the size at least macroscopically; the perilipin staining, which can label only alive adipocytes enabled to clearly discriminate alive and dead adipocytes. Thus, we first found that adipose-tissue remodeling after non-vascularized grafting was conducted very slowly by replacement of new generation donor-derived adipocytes and finished by 8–12 weeks (Fig. 7.7). After finishing the adipose remodeling, which means that no more small new adipocytes are observed, there are still proliferating cells observed depending on the number and amount of oil cysts (non-yet-scavenged lipid droplets); most of the proliferating cells were found to be macrophages (Fig. 7.10). These recent findings for non-vascularized fat grafting seem to show common cellular events after any type of non-vascularized tissue grafting including skin graft.

7.10 Supplementation of Adipose Progenitor Cells in Micro Fat Grafting

As discussed above, aspirated fat tissue has a significantly lower progenitor/mature cell ratio [12, 20]. This low ASC/adipocyte ratio may be the main reason for

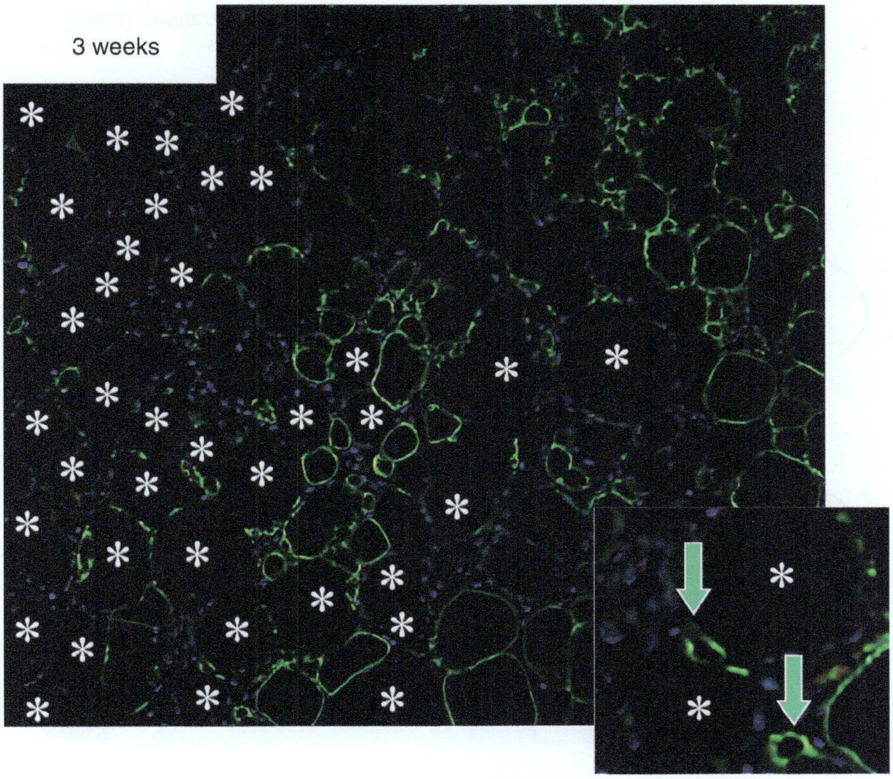

Fig. 7.9 Histology of grafted adipose tissue at 3 weeks. A section of adipose graft at 3 weeks stained with perilipin (alive adipocytes: *green*) and Hoechst (nuclei: *blue*) indicated that many adipocytes are dead (perilipin negative), which means they are lipid droplets (some of the dead adipocytes are indicated by *asterisks*). On the other hand, small new adipocytes (perilipin positive: *arrows*) are emerging to replace the oil droplets. This remodeling was progressing very slowly; the oil droplets stayed for a long time without scavenging or absorption and thus the adipose graft tissue can keep its volume even during the remodeling phase. This phenomenon cannot be detected by H & E staining, because it cannot discriminate alive adipocytes from dead ones

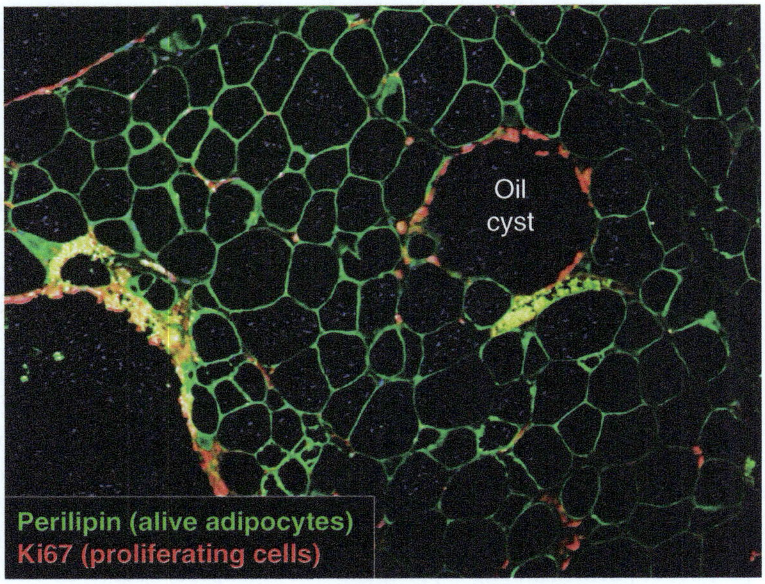

Fig. 7.10 Histology of grafted adipose tissue at 8 weeks. A section of adipose graft at 8 weeks stained with perilipin (alive adipocytes: *green*) and Hoechst (nuclei: *blue*) indicated that there are no more small new adipocytes (perilipin positive) or dead adipocytes (perilipin negative); there remain regenerated alive adipocytes and larger oil cysts. The oil cysts are surrounded by Ki67-positive proliferating cells, which are mainly infiltrated macrophages

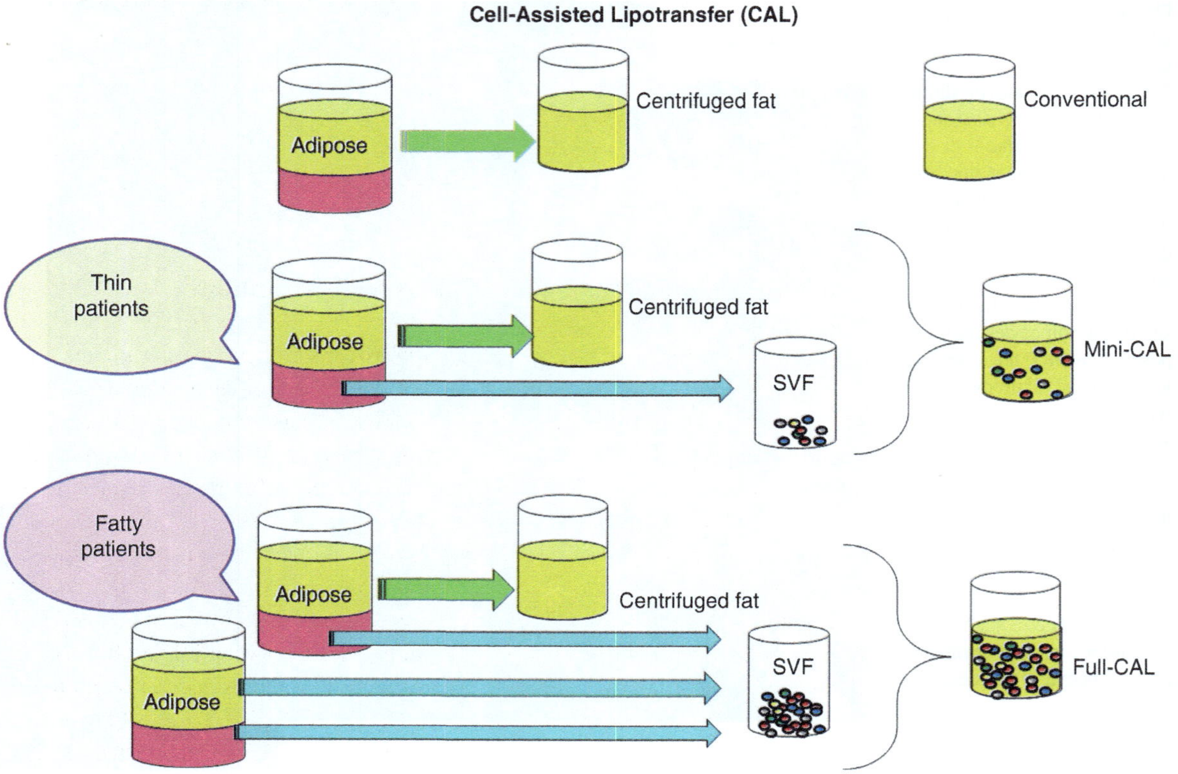

Fig. 7.11 Concept of cell-assisted lipotransfer (*CAL*). The adipose portion (*yellow*) of liposuction aspirates is centrifuged and used as injection material in conventional lipoinjection. In CAL, relatively progenitor-poor aspirated fat tissue is converted to progenitor-enriched fat tissue by supplementation with the stromal vascular fraction (*SVF*). Freshly isolated (non-cultured) SVF cells are attached to the aspirated fat tissue, which acts a scaffold in this strategy. The SVF can be obtained from adipose and fluid portions of liposuction aspirates. The SVF from the adipose portion obtained through collagenase digestion contains 10–40% of adipose progenitor cells (*ASCs*) (CD34+ CD31− CD45−), some of which have multipotency and can differentiate into several lineages in vitro. In mini-CAL, which is used in thin patients, only the fluid portion (*pink*) of the liposuction aspirate is used for the isolation of the SVF, while another volume of liposuction aspirate is additionally harvested for SVF isolation in full-CAL, which is performed in fatty patients. A huge body of basic and translational research using adipose-derived stem/progenitor cells has been conducted and ASCs are currently being used in some clinical trials, including treatments for bone defects (autologous fresh ASCs) [12], rectovaginal fistula (autologous cultured ASCs) [13], graft-versus-host disease (non-autologous ASCs) [14], and soft tissue augmentation by progenitor-enriched fat tissue grafting (autologous fresh ASCs) [15–18]. ASCs have been found to have potential similar to bone marrow-derived mesenchymal stem cells and are now of great interest as a tool for cell therapies

long-term atrophy of transplanted adipose tissue, because ASCs are supposed to contribute to adipose-tissue remodeling and subsequent turnover by replacing old adipocytes with next-generation adipocytes [48]. Based on the findings and hypothesis, we designed a novel therapeutic strategy called cell-assisted lipotransfer (CAL) in which progenitor-poor adipose tissue was converted to progenitor-enriched one by supplementing SVF containing ASCs freshly isolated from extra volume of aspirated adipose tissue [20, 46] (Fig. 7.11). There are at least three experimental studies [19, 20, 23] demonstrating supplementation of adipose progenitor cells enhances the volume or weight of surviving adipose tissue. In CAL [20, 43, 46, 47, 49], freshly isolated SVF cells are attached to the aspirated fat tissue, which acts as a living bioscaffold (Fig. 7.5).

There are four possible roles for ASCs in CAL treatment, which were partly confirmed in pre-clinical studies [19, 20, 23]. First, ASCs can differentiate into adipocytes and contribute to regeneration of adipose tissue. Second, ASCs can differentiate into VECs and also probably into vascular mural cells [2, 7, 20–22], resulting in the promotion of angiogenesis and graft survival. Third, ASCs are known to release angiogenic growth factors such as hepatocyte growth factor (HGF) and stromal cell-derived factor 1 (SDF-1) in response to injury, hypoxia, and other conditions [17, 26, 29, 37, 40] and these factors influence surrounding host tissue. The fourth, ASCs survive as undifferentiated ASCs and contribute to physiological turnover or incidental remodeling in the future [20]. Although adipocyte life span is very long (2–10 years) [33, 34], most adipocytes likely die soon after fat grafting and either are replaced with new adipocytes derived from ASCs or failed in the replacement.

Clinical outcomes over 600 patients have been encouraging; average volume retention was superior to that of conventional fat grafting, though results were relatively variable among patients [43, 46–49]. ASC supplementation appeared to minimize adipose atrophy after transplantation, because tissue volume reduction usually seen after 3 months in conventional fat grafting was minimal.

In the clinical trial, ectopic fibrogenesis was observed in two patients injected with SVF cells as a cell suspension separately from fat grafts [45]. In cell-based therapies using adherent mesenchymal stem cells, unfavorable behaviors such as differentiation into myofibroblasts have been reported [31]. This possibility of unexpected behaviors should be taken into account even if cells are derived from adult tissue and have not been substantially manipulated. It was suggested that ASCs should be adhered to cells, tissue, extracellular matrix, or biological scaffold before administration to avoid unexpected migration or differentiation.

7.11 Stem Cells and Signals for Tissue Regeneration

The fate of ASCs, like any other stem cells, is determined by microenvironments, which include cell death, ECM disruption, bleeding, inflammation, hypoxia, cytokines, chemokine, tissue injury, mechanical force, etc. Stem cells are usually in a silent state at their niche and progenitor cells are likely the same because adipose tissue is slow-cycling. Only upon emergency such as injury, the sleeping stem cells are activated and act according to commitments from the microenvironment; they may reconstruct adipose tissue by differentiating into adipocytes and capillaries, while they may induce fibrosis and/or calcification under unfavorable conditions.

For therapeutic tissue (re)generation, stem cells and signals are two critical requisites. Stem cells are actual players for tissue generation, maintenance, and repair, while signals (microenvironments) are assignments (commitments) to stem cells and stem cells can neither be activated nor differentiate without any signals. We need to control stem cell fate to avoid unexpected behavior of stem cells such as differentiation into unwanted direction, regardless that they are resident stem cells, recruited stem cells mobilized from bone marrow or administered stem cells.

Many efforts have been done to identify specific signals useful for (re)generation of each organ, but there are few signals established so far for therapeutic uses. It was suggested that a dying (apoptotic) cell release specific signals to activate stem/progenitor cells for providing a new cell to replace the dying cell; this is called "compensatory proliferation" [14]. Providing a dying cell as a microenvironment, which is the target cell type to (re)generate, appears to be safest and most reliable way at present to generate tissue, because this is the natural way for compensatory proliferation.

The balance between the stem cells and signals affects the efficiency of the tissue regeneration. In non-vascularized fat grafting, there are many signals such as injury or bleeding, but dying adipocytes seem to be crucial signals to induce ASC differentiation into adipogenic lineage; thus dying adipose tissue would be the key signal to adipose tissue regeneration after fat grafting. If the dying tissue does not contain enough stem cells, the tissue regeneration would not be efficient. On the other hand, if only stem cells are transplanted, stem cells would not be either activated or differentiated sufficiently. So, if the tissue is stem-cell-deficient in number, it would be reasonable to normalize stem cell density in the tissue by supplementing the stem cells.

7.12 Conclusion

Literature regarding ASCs has increased rapidly in the last 5 years; ASCs have been used as a potent therapeutic tool and an alternative to bone marrow-derived adult stem cells in a variety of translational research. A number of clinical trials using freshly isolated or cultured ASCs are ongoing in more than ten countries. Given the physiological functions of ASCs, its homologous use, namely promoting adipogenesis/angiogenesis, would be the easiest and most promising way to perform in clinical settings. Identification of adipocyte-releasing factors upon apoptosis/necrosis would be a breakthrough to step up to the next stage for adipose-tissue regeneration. Careful design of microenvironment activating ASCs, cell delivery protocol to avoid unexpected behavior and induce maximal potential of ASCs, and selection of target diseases, will be critical to the success of clinical applications.

References

1. Aiba-Kojima E, Tsuno NH, Inoue K et al (2007) Characterization of wound drainage fluids as a source of soluble factors associated with wound healing: comparison with platelet-rich plasma and potential use in cell culture. Wound Repair Regen 15:511–520
2. Alessandri G, Girelli M, Taccagni G et al (2001) Human vasculogenesis ex vivo: embryonal aorta as a tool for isolation of endothelial cell progenitors. Lab Invest 81:875–885
3. Amos PJ, Shang H, Bailey AM et al (2008) IFATS collection: the role of human adipose-derived stromal cells in inflammatory microvascular remodeling and evidence of a perivascular phenotype. Stem Cells 26:2682–2690
4. Aoi N, Yoshimura K, Kadono T et al (2009) Ultrasound assessment of deep tissue injury in pressure ulcers: possible prediction of pressure ulcer progression. Plast Reconstr Surg 124:540–550
5. Brahimi-Horn MC, Pouysségur J (2007) Oxygen, a source of life and stress. FEBS Lett 581:3582–3591
6. Braun RD, Lanzen JL, Snyder SA et al (2001) Comparison of tumor and normal tissue oxygen tension measurements using OxyLite or microelectrodes in rodents. Am J Physiol Heart Circ Physiol 280:H2533–H2544
7. Cao Y, Sun Z, Liao L et al (2005) Human adipose tissue-derived stem cells differentiate into endothelial cells in vitro and improve postnatal neovascularization in vivo. Biochem Biophys Res Commun 332:370–379
8. Caplan AI (2007) Adult mesenchymal stem cells for tissue engineering versus regenerative medicine. J Cell Physiol 213:341–347
9. Crisan M, Yap S, Casteilla L et al (2008) A perivascular origin for mesenchymal stem cells in multiple human organs. Cell Stem Cell 3:301–313
10. di Bonzo LV, Ferrero I, Cravanzola C et al (2008) Human mesenchymal stem cells as a two-edged sword in hepatic regenerative medicine: engraftment and hepatocyte differentiation versus profibrogenic potential. Gut 57:223–231
11. Erecińska M, Silver IA (2001) Tissue oxygen tension and brain sensitivity to hypoxia. Respir Physiol 128:263–276
12. Eto H, Suga H, Matsumoto D et al (2009) Characterization of adipose tissue structure and cellular components: differences between aspirated adipose tissue and excised adipose tissue. Plast Reconstr Surg 124:1087–1097
13. Eto H, Suga H, Aoi N et al (2011) Adipose injury-associated factors activate adipose stem/stromal cells, induce neoangiogenesis, and mitigate hypoxia in ischemic tissues. Am J Pathol 178:2322–2332
14. Fan Y, Bergmann A (2008) Apoptosis-induced compensatory proliferation. The cell is dead. Long live the cell! Trends Cell Biol 18:467–473
15. Gago N, Pérez-López V, Sanz-Jaka JP et al (2009) Age-dependent depletion of human skin-derived progenitor cells. Stem Cells 27:1164–1172
16. Kato H, Suga H, Eto H et al (2010) Reversible adipose tissue enlargement induced by external tissue suspension: possible contribution of basic fibroblast growth factor for preservation of enlarged tissue. Tissue Eng A 16: 2029–2040
17. Kondo K, Shintani S, Shibata R et al (2009) Implantation of adipose-derived regenerative cells enhances ischemia-induced angiogenesis. Arterioscler Thromb Vasc Biol 29:61–66
18. Kuroda Y, Kitada M, Wakao S et al (2010) Unique multipotent cells in adult human mesenchymal cell populations. Proc Natl Acad Sci USA 107:8639–8643
19. Masuda T, Furue M, Matsuda T (2004) Novel strategy for soft tissue augmentation based on transplantation of fragmented omentum and preadipocytes. Tissue Eng 10: 1672–1683
20. Matsumoto D, Sato K, Gonda K et al (2006) Cell-assisted lipotransfer: supportive use of human adipose-derived cells for soft tissue augmentation with lipoinjection. Tissue Eng 12:3375–3382
21. Miranville A, Heeschen C, Sengenes C et al (2004) Improvement of postnatal neovascularization by human adipose tissue-derived stem cells. Circulation 110:349–355
22. Moon MH, Kim SY, Kim YJ et al (2006) Human adipose tissue-derived mesenchymal stem cells improve postnatal neovascularization in a mouse model of hindlimb ischemia. Cell Physiol Biochem 17:279–290
23. Moseley TA, Zhu M, Hedrick MH (2006) Adipose-derived stem and progenitor cells as fillers in plastic and reconstructive surgery. Plast Reconstr Surg 118:121S–128S
24. Muthukrishnan L, Warder E, McNeil PL (1991) Basic fibroblast growth factor is efficiently released from a cytolsolic storage site through plasma membrane disruptions of endothelial cells. J Cell Physiol 148:1–16
25. Myolotte LA, Duffy AM, Murphy M et al (2008) Metabolic flexibility permits mesenchymal stem cell survival in an ischemic environment. Stem Cells 26:1325–1336
26. Nakagami H, Maeda K, Morishita R et al (2005) Novel autologous cell therapy in ischemic limb disease through growth factor secretion by cultured adipose tissue-derived stromal cells. Arterioscler Thromb Vasc Biol 25:2542–2547
27. Nishimura S, Manabe I, Nagasaki M et al (2009) CD8+ effector T cells contribute to macrophage recruitment and adipose tissue inflammation in obesity. Nat Med 15:914–920
28. Pasarica M, Sereda OR, Redman LM et al (2009) Reduced adipose tissue oxygenation in human obesity: evidence for rarefaction, macrophage chemotaxis, and inflammation without an angiogenic response. Diabetes 58:718–725
29. Rehman J, Traktuev D, Li J et al (2004) Secretion of angiogenic and antiapoptotic factors by human adipose stromal cells. Circulation 109:1292–1298

30. Rodeheffer MS, Birsoy K, Friedman JM (2008) Identification of white adipocyte progenitor cells in vivo. Cell 135: 240–249
31. Russo FP, Alison MR, Bigger BW et al (2006) The bone marrow functionally contributes to liver fibrosis. Gastroenterology 130:1807–1821
32. Soulez M, Sirois I, Brassard N et al (2010) Epidermal growth factor and perlecan fragments produced by apoptotic endothelial cells co-ordinately activate ERK1/2-dependent antiapoptotic pathways in mesenchymal stem cells. Stem Cells 28:810–820
33. Spalding KL, Arner E, Westermark PO et al (2008) Dynamics of fat cell turnover in humans. Nature 453:783–787
34. Strawford A, Antelo F, Christiansen M et al (2004) Adipose tissue triglyceride turnover, de novo lipogenesis, and cell proliferation in humans measured with 2H2O. Am J Physiol Endocrinol Metab 286:E577–E588
35. Suga H, Matsumoto D, Inoue K et al (2008) Numerical measurement of viable and nonviable adipocytes and other cellular components in aspirated fat tissue. Plast Reconstr Surg 122:103–114
36. Suga H, Eto H, Inoue K et al (2009) Cellular and molecular features of lipoma tissue: comparison with normal adipose tissue. Br J Dermatol 161:819–825
37. Suga H, Eto H, Shigeura T et al (2009) IFATS collection: FGF-2-induced HGF secretion by adipose-derived stromal cells inhibits post-injury fibrogenesis through a JNK-dependent mechanism. Stem Cells 27:238–249
38. Suga H, Eto H, Aoi N et al (2010) Adipose tissue remodeling under ischemia: death of adipocytes and activation of stem/progenitor cells. Plast Reconstr Surg 126(6):1911–1923
39. Tang W, Zeve D, Suh JM et al (2008) White fat progenitor cells reside in the adipose vasculature. Science 322:583–586
40. Thangarajah H, Vial IN, Chang E et al (2009) IFATS collection: adipose stromal cells adopt a proangiogenic phenotype under the influence of hypoxia. Stem Cells 27:266–274
41. Traktuev DO, Merfeld-Clauss S, Li J et al (2008) A population of multipotent CD34-positive adipose stromal cells share pericyte and mesenchymal surface markers, reside in a periendothelial location, and stabilize endothelial networks. Circ Res 102:77–85
42. Yang YI, Kim HI, Choi MY et al (2010) Ex vivo organ culture of adipose tissue for in situ mobilization of adipose-derived stem cells and defining the stem cell niche. J Cell Physiol 224:807–816
43. Yoshimura K, Asano Y (2010) Fat injection to the breasts: cosmetic augmentation, implant replacement, inborn deformity, and reconstruction after mastectomy. In: Hall-Findlay EJ, Evans GRD (eds) Aesthetic and reconstructive surgery of the breast. Elsevier, London, pp 405–420
44. Yoshimura K, Shigeura T, Matsumoto D et al (2006) Characterization of freshly isolated and cultured cells derived from the fatty and fluid portions of liposuction aspirates. J Cell Physiol 208:64–76
45. Yoshimura K, Aoi N, Suga H et al (2008) Ectopic fibrogenesis induced by transplantation of adipose-derived progenitor cell suspension immediately after lipoinjection. Transplantation 85:1868–1869
46. Yoshimura K, Sato K, Aoi N et al (2008) Cell-assisted lipotransfer (CAL) for cosmetic breast augmentation - supportive use of adipose-derived stem/stromal cells. Aesthetic Plast Surg 32:48–55
47. Yoshimura K, Sato K, Aoi N et al (2008) Cell-assisted lipotransfer for facial lipoatrophy: efficacy of clinical use of adipose-derived stem cells. Dermatol Surg 34:1178–1185
48. Yoshimura K, Suga H, Eto H (2009) Adipose-derived stem/progenitor cells: roles in adipose tissue remodeling and potential use for soft tissue augmentation. Regen Med 4:265–273
49. Yoshimura K, Asano Y, Aoi N et al (2010) Progenitor-enriched adipose tissue transplantation as rescue for breast implant complications. Breast J 16:169–175
50. Zengin E, Chalajour F, Gehling UM et al (2006) Vascular wall resident progenitor cells: a source for postnatal vasculogenesis. Development 133:1543–1551
51. Zuk PA, Zhu M, Ashjian P et al (2002) Human adipose tissue is a source of multipotent stem cells. Mol Biol Cell 13:4279–4295

8 Storage of Adipose Stem Cells

Giorgio Bronz and Gianni Soldati

Contents

8.1 **Introduction** 83

8.2 **Stem Cell Bio-Banking** 84

8.3 **Bringing ACTPs to Man: A Strategy** 85

8.4 **The Cellular Component: Mesenchymal Stem Cells (MSCs)** 86

8.5 **Adipose Tissue as a Source of Mesenchymal Stem Cells** 88

8.6 **Storage of Adipose-Derived MSCs** 89

References 92

G. Bronz (✉)
Clinica Giorgio Bronz, Viale Castagnola 21f,
6900 Lugano, Switzerland
e-mail: info@clinicagiorgiobronz.ch; gbronz@swissonline.ch

G. Soldati
Swiss Stem Cell Foundation, Via Tesserete 48,
6900 Lugano, Switzerland
e-mail: gianni.soldati@ldm.ch

8.1 Introduction

The goal of the alchemists of the sixteenth century was the transmutation of common metals into gold and the creation of an elixir of life, a remedy that, it was supposed, would cure all diseases and prolong life. The dream of transforming a common or less regarded material into a very valuable one has been probably accomplished by the scientists of the twenty-first century with human adipose-derived stem cells or mesenchymal stem cells (MSCs).

Stem cells can be stored in bio-banks, intended as organized collections of biological samples and their related data. They are established in order to contain samples with research significance or potential therapeutic use [3]. Biological samples stored in bio-banks can include organs, tissues, cells of different type, body fluids, or recently discovered bioengineered cells. Thus, bio-banks represent an important resource for identifying the causes and mechanisms of a large number of diseases. In particular, stem cell bio-banks are intended as storage facilities for stem cells of different type having a possible future therapeutic application [17]. Given the spreading of chronic degenerative diseases and aging demographics, the development of efficient approaches for tissue repair is required. A vast and growing body of evidence indicates that cell-based strategies have promising therapeutic potential. The goals of stem cell–based medicine are to utilize autologous as well as allogeneic natural or bioengineered stem cells to enhance or activate innate healing processes in order to supplement repair deficiencies. Many stem cell platforms can be stored in bio-banks and offer therefore a wide range of potential lineages

Y.-G. Illouz and A. Sterodimas (eds.), *Adipose Stem Cells and Regenerative Medicine*,
DOI: 10.1007/978-3-642-20012-0_8, © Springer-Verlag Berlin Heidelberg 2011

ranging from pluripotent to oligopotent cytotypes. Like bio-banks in general, stem cell bio-banks may be operated by public institutions, such as university departments, individuals, or private companies. Irrespective of the responsible institution, they may get funding from public or private sources. Indeed, the importance and the role of adipose tissue has been greatly revaluated, after the finding that adipose tissue is the largest endocrine organ, able to interact with all major organs via the production of a wide range of hormones and cytokines. Furthermore, many groups working independently have shown that adult stem cells derived from white adipose tissue can differentiate along multiple pathways raising great hope in regenerative medicine as adipose tissue can be an abundant source of therapeutic cells. Moreover, human MSCs were recently reprogrammed successfully into embryonic stem cell-like colonies (induced pluripotent stem cell, iPS) faster and more efficiently than adult human fibroblasts using the strategy developed by Yamanaka and co-workers [25]. These results make definitively MSCs as major candidates for tissue-engineering applications.

MSCs cells are also increasingly appreciated in the plastic and reconstructive surgical procedures, and in tissue-engineering strategies. Currently available reconstructive surgery using synthetic materials or autologous fat transplants is often unsatisfactory, due to the long-term unpredictability of volume maintenance. Transplanted MSCs may overcome the problems and offer the possibility of fulfilling the key principle of replacing like with like. Similar applications of MSCs are also widely desired for cosmetic purposes. So, the use of autologous MSCs clinically has been reported across the globe. For example, a novel transplantation strategy, termed cell-assisted lipo-transfer (CAL), which involves the enrichment of MSCs in graft, has been used with promising results by Yoshimura and co-workers [29]. In Europe, Rigotti and co-workers reported the use of MSCs in 20 patients to treat the late side effects of radiotherapy [22].

Due to the increasing importance given to these cells it becomes crucial to store MSCs in order to use them later in life to repair, substitute, or regenerate tissues that could be damaged during the individual's life. There are unfortunately today no such bio-banks for adipose tissues that respond to the GMP (Good Manufacturing Practices) guidelines given by the Authority. In particular, Swissmedic, the Swiss authority for medicinal products, has put clearly the point of storage into the GMP procedures because liquid nitrogen–stored MSCs are effectively a drug and by consequence they have to be treated exactly as a medicinal product following GMP guidelines in order to be used later, when thawed, as advanced cell therapy products (ACTPs).

The purpose of this chapter is to show how storage of adipose stem cells could be done in the near future and what kind of experience has been cumulated in our Institute in order to store MSCs safely and usefully for later human cell therapies.

8.2 Stem Cell Bio-Banking

Although dramatically successful in some cases, transplant medicine remains an option for a limited number of patients mostly because of the shortage of donated organs, the epidemic of chronic diseases, and the high costs associated with patient transplantation and follow-up. For this reason, alternative solutions to advance transplant medicine are strongly needed [19]. Novel approaches involving the replacement of the damaged tissues with "new" tissues derived from stem cell–based regeneration are rapidly coming to clinics [20]. As they are naturally propagated throughout development, stem cells provide the foundation for *de novo* tissue formation. Thus, stem cells are isolated from either natural sources or even bioengineered to constitute a large spectrum of progenitor cells with peculiar characteristics for different therapeutic repair applications. In this way, embryonic stem cells, umbilical cord–derived stem cells, and adult stem cells, just to mention a few, can be really effective. The new frontier is then represented by bioengineered stem cells obtained by the conversion of somatic donor cells into pluripotent progenitors by "therapeutic cloning" and nuclear reprogramming (iPS cells) to derive patient-specific progenitor cells [21]. All these stem cell platforms are currently available for "customized" clinical applications. The ability of storing stem cells of different types making them available at the right time for therapeutic use is nowadays of unquestionable importance. Thus, stem cell bio-banks are established worldwide in order to provide high-quality units to transplant centers and patients able to receive them. In general, stem cell units can be cryopreserved in public and private banks.

Actually, stem cell banks operate under strict guidelines. For umbilical cord blood banks, these guidelines have been established, released, and continuously reviewed by NetCord/FACT (Foundation for the Accreditation of Cellular Therapy).

As for existing stem cell bio-banks, adipose-derived stem cell bio-banks will follow these guidelines. Furthermore, the new legislation operative since July 2007 in Switzerland states that every ACTP must comply with GMP procedures. This implies that ACTPs are considered as a matter of fact a drug or a medicinal product and must thus be produced following these guidelines. In particular, every manipulation of cells destined to human therapy has to be accomplished in a clean room, a specially designed work area, where the laboratories are under a constant control of a number of parameters like air purity, personnel clothing, cleaning procedures, and so on. Bio-banks based on adipose-derived stem cells or MSCs will thus be considered as GMP facilities or better as cell factories strictly controlled and authorized by the Country Regulatory Offices.

These strict rules represent a high-quality system to which a stem cell bank should comply to be accepted as a source of transplantable units worldwide. These guidelines and pertaining authorities' inspections guarantee standardization of the cellular products derived from a donor tissue and comparable quality of the stored units for long-term storage and possible subsequent release. Clearly, when storing samples in a stem cell bio-bank, standard operative procedures must necessarily be applied for sample collection, transport, identification, processing, storage at the proper temperature, and for data collection and processing [17]. In general, what is really needed for a stem cell bank to be operative are: the implementation of a quality management system to make the banking facility function properly in every step of its activity; the standardization of procedures, and therefore the use of protocols for sample collection, transport, processing, testing, and storage and release. The standardization will guarantee that all the samples will share the same conditions and same quality controls. Then there must be sample traceability, meaning proper sample coding and identification; computer management and paper copies of all applications and the processes to which samples and associated data have been submitted; employment of qualified personnel who will be exclusively dedicated to the different operations of the bio-bank. Aside from the organizational plan and the positive aspects that bio-banks can contribute, their start-up is not free from difficulties such as a big initial investment on new equipment and proper instrumentation, hiring qualified personnel, and last but not least, organizing space for manipulation and storage of samples. This is a huge initial effort. Being aware that greater reliability of stem cell bio-banks are essential for the success of any transplantation strategy, if a stem cell bank adopts the proposed quality standards, this should provide clinicians high-quality information and properly manipulated and stored units for a safer transplant procedure in therapeutic use.

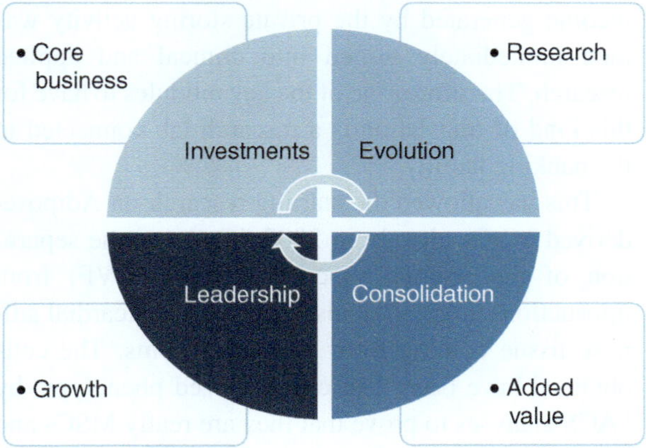

Fig. 8.1 The working model of Swiss Stem Cell Bank. Core business funds research which creates an added value and a total growth for the bio-banking facility

8.3 Bringing ACTPs to Man: A Strategy

It is widely known that the use of stem cells in cell-based therapies is increasing worldwide. Despite the fact that every country has almost its own rules on this issue, cell-based therapies are based on the indivisible binomial: clinical research and therapeutic application. One goes with the other and vice versa. For this reason, it was most likely a wise choice when it was decided in our stem cell bio-bank upon establishment that the return of investment obtained by the banking activity would then be turned into funding clinical research activity, making therefore research self-funded as in Fig. 8.1 where the working model of Swiss Stem Cell Bank is shown. Our private stem cell bank was established by a big initial investment's effort in 2005 in Switzerland; a quality management system was set up and then the bank ran quite easily and the income generated by the private storing activity was then immediately turned into clinical and applied research. Therefore, one of the key modules to have for this kind of translation is a research lab connected to the banking facility.

This has allowed the starting research on Adipose-derived MSCs already in 2007, involving the separation of the stromal vascular fraction (SVF) from liposuction samples but also from peri-pericardial adipose tissue coming from operative rooms. The cells obtained have been then characterized phenotypic by FACS analyses to prove that they are really MSCs and they have been cultured for long periods of time in flasks to check whether they could grow in culture and maintain their original phenotype after many passages with a new serum-free culture medium.

All these research projects were necessary to step forward from basic research to clinical research where patients are involved and as an example, cells must be cultivated without animal-derived cell culture media. In this regard, we had to develop a serum-free culture medium to be able to separate, grow, freeze, and eventually thaw cells for clinical use.

The SVF cells are now being investigated for their ability to grow on 3D scaffolds for bone and tenocyte regeneration.

Thus, the step forward from clinical research to clinical application is possible only when particular conditions are realized and having set up in collaboration with Cardiocentro Ticino, a clean room producing clinical-grade stem cell preparations immediately available was strategic for our vision. Actually, ours is the only clean room accredited in Switzerland for the production of cell therapy products and logistically everything is located within the same Institute. Having a daily contact between researchers, physicians, and in later times engineers, our research projects can move easily and directly from applied research to clinics. Currently, for example, we are focusing on human bone regeneration and cartilage reconstruction and last but not least cardiovascular diseases. This is what we would call "discovery translation" to clinics. The personnel composition was also determined not by chance: Three groups of people self-connect every day for discussion. Biologists for the research aspects, physicians for clinical ones, and engineers for developments into real-world devices, making this group quite heterogeneous but unique in its effort to make things clinically practical.

8.4 The Cellular Component: Mesenchymal Stem Cells (MSCs)

The emergence of regenerative medicine calls for substantial sources and amounts of stem cells as a first line but also for growth factors and biomaterials in driving the correct differentiation. A stem cell is defined as a cell that can self-renew and differentiate toward multiple tissues. Among them bone tissues are the most prominent and advanced for cell therapy. Osteoblasts, the precursors of osteocytes, can be easily differentiated from MSCs.

Embryonic stem cells isolated from the inner cell mass in the blastocyst are totipotent cells and have unlimited self-renewal ability [16]. Several experiments in vivo and in vitro have clearly demonstrated their ability to differentiate in bone tissues [5, 15]. If on the one hand, these cells are very attractive for basic research, on the other hand, the predisposition of animals to develop teratoma raise new concerns on the use of these cells in regenerative medicine. Furthermore, embryonic stem cells are at the center of a social, ethical, and political debate. In the last few years, indeed we are assisting the development of new strategies based on the use of cellular components that could represent new solutions to the unconformity of the actual procedures for bone reconstruction. Whether a particular cell population will maintain a proliferative capacity or not after isolation from its original tissue is today a matter of concern in the scientific community. But not only proliferative ability is needed, but also sufficient amount of cells obtained in the isolation procedure keeping the patient on the safe side with easy and painless extraction procedures.

For these purposes, researchers are focusing on the identification of precursor cell populations in postnatal tissues that comply with applications in tissue engineering. From these efforts, MSCs are coming out as the dominant cell population in current translational research and in particular in the development of new ACTPs. Since adult stem cells reside in adult tissues (adult is considered from birth on), they are not concerned with ethical or social discussions as embryonic stem cells. Furthermore, many new adult tissues have been identified to hide stem cells within differentiated cells and they have been characterized as efficient sources for MSCs which include bone marrow, placenta, muscle tissues, adipose tissues, amniotic fluid, skin, cartilage, and peripheral blood among others. For this reason, the

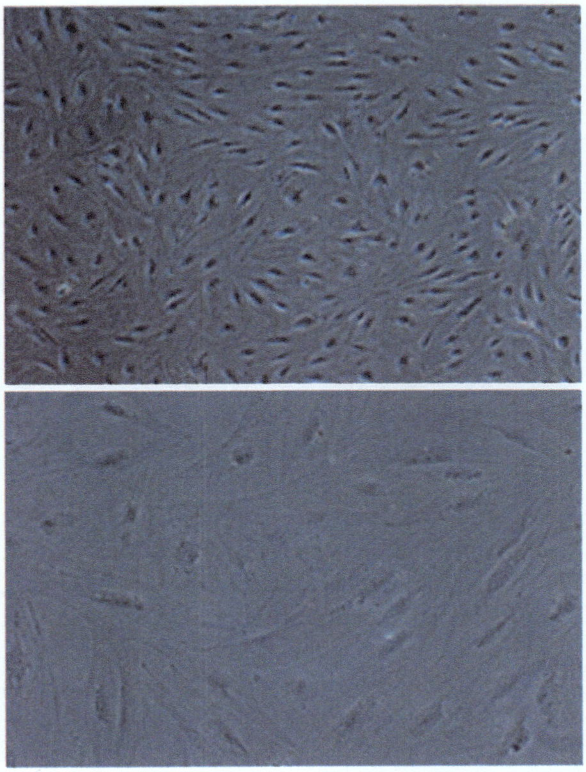

Fig. 8.2 MSCs grown at confluence after 7 days in culture at 100× and 200×. Observe the typical morphologic pattern of adherent fibroblast-like cells

initial enthusiasm that surrounded MSCs in the first years is now flourishing even more in the last 2 years.

Our knowledge of MSCs in stromal tissues arised 40 years ago. Indeed in 1966, Friedenstein and co-workers were the first to describe stromal cells that would be identified as MSCs [10]. In this published work, the identified cells were defined as having the ability to adhere to plastic in cell cultures, to possess a fibroblastoid phenotype, and to be able to form colonies. Today, MSCs are defined to possess a mesenchymal phenotype if they meet the minimal criteria stated by the International Society for Cellular Therapy [8]. A picture of MSCs in culture is shown in Fig. 8.2, where the adherence to plastic is the main macroscopic characteristic of these cells.

Additionally, it was in this paper that the authors hypothesized a potential in the osteogenic differentiation of these MSCs. From this work on, many studies have focused on the fine characterization of this promising population of MSCs, in particular on surface antigen expression that better identify these cells. A particular positivity was observed for Thy-1 (CD90), TGF-beta receptor type III endoglin (CD105), hyaluronic acid receptor CD44, integrin alpha-1 subunit CD29, CD133, activated leukocite-cell adhesion molecules (ALCAM, CD166). In this cell population, common hematopoietic markers CD34, CD45, and CD14 are

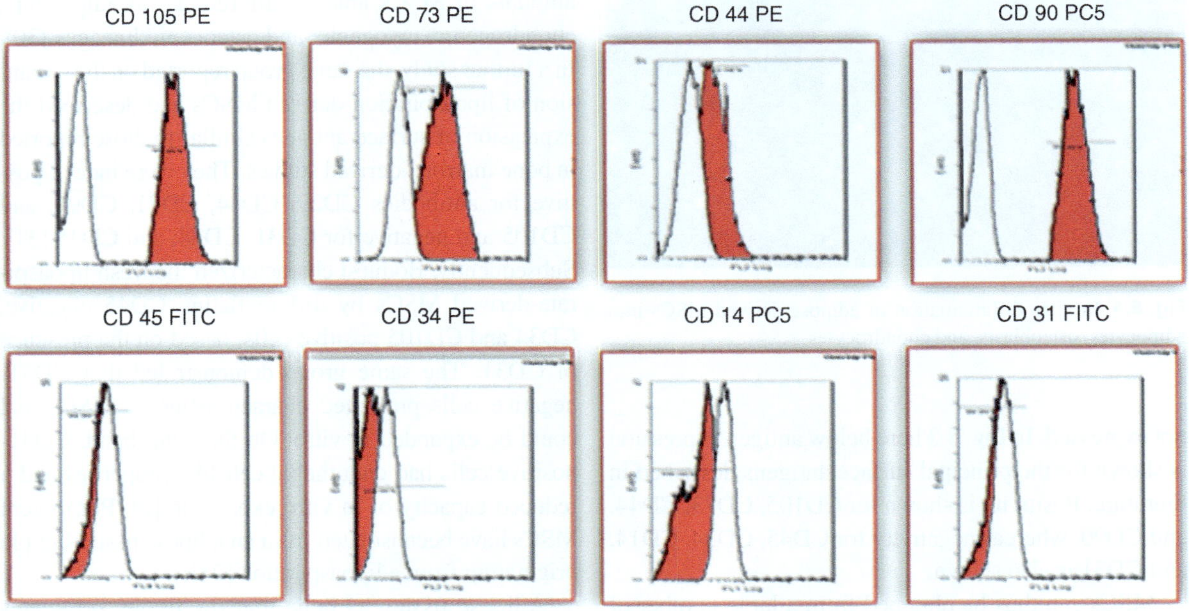

Fig. 8.3 *Upper panel* shows the presence of antigens on adipose-derived MSCs. *Lower panel* shows that MSCs are negative for the described antigens

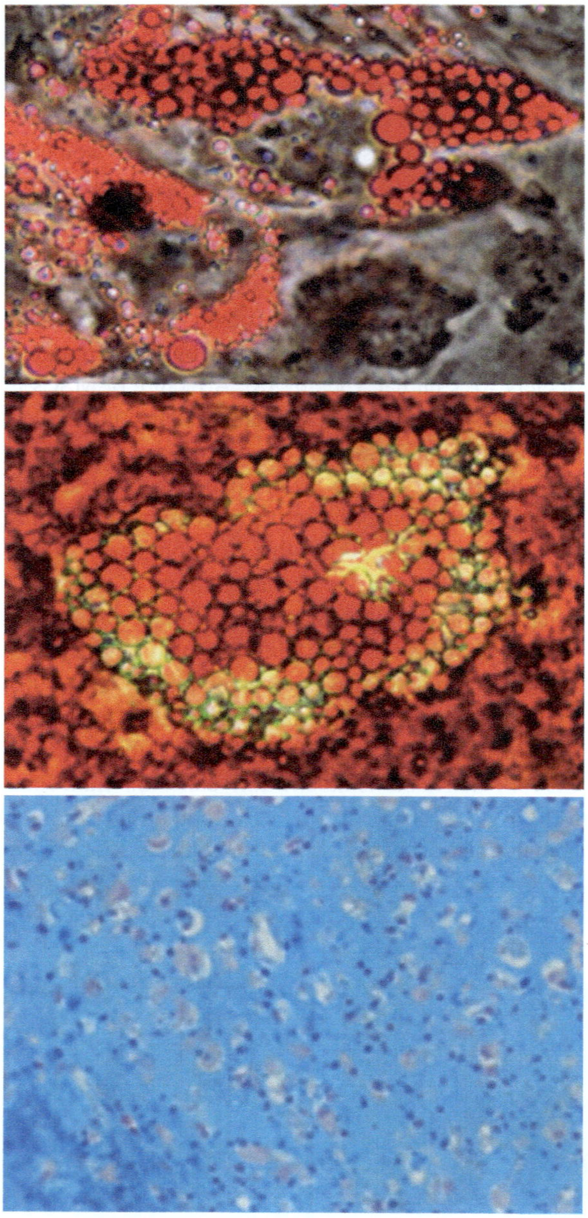

Fig. 8.4 In vitro differentiation of adipose-derived MSCs into adipocytes, osteoblasts, and chondrocytes

not expressed. In Fig. 8.3 here below antigen expression is shown for the principal surface antigens described in literature. Positivity is shown for CD105, CD73, CD44, and CD90, whereas negativity for CD45, CD34, CD14, and CD31 is also shown.

MSCs can thus be obtained from a large number of tissues including heart, bone marrow, and adipose tissues among others. In particular, they possess a wide spectrum of differentiation including bone tissues, cartilage, cardiomyocytes, hepatocytes, and neurons. They represent thus a preferential source for use in human advanced cell therapies, given the availability of autologous tissues, in particular adipose tissue, as a source of stem cells, their efficacy in the MSCs isolation, and their safety if processed in certified structures. Figure 8.4 shows the adherent cell population derived after cell culture of the SVF expanded in vitro and differentiated in adipocytes, osteoblasts, and chondrocytes.

8.5 Adipose Tissue as a Source of Mesenchymal Stem Cells

Recently adipose tissue has provoked a rise in interest as an alternative source, in particular versus bone marrow, of MSCs. The mononucleated phase of this tissue, named stromal vascular fraction (SVF), was first described as a source of adipocyte precursors [14]. These cells morphologically are similar to fibroblasts and can differentiate into pre-adipocytes and thus generate in vitro adipose tissue [11]. Despite the proposal of other authors that the differentiation of SVF can happen only under specific conditions [13], the concept of "adipose-derived stem cell" was introduced for the first time in literature in 2001 in Tissue Engineering, where group leader Patricia Zuk demonstrated the presence of large amounts of MSCs able to differentiate to adipogenic, chondrogenic, myogenic, and osteogenic lineages [30]. In a further study, the same group reported on the expansion of lipoaspiration-derived MSCs and described the expression of surface antigens similar to those reported in bone marrow–derived studies. They were indeed positive for antibodies CD29, CD44, CD71, CD90, and CD105 and negative for CD31, CD34, and CD45 [31]. Subsequently, Boquest characterized the fresh lipoaspirate-derived MSCs by differentiating CD45 negative, CD34 and CD105 positive cells, based on the presence of CD31. The same group demonstrated that CD31-negative cells presented characteristics of MSCs and could be expanded in vitro. On the other hand, CD31-positive cells had endothelial cell-like properties and a reduced capacity of in vitro expansion [6]. Pluripotent MSCs have been isolated from an adipose tissue sample originating from a lipoaspiration [28].

Adipose tissue, as bone marrow tissue, originates from the embryonic mesenchima and contains a tissue called stroma, that can be easily isolated.

Many recent works indicate that MSCs from bone marrow, adipose tissue, or cord blood do not show any difference in the fibroblastoid morphology, immunophenotype, isolation ability, Colony Forming Unit (CFU-F), and differentiation capacity. However, Gimble and co-workers [12] suggested that a stem cell useful for applications in regenerative medicine should have the following characteristics:

1. Can be found in consistent amounts (millions/billions of cells)
2. Can be isolated with relatively noninvasive procedures
3. Can differentiate in many cell lines in a regular and reproducible way
4. Can be easily transplanted
5. Can be processed with the actual GMP guidelines (Good Manufacturing Practices)

Adipose tissue satisfies all these criteria. Consider also that, with the increasing phenomenon of obesity, adipose tissue has become abundant and easily accessible and the liposuction technique is a procedure that is far less invasive than bone marrow aspiration. Generally we can say that this technique produces less pain and discomfort in the patient and does not represent a pathological condition. Small amounts of adipose tissue (100–200 ml) can be easily obtained by means of a simple local anesthesia. Furthermore, in a single gram of adipose tissue can be found 5,000 stem cells, namely adipose-derived stem cells (ASCs), 500 more abundant than those found in the same volume of bone marrow. For this reason, adipose tissue can be considered a rich resource of MSCs that can be useful in many different domains of application. This cellular population, named SVF, can be isolated in big amounts from adipose tissue and easily maintained and expanded in culture, showing stable growth and proliferation curves. Figure 8.5 shows the growth curve of SVF cells as function of time in a regular human serum and in our new serum-free medium.

The processing of adipose tissue is carried out by washing initially with sterile PBS the lipoaspirate to remove erythrocytes and local anesthetic in excess. The fresh tissue is then submitted to enzymatic digestion, a collagenase mixture for 30 min at 37°C to release the cellular component from the fibrosis component of the tissue. Cells obtained from this process can adhere to the plastic and show a fibroblastoid morphology. They can also differentiate to the adipogenic, chondrogenic, myogenic, neurogenic, and osteogenic

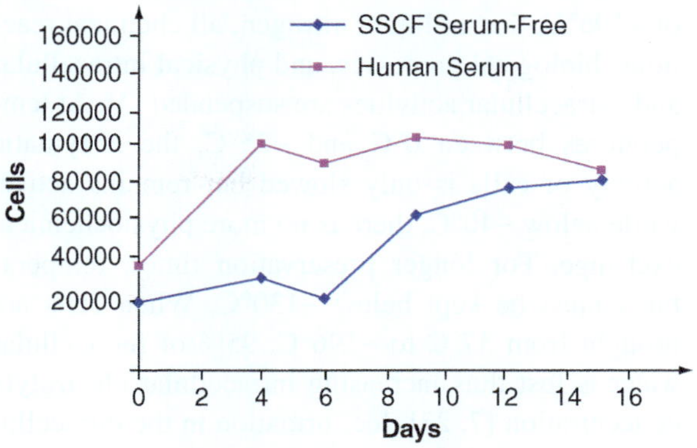

Fig. 8.5 The nucleated cells extracted from the adipose tissue (SVF) and put in culture with our serum-free medium need more time (roughly a week) to adapt to the new conditions. After this period, they grow very well, and after 15 days they reach a density similar to the one obtained with cells cultured in serum-containing medium

cell lines. Following particular conditions of culture, these cells show that they are multipotent. Generally, they show a duplication time of 2–4 days, following the age of the donor, the type of adipose tissue (brown or white), and the localization (subcutaneous or visceral). Furthermore, other conditions have to be taken into account, in particular the type of surgery technology used, condition of culture, density of plating, and the composition of the cell culture medium.

The antigen expression has been widely studied by many research groups and, despite different isolation techniques, number of cell passages, and culture conditions, the immunophenotypic profiles obtained are very similar to those obtained for MSCs from bone marrow. Another aspect that has to be considered is that MSCs, when in culture, secrete various growth factors, like VEGF, HGF, FGF-2, and IGF-1.

The next step in clinical research is now application of MSCs to cell therapy in man, where many groups are now experimenting or even entering phase I trials to transplant scaffolds or matrixes inside which MSCs have been cultured and grown. These experiments will lead to a number of ACTPs ready to be used in cell therapies.

8.6 Storage of Adipose-Derived MSCs

By definition, cryopreservation is the maintenance of biological tissues in a living state of suspended animation at cryogenic temperatures. At a temperature of −196°C, that of liquid nitrogen, all chemical reactions, biological processes, and physical intracellular and extracellular activities are suspended [4]. At temperatures between 0°C and −25°C, the enzymatic activity of cells is only slowed but remains active, while below −40°C, there is no more physiochemical exchange. For longer preservation times, temperatures must be kept below −130°C. When cells are brought from 37°C to −196°C, 95% of intracellular water is lost thus increasing intracellular electrolyte concentration [7, 23]. Ice formation in the intracellular spaces can deform and compress cells and even destroy them.

Thus, a convenient freezing protocol should take into account cellular dehydration, mechanical stress, and intracellular crystallization. Freezing speed is another important variable in this process, considering that when cells are cooled slowly, the flow of water in the extracellular space is increased, whereas with higher freezing rates, diffusion of water is relatively slow. To partially solve this complex problem, cryoprotective substances help in protecting the cells during the process of cryopreservation. Cryoprotectors protect cellular proteins from denaturation [2, 26], in particular those of the plasma membrane by DMSO, through electrostatic interactions [1]. It is accepted that the presence of high concentration of cryoprotectors like DMSO considerably reduces nucleation and aggregation of crystals that can increase the number of cell deaths in the process, whereas the same high concentration may induce cellular lesion via direct toxic effects [9]. This risk can be limited by reducing the time of exposure of cells to cryoprotectors at ambient temperatures.

Only a few studies have examined the role of frozen storage of adipose tissue. One of them was described as a domestic −18°C storage of adipose tissue for 2 weeks. Injection of fat tissue in nude mice demonstrated the survival of this tissue as compared to a control group of non-frozen tissue [24]. A simple freezing technique was used recently by storing fat tissues at −196°C in liquid nitrogen for up to 8 days demonstrating a good maintenance of mitochondrial metabolic activity in the frozen grafts [18]. Remarkably, in both experiments, fat tissue samples were frozen without the addition of a cryoprotective agent. Another study reported the use of a cryoprotective agent to better save and keep viable tissues after thawing [27]. Nevertheless, one has to consider that adipose tissue is the source of MSCs and these cells are responsible for the biological effect that we attribute to them concerning regenerative medicine. Thus, we could consider to store only the adipose-derived stromal vascular fraction (SVF) by isolating it from the carrier tissue. Indeed, the vast majority of studies report the separation, growth, and differentiation of the SVF and all clinical trials to date using adipose-derived MSCs have been designed on this particular fraction, inside which a large number of stem cells have been found. As reported above, the adipose-derived MSCs are pluripotent and can thus give rise to many target tissues, like bone, tendons, cartilages, heart, and nerves, opening the door to real-world Advanced Therapy Products that, in a first time, will be autologous-based but could in the near future be engineered to everyone's need.

One main obstacle related to the application of standardized protocols with ACTPs is legal, and the European and Swiss regulatory instances are developing new criteria to better control this new and expanding field. In particular, Swissmedic, the regulatory body for Switzerland, has clearly put the obstacle at a very high quality expectation. Indeed, all processes relating adipose-derived SVF have to be completed under GMP (Good Manufacturing Practice) guidelines practically meaning that transport, separation, and freezing of cells have to be performed in a clean room. (for more details, see http://www.swissmedic.ch/produktbereiche/00451/00466/index.html?lang=it). Advanced Cell Therapy products will thus need clean rooms or cell factories for their practical application in patients.

The rationale behind this is that if ACTPs are considered as drugs, they also have to be produced as drugs, and since July 2007, they are indeed to be considered as drugs. For storage purposes, this implies that the freezing procedure has also to be performed in GMP and, as a consequence, the final liquid nitrogen storage, has to be qualified in GMP. The first consequence is that a cell factory must validate every protocol or product entering the facility before being able to process, store, and eventually release an ACTP.

To this aim, we designed and validated a protocol to extract and freeze SVF stem cells from liposuction adipose tissues. During this process, the following steps have been evaluated and performed:

1. SVF characterization
2. Freezing procedures and storage
3. Thawing of cells and characterization

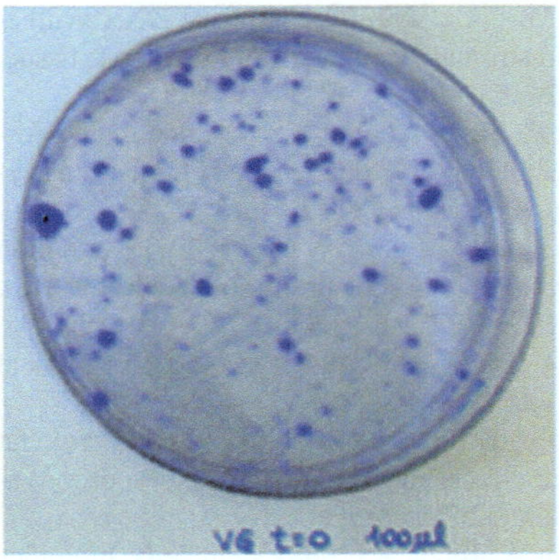

Sample number	Sample weight	Number of cells/g in the SVF	CFU-F (5000 cells plated)	% of CFU-F
1	50 g	36000	63	1/79
2	50 g	25000	171	1/29
3	50 g	10000	32	1/156

Fig. 8.6 Picture of a CFU-F plate for colony counting and table for number of cells extracted per gram of tissue and enumeration of CFU-F

Table 8.1 Results of the freezing/thawing procedure on three SVF samples

Sample no.	No. of cells frozen	Alive cells after thawing (Trypan blue)	% of cell loss	Number of adherent cells after thawing
1	670,000	530,000	18	300,000
2	650,000	500,000	24	280,000
3	650,000	500,000	24	300,000

Here below we describe the procedures and results of these procedures and the final evaluation of SVF-derived stem cells for their potential use as ACTPs.

1. SVF characterization

A sterile syringe containing 50 g of adipose tissue coming from the operating room was used as a device to collect and transport liposuction samples. Isolation of SVF from 3 liposuction samples was immediately performed in sterile conditions by washing 2 times the sample in sterile PBS with Calcium and Magnesium. Washed adipose tissue was then digested with a collagenase cocktail 30 min at 37°C under agitation. PBS was then added again to further wash the sample and after phase separation (lipophilic and hydrophilic), the hydrophilic part of the sample was collected in a sterile 50-ml tube. The digested sample was centrifuged at 15°C, 400 g for 5 min and the pellet (SVF) was washed again two times with PBS. Stromal vascular fraction was re-suspended in 5% human albumin. The following parameters were analyzed before freezing the sample:

- Number of cells/g of tissue
- Ability of CFU-F formation (%) expressed as colonies number over the total cell counted

Figure 8.6 shows a picture of CFU-F colonies after 7 days in culture. Cells are then counted by eye and number of CFU-F colonies is calculated over the total number of cells in the SVF.

2. Freezing procedures and storage

After counting the fresh SVF, a fixed number of cells were frozen (see Table 8.1), by keeping the re-suspended SVF at 4°C under sterile conditions and slowly adding 2.5 mL of DMSO in a 25-ml cryobag and put overnight at −80°C. The bag was then transferred in a vapor liquid nitrogen tank for 2 days.

3. Thawing and characterization of cells

Cells were thawed by putting immediately the frozen bag at 37°C for 3–5 min, re-suspended, and put in culture for further characterization. Alive cells after thawing were counted by Trypan Blue exclusion. Since the first criteria for identification of mesenchymal stem cells is adherence to plastic, cells were then plated on Petri dishes. Table 8.1 shows the results of the three analyzed samples.

Overall, our experience shows that the SVF can be easily frozen following standard conditions for cell freezing. The yield after the procedure is comparable to other cell freezing procedures and can thus be safely used for banking purposes.

Of course, the results of these experiments need further investigations and we are actively working on the validation of many other adipose tissue samples, but the amount of adipose tissue (50–100 g), the isolation protocols, and the freezing and thawing protocols established in our Institute demonstrate that adipose-derived mesenchymal stem cells can be treated as many other cell types with respect to these procedures and represent a useful source of material to proceed to the next step in this domain, expansion for advanced cell therapies.

References

1. Anchordoguy TJ, Cecchini CA, Crowe JH et al (1991) Insights into the cryoprotective mechanism of dimethyl sulfoxide for phospholipid bilayers. Cryobiology 28:467–473
2. Arakawa T, Carpenter JF, Kita YA et al (1990) The basis for toxicity of certain cryoprotectants: a hypothesis. Cryobiology 27:401–415
3. Auray-Blais C, Patenaude J (2006) A biobank management model applicable to biomedical research. BMC Med Ethics 7:E4
4. Bakhach J (2009) The cryopreservation of composite tissues. Organogenesis 5(3):119–126
5. Bielby RC, Boccaccini AR, Polak JM et al (2004) In vitro differentiation and in vivo mineralization of osteogenic cells derived from human embryonic stem cells. Tissue Eng 10:1518–1525
6. Boquest AC, Shahdadfar A, Fronsdal K et al (2005) Isolation and transcription profiling of purified uncultured human stromal stem cells: alteration of gene expression after in vitro cell culture. Mol Biol Cell 16:1131–1141
7. Diller KR, Raymond JF (1990) Water transport through a multicellular tissue during freezing: a network thermodynamic modeling analysis. Cryo Lett 11:151–162
8. Dominici M, Le Blanc K, Mueller I et al (2006) Minimal criteria for defining multipotent mesenchymal stromal cells. The International Society for Cellular Therapy position statement. Cytotherapy 8(4):315–317
9. Fahy GM (1986) The relevance of cryoprotectant "toxicity" to cryobiology. Cryobiology 23:1–13
10. Friedenstein AJ, Piatetzky S II, Petrakova KV (1966) Osteogenesis in transplants of bone marrow cells. J Embryol Exp Morphol 16:381–390
11. Gaben-Cogneville AM, Aron Y, Idriss G et al (1983) Differentiation under the control of insulin of rat preadipocytes in primary culture. Isolation of homogeneous cellular fractions by gradient centrifugation. Biochim Biophys Acta 762:437–444
12. Gimble JM, Katz AJ, Bunnell BA (2007) Adipose-derived stem cells for regenerative medicine. Circ Res 100(9):1249–1260
13. Glick JM, Adelman SJ (1983) Established cell lines from rat adipose tissue that secrete lipoprotein lipase. In Vitro 19:421–428
14. Hollenberg CH, Vost A (1969) Regulation of DNA synthesis in fat cells and stromal elements from rat adipose tissue. J Clin Invest 47:2485–2498
15. Karner E, Unger C, Sloan AJ et al (2007) Bone matrix formation in osteogenic cultures derived from human embryonic stem cells in vitro. Stem Cells Dev 16:39–52
16. Klimanskaya I, Chung Y, Becker S et al (2007) Derivation of human embryonic stem cells from single blastomeres. Nat Protoc 2(8):1963–1972
17. Liaño F, Torres A (2009) Biobanks: a new tool for clinical research. Nefrologia 29(3):193–195
18. MacRae JW, Tholpady SS, Ogle RC et al (2004) Ex vivo fat graft preservation: effects and implications of cryopreservation. Ann Plast Surg 52:281–283
19. Nelson TJ, Behfar A, Terzic A (2008) Stem cells: biologics for regeneration. Clin Pharmacol Ther 84(5):620–623
20. Nelson TJ, Behfar A, Terzic A (2008) Strategies for therapeutic repair: the "R3" regenerative medicine paradigm. Clin Transl Sci 1(2):168–171
21. Nelson TJ, Behfar A, Yamada S et al (2009) Stem cell platforms for regenerative medicine. Clin Transl Sci 2(3): 222–227
22. Rigotti G, Marchi A, Galiè M et al (2007) Clinical treatment of radiotherapy tissue damage by lipoaspirate transplant: a healing process mediated by adipose-derived adult stem cells. Plast Reconstr Surg 119(5):1409–1422
23. Rubinsky B (1989) The energy equation for freezing of biological tissue. J Heat Transfer 111:988–997
24. Shoshani O, Ullmann Y, Shupak A et al (2001) The role of frozen storage in preserving adipose tissue obtained by suction-assisted lipectomy for repeated fat injection procedures. Dermatol Surg 27:645–647
25. Takahashi K, Okita K, Nakagawa M et al (2007) Induction of pluripotent stem cells from fibroblast cultures. Nat Protoc 2(12):3081–3089
26. Timasheff SN (1982) Preferential interactions in protein-water co-solvent systems. In: Franks F, Mathias S (eds) Biophysics of water. Wiley, New York, pp 70–72
27. Wolter TP, Heimburg DV, Stoffels I et al (2005) Cryopreservation of mature human adipocytes: in vitro measurement of viability. Ann Plast Surg 55:408–413
28. Yoshimura K, Shigeura T, Matsumoto D et al (2006) Characterization of freshly isolated and cultured cells derived from the fatty and fluid portions of liposuction aspirates. J Cell Physiol 208:64–76, 27
29. Yoshimura K, Asano Y, Aoi N et al (2010) Progenitor-enriched adipose tissue transplantation as rescue for breast implant complications. Breast J 16(2):169–175
30. Zuk PA, Zhu M, Mizuno H et al (2001) Multilineage cells from human adipose tissue: implications for cell-based therapies. Tissue Eng 7:211–228
31. Zuk PA, Zhu M, Ashjian P et al (2002) Human adipose tissue is a source of multipotent stem cells. Mol Biol Cell 13:4279–4295

Engineered Scaffolds and Matrices: Tailored Biomaterials for Adipose Stem Cell Engineering

9

Karen J.L. Burg, Nichole Myers Cavin, and Katherine Neser

Contents

9.1 Introduction 93

9.2 Injectable Composite: A Tailorable Biomaterials System 94

9.3 Using Bead Chemistry and Morphology to Drive Cellular Behaviors 95

9.4 Driving Cellular Behaviors: How and Why? 98

9.5 The Biomaterial: An Interactive, Dynamic Tissue-Engineering Tool 99

9.6 Conclusions 100

References 100

K.J.L. Burg (✉)
Department of Bioengineering, Institute for Biological Interfaces of Engineering, Clemson University, 401 Rhodes Research Center, Clemson, SC 29634-0119, USA
e-mail: kburg@clemson.edu

N.M. Cavin • K. Neser
Department of Bioengineering, Clemson University, 401 Rhodes Research Center, Clemson, SC 29634-0905, USA

9.1 Introduction

Tissue-engineering methods are continually being investigated for reconstruction of soft-tissue defects [2, 14, 24, 26]. Many patients opting to undergo a soft-tissue reconstruction procedure are breast cancer patients; breast cancer is the second most prevalent cancer in the United States [22]. Although the principles in this chapter may be applied to a range of soft and hard tissue applications, the focus for the purpose of this discussion will center on breast reconstruction. Breast cancer treatments include surgeries (lumpectomies or mastectomies) and/or therapeutic treatments (radiation, chemotherapy, immunotherapy, or hormone therapy) [22]. Over 20% of U.S. breast cancer mastectomy patients will undergo subsequent breast reconstruction [1]. Breast reconstruction procedures offer an aesthetic rather than functional breast replacement. Current breast reconstruction options include soft-tissue transplant, tissue flaps, and implants. Due to the negative side effects associated with these reconstruction options, new alternatives must be developed.

Adipose tissue is a component of breast tissue that is removed during a mastectomy. Stem cells, preadipocytes, and adipocytes have been used as fillers in the replacement of excised tissue [13, 24, 26]. Several strategies have been investigated to engineer adipose tissue using biomaterials or cells [13]. The long-term objective is to develop normal, healthy tissue that will integrate with native tissue and will not require follow-up surgery or treatment in the long-term. Toward this end, tissue-engineering strategies combining cells and biomaterials have also been scrutinized. The traditional

Y.-G. Illouz and A. Sterodimas (eds.), *Adipose Stem Cells and Regenerative Medicine*,
DOI: 10.1007/978-3-642-20012-0_9,

tissue-engineering concept is focused on combining healthy patient cells on/in a biomaterial scaffold that is designed to provide physicochemical and mechanical cues to the cells [17]. The scaffold is designed to be degradable so that it is absorbed as the cells grow and form tissue. Research has focused on seeding cells in a laboratory setting and cultivating the cellular scaffolds in that setting; this preparatory step is thought as necessary prior to implantation. The realities of the clinic suggest that, although this approach may be scientifically optimal, it is far more logistically appealing from a clinical and regulatory perspective, to retrieve cells in the operating room, combine them with a material on site, and immediately implant the cellular material.

Hence, for a clinically relevant approach to be successful, far more emphasis must be placed on designing the implant to be tunable to meet the needs of a particular host-tissue environment. Soft tissue is comprised of many cell types, including adipocytes, stem cells, pre-adipocytes, as well as vascular cells. The complex signaling in the cellular microenvironment is impossible to replicate in a single cell system; hence, traditional tissue-engineering approaches, where one specific parenchymal cell type is isolated, likely add unnecessary processing and handling challenges and do not provide the necessary microenvironmental stromal cues to allow normal tissue development. The traditional approach has largely presented the biomaterial as a relatively independent "scaffold" or housing complex for cells, rather than an interactive, dynamic, tailorable part of an implant that can be designed to drive cellular behaviors.

9.2 Injectable Composite: A Tailorable Biomaterials System

Accordingly, one approach is to combine an injectable composite system with a mix of cells taken from the patient [6]. By definition, a composite consists of multiple components that can be configured in a wide range of manners to meet a wide range of clinical needs. The composite system consists of cells loaded on absorbable beads distributed in a degradable hydrogel carrier that will be injected at the defect sight (Fig. 9.1). Assuming the implanted cells are derived from an adipose depot, the implanted cells are a mix of mature and immature, undifferentiated cells; hence, the biomaterial can be designed to modulate a host of cellular

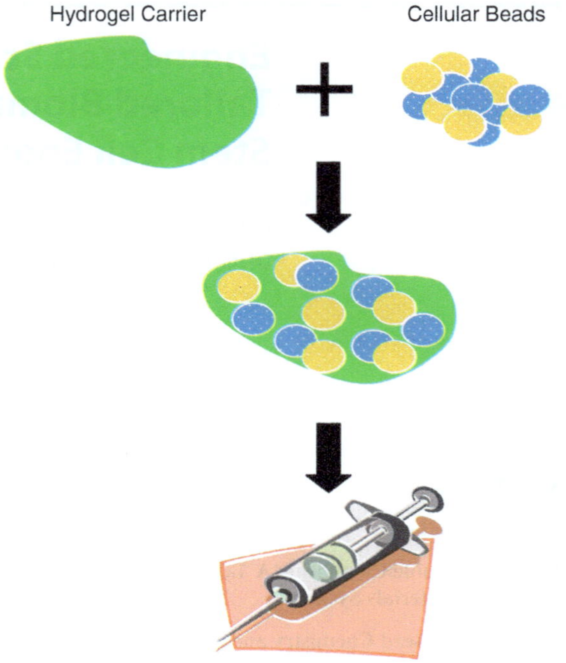

Fig. 9.1 The injectable tissue-engineered composite is modular and may be tuned to suit a particular application

behaviors. Preadipocytes are committed to the adipose tissue lineage through differentiation, which can be induced in laboratory culture using dexamethasone, insulin, and 3-isobutyl-1-methylxanthine (IBMX) and which can be induced *in vivo* through cellular signaling and/or biomaterial stimulation in the host microenvironment. Through a series of cellular and nuclear events, these inducers cause phenotype changes within the cell, producing a mature adipocyte.

Stem cells, preadipocytes, and adipocytes are anchorage-dependent cells and must be seeded on mechanically appropriate substrates for cell proliferation and differentiation to occur [2, 13]. Hence, a composite material containing absorbable beads can provide the appropriate environment for mechanotransduction by anchorage-dependent cells, which the typical soft, gelatinous material simply cannot provide. We have shown, for example, that mesenchymal stem cells will remain viable in collagen gels, but will not differentiate, even in the presence of adipocytic inducers, whereas once seeded on collagen beads of similar chemistry and in culture with adipocyte cocktail, these cells will differentiate, expressing the aP2 gene, indicative of mature fat. The composite format allows adjustment to different applications (e.g., orthopedic to soft-tissue applications) as well as to specific needs of a particular patient. For example, a mix of beads may be used to allow loading of cells as well as encapsulation of therapies, also to allow timed and/or gradual degradation of beads and correlated mass loss. Bead chemistry may be selected to suit the specific need – e.g., cell affinity, drug release – and the ratio of bead chemistries in a given composite may be selected to suit the situation.

An injectable composite is a minimally invasive solution, and has the added advantage of conforming to the geometric variations of a range of defects, where no two defects will have exactly the same geometries. The viscosity of the delivery carrier may similarly be selected to suit the need – long or short-term retention in the defect, release of therapeutic, etc. First and foremost, however, the carrier plays a very important role in delivering the cellular beads from a syringe to a defect site. Without the carrier, the beads would be compressed upon delivery, and difficult to impossible to inject, depending on the particular bead characteristics. A higher viscosity gel is beneficial, for example, in maintaining separation between the beads and allowing good distribution of cellular components upon implantation. One of the major limiting factors in soft-tissue engineering is the inability to vascularize an implant. Again, an injectable composite provides a distinct advantage as small volumes may be injected serially over time to allow the development of vascularity. Additionally, vascular enhancing factors may be added to the composite, either to the bead or carrier component, to enhance development of vascularity. The carrier provides paths of least resistance for infiltrating cells and transport of nutrients and waste so the problems of necrosis and avascularity associated with bulk, continuous cellular scaffolds are averted. Overall, the composite design allows the implant to be tuned most appropriately to the particular circumstance; it is a so-called platform technology.

9.3 Using Bead Chemistry and Morphology to Drive Cellular Behaviors

In a proof-of-concept case study, cellular attachment, proliferation, and differentiation were assessed on biomaterials of varying textures and surface chemistry to demonstrate the ability of surface chemistry and topography in guiding cellular behavior. The purpose to the case study was to highlight the potential of material selection and design in a modular composite design and not to promote specific materials. Specifically, two synthetic materials and one naturally derived material, that is, polylactide (PL) beads, gelatin beads, and polystyrene flat cell culture surfaces, were examined. Polystyrene is a conventional material used in 2-D cell culture and thus serves as an excellent control material. Polylactide has widely been studied, has tunable physiochemical properties, and has been approved for specific use in human medicine [18]. Gelatin is commonly used as a support structure for cells and to expand chondrocytes *in vitro* [20].

Gelatin and PL beads were prepared, the latter using a water–oil emulsion process with subsequent hydrolysis in ethanol. Mouse 3T3-L1 preadipocytes were seeded at a density of 3×10^5 cells/100 μL of scaffolds, for gelatin beads and PL beads, and on polystyrene control plates. A differentiation medium, including dexamethasome, insulin, and IBMX was added on Day 4 to half of the wells, the other half remained inducer-free, and all samples were cultured for 18 days.

Scanning electron microscopy (SEM) results show the cellular attachment to the PL and gelatin beads after 18 days in culture (Fig. 9.2) and highlight the very different surface topologies. Differential scanning calorimetry results indicate that the as-received polylactide was amorphous. The beads that were hydrolyzed and the ones that were maintained in culture did not have uniform melting temperatures, but rather a range of melting temperatures (115–190°C). The calculated percent crystallinity for the hydrolyzed and cultured scaffolds ranged from $1.24 \pm 0.02\%$ to $2.12 \pm 0.06\%$. Gel-permeation chromatography results (Fig. 9.3) indicate that the weight average molecular weight of the as-received PL pellet decreased after hydrolysis and decreased furthermore after 18 days in culture. The results emphasize a larger message, that is, that the PL characteristics can be readily modified by design, and small changes in crystallinity and/or molecular weight can have a pronounced effect on cellular behavior.

Cumulative increases in lactic acid and glucose in the medium indicated that the cells were metabolically active. The difference in lactic acid measurements between Day 2 and Day 18 were significant for cells on all scaffolds or surfaces. The difference in glucose measurements between Day 2 and Day 18 was also significant for the cells on all scaffolds or surfaces.

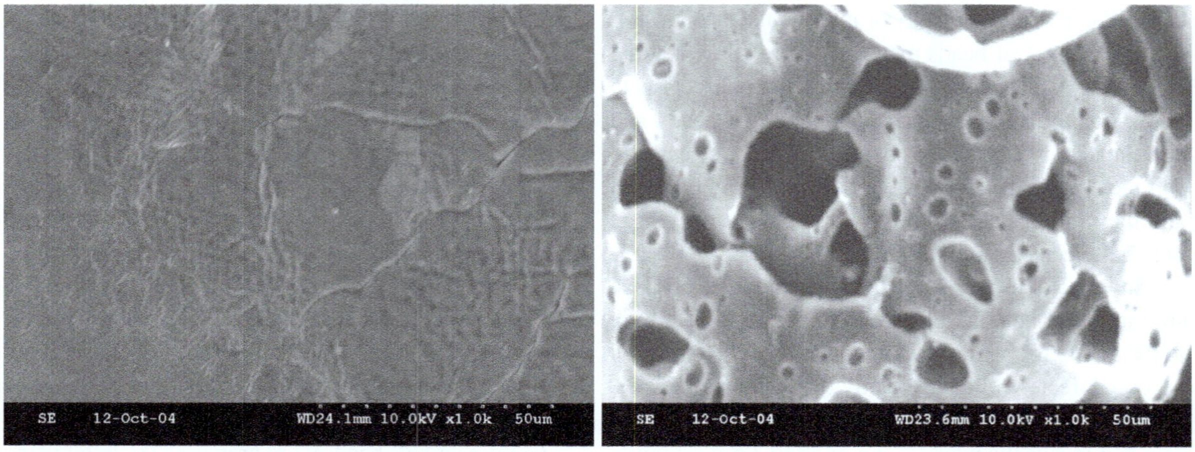

Fig. 9.2 SEM image of PL bead surface (*left*, 400×) and gelatin bead surface (*right*, 450×) after 18 days in culture

Fig. 9.3 Weight average molecular weight of PL scaffolds. *DI* denotes differentiation medium. *Asterisk* (*) indicates statistical difference ($p<0.05$) between PL Day 0 (dry) scaffolds and as-received PL and hydrolyzed scaffolds. *Pound* (#) indicates statistical difference ($p<0.05$) between PL Day 0 (dry) scaffolds and scaffolds in culture for 18 days. *Error bars* denote standard deviation

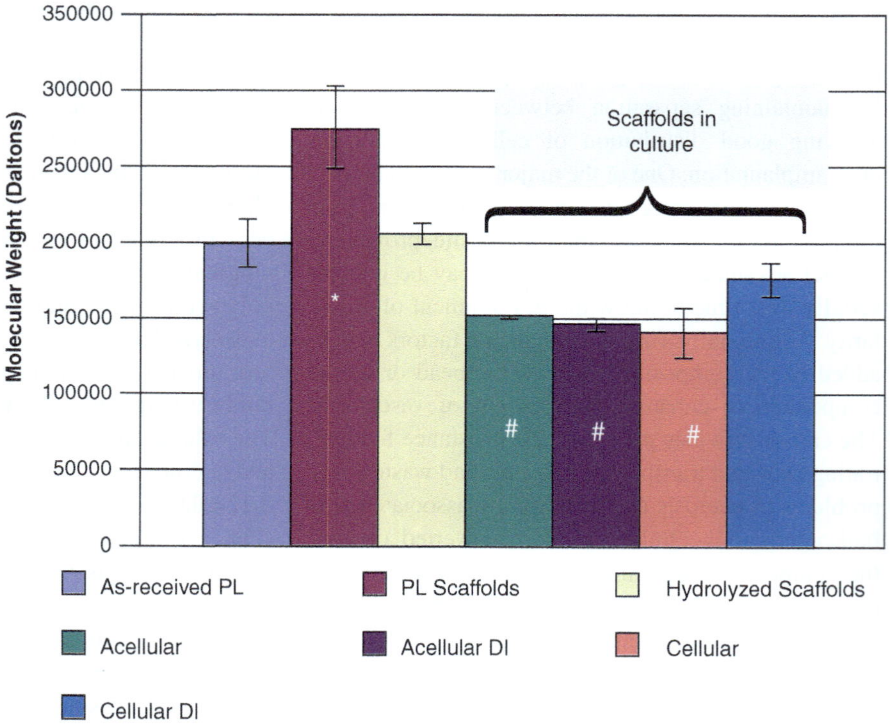

Histology results implied that cells seeded on the PL beads did not form a confluent layer around the circumference of the beads by Day 4, whereas at this time point, the gelatin beads were completely covered by the cells, which appeared to have grown in multiple layers on the beads. By Day 18, the cells cultured on PL beads in the absence of differentiation inducers were present on the beads, but appeared to be beginning to detach. The cells cultured on PL beads in the differentiation medium did not appear to be attached to the PL beads at this time point. The cells seeded on the gelatin beads cultured with or without differentiation medium for 18 days remained attached to the scaffolds and grew into the porous structures. In contrast, live/dead images illustrate that both the PL and gelatin beads had nearly 100% cellular coverage by Day 4. The number of cells cultured in the absence of differentiation inducers on the PL beads, after 18 days, appear to be less than that at Day 4; there was clearly an increase in the presence of dead cells after 18 days of culture. By Day 18, there was also an increase in dead cells in the PL samples cultured in the differentiation medium. The cells cultured on the gelatin beads in both mediums had mostly live cells, and there was no evidence of cellular detachment. The differences between histology and microscopy results are to be expected given the more extensive handling required during histological processing [16].

Cells seeded on polystyrene control plates stained with Oil Red O [9], indicating lipid was present in these cells. By Day 4, minimal lipid was evident (Fig. 9.4), but by Day 18, the cells cultured without differentiation inducers had approximately 1% lipid surface area coverage (Fig. 9.4). By Day 18, the cells cultured in the differentiation medium had approximately 18% lipid surface area coverage (Fig. 9.4).

Although chemistry effects were not isolated from spatial effects (2-D vs. 3-D cultures), the results from these and other studies strongly suggest that information gleaned from 2-D culture systems may be radically different from results obtained from 3-D systems of similar chemistry. Hence, if one is attempting to predict *in vivo* behavior, the *in vitro* biomaterials system must be designed accordingly. The cells cultured on the gelatin beads in the differentiation medium produced more triglyceride than the other cell populations cultured in the differentiation medium [31]. Cells seeded on the gelatin beads with differentiation inducers produced 142.5 ± 16.3 μM triglycerides by Day 18, whereas cells seeded on the 2-D surface cultured in the differentiation medium produced 39.0 ± 26.5 μM of triglycerides, and the cells cultured in the differentiation medium seeded on PL beads produced 29.8 ± 20.9 μM of triglycerides. The concentration of triglycerides produced by cells cultured in the differentiation medium on the gelatin scaffolds was significantly different ($p < 0.05$) from the concentrations of triglycerides produced by cells cultured in the differentiation medium seeded on the PL

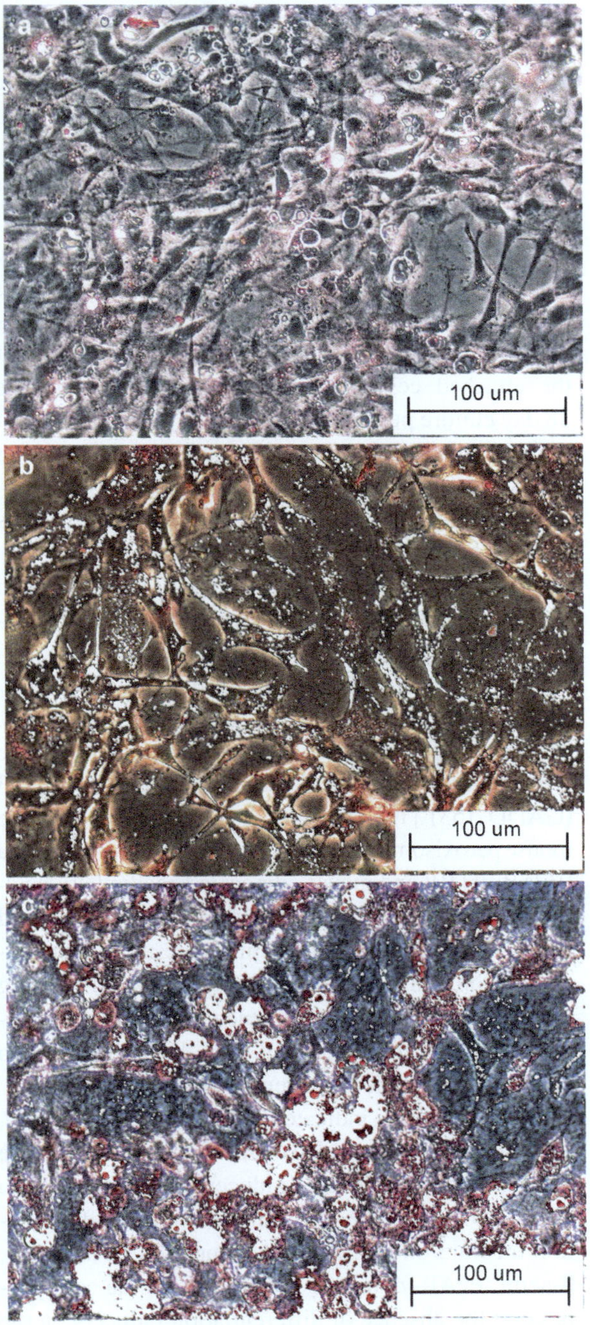

Fig. 9.4 Oil Red O staining. Cells cultured in absence of differentiation inducers on a 2-D surface at Day 4 (*top*), cells cultured for 18 days in absence of differentiation inducer (*bottom, left*) and cells cultured for 18 days in presence of differentiation medium (*bottom, right*), 32× objective

microspheres and the 2-D surfaces. These results emphasize the importance of biomaterial physicochemical specifications in defining the cellular path. The lower density of cells on PL beads indicated by histology is very likely an artifact of processing. Nonetheless, the results point to the fragile interface between the cells and the PL beads, a point that is important but not readily apparent via live/dead imaging.

Using a Pico Green® assay, it was determined that approximately 300,000 cells were seeded in each well containing beads at the beginning of the study, but by Day 4, there were less than 150,000 cells per bead type indicating that less than 50% of the seeded cells attached to the bead surfaces. The results indicate that the attached cells increased in number during the 18 day culture period. Statistical analysis revealed that the cell number for the cells seeded on the gelatin beads with or without differentiation inducers was significantly different ($p < 0.05$) from the number of cells attached to the PL beads in both media.

Reverse transcriptase polymerase chain reaction (RT-PCR) was performed to determine at what timepoint cells expressed aP-2 after seeding onto 3-D substrates, and what additional selected preadipocyte and/or adipocyte genes were expressed in this timeframe. The preadipocyte primer selected for use was mouse growth arrest and DNA-damage inducible protein 153 (GADD 153) [11]. The adipocyte primers used were mouse peroxisome proliferator activator receptor gamma (PPAR-γ) [10] and mouse adipocyte-specific fatty-acid-binding protein 2 (aP-2) [25]. Mouse beta-actin served as the control gene. Semi-quantitative results were determined from band intensity using a Kodak gel imaging system. The band intensity of each sample was normalized to the corresponding β-actin band intensity to obtain the relative level of gene expression.

RT-PCR results indicate that on Day 10 of culture, the cell populations seeded on the PL beads without the addition of differentiation inducers expressed β-actin and GADD 153. The cells cultured in the differentiation medium on the PL beads expressed GADD 153, β-actin, PPAR-γ, and aP-2. On Day 18, the cells cultured in the absence of differentiation inducers on the PL beads expressed GADD 153, aP-2, and β-actin, while at this time point, the cells cultured in the differentiation medium expressed β-actin, PPAR-γ, and aP-2. The cells seeded on the gelatin beads with or without differentiation inducers expressed GADD 153, β-actin, PPAR-γ, and aP-2, that is, all the genes targeted, by the earlier time point of Day 10. The cells cultured on polystyrene without differentiation inducers expressed GADD 153, β-actin, and PPAR-γ on Day 10, while the cells cultured in the differentiation medium on polystyrene expressed all four genes of interest by Day 10. On Day 18, the cells cultured on polystyrene without differentiation inducers expressed GADD 153, β-actin, and PPAR-γ but still did not express aP-2.

9.4 Driving Cellular Behaviors: How and Why?

The objective of this proof-of-concept case study was to differentiate and assess the behavior and gene expression of preadipocytes cultured on scaffolds of differing surface textures and chemistries. The scaffolds used were gelatin beads and PL beads and they were compared to control scaffolds of polystyrene. As expected, the scaffold material and/or surface topography influenced cellular attachment and behavior. The preadipocytes cultured on the gelatin beads, for example, were able to continue proliferating, remain attached to the scaffold for the entire culture period, produce more triglycerides, and yield more viable cells than dead cells as compared to the preadipocytes cultured on the PL beads.

The significant increases between Day 2 and Day 18 in lactic acid suggest that the cells were metabolically active, but these measurements alone are not able to indicate specific changes in cellular behavior, distinguishing differentiation from proliferation. The cells on the PL beads and polystyrene, cultured in the presence of differentiation medium, did not have a significant increase in lactic acid over the course of the study. The significant changes in glucose between Day 2 and Day 18 suggest that the cells were taking in a larger amount of glucose, likely due to cell division, as suggested by results from the other assays. Indeed, when preadipocytes begin to differentiate, these cells accumulate triglycerides [30]. The triglyceride results show that the cells cultured in differentiation medium on the gelatin scaffolds produced a significantly greater triglyceride concentration than cells cultured in the same medium on PL beads and the 2-D surfaces by Day 18. These results indicate that the cells on the gelatin beads were in the beginning stages of differentiation.

The Pico Green® results indicate that the cell number increased during the entire culture period regardless of medium. The increase in cell number suggests that even though most cells were differentiating, others were still in the proliferation stages, signifying that the cell populations were heterogeneous. The results for Pico Green® also show a 2.5-fold increase in cell number on PL beads while the gelatin beads had a 2.2-fold increase. Cells attached readily to the gelatin beads and were able to differentiate; therefore, less cells likely proliferated. The cells on the PL beads bound by non-specific attachment and were therefore slower to proliferate; the less involved attachment likely allowed the extended proliferation phase. Cellular affinity for the gelatin beads, as compared with the PL beads stems from both the differences in surface texture and surface chemistry. For example, the gelatin beads are porous, increasing the surface area for cellular attachment and permitting more cellular attachment per bead [20]. Surface roughness has also been reported to increase cellular attachment [19, 29]. The PL beads initially had a smooth surface, but were hydrolyzed to create a rough surface to promote cellular attachment. The change in the surface of these beads evidently was not significant enough to allow the cells to stop proliferating and start differentiating.

Surface chemistry is another reason for the difference in cell number on the two scaffolds. Gelatin, which is denatured collagen, is a natural biomaterial that allows cells to naturally attach to its surface. PL is a synthetic biomaterial and does not have functional groups on the surface to encourage cell attachment. Cells are able to attach to the PL bead surfaces because serum added to the medium coats the surface of the beads, just as they would be coated in the body. The serum in the medium coats the PL surface, and the cells attach to the hydrophilic coating rather than to the PL hydrophobic surface [5, 12].

The results indicate that the cell population seeded on the gelatin beads are producing lipid, have phenotype characteristics of mature adipocytes, and are the better scaffold choice for supporting cellular proliferation and differentiation. GADD 153 is a growth arrest gene that is expressed during the cell's exit from proliferation and entry into differentiation. Tang and Lane report that GADD 153 is expressed during growth arrest in preadipocytes and down-regulated thereafter [28]. GADD 153 has been reported as being down-regulated as preadipocytes enter the differentiation cycle [7, 11]. However, Shugart and coworkers suggest that cells committed to a particular lineage continue to proliferate but at a much slower rate [27]. Darlington and coworkers propose similar findings to Shugart and coworkers. Darlington and coworkers suggest that GADD 153 is suppressed during clonal expansion preceding differentiation, but that GADD 153 is then elevated on Day 4 of differentiation and remains elevated throughout differentiation [8]. Our results suggest that GADD 153 is expressed throughout differentiation, which support findings by Darlington and coworkers.

9.5 The Biomaterial: An Interactive, Dynamic Tissue-Engineering Tool

The results from these types of case studies will serve in selecting bead chemistries and morphologies for specific clinical applications. The beads tested could be of use in the injectable composite system. The results suggest that the gelatin beads and perhaps other select scaffolds made from natural materials with porous structures are optimal for stem cell and/or preadipocyte attachment and differentiation. Both scaffold types tested had varying diameters; the gelatin beads are prefabricated with a diameter that can be ejected through a 14-gauge needle and the PL scaffolds can be constructed with a smaller diameter by varying the design parameters. Differentiation of the cells on the PL scaffolds could be enhanced by creating a rougher surface or by introducing pores into the scaffold structure, or by absorbing a natural polymer to the surface. Alternatively, PL may simply be better suited for other functions in the injectable composite, such as drug delivery or maintenance of volume, for example, avoiding fast biomaterial absorption and subsequent development of scar tissue.

Indeed, although the cells did not remain attached to the PL beads for the entire culture period or appear to grow to confluence on the scaffolds, the PL beads may serve other purposes in the injectable scaffold system. The fabrication, storage, cellular response, tissue reaction, and degradation of PL scaffolds has been investigated by many researchers and is well documented [3, 4, 12, 15, 21, 23]. The degradation of these scaffolds can be controlled based on scaffold design parameters. That is, the PL beads may be used as drug delivery agents or for structural support in the injectable system, with gelatin or other beads being used as cell carriers. PL scaffolds may be better used in the injectable scaffold system as acellular support structures coupled with the cellular gelatin microspheres. The gelatin scaffolds show more promise for use in the injectable composite system as cellular support devices than the PL scaffolds for an 18-day culture period. Cells seeded on the gelatin microspheres were more proliferative and produced more triglycerides than those seeded on the PL microspheres. Both scaffolds can be used in combination to form an injectable composite system, the PL microspheres serving as a drug delivery vehicle or cellular support device, and the gelatin microspheres functioning as cellular carriers.

The global purpose of the composite injectable is to provide a tailorable cellular polymeric system that may be readily adjusted as the need dictates. The injectable composite system would allow, for example, breast cancer patients the option of reconstructive surgery following a lumpectomy.

9.6 Conclusions

The biomaterial is indeed a powerful tool influencing cellular behavior and, if carefully characterized and designed in appropriate form, can provide the means for directing stem cell proliferation and differentiation for, among other applications, adipose-tissue engineering. Future research should result in the development of a toolbox of methods and techniques to produce the most optimal composite features or a specific application. This work should include developing methods to tailor absorption, texture, and drug delivery capacity for a range of bead chemistries, pinpointing appropriate culture methods, exploring co-culture bench top systems, and facilitating the selection of a range of appropriate hydrogel carriers for the cellular composites in the injectable composite system.

Acknowledgments The authors gratefully acknowledge technical support from Dina Basalyga, Chuck Thomas, Kim Ivey, Joan Hudson, and Larry Grimes. Funding for work presented in this chapter was provided by the National Science Foundation PECASE (BES 0093805) grant.

References

1. American Society of Breast Surgeons (2010) http://www.breastsurgeons.org/ Accessed on 5/24/2011
2. Bauer-Kreisel P, Goepferich A, Blunk T (2010) Cell-delivery therapeutics for adipose tissue regeneration. Adv Drug Deliv Rev 62(7–8):798–813
3. Burg KJL, Shalaby SW (1997) Physicochemical changes in degrading polylactide films. J Biomater Sci Polym Ed 9(1):15–29
4. Burg KJL, Shalaby SW (1997) Water fugacity in absorbing polymers. J Biomed Mater Res 38(4):337–341
5. Burg KJL, Shalaby SW (1999) Biodegradable materials. In: Zilla P, Greisler HP (eds) Tissue engineering of prosthetic vascular grafts. R.G. Landes, Austin, p 505
6. Burg KJL, Austin CE, Culberson CR, Greene KG, Halberstadt CR, Holder WD Jr, Loebsack AB, Roland WD (2000) A novel approach to tissue engineering: injectable composites. Transactions of the 2000 World Biomaterials Congress, Kamuela, 2000
7. Burton GR, Guan Y, Nagarajan R, McGehee RE Jr (2002) Microarray analysis of gene expression during early adipocyte differentiation. Gene 293(1–2):21–31
8. Darlington GJ, Ross SE, MacDougald OA (1998) The role of C/EBP genes in adipocyte differentiation. J Biol Chem 273(46):30057–30060
9. Energy Beam Sciences (2004) Oil red O stain for glycol methacrylate sections. http://www.ebsciences.com/histology/gma_oilredo.htm Accessed on 5/24/2011
10. Fischbach C, Seufert J, Staiger H, Hacker M, Neubauer M, Gopferich A, Blunk T (2004) Three-dimensional *in vitro* model of adipogenesis: comparison of culture conditions. Tissue Eng 10(1–2):215–229
11. Fleming JV, Fontanier N, Harries DN, Rees WD (1997) The growth arrest genes GAS5, GAS6, and CHOP-10 (GADD153) are expressed in the mouse preimplantation embryo. Mol Reprod Dev 48(3):310–316
12. Griffith LG (2002) Emerging design principles in biomaterials and scaffolds for tissue engineering. Ann NY Acad Sci 961:83–95
13. Gomillion CT, Burg KJL (2006) Stem cells and soft tissue engineering. Biomater 27:6052–6063
14. Gomillion CT, Parzel CA, White RL Jr, Burg KJL (2007) Tissue engineering: breast. Encyclopedia of biomaterials and biomedical engineering. Informa Healthcare, Taylor & Francis, New York
15. Jain RA (2000) The manufacturing techniques of various drug loaded biodegradable poly (lactide-co-glycolide) (PLGA) devices. Biomater 21(23):2475–2490
16. James R, Jenkins L, Ellis SE, Burg KJL (2004) Histological processing of hydrogel scaffolds for tissue engineering applications. J Histotechnol 27(2):133–139
17. Langer R (1997) Tissue engineering: a new field and its challenges. Pharm Res 14:840–841
18. Lu L, Peter SJ, Lyman MD, Lai HL, Leite SM, Tamada JA, Vacanti JP, Langer R, Mikos AG (2000) *In vitro* degradation of porous poly (L-lactic acid) foams. Biomater 21(15):1595–1605
19. Lydon MJ, Clay CS (1985) Substratum topography and cell traction on sulfuric-acid treated bacteriological-grade plastic. Cell Biol Int Rep 9(10):911–921
20. Malda J, Kreijveld E, Temenoff JS, van Blitterswijk CA, Riesle J (2003) Expansion of human nasal chondrocytes on macroporous microcarriers enhances redifferentiation. Biomater 24(28):5153–5161
21. Migliaresi C, Fambri L, Cohn D (1994) A study on the in-vitro degradation of poly (lactic acid). J Biomater Sci Polym Ed 5(6):591–606

22. National Cancer Institute (2010) http://www.cancer.gov/cancertopics/types/commoncancers Accessed on 5/24/2011
23. Park A, Griffith Cima L (1996) In vitro cell response to differences in poly-L-lactide crystallinity. J Biomed Mater Res 31(1):117–130
24. Patrick CW Jr, Chauvin PB, Robb GL (1998) Tissue engineered adipose. In: Patrick CW Jr, Mikos AG, McIntire LV (eds) Frontiers in tissue engineering. Elsevier, Houston, p 369
25. Qui J, Ogus S, Lu R, Chehab FF (2001) Transgenic mice overexpressing leptin accumulate adipose mass at an older, but not younger, age. Endocrinol 142(1):1348–1358
26. Shenaq SM, Yukse E (2002) New research in breast reconstruction – adipose tissue engineering. Clin Plast Surg 29: 111–125
27. Shugart EC, Levenson AS, Constance CM, Umek RM (1995) Differential expression of GAS and GADD genes at distinct growth arrest points during adipocyte development. Cell Growth Differ 6(12):1541–1547
28. Tang QQ, Lane MD (2000) Role of C/EBP homologous protein (CHOP-10) in the programmed activation of CCAAT/enhancer-binding protein-beta during adipogenesis. Proc Natl Acad Sci USA 97(23):12446–12450
29. von Recum AF, Shannon CE, Cannon CE, Long KJ, van Kooten TG, Meyle J (1996) Surface roughness, porosity, and texture as modifiers of cellular adhesion. Tissue Eng 2(4):241–253
30. Warne JP (2003) Tumour necrosis factor alpha: a key regulator of adipose tissue mass. J Endocrinol 177(3):351–355
31. Zen-Bio, Inc., Research Triangle Park, NC. Cultured human adipocyte differentiation assay kit. http://www.zen-bio.com Accessed on 5/24/2011

Fundamentals and Principles of Biomolecules in Adipose Stem Cell Engineering

10

Michael Magarakis, Sachin M. Shridharani, Navin K. Singh, and Richard J. Redett

Contents

10.1 **Introduction** 103

10.2 **Biomolecules and Adipose Stem Cells** 104
10.2.1 Angiogenic Factors 104
10.2.2 Hematopoietic and Pro-inflammatory Factors 106

10.3 **Conclusions** 108

References 108

M. Magarakis (✉) • S.M. Shridharani •
N.K. Singh • R.J. Redett
Department of Plastic, Reconstructive and Maxillofacial Surgery, The Johns Hopkins University School of Medicine, 601 N. Caroline St, JHOC # 8, Baltimore, MD 21287, USA
e-mail: mmagarakis@yahoo.com; sshridh1@jhmi.edu; doctor@washingtonianplasticsurgery.com; rredett@jhmi.edu

10.1 Introduction

Human adipose tissue is comprised of three main fat deposits – visceral white fat, subcutaneous white fat, and brown fat – each with its own unique properties. In particular, white adipose tissue is associated with energy storage and hormone production, while brown adipose tissue is mainly responsible for heat production through energy expenditure (thermogenesis) [29]. Although many informative studies have been performed on cultured adipocytes, there are still some aspects of adipocyte function that require further investigation. For instance, the regulation of adipose tissue metabolism is controlled by activation of the autonomic nervous system, delivery of a complex mixture of substrates and hormones to adipose tissue, feedback from autocrine and paracrine effectors secreted by adipocytes, and the vascularity of the adipocytes [7]. Humans are born with a specific numeric amount of adipocytes that multiply and develop until puberty, subsequently remaining constant thereafter. Irrespective of exercise and/or strict dietary modification, humans cannot reduce the number of fat cells. Nonsurgical treatment such as aerobic exercise and balanced diet will eventually decrease adipose cell mass; however, the actual number of those cells will remain constant [3]. Adipose tissue contains adipose-derived stem cells, which possess the ability to differentiate into multiple cellular lineages, a property that represents the key to regenerative medicine. By definition, stem cells are characterized by their ability to undergo multilineage differentiation and form terminally differentiated cells. Guilak et al. assessed this potential by culturing and ring cloning to select cells derived from one progenitor cell. Forty-five

Y.-G. Illouz and A. Sterodimas (eds.), *Adipose Stem Cells and Regenerative Medicine*,
DOI: 10.1007/978-3-642-20012-0_10, © Springer-Verlag Berlin Heidelberg 2011

clones were expanded through four passages and then induced for adipogenesis, osteogenesis, chondrogenesis, and neurogenesis using lineage-specific differentiation media. The authors found that 81% of adipose stem cell (ASC) clones differentiated into at least one of the lineages [9]. An ideal stem cell, one that can potentially be used in regenerative medicine, should have the following characteristics: (a) found in large quantities, (b) easily collected or harvested, (c) is differentiated into multiple cell lineage pathways in a reproducible manner, and (d) can be easily transferred to an autologous or even allogeneic host [4]. Tissue-specific stem cells originate from specific organs such as: brain, gut, lung, liver, bone marrow, and adipose tissue [30]. It is well known that these stem cells persist in adults; however they represent a rare population "hidden" amongst other cell populations [31]. ASC have a broad differentiation potential, but their ability to develop is limited compared to embryonic stem cells. They can be isolated from either bone marrow or adipose tissue. This population was initially thought to differentiate only to their tissue of origin; however, it has been shown that ASC have the capacity to differentiate into cells of mesodermal, endodermal, and ectodermal origin. Furthermore, they cross-lineage barriers and acquire the phenotype and biochemical properties of cells that are unique to other tissues [2, 10, 17, 20–22]. Adipocytes develop from mesenchymal cells through a combination of transcriptional and nontranscriptional events that occur throughout human life. Adipocyte differentiation is a complex process accompanied by simultaneous changes in cell morphology, hormone sensitivity, and gene expression [4]. Although, for many years, ASC have been described as pre-adipocytes [5, 11], today they are appreciated as multipotent cells with a chondrogenic, neurogenic, and osteogenic potential [5, 9, 11, 32]. Sedentary lifestyle and limited time for exercise have contributed to irregularities in body contour and excess adipocyte mass that is often resistant to the most strenuous exercise or weight loss efforts. The significant accumulation of subcutaneous fat among individuals in the United States and indeed world-wide in developed nations makes adipose tissue an abundant source of ASC. Approximately 400,000 liposuction procedures are performed in the United States each year, and these procedures yield anywhere from 100 mL to >3 L of adipocyte tissue [14]. Today, most of this lipoaspirate, which contains a significant amount of ASC with a wide range of therapeutic potential, is discarded.

10.2 Biomolecules and Adipose Stem Cells

Biomolecules refer to the biological materials which serve as the structural integrity of tissue-engineered constructs and regulate their components. The main components of biomolecules are the following cellular factors: growth, differentiation, angiogenic, pro-inflammatory, and gene modulated. The specific factors to be used for each tissue-engineered construct can be provided either exogenously or by local or systemic delivery. Adipose tissue is a dynamic "player" in endocrine physiology and serves as a source of cytokine secretion. In the clinical setting, it has been shown that individuals with large volumes of adipose tissue are more likely to have increased levels of pro-inflammatory cytokines such as interleukin (IL) 6, IL-8, and tumor necrosis factor alpha (TNF-α). Furthermore, adipose tissue expresses hematopoietic growth factor and macrophage colony-stimulating factor (M-CSF), whose expression can lead to adipose tissue volume expansion [15].

ASC are multipotent and can potentially differentiate in various pathways in response to growth factors and environmental agents [8]. There is evidence that ASC can promote tissue recovery through the delivery and localized secretion of cytokines. Recent in vivo studies support this hypothesis. Intravenous infusion of ASC improved recovery of limb function in mice following ischemic injury [19]. The positive effects of ASC in ischemia are most likely secondary to their ability to secrete angiogenic cytokines, such as hepatocyte growth factor (HGF) and vascular endothelial growth factor (VEGF).

In this chapter the authors reviewed the endocrine function and cytokine profile of ASC, and focused on elucidating the basic principles, as well as interactions, between adipose stem cells and cytokines, adipokines, or biomolecules in general.

10.2.1 Angiogenic Factors

10.2.1.1 Hepatocyte Growth Factor (HGF)

The role of implanting ASC into ischemic cardiac tissue as a means to increase angiogenesis is an emerging therapeutic approach [6, 27]. Most of the

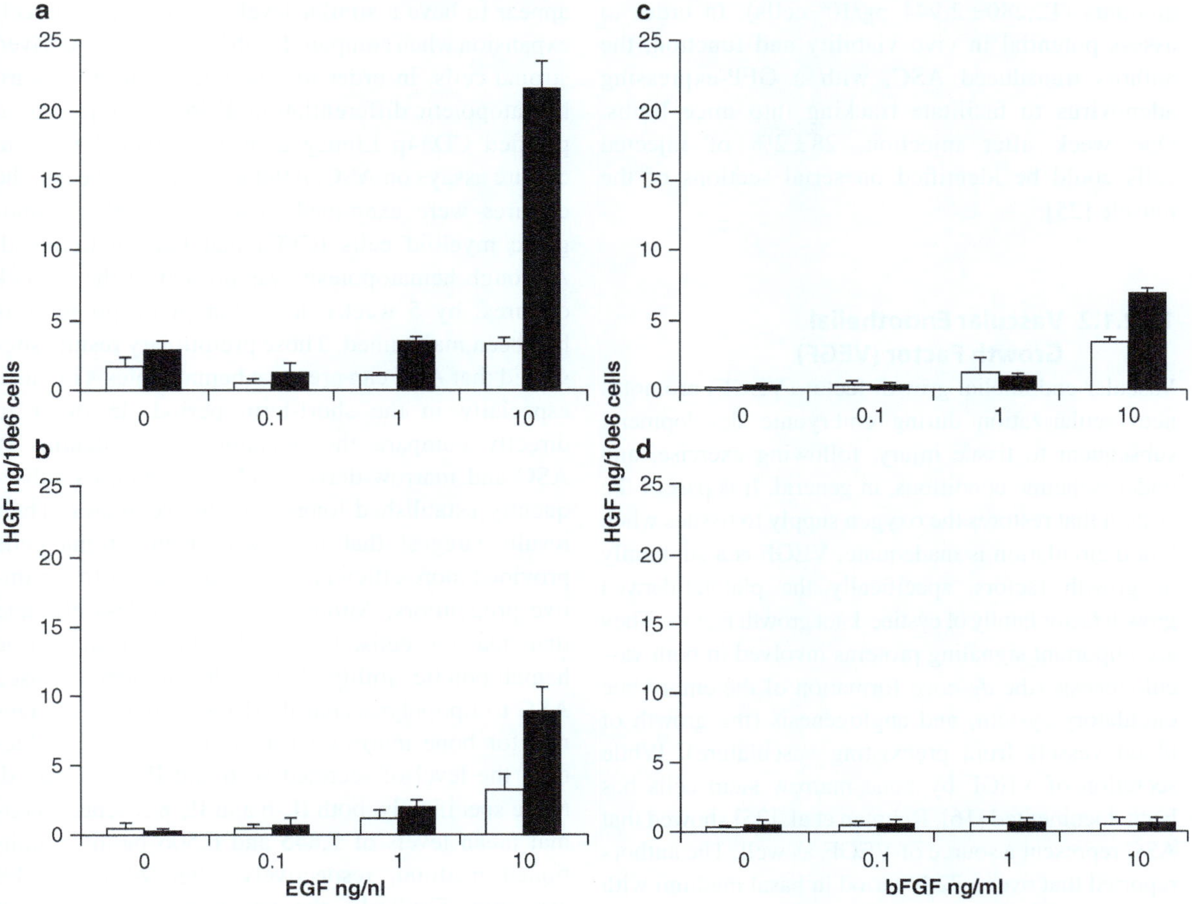

Fig. 10.1 Hepatocyte growth factor (*HGF*) secretion. The secretion of HGF was determined by ELISA on conditioned medium from undifferentiated (**a**, **c**) and adipocyte-differentiated (**b**, **d**) ASC following exposure to epidermal growth factor (*EGF*) (**a**, **b**) or basic fibroblast growth factor (*bFGF*) (**c**, **d**) in the absence (*white bars*) or presence (*solid bars*) of varying concentrations of 2-sodium ascorbic acid. The values represent the mean (ng/10^6 cells) ± S.D. of $n=3$ ASC donors (Reprinted with permission from the publisher from Kilroy et al. [15])

clinical studies have used bone marrow cells which are only available in limited quantities and cannot be easily isolated. There are data to support that ASC secrete HGF, thus representing a potential source for cells to be utilized in cardiovascular cell therapy [15, 24, 25]. In vitro studies have depicted a link between ASC-derived HGF and physiologic or pathologic processes. In particular, secretion of HGF by ASC has been shown to have a positive effect on tubule formation by vascular endothelial cells. This action was found to be independent of VEGF [26]. Unfortunately, Rahimi et al. showed that HGF secreted by ASC promoted the proliferation of mammary tumor epithelial cells [23]. Kilroy et al. reported the constitutive and inducible secretion of HGF by ASC in vitro. The authors showed that this property was dependent on the level of ASC differentiation. In particular, the adipocyte-differentiated ASC appear to lose their responsiveness to basic fibroblast growth factor (b-FGF) and failed to induce HGF expression. On the other hand, treatment of undifferentiated ASC with either b-FGF or EGF was associated with increased levels of HGF release. Finally, it appears that the addition of ascorbic acid increased the increased HGF secretion by a factor of twofold or greater (Fig. 10.1) [15].

In a similar manner, Rehman et al. reported the secretion of HGF by human ASC in significant amounts (12,280 ± 2,944 pg/10^6 cells). In order to assess potential in vivo viability and function, the authors transduced ASC, with a GFP-expressing adenovirus to facilitate tracking into mice limbs. One week after injection, 28 ± 2% of injected cells could be identified on serial sections of the muscle [25].

10.2.1.2 Vascular Endothelial Growth Factor (VEGF)

Vascular endothelial growth factor (VEGF) promotes neovascularization during embryonic development, subsequent to tissue injury, following exercise, and under ischemic conditions, in general. It is part of the system that restores the oxygen supply to tissues when blood circulation is inadequate. VEGF is a subfamily of growth factors, specifically the platelet-derived growth factor family of cystine-knot growth factors. They are important signaling proteins involved in both vasculogenesis (the *de novo* formation of the embryonic circulatory system) and angiogenesis (the growth of blood vessels from preexisting vasculature). While secretion of VEGF by bone marrow stem cells has been documented [16], Rehman et al. [25] showed that ASC represent a source of VEGF, as well. The authors reported that over a 72-h period in basal medium with 5% fetal bovine serum and no additional growth factors under normoxic conditions, ASC secreted significant amounts of VEGF (1,203 ± 254 pg/10^6 cells). Interestingly, when ASC were cultured in hypoxic conditions, there was a fivefold increase in the secretion of VEGF from 1,203 ± 254 to 5,980 ± 1,066 pg/10^6 cells ($p = 0.0016$, paired t-test, $n = 7$). The property of ASC to react to a stimulus such as hypoxia shows that they can adapt to the environment into which they are placed (ischemic myocardium), by increasing the production of VEGF in response to ischemia and thus, induce neovascularization.

10.2.2 Hematopoietic and Pro-inflammatory Factors

One of the most clinically relevant properties of bone marrow-derived mesenchyme is the ability to provide long-term hematopoietic support. ASC appear to have a similar level of hematopoietic cell expansion when compared with bone marrow-derived stroma cells. In order to assess their ability toward hematopoietic differentiation, Kilroy et al. [15] used purified CD34p Linneg cells to initiate long-term culture assays on ASC. After either 3 or 5 weeks, the cultures were examined to assess whether clonogenic myeloid cells (CFC) had been maintained. Although hematopoiesis was present in the 3-week cultures; by 5 weeks, less clonogenic progenitors had been maintained. Those preliminary results suggested that ASC can preserve hematopoiesis in vitro, especially in the short-term period. In order to directly compare the hematopoiesis potential of ASC and marrow-derived cells, the authors subsequently established long-term culture assays. Their results suggest that marrow-derived stroma cells provided more efficient long-term support for primitive progenitors. Although ASC were less efficient than marrow cells, they still exhibited some true hematopoietic ability. When the authors exposed ASC to lipopolysaccharide (LPS), which is an agonist for bone marrow stromal cell cytokine induction, the level of secreted IL-6 and IL-8 increased. More specifically, both IL-6 and IL-8 reached maximal mean levels of 7,845 and 6,506 pg/mL conditioned medium, respectively, after 24 h of LPS exposure. Similarly, the hematopoietic cytokines: macrophage colony-stimulating factor (M-CSF) and granulocyte–macrophage colony-stimulating factor (GM-CSF) reached maximal mean levels of 976 and 52 pg/mL, respectively, at 24 h. TNF-α however, reached its peak mean level of 112 pg/mL after 8 h of LPS exposure. IL-7 and the pro-inflammatory cytokine IL-11 were low. They displayed a significant induction by ELISA, reaching maximal mean levels 24 h after LPS exposure of 3.4 and 12.7 pg/mL, respectively (Fig. 10.2).

Consistent with the ELISAs, the steady-state levels of mRNAs for representative cytokines were elevated within 4 h following LPS exposure based on RT-PCR. IL-1a, IL1b, and IL-12 protein were not detected in the conditioned medium from undifferentiated ASC following LPS exposure. The data produced by this study indicate that ASC may have clinical value for the patient population undergoing hematopoietic stem cell transplantation following high-dose chemotherapy. Conclusively, there is potential of co-infusing ASC with hematopoietic stem cells as a means to optimize

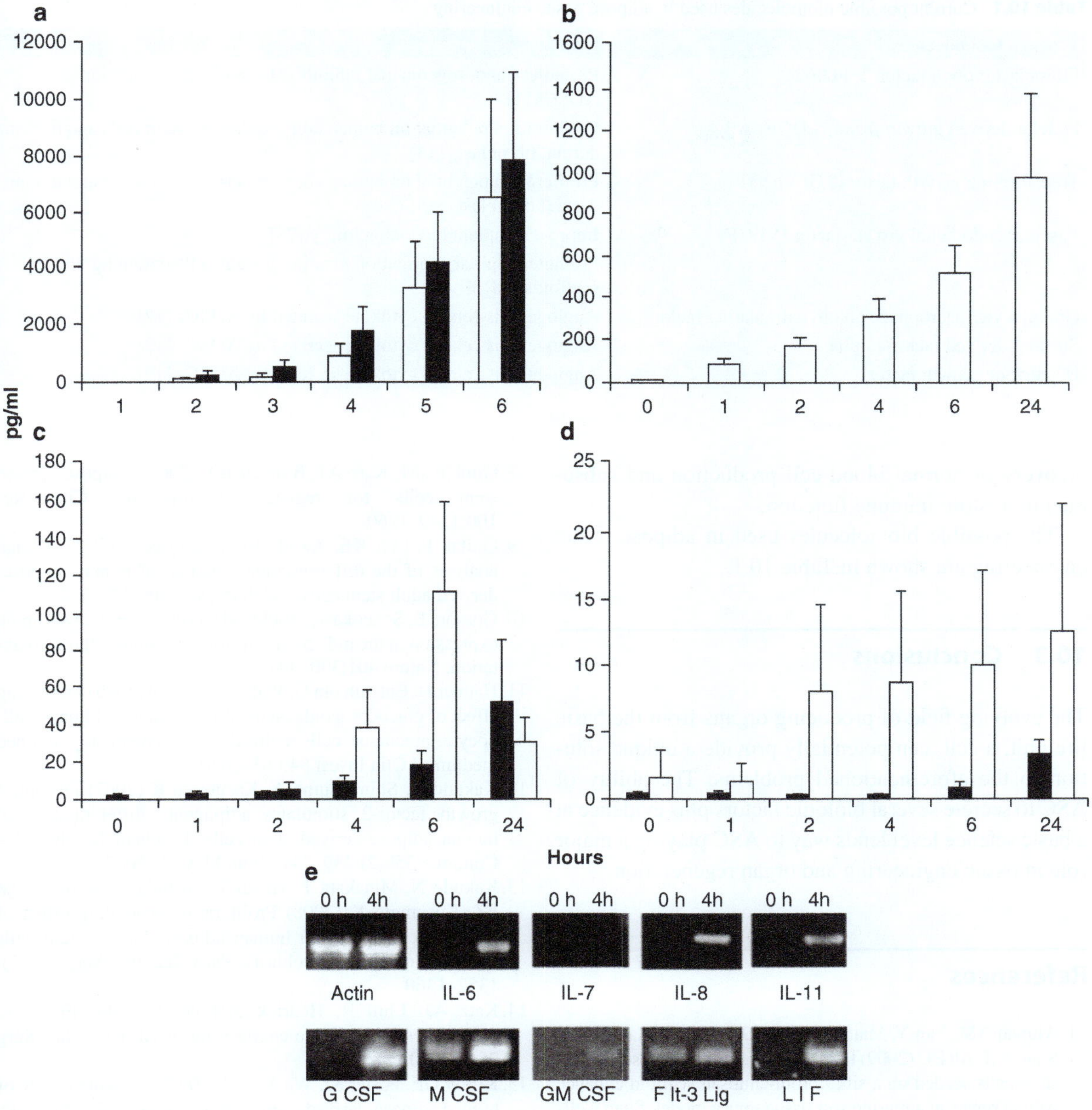

Fig. 10.2 Pro-inflammatory and hematopoietic cytokine secretion. The conditioned medium from undifferentiated ASC was assayed for secretion of selected cytokines at varying times following exposure to LPS (100 ng/mL) for periods of 0–24 h; (**a**) IL-6 (*solid bar*) and IL-8 (*clear bar*); (**b**) M-CSF; (**c**) GM-CSF (*clear bar*) and TNF (*solid bar*); (**d**) IL-7 (*clear bar*) and IL-11 (*solid bar*). The values represent the mean (pg/mL) ± S.E.M. of $n = 6–8$ ASC donors. (**e**) The mRNA levels of selected cytokines in ASC from a representative donor were assayed by PCR analysis following exposure to LPS (100 ng/mL) for 0 or 4 h (Reprinted with permission from the publisher from Kilroy et al. [15])

Table 10.1 Current possible biomolecules used in adipose tissue engineering

Types of biomolecules	Properties
Fibroblast growth factor-2 (FGF-2)	Promotes chondrogenic and inhibits osteogenic differentiation of ADSCs [1]
Platelet-derived growth factor (PDGF)-AB	Proliferation potential on human adipose-derived stem cells and human dermal fibroblasts [13]
Transforming growth factor (TGF)-beta1	Proliferation potential on human adipose-derived stem cells and human dermal fibroblasts [13]
Vascular endothelial growth factor (VEGF)	Improves implant biocompatibility [28] Promotes capillary formation in adipose stem cell containing tubular scaffolds [18]
Granulocyte/macrophage colony-stimulating factor	Angiogenesis-related cytokine secreted by ADSCs [12]
Stromal-derived factor-1alpha	Angiogenesis-related cytokine secreted by ADSCs [12]
Hepatocyte growth factor	Angiogenesis-related cytokine secreted by ADSCs [12]

recovery of normal blood cell production and subsequently restore immune function.

The possible biomolecules used in adipose tissue engineering are shown in Table 10.1.

10.3 Conclusions

The evolving field of producing organs from the basic life unit, a cell, can potentially provide a unique solution to the aforementioned problems. The ability of ASC to secrete several biologic factors plus evidence at a basic science level lends way to ASC playing a major role in tissue engineering and organ regeneration.

References

1. Altman AM, Yan Y, Matthias N, Bai X, Rios C, Mathur AB, Song YH, Alt EU (2009) IFATS series: human adipose-derived stem cells seeded on a silk fibroin-chitosan scaffold enhance wound repair in a murine soft tissue injury model. Stem Cells 27(1):250–258, Sept 25, 2008 [Epub ahead of print]
2. Beltrami AP, Urbanek K, Kajstura J et al (2001) Evidence that human cardiac myocytes divide after myocardial infarction. N Engl J Med 344:1750–1757
3. Bjorntorp P, Ostman J (1971) Human adipose tissue. Dynamics and regulation. Adv Metab Disord 5:277
4. Bunnell BA, Flaat M, Gagliardi C et al (2008) Adipose-derived stem cells: isolation, expansion and differentiation. Methods 45:115–120
5. Deslex S, Negrel R, Vannier C et al (1987) Differentiation of human adipocyte precursors in a chemically defined serum-free medium. Int J Obes 11:19–27
6. Freedman SB, Isner JM (2002) Therapeutic angiogenesis for coronary artery disease. Ann Intern Med 136:54–71
7. Frayn KN, Karpe F, Fielding BA (2003) Integrative physiology of human adipose tissue. Int J Obes 27:875–888
8. Gimble JM, Katz AJ, Bunnell BA (2007) Adipose-derived stem cells for regenerative medicine. Circ Res 100:1249–1260
9. Guilak F, Lott KE, Awad HA, Cao Q et al (2006) Clonal analysis of the differentiation potential of human adipose-derived adult stem cells. J Cell Physiol 206:229–237
10. Gussoni E, Soneoka Y, Strickland CD et al (1999) Dystrophin expression in the mdx mouse restored by stem cell transplantation. Nature 401:390–394
11. Hauner H, Entenmann G, Wabitsch M et al (1989) Promoting effect of glucocorticoids on the differentiation of human adipocyte precursor cells cultured in a chemically defined medium. J Clin Invest 84:663–1670
12. Kakudo N, Shimotsuma A, Kusumoto K (2007) Fibroblast growth factor-2 stimulates adipogenic differentiation of human adipose-derived stem cells. Biochem Biophys Res Commun 359(2):239–244, Epub May 21, 2007
13. Kakudo N, Minakata T, Mitsui T, Kushida S, Notodihardjo FZ, Kusumoto K (2008) Proliferation-promoting effect of platelet-rich plasma on human adipose-derived stem cells and human dermal fibroblasts. Plast Reconstr Surg 122(5): 1352–1360
14. Katz AJ, Llull R, Hedrick MH et al (1999) Emerging approaches to the tissue engineering of fat. Clin Plast Surg 26:587–603
15. Kilroy GE, Foster SJ, Wu X et al (2007) Cytokine profile of human adipose-derived stem cells: expression of angiogenic, hematopoietic, and pro-inflammatory factors. J Cell Physiol 212:702–709
16. Kinnaird T, Stabile E, Burnett MS et al (2004) Marrow-derived stromal cells express genes encoding a broad spectrum of arteriogenic cytokines and promote in vitro and in vivo arteriogenesis through paracrine mechanisms. Circ Res 94(5):678–685
17. Kotton DN, Fine A (2003) Derivation of lung epithelium from bone marrow cells. Cytotherapy 5:169–173
18. Liu ZJ, Zhuge Y, Velazquez OC (2009) Trafficking and differentiation of mesenchymal stem cells. J Cell Biochem 106(6):985–991, Feb 19, 2009 [Epub ahead of print]
19. Miranville A, Heeschen C, Sengenes C et al (2004) Improvement of postnatal neovascularization by human adipose tissue-derived stem cells. Circulation 110: 349–355

20. Petersen BE, Bowen WC, Patrene KD et al (1999) Bone marrow as a potential source of hepatic oval cells. Science 284:1168–1170
21. Pittenger MF, Mackay AM, Beck SC et al (1999) Multilineage potential of mesenchymal stem cells. Science 284: 143–147
22. Prockop DJ (1997) Marrow stromal cells as stem cells for nonhematopoietic tissues. Science 276:71–74
23. Rahimi N, Saulnier R, Nakamura T et al (1994) Role of hepatocyte growth factor in breast cancer: a novel mitogenic factor secreted by adipocytes. DNA Cell Biol 13: 1189–1197
24. Rehman J, Considine RV, Bovenkerk JE et al (2003) Obesity is associated with increased levels of circulating hepatocyte growth factor. J Am Coll Cardiol 41:1408–1413
25. Rehman J, Traktuev D, Li J et al (2004) Secretion of angiogenic and antiapoptotic factors by human adipose stromal cells. Circulation 109:1292–1298
26. Saiki A, Watanabe F, Murano T et al (2006) Hepatocyte growth factor secreted by cultured adipocytes promotes tube formation of vascular endothelial cells in vitro. Int J Obes (Lond) 30:1676–1684
27. Strauer BE, Kornowski R (2003) Stem cell therapy in perspective. Circulation 107:929–934
28. Suga H, Eto H, Shigeura T, Inoue K, Aoi N, Kato H, Nishimura S, Manabe I, Gonda K, Yoshimura K (2009) IFATS series: FGF-2-induced HGF secretion by adipose-derived stromal cells inhibits post-injury fibrogenesis through A JNK-dependent mechanism. Stem Cells 27:238–249, Sept 4, 2008 [Epub ahead of print]
29. Tran TT, Kahn CR (2010) Transplantation of adipose tissue and stem cells: role in metabolism and disease. Nat Rev Endocrinol 6:195–213
30. Wei G, Schubiger G, Harder F et al (2000) Stem cell plasticity in mammals and transdetermination in Drosophila: common themes? Stem Cells 18:409–414
31. Woodbury D, Reynolds K, Black IB (2002) Adult bone marrow stromal stem cells express germline, ecto-dermal, endodermal, and mesodermal genes prior to neurogenesis. J Neurosci Res 96:908–917
32. Zuk PA, Zhu M, Ashjian P et al (2002) Human adipose tissue is a source of multipotent stem cells. Mol Biol Cell 13:4279–4295

Gene Therapy Used for Adipose Stem Cell Engineering

11

Ulrich R. Goessler

Contents

11.1 **Introduction** 111

11.2 **Tissue Engineering: Definition and Limitations** 112

11.3 **Concepts and Strategies of Gene Therapy** 112

11.4 **Stem Cells and Regenerative Medicine** 114

11.5 **Adipose Stem Cells as a Valuable Target for Research** 114

11.6 **Gene Therapy and Stem Cells** 115

11.7 **Practical Implications and Clinical Strategies in the Usage of Adipose Stem Cells** 117

11.8 **Conclusion** 118

References 118

U.R. Goessler
Department of Otolaryngology, Head and Neck Surgery, Universitäts-HNO-Klinik Mannheim, Theodor-Kutzer-Ufer, DE-68135 Mannheim, Germany
e-mail: ugoessler@web.de

11.1 Introduction

Each year, millions of patients suffer from organ failure or tissue loss. The causes include injury, disease or congenital malformation and the resulting morbidity is one of the most frequent, devastating, and costly problems in health care. Tissue engineering and regenerative medicine are evolving interdisciplinary fields in medicine that attribute the principles of biology and engineering to the development of viable substitutes that restore, maintain, or even improve the function of human tissues and organs. Tissue engineering has grown from an in vitro basic science discipline into a clinically oriented specialty in medicine and has provided researchers with critical new insights.

However, many questions need to be addressed and solved. The influence of the local milieu cannot be underestimated; different cell types often require unique culture environments, making it difficult to design a multilayer tissue-engineered construct. When embryonic or pluripotent stem cell lines are used, their differentiation mechanisms must be under guidance and constant surveillance in order to accomplish permanent differentiation. There is much to learn regarding the use of growth factors. The appropriate dosing or the complex cascades and coordinated sequences of their elaborations are not completely understood.

Genetic engineering and the use of adult stem cells may hold the key to future development of tissue-engineered constructs. The identification or perhaps deletion of specific genetic sequences might be able to identify and modify genes critical to tissue development.

Y.-G. Illouz and A. Sterodimas (eds.), *Adipose Stem Cells and Regenerative Medicine*,
DOI: 10.1007/978-3-642-20012-0_11, © Springer-Verlag Berlin Heidelberg 2011

11.2 Tissue Engineering: Definition and Limitations

Tissue engineering combines the usage of cells coupled with biological or artificial matrices to influence and guide the cells during tissue repair or regeneration [15]. This guidance may occur by specific bioactive molecules, ex vivo gene transfer, or other physical factors to form neotissues in vitro for future reimplantation in vivo. The fundamental premise of tissue engineering is the regeneration of tissues followed by restoration of function of organs through implantation of cells or tissues grown outside the body [36]. Tissue engineering has evolved from the use of biomaterials to repair or replace diseased or damaged tissue toward using controlled three-dimensional scaffolds in which the cells can be seeded before implantation. This living tissue construct should be functionally, structurally, and mechanically equal to the tissue it has been designed to replace.

11.3 Concepts and Strategies of Gene Therapy

The fundamental principal of gene therapy is the transfer of genetic material into individuals for therapeutic purposes by altering cellular function or structure at the molecular level. The genetic alteration ultimately leads to the production of a therapeutic protein that is secreted into the surrounding tissue milieu, is expressed on the cell surface or acts as a signaling molecule to influence cell or tissue behavior. As a result, the techniques of gene therapy can be used therapeutically to produce proteins to treat and potentially cure acute and chronic conditions [35]. There are two general ways that gene therapy can be performed: The direct method involves transferring the genetic material into the target somatic cells in vivo [10]. The indirect technique comprises removal of cells from the patient followed by genetic modification of the cells ex vivo and return of the cells to the patient. Of the two approaches, the in vivo method is technically simpler to perform in a clinical setting giving it greater potential utility. The ex vivo techniques may be more complex, but are relatively safer. To choose a suited method, one has to take into account the disease to be treated, the gene to be delivered to treat the disease, and the vector used to deliver the gene [30].

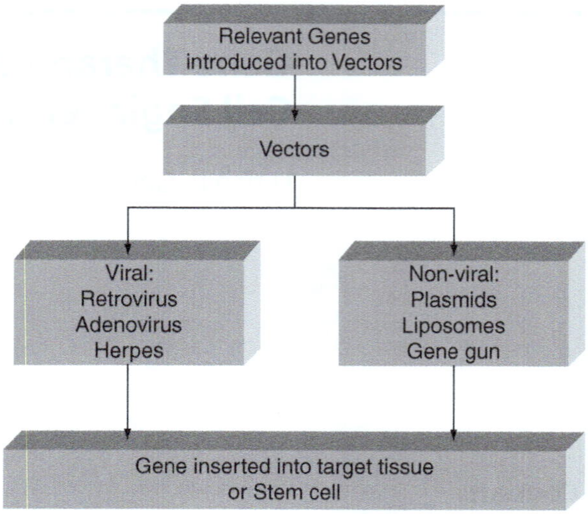

Fig. 11.1 Principles of gene therapy and different vectors

The deliverance of the genetic information to the nucleus leading to production of proteins is called transfection and can occur via two techniques (Fig. 11.1): Viruses are ideal vehicles to perform the task of transfection, but this method yields higher technical demands and increased risk of virus-associated toxicity [35]. These downsides are usually outweighed by the viral ability to efficiently infect cells and in the process transfer DNA to the host without invoking an immune response [25].

Retroviruses are the most widely used tools for gene therapy applications. This is due to their inability to infect nondividing cells [11, 34]. Adenovirus, herpes simplex virus, and adeno-associated virus are candidates for direct in vivo or ex vivo delivery [30]. Retroviruses are RNA viruses that carry a gene for a reverse transcriptase that is able to transcribe the viral genetic material into a double-stranded DNA intermediate which is then incorporated into the host DNA allowing the host cell machinery to produce all the necessary viral components. As the viral genome has been integrated into the host DNA, any modification made will be passed on to all daughter cells that are derived from the transfected cell [18, 30].

Currently, the most commonly used retrovirus is derived from the murine leukemia virus. The majority of clinical trials have utilized vectors based on the murine leukemia virus [28, 29]. Murine leukemia virus offers a number of characteristics that make it attractive as a gene therapy vector: It can be considered a fairly safe candidate, since Murine leukemia virus is nonpathogenic in humans. Additionally, because it has little homology with human retroviruses, the risk of recombination between the vector and any resident human viruses is low [11].

In contrast to retroviruses, adenovirus does not integrate its genome into the host genome. Instead, the adenoviral genome persists within the nucleus as an episomal element after infection of the host cell. The advantages common to all adenoviral vectors include the ease of purification and concentration and the high efficiency rate of host cell infection or various cell types, dividing or nondividing [30]. These advantages make adenoviral vectors a good candidate for direct in vivo gene transfer. The advantages of herpes simplex virus are its large size, wide spectrum of action, and continuous expression of genes [30]. Unfortunately, herpes simplex virus also has its limitations, which include low infection efficiency, wild-type breakthrough, and a large genome size that makes it more difficult to manipulate than other viral vectors [23, 24].

Nonviral vectors can be categorized into three groups and comprise variations of incorporation of genetic material into the cell. These injections can be injections of naked DNA (usually plasmids), liposomes, or particle-mediated gene transfer ("the gene gun"). To ameliorate DNA-uptake in the target tissue, the genetic material can be placed into liposomes or can be coated to micro-projectiles (e.g., gold, tungsten). With the gene-gun, these particles are then accelerated by either helium pressure or a high-voltage electrical discharge thus carrying enough energy to penetrate the cell membrane [30]. Nonviral vectors are much cheaper and easier to produce in large amounts. These vectors have a limited immunogenicity, which allows for potential redosing and they are considered the safer option, as there is no possibility of recombination that would result in a competent virus that could potentially cause disease [30]. However, the gene transfer rate is significantly lower compared to viral vectors [35].

A novel strategy of nonviral gene transfer is to load cDNA onto a porous biomaterial scaffold and pack it directly into a wound with subsequent transfer of the gene into endogenous cells migrating into the site [2, 3, 38]. This technique is called gene-activated matrix (GAM) and is an extension of research producing biodegradable polymers appropriate for tissue engineering [4].

Gene therapy has its limitations in treatment efficiency and safety. While gene therapy may represent a "last resort" treatment option for severe disorders such as cancer or cystic fibrosis, the risk of side effects may be unacceptable in elective reconstructive surgery. In addition, integration of viral vectors into the host genome carries the risk of insertional mutagenesis [10]. Abnormal regulation of cell growth, toxicity from chronic overexpression of the growth factor and cytokines and malignancy are all theoretically possible, but no cases have been reported yet. However, there is no guarantee that integrated DNA sequences will not cause mutations or malignancies years later. Most clinical trials of gene therapy are using the ex vivo approach, so the virus is not directly introduced into patients and cells can be extensively tested before implantation.

Loss of expression of the transferred gene after a few weeks is a common and not fully understood phenomenon. However, temporary and self-limiting gene expression could be useful in the treatment of musculoskeletal injuries, in which only transient high levels of growth factors are needed to promote healing response. Present research is also focusing on the development of specific inducible DNA sequences that are adjacent to the functional genes and are required for expression and regulation of gene transcription.

Cellular therapy has become an important strategy of regenerative medicine. The cellular component of the tissue-engineering paradigm forms the cornerstone to the complex task of repairing damaged or diseased tissue. A prerequisite for clinical strategies is the need for reliable sources of multipotent cells that can be obtained with limited morbidity and can be precisely influenced, shaped, and integrated into tissue. The adult stem cell population may be well suited for this task. Ahead lies the challenge to master the creation of a defined local milieu with regard to the unique culture environments for different cell types.

Genetic engineering may hold the key to identifying and modifying genes critical to cellular development and differentiation. Approved therapeutic techniques have not made the transition from bench to bedside yet. However, a great potential exists for the treatment of musculoskeletal injuries in the future. Currently, only a few effective therapeutic gene therapy techniques for cartilage reconstruction have been tested in human joints [13]. At the experimental level, many studies have been performed successfully to prove the feasibility of gene delivery into different tissues of the musculoskeletal system. Beyond this stage, initial experimental studies demonstrated positive effects of transduced genes (especially BMP-2, IGF-1, TGF-beta) in vitro and in vivo. The main obstacle today seems to be the availability of vectors carrying effective genes, and some concern with the safety of viral vectors.

11.4 Stem Cells and Regenerative Medicine

The term mesenchymal stem cell (MSC) has first been described by Bruder et al. 1994 [5]. Stem cells are the "cellular motors" of development and regeneration and are defined as being clonogenic, self-renewing throughout lifetime, and being able to differentiate into various cell lineages. Their differentiation pathway is unidirectional, passing through the stage of lineage commitment and finally generating terminally differentiated cells. Adult stem cell differentiation is traditionally believed to be restricted to the tissue in which the stem cells reside [22]. The concept of adult stem cells being restricted to their own tissue has been challenged over the past 5 years by numerous reports that adult stem cells can jump lineage barriers and differentiate into cells outside their own tissue, a process called stem cell transdifferentiation [22].

The enormous recent interest in multipotential cells stems from the need to ensure a sufficient amount of tissue-specific cells with limited donor site morbidity. As stem cells are multipotent and self-regenerating, they carry the unique potential to maintain normal cellular homeostasis in functional tissues for the lifetime of the organism. However, the biology and behavior of a stem cell is complex and at best poorly understood. The process of stem cell tissue engineering can be simplified into a two-stage process. First, the stem cell must be differentiated into the lineage-committed cell types of which the organ or tissue is composed. Next, the stem cells must assemble in a three-dimensional architecture dictated by the tissue type to be generated. Adding to the complexity, each process is typically controlled by distinct regulating factors unique to the tissue and the temporal sequence.

The source of stem cells for regenerative medicine can be grouped into two major sources: embryonic or adult tissues. The multilineage potential of adult stem cells, primarily from bone marrow, has been characterized extensively [16].

Stem cells are thought to have self-replicating potential and the ability to give rise to terminally differentiated cells of multiple lineages [19]. The traditional view of adult stem cell differentiation believed that stem cell differentiation progressed in a linear, irreversible fashion and restricted their fate to within a germ line. Recently, newer ideas suggest that stem cells can differentiate in a more graded manner [1, 14, 19]. These analyses may be the beginning of a paradigm shift. It is possible that stem cells, unlike more committed precursors, are capable of switching phenotypes at a "late" stage of development. This plasticity, coupled with the ability of stem cells to cross germ layers, offers exciting possibilities and the definition of a stem cell may have to be changed. Adding to the paradigm shift is the emerging concept that stem cells may be found in multiple organs (e.g., muscle, heart, and liver [27]) and tissues (skin [37], fat [40]). Recent studies suggest that adipose tissue may be another source of pluripotent stem cells with multi-germline potential.

11.5 Adipose Stem Cells as a Valuable Target for Research

Although bone marrow has been the main source for the isolation of multipotent stem cells in the previous years and the bone marrow-derived stem cells are well characterized and safe in handling, the harvest of bone marrow is a highly invasive procedure and the number, differentiation potential, and maximal life span of stem cells from bone marrow decline with increasing age. Therefore, alternative sources from which to isolate mesenchymal stem cells are subject to intensive investigation (Fig. 11.2). Adipose tissue is an alternative source that can be obtained by a less invasive method and in larger quantities than bone marrow. It has been demonstrated that adipose tissue contains stem cells similar to bone marrow-derived stem cells, which are termed processed lipoaspirate cells [40, 41]. These cells can be isolated from cosmetic liposuctions in large numbers and grown easily under standard tissue culture conditions [40, 41]. Further analysis has been performed to confirm the stem cell character of the lipoasirate cells. Usually, the results found within the lipoaspirate population are compared to expression patterns in bone marrow-derived stem cells as they are longstanding and well-characterized cell populations. The analysis included morphologic characteristics as

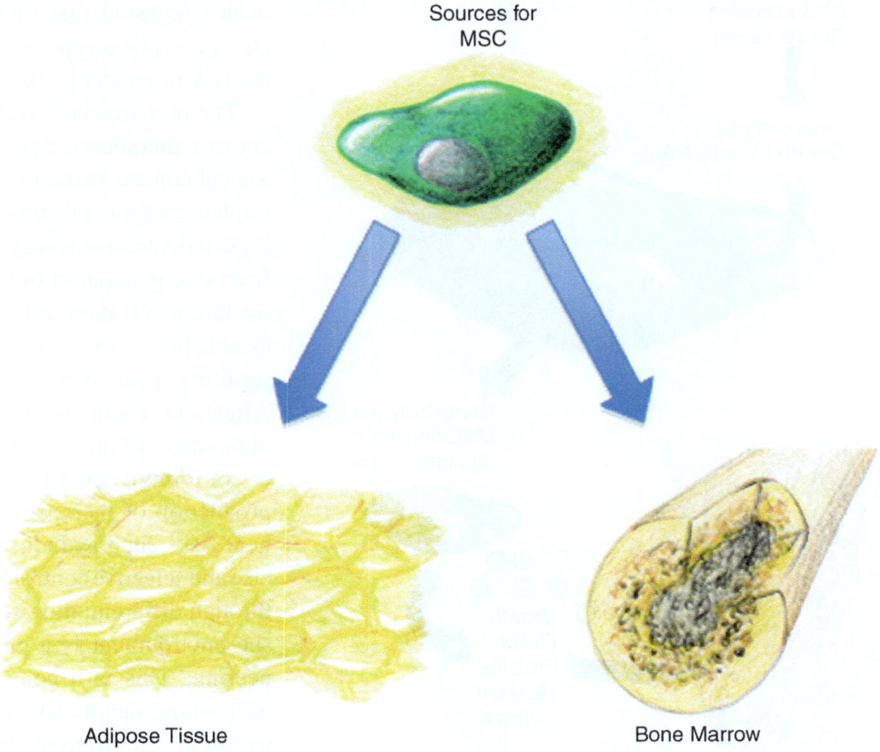

Fig. 11.2 Sources for mesenchymal stem cells: bone marrow and alternatively adipose tissue

well as functional parameters. The morphologic analysis included cell-specific proteins and CD markers [6] like other MSCs, adipose stem cells expressed CD29, CD44, CD71, CD90, CD105/SH2 and SH3. However, they were not found to express CD31, CD34, and CD45 [32]. Lipoaspirate cells were found to express CD13, whereas no expression of CD14, 16, 56, 62e, or 104 could be revealed [32]. These similarities in expression patterns differed in two CD markers as adipose stem cells expressed CD49d and did not express CD106. MSCs derived from bone marrow showed the inverse expression patterns.

The similarities between the two populations lend support to the theory that stem cells can be found within adipose tissue. However, there is a possibility that lipoaspirate cells might be a clonal variant of circulating MSCs that just reside in the specific tissue.

Functional and proliferative analyses included the multilineage differentiation capacity of adipose stem cells – this capacity has clearly been proven in previous studies: osteogenic, myogenic, chondrogenic, and neurogenic differentiation have been shown [40, 41].

11.6 Gene Therapy and Stem Cells

Recent research tried to combine the advantages of tissue engineering and gene therapy resulting in stem cell-based tissue engineering in conjunction with gene therapy to enhance tissue regeneration by providing the stem cells with an environment of optimal protein expression (Fig. 11.3) [17].

Adult mesenchymal stem cells are a promising target for further research as they mediate the reproduction and transmission of genetic information to subsequent cellular generations. At present, gene therapy of embryonal stem cells is not a practicable option due to its complex technical nature and ethical considerations. Somatic cell gene therapy on the other hand offers the advantage that the changed genetic information cannot be passed on to the next generation and has become a major focus of stem cell research. A central issue of practicability in stem cell engineering is the specific methodology used to introduce therapeutic genes into the progenitor cells. These therapeutic genes are the central factor as they

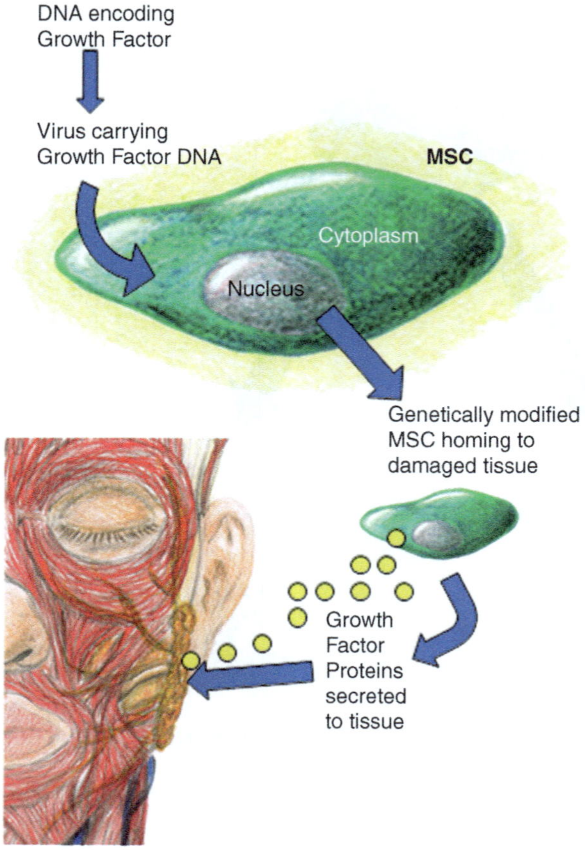

Fig. 11.3 Possible approach to genetically engineered stem cells

carry the relevant information. The algorithm of gene therapy-modulated stem cell engineering encompasses identification of candidate genes, isolation of these genes, introduction into viral genomes or non-viral vectors, transfection and introduction into stem cells and again controlled expression in the surrounding matrix without harmful side effects. As mentioned above, retroviral vectors have been used in many settings for the transfer of genes into stem cells.

The potential of genetically modified stem cells lies in the correction of genetically caused diseases such as muscle dystrophy, hemophilia, or cystic fibrosis. On top of that, stem cell-based gene therapy approaches are also being investigated to treat acquired diseases that do not have a genetic basis. These include cancer, diabetes, AIDS, Parkinson's, or Alzheimer's diseases. However, there are a number of risk factors to be taken into account. For example, retroviruses tend to insert into active genes and it has been suggested that their use in stem cells and the change of phenotype in the host cell may also increase the risk of cancer [39].

The prerequisite for efficient gene therapy is delivery of a therapeutic gene product into the correct biological context with minimal harmful side effects. The implementation of stem cells into this concept is a logical deduction as they can pass on the genetic codes from one generation to the next. However, the implementation of stem cells into gene therapy demands establishing novel strategies in order to make the resulting gene expression predictable and reliable. After attainment of this goal, the next step will be obtainment of methods for the efficient and safe delivery of foreign genes into stem cells as an uncontrolled production of therapeutic gene products carries eminent risks such as induction of malignancy. The precise and selective control of therapeutic gene expression through differentiating stem cells within a defined tissue environment is a prerequisite in stem cell engineering. This is a complex approach with multiple influencing factors which could in future help to control stem cell differentiation into specific lineages, the maintenance of their undifferentiated state for later transplantation, moderation of proliferation, or the regulation of expression of therapeutic genes. Target genes would be regulators of cell cycle, suicide genes, cytokines, or growth factors.

Various methodologies have been employed for engineering gene delivery and their expression in adult stem cells. Stem cells offer the potential to provide cellular therapies, the strategy for gene therapy is expression of a specific repertoire of genes, thereby modifying its own identity to maintain, replace, or rescue a particular tissue.

Stem cells function as the cellular generators that drive the renewal of adult mammalian tissues. They continuously proliferate throughout life to produce new progenitors that undergo a program of differentiation and maturation to replace older tissue cells. The same cell turnover program is thought to provide a source of cells for the repair and regeneration of adult tissues.

Stem cells derived from bone marrow may serve as a potential vehicle for cell and gene specific therapy against disease. In addition to bone marrow, other potential sources of stem cells for therapy include the peripheral blood, CNS, liver, pancreas, muscle, skin lung, intestine, heart, and fat [21]. Ideal characteristics of a candidate stem cell source should be easy accessibility, ease of harvest with minimal risk to the patient, and sufficient quantity. With these issues in mind, fat tissue represents a promising tissue source [31]. Adipose-derived stem cells share growth kinetics with bone marrow-derived stem cells; additionally, characteristics regarding cell senescence, gene transduction efficiency, CD surface marker expression, and gene transcription profiles parallel those of bone marrow-derived stem cells [12].

11.7 Practical Implications and Clinical Strategies in the Usage of Adipose Stem Cells

Adipose stem cells offer a variable target for further experiments in combination with gene therapy for several reasons: They are readily harvested in a minimal invasive and technically not very demanding fashion. They are easily harvested in sufficient numbers. They offer safe and reliable differentiation in different cell lines. The implementation of gene therapy methods offers exciting new possibilities in the treatment of disease. Stem cells possess the ability to differentiate into diverse tissues and, due to their inherent ability to home to damaged tissue, they have the potential to deliver therapeutic genes to specific tissue environments. By using tissue-specific promoters and markers, gene expression can be selectively tailored to the needs of damaged areas. Diseases where the application of genetically engineered stem cells shows promise include those where a protein or an entire enzyme is missing or nonfunctional. Another setting of disease includes diminished function of a protein in a specific tissue matrix. Diseases where stem cell-based gene therapy would be most promising include cancer, neurodegenerative disorders (Parkinson's or Alzheimer's disease), ischemic heart disease, and muscle dystrophies.

Innovative new treatment strategies might change the approach to cancer treatment. A recently developed approach makes use of the ability of stem cells to be recruited by tumor vessels and then undergo differentiation into endothelial-like cells. Averagely one-third of new vascular endothelial cells in tumors could be derived from bone marrow progenitors [20] and the use of genetically modified progenitor cells recruited from the peripheral circulation may represent a potential vehicle for selective gene therapy of tumors [33].

The treatment of Parkinson's disease has been in the focus of research. Treatment approaches have used either cell transplantation or gene therapy approaches [7]. Limited studies to date have attempted to combine the two approaches.

Gene therapy could also offer amelioration of ischemic heart disease. Stem cell engineering has been used to transfect stem cells with human angiopoietin-1 and VEGF. The application of these stem cells led to decreased infarct size and significantly increased capillary density, as well as improved long-term cardiac performance [9].

Muscle dystrophies are chronic diseases characterized by progressive muscle wasting. To date no adequate treatment modality exists for these patients. MSC and embryonic stem cells have not shown much promise in the treatment of dystrophies.

Fanconi anemia is a rare hereditary disease that has been the subject of intensive research directed toward its treatment by gene therapy. Disease presents with bone marrow failure and developmental anomalies leading to a high incidence of myelodysplasia, acute nonlymphocytic leukemia, and solid tumors.

The genetic basis for Fanconi anemia lies in selective mutations in any one of the characterized Fanconi anemia genes and 12 genetic subtypes have been described. The Fanconi anemia proteins are thought to be functionally linked to the repair of DNA interstrand cross-links, which block the progression of DNA-replication forks [26]. The defective proteins may be targeted via stem cell-based gene therapy.

Successful gene therapy strategies implemented into practical use was demonstrated with the example of osteogenesis imperfecta. The molecular basis of the disease is mutations in the collagen-I-encoding genes. MSCs from the bones of osteogenesis imperfecta patients were analyzed and point mutations in the COL1A1 gene could be identified [8]. MSCs were successfully infected with an adeno-associated virus to target and deactivate the mutated COLIA1 gene. The corrected MSCs were then transplanted into immunodeficient mice and damaged cells demonstrated improved stability and collagen processing [8].

11.8 Conclusion

Genetically engineered stem cells represent a promising therapeutic approach. Even with the latest methods it is still difficult to transfect stem cells at high efficiency and still retains a multipotent phenotype. The overall safety of the various gene delivery systems is also an important consideration. Clearly, many hurdles remain to be addressed before these approaches can be widely applied as a common therapeutic strategy.

References

1. Bennett JH et al (1991) Adipocytic cells cultured from marrow have osteogenic potential. J Cell Sci 99(pt 1):131–139
2. Bonadio J (2000) Tissue engineering via local gene delivery: update and future prospects for enhancing the technology. Adv Drug Deliv Rev 44(2-3):185–194
3. Bonadio J (2002) Genetic approaches to tissue repair. Ann NY Acad Sci 961:58–60
4. Bonadio J et al (1999) Localized, direct plasmid gene delivery in vivo: prolonged therapy results in reproducible tissue regeneration. Nat Med 5(7):753–759
5. Bruder SP, Fink DJ, Caplan AI (1994) Mesenchymal stem cells in bone development, bone repair, and skeletal regeneration therapy. J Cell Biochem 56(3):283–294
6. Bruder SP, Jaiswal N, Haynesworth SE (1997) Growth kinetics, self-renewal, and the osteogenic potential of purified human mesenchymal stem cells during extensive subcultivation and following cryopreservation. J Cell Biochem 64(2):278–294
7. Burton EA, Glorioso JC, Fink DJ (2003) Gene therapy progress and prospects: Parkinson's disease. Gene Ther 10(20):1721–1727
8. Chamberlain JR et al (2004) Gene targeting in stem cells from individuals with osteogenesis imperfecta. Science 303(5661):1198–1201
9. Chen HK et al (2005) Combined cord blood stem cells and gene therapy enhances angiogenesis and improves cardiac performance in mouse after acute myocardial infarction. Eur J Clin Invest 35(11):677–686
10. Crystal RG (1995) Transfer of genes to humans: early lessons and obstacles to success. Science 270(5235):404–410
11. Danos O, Heard JM (1992) Recombinant retroviruses as tools for gene transfer to somatic cells. Bone Marrow Transplant 9(suppl 1):131–138
12. De Ugarte DA et al (2003) Comparison of multi-lineage cells from human adipose tissue and bone marrow. Cells Tissues Organs 174(3):101–109
13. Evans CH, Robbins PD (1995) Possible orthopaedic applications of gene therapy. J Bone Joint Surg Am 77(7):1103–1114
14. Galotto M et al (1994) Hypertrophic chondrocytes undergo further differentiation to osteoblast-like cells and participate in the initial bone formation in developing chick embryo. J Bone Miner Res 9(8):1239–1249
15. Goessler UR, Hormann K, Riedel F (2004) Tissue engineering with chondrocytes and function of the extracellular matrix (review). Int J Mol Med 13(4):505–513
16. Goessler UR, Hormann K, Riedel F (2005) Tissue engineering with adult stem cells in reconstructive surgery (review). Int J Mol Med 15(6):899–905
17. Goessler UR et al (2006) Perspectives of gene therapy in stem cell tissue engineering. Cells Tissues Organs 183(4):169–179
18. Goff SP, Lobel LI (1987) Mutants of murine leukemia viruses and retroviral replication. Biochim Biophys Acta 907(2): 93–123
19. Hall PA, Watt FM (1989) Stem cells: the generation and maintenance of cellular diversity. Development 106(4):619–633
20. Hammerling GJ, Ganss R (2006) Vascular integration of endothelial progenitors during multistep tumor progression. Cell Cycle 5(5):509–511
21. Korbling M, Estrov Z (2003) Adult stem cells for tissue repair–a new therapeutic concept? N Engl J Med 349(6): 570–582
22. Korbling M, Estrov Z, Champlin R (2003) Adult stem cells and tissue repair. Bone Marrow Transplant 32(suppl 1):S23–S24
23. Krisky DM et al (1998) Development of herpes simplex virus replication-defective multigene vectors for combination gene therapy applications. Gene Ther 5(11): 1517–1530
24. Krisky DM et al (1998) Deletion of multiple immediate-early genes from herpes simplex virus reduces cytotoxicity and permits long-term gene expression in neurons. Gene Ther 5(12):1593–1603
25. Krougliak V, Graham FL (1995) Development of cell lines capable of complementing E1, E4, and protein IX defective adenovirus type 5 mutants. Hum Gene Ther 6(12):1575–1586
26. Levitus M, Joenje H, de Winter JP (2006) The Fanconi anemia pathway of genomic maintenance. Cell Oncol 28(1–2):3–29
27. Lucas WT, Youngner JS (1992) The use of hybrid-selected template increases the specificity of the polymerase chain reaction. PCR Meth Appl 2(1):41–44
28. Marshall E (2000) Improving gene therapy's tool kit. Science 288(5468):953
29. Marshall E (2002) Clinical research. Gene therapy a suspect in leukemia-like disease. Science 298(5591):34–35
30. Oligino TJ et al (2000) Vector systems for gene transfer to joints. Clin Orthop Relat Res 379(suppl):S17–S30
31. Parker AM, Katz AJ (2006) Adipose-derived stem cells for the regeneration of damaged tissues. Expert Opin Biol Ther 6(6):567–578
32. Pittenger MF et al (1999) Multilineage potential of adult human mesenchymal stem cells. Science 284(5411):143–147
33. Reyes M et al (2002) Origin of endothelial progenitors in human postnatal bone marrow. J Clin Invest 109(3): 337–346
34. Robbins PD, Ghivizzani SC (1998) Viral vectors for gene therapy. Pharmacol Ther 80(1):35–47
35. Salyapongse AN, Billiar TR, Edington H (1999) Gene therapy and tissue engineering. Clin Plast Surg 26(4):663–676, x

36. Stock UA, Vacanti JP (2001) Tissue engineering: current state and prospects. Annu Rev Med 52:443–451
37. Toma JG et al (2001) Isolation of multipotent adult stem cells from the dermis of mammalian skin. Nat Cell Biol 3(9):778–784
38. Warren SM et al (2002) New directions in bioabsorbable technology. J Neurosurg 97(4 suppl):481–489
39. Young LS et al (2006) Viral gene therapy strategies: from basic science to clinical application. J Pathol 208(2): 299–318
40. Zuk PA et al (2001) Multilineage cells from human adipose tissue: implications for cell-based therapies. Tissue Eng 7(2):211–228
41. Zuk PA et al (2002) Human adipose tissue is a source of multipotent stem cells. Mol Biol Cell 13(12):4279–4295

Stem Cells from Human Adipose Tissue: A New Tool for Pharmacological Studies and for Clinical Applications

12

Claude A. Dechesne and Christian Dani

Contents

12.1 **Introduction** 121

12.2 **hMADS Cells Are Powerful Tools for Pharmacological Studies** 123

12.3 **hMADS Cells, a New Tool for Clinical Applications** 123

12.3.1 Muscular Dystrophies and Mesenchymal Stem Cells 124

12.3.2 In Vivo Myogenic Potential of Adipose-Derived Stem Cells 124

12.3.3 In Vitro Myogenic Potential of Adipose-Derived Stem Cell 125

12.3.4 Autonomous Myogenic Potential 125

12.3.5 Myogenic Potential of Adipose-Derived Stem Cells Cultured with Myoblasts.......... 126

12.3.6 Genetic Modification of Adipose-Derived Stem Cells 126

12.3.7 In Vitro Myogenic Differentiation Potential of MyoD-hMADS Cells.................... 126

12.3.8 In Vivo Contribution of MyoD-hMADS Cells to Muscle Repair 128

12.4 **Conclusions**.......................... 130

References.................................... 131

C.A. Dechesne • C. Dani (✉)
Institute of Developmental Biology and Cancer, University of Nice Sophia-Antipolis, CNRS, UMR6543 Nice, France
e-mail: dechesne@unice.fr; dani@unice.fr

12.1 Introduction

Up to now, preadipocyte clonal lines from rodents have been mainly used to gain insight into cellular and molecular mechanisms of adipogenesis [22]. Much less is known about the molecular regulation of human adipogenesis, partly due to the absence of appropriate human cellular models. Primary cultures of preadipocytes derived from stromal-vascular fraction (SVF) of adipose tissue, although being able to differentiate into adipocytes in vitro, undergo a dramatic decrease in their ability to differentiate before growth arrest and replicative senescence with serial subculturing, making it difficult to investigate effects of compounds in a fully reproducible manner. This severe limitation has been partly circumvented with cells that are immortalised either genetically or spontaneously [4, 34, 35, 39]. Upon differentiation, cells from these human clonal cell lines expressed only some of the characteristic markers of human adipocytes, and the lipolytic responses specific of human adipocytes and secretion of adipocytokines have not been reported. Recently, we have isolated multipotent stem cells from the SVF of infant adipose tissues, cells called human multipotent adipose-derived stem (hMADS) cells [21]. Diverse terms, from adipose mesenchymal stem cells to preadipocytes, have been used to name stem cell populations isolated from adipose tissue. The International Fat Applied Technology Society reached a consensus to adopt the term "adipose-derived stem cells (ASCs)." However, the name hMADS cells has been maintained as these cells display additional and specific characteristics. hMADS cells are isolated from infant adipose tissues and hMADS cell lines can be established. Cells

Y.-G. Illouz and A. Sterodimas (eds.), *Adipose Stem Cells and Regenerative Medicine*,
DOI: 10.1007/978-3-642-20012-0_12, © Springer-Verlag Berlin Heidelberg 2011

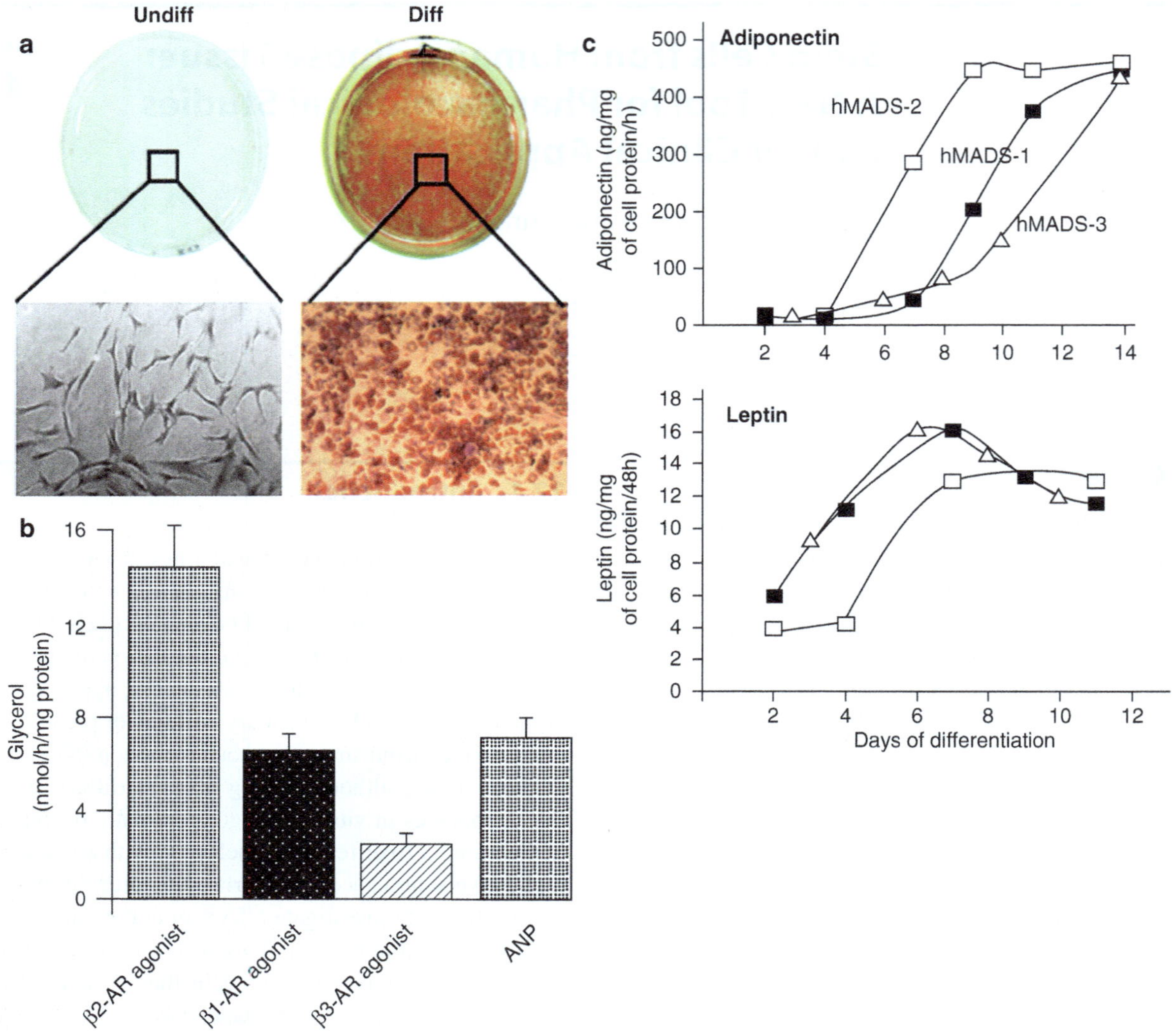

Fig. 12.1 Phenotypic characteristics of hMADS cells differentiated into adipocytes. (**a**) Undifferentiated and differentiated (day 14) hMADS cells are stained with Oil Red O for triglycerides. (**b**) Lipolytic response of hMADS cells differentiated into adipocytes. hMADS cells exhibit after differentiation the panoply of lipolytic responses, which are characteristic of human adipocytes. After stimulation at an optimal concentration of 1 mM with an agonist known to be specific for b1-adrenoreceptor (dobutamine), or b2-AR (terbutaline), or b3-AR (CL31643) [10, 25], the rank order of maximal potency is terbutaline > dobutamine > CL316243. (**c**) Secretion of adipocytokines during differentiation of hMADS cells. Secretion of leptin and of adiponectin is shown during differentiation of three different hMADS cell lines

exhibit the characteristics of mesenchymal stem cells, i.e., the capacity for self-renewal, as cells can be expanded in vitro for more than 160 population doublings (i.e., around 30 passages) while maintaining a normal diploid karyotype and the potential to undergo differentiation into adipocytes, osteoblasts, and chondrocytes at the single cell level [21, 37]. When transplanted into animal models for muscular dystrophies, hMADS cells participate to muscle regeneration emphasising their therapeutic potential, as discussed latter in this chapter. hMADS cells can be frozen and thawed as a usual clonal cell line, making them a very convenient tool for in vitro studies. We and others have now demonstrated that hMADS cells are appropriate to study human white adipocyte differentiation [3, 13, 19, 20, 30]. In vitro, hMADS cells enter the adipose lineage at a high rate, differentiation yield is estimated at more than 80% (Fig. 12.1a), and differentiate into cells that display a unique combination of properties similar, if not identical, to those of native human adipocytes. Within 10 days after induction of adipocyte differentiation under serum-free adipogenic condition, cells express the major molecular markers, key transcription factors and nuclear receptors of human white adipocytes. hMADS cells exhibit after differentiation the panoply of lipolytic responses, which are characteristic of human adipocytes (Fig. 12.1b). Interestingly, hMADS cells respond to the atrial natriuretic peptides (ANP), a unique characteristic both in vitro and in vivo of adipocytes from primates [14]. Moreover, compared to other preadipocyte cell lines so far described, an important feature of differentiated hMADS cells is their ability to secrete leptin and adiponectin (Fig. 12.1c) within values reported for isolated human adipocytes [20].

12.2 hMADS Cells Are Powerful Tools for Pharmacological Studies

We would like to emphasise, with the following examples, the importance for working with human cells for pharmacological studies. The development of HIV protease inhibitor (PI) antiretroviral therapy has dramatically improved the lifespan of HIV-infected patients. However, lipodystrophy is a major side effect of this therapy: HIV-infected patients treated with PIs develop dyslipidemia, systemic insulin resistance and it has been shown that PIs interfere directly with adipocyte differentiation. Therefore, it was important to investigate the effects of PIs on the development of adipose cells. However, we demonstrated that the effects of PIs were dependent on the cell models when using murine preadipose cell lines [29], indicating the need of a human cell model for such studies. Latter, using hMADS cells, we demonstrated that HIV drugs enter into undifferentiated and differentiated cells and that some of them, not all, inhibit adipocyte differentiation with various effects on adipocytokine gene expression [31]. Therefore, hMADS cells allowed us to propose that PIs participate in insulin resistance through a direct effect on adipocyte. The second example is related to the Hedgehog pathway and osteogenesis. hMADS cells express critical factors for bone formation such as bone morphogenetic proteins (BMP) and BMP receptors, and they can form functional osteoblast and osteocytes in vitro and in vivo after transplantation into nude mice [6]. The ability of hMADS cells to undergo both adipocyte and osteoblast differentiations allowed us to identify oxytocin as a critical regulator of bone formation and as a potential novel therapeutic molecule against osteoporosis [7]. Altogether, these data indicate that hMADS cell lines represent also a faithful model for investigating human osteogenesis. It has been reported that activation of Hedghog pathway promotes osteogenesis in murine models. However, we have shown that in hMADS cells activation of this pathway inhibits osteogenesis [18]. The molecular reasons of this inter-species difference is not understood, but these data demonstrate the requirement, when it is possible, to use human models for drug screenings. Indeed, two adipose tissues with different functions coexist in humans, i.e. white and brown adipose tissues. White adipose tissue (WAT) is mainly involved in energy storage and mobilisation. WAT is localised in various sites of the body, has an enormous capacity for expansion and excess of fat accumulation is associated with metabolic disorders. WAT not only stores lipids, it is also a secretory organ. Adipocytes secrete cytokines, named adipokines, pro-inflammatory cytokines, and many other factors [1]. Brown adipose tissue (BAT) is specialised in energy expenditure. It is a key thermogenic organ and brown adipocytes burn fat. In humans, brown adipocytes were considered to be present in newborn only but within the last 2 years a series of papers have reported that active deposits of BAT can be active in adult healthy individuals [17, 28, 33]. These observations open new therapeutic avenues to treat obesity. We have reported culture conditions to turn hMADSc-white adipocytes into brown adipocytes. Upon chronic exposure to a specific PPARg agonist, but not to a PPARb/d or PPARa agonists, white adipocytes derived from hMADS cells are able to switch to a functional brown phenotype by expressing UCP1 protein. This switch is accompanied by an increase in oxygen consumption and uncoupling [8], opening the opportunity to screen for drugs stimulating the formation and/or the uncoupling capacity of human brown adipocytes.

12.3 hMADS Cells, a New Tool for Clinical Applications

In October 2010, 32 clinical trials using adipose stem cells are listed in the registry of federally and privately supported trials conducted in the United States and around the world, as seen on the internet site: http://clinicaltrials.gov/. The amount of clinical trials have risen rapidly the last 3 years, and the clinical applications may potentially concern a much broader panel of disorders than those targeted by the current trials. Among these diseases, muscular dystrophies are subjected to intense research investigations, and we will assess here the potential of human stem cells derived from adipose tissue for their use in cell-based therapy of myopathies.

12.3.1 Muscular Dystrophies and Mesenchymal Stem Cells

Skeletal muscle is a tissue that benefits from a strong power of regeneration that allows recovering from rather severe injuries. Most if not all of the regenerated muscle fibres are generated by the satellite cells, which are the muscle progenitor cells. They lay along the fibres, under the basal lamina, in a quiescent status and are recruited when muscles are subjected to injury. Upon different signals generated at the damage site, they proliferate and the progeny constitutes the population of myoblasts that either fuse with pre-existing fibres or fuse together into new multinucleated fibres. In the meantime, the satellite cells replenish their original niche. A similar process occurs in muscle dystrophies to replace the necrotic fibres. This is the case for instance for the Duchenne muscular dystrophy (DMD), a severe and frequent disease, due to the absence of dystrophin because of spontaneous or inherited mutations/deletions in the dystrophin gene. This lethal muscle-wasting disease is the most prevalent hereditary muscle disorder which afflicts approximately 1 in every 3,500 boys. Dystrophin is a subsarcolemmal protein that connects cytoskeletal actin to sarcolemma and extracellular matrix through a large protein complex. After too many necrosis/regeneration cycles, the repair potential of satellite cells is exhausted, muscle fibres are atrophied and more or less replaced by fibrosis and fat. The subsequent loss of function creates severe pathological conditions. No treatment capable of stopping the deleterious evolution is yet available, although the origin of DMD has been described more than 20 years ago. Much effort is devoted to develop various therapeutic strategies including drug-, gene- and cell-based therapies. The use of heterologous or autologous cells aims at repopulating or inducing regeneration of damaged muscles. In this context, identification of any cell population that could efficiently regenerate muscle has special interest to assess new cell-based therapies. Several cellular types are under investigation including myoblasts, mesoangioblasts and muscle or blood $CD133^+$ cells (for a recent review see [9]). Mesenchymal stem cells (MSCs) are also promising candidates because of their differentiation potential and their possibility of extensive multiplication in culture related to their capacity for self-renewal.

Before 1995, MSCs were known as multipotential progenitors but expression of myogenic properties had not been reported. Then, Arnold Caplan's lab described that rat bone marrow-derived MSCs exposed to the DNA-demethylating compound 5-azacytidine can differentiate into multinucleated myotubes, suggesting that these cells are a source of myo-progenitor cells [36]. The same group hypothesised that under suitable conditions MSCs can be induced to express a myogenic phenotype. Such environment might be found in regenerating muscles. They injected mouse bone marrow MSCs into muscles from *mdx* mice. *Mdx* mice are a convenient animal model for DMD, since their muscles are essentially composed of dystrophin-negative myofibres undergoing cycles of degeneration/regeneration, although these mice do not suffer from the severe features of myopathy. They provide an excellent natural model to study muscle regeneration. Significant increase of the number of dystrophin-positive myofibres was found, strongly suggesting that implanted MSCs differentiate into myogenic cells [23]. These results extended the differentiation potential of MSCs to the myogenic lineage and raised new important questions including the physiological relevance of the myogenic capacity of these cells. A new field of investigation was opened on the evaluation of MSC potential for muscle repair.

12.3.2 In Vivo Myogenic Potential of Adipose-Derived Stem Cells

As in other areas of research, MSCs from adipose tissue were also considered for myogenic differentiation, after the pioneer work using bone marrow MSCs. The myogenic potential of adipose-derived stem cells (ASCs) has been investigated since the early 2000s.

Several studies reported an engraftment of ASCs in animal recipient's muscles. Bacou and collaborators were the first to investigate the potential of cells from the adipose SVF, which represents the non-

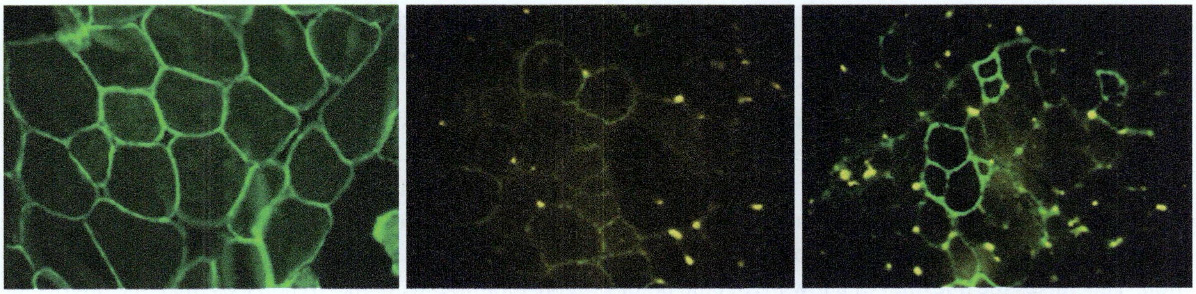

Fig. 12.2 Contribution of hMADS cells to muscle regeneration. Expression of dystrophin, detected with an anti-dystrophin antibody revealed with a fluorescent second antibody (*green*) is shown on cryostat transversal sections of *Tibialis anterior* muscles of wild-type, *mdx* and hMADS cells-injected *mdx* mice. Approximately 500,000 hMADS cells were injected in the muscles of *mdx* mice, which undergo continuous degeneration/regeneration cycles consequently to the genetic lack of dystrophin. Partial restoration of dystrophin was observed 2 weeks after hMADS cells engraftment

differentiated adipocyte cell population. Autologous cells freshly harvested and labelled with LacZ were transferred into rabbit muscles. They were found to contribute to 10% of fibres of regenerating muscles, which had been injured before cell transfer [2]. Two months after cell transfer, muscles were heavier and developed an increased maximal force. Thus, this group established that a myogenic potential is supported by at least a fraction of adipose SVF. Similarly, mouse uncultured ADSCs injected into *mdx* mouse muscles or mouse non-dystrophic muscles, which had been damaged by femoral artery removal, were respectively found in up to 10% and 20% of recipient's muscle fibres [5]. However, it is interesting to note that only a few muscle fibres derived from mouse donor's cells were found in *mdx* mice injected with ASCs after in vitro expansion [38]. Finally, human ASCs were also found incorporated in regenerating mouse muscle fibres. hMADS cells were found to contribute to regenerating muscles from *mdx* and pre-injured immunodeficient mice (Fig. 12.2) [21]. hMADS cell-derived myofibres were detected by labelling mouse muscle sections with human-specific antibodies. More human-specific muscle markers could be amplified by RT-PCR experiments from RNA extracted from mouse muscle injected with hMADS cells. Similar results have been reported with ASCs prepared from liposuction procedures engrafted in muscles of SJL mouse, a murine model for limb girdle muscular dystrophy type 2B [32]. Therefore, ASCs transplanted in a regenerating muscle environment exhibit a significant potential to differentiate into skeletal muscle fibres.

12.3.3 In Vitro Myogenic Potential of Adipose-Derived Stem Cell

To study further the myogenic commitment of ASCs, many experiments have been conducted in cell culture dishes.

12.3.4 Autonomous Myogenic Potential

ASC autonomous myogenic differentiation has been investigated by several groups, and the general point of view is that the myogenic differentiation is very limited in culture, without any muscle-like environment. As reference, the full differentiation of myoblasts is obtained under appropriate culture conditions including the withdrawal of serum and the presence of insulin, after the cells have reached sub-confluency. Then the myoblasts stop proliferating and express myogenic determination factors (MyoD, Myf5) followed by other myogenic regulatory factors (myogenin, MRF4). They start to fuse into elongated and multinucleated myotubes, which will express most of the muscle terminal markers. Myotubes is the most differentiated stage obtained in culture; fully organised myofibres need innervation that exists in vivo. From all the reports devoted at the myogenic differentiation of ASCs, it emerges that, although ASCs can express early and even late muscle markers under myogenic culture conditions, fusion into myotubes is rarely observed. This has been published for human ASCs isolated from lipoaspirates as well as from excised adipose tissues

[16, 40, 41]. hMADS cell myotubes were found at a very low frequency (<0.1% of total plated cells) and only with cells from a few donors, even after multiple experiments to optimise culture conditions. Similar observations were reported in mouse with ADSCs prepared from subcutaneous or inguinal fat pads [5]. Thus, the yield of ASC myogenic differentiation is far less than found for adipogenic or osteogenic differentiation. This raises the question of a low myogenic potential shared by all the ASCs population or a myogenic potential restricted to a minor subset of cells, responsible for the in vivo muscle-regenerating capacities.

12.3.5 Myogenic Potential of Adipose-Derived Stem Cells Cultured with Myoblasts

The very low autonomous myogenic potential of ASCs but their capacity to contribute to regenerating myofibres in vivo points out the importance of muscle environment to achieve complete myogenic differentiation. In vitro this question has been addressed through co-cultures of ASCs with myoblasts or culture media conditioned by myoblasts or myotubes. Human or mouse ASCs traced with GFP were mixed with C2C12 murine myoblasts [15] or mouse primary myoblasts [5] and cultured under myogenic differentiation conditions. GFP-positive myotubes were detected in both experiments and the human GFP-positive myotubes expressed human nestin, known to be significantly expressed in muscle cells. This indicated that ASCs can fuse with muscle cells to form myotubes. In addition, Di Rocco and collaborators found that differentiating myogenic cells are a source of soluble factors involved in ASC myogenic differentiation. The co-culture experiments indicated that ASCs share with muscle cells the capacity to fuse into myotubes and then to express muscle proteins. However, the yield was always very low, at the best in a 1% range of ASCs integrated into myotubes. Moreover, it is difficult to conclude that the co-culture with muscle cells quantitatively increases the level of myogenic differentiation of ASCs alone. The accuracy of the measurements of very low percentages has to be considered and it cannot be ruled out that a fraction of ASCs might undergo non-significant fusion with myoblast cells. A few non-myogenic cells, physically in contact with several muscle cells, may have been enrolled in the massive wave of fusion of differentiating muscle cells.

12.3.6 Genetic Modification of Adipose-Derived Stem Cells

The above data show that the muscle repair potential of ASCs should be improved for clinical application perspectives. Two directions can be followed for this purpose. It can be tried to identify more myogenic sub-populations of adipose SVF cells, for instance by cell cloning, with the hypothesis that the myogenic potential is not evenly shared by all SVF cells. To date this hypothesis has not been confirmed. In addition, difficulties can arise if candidate sub-populations represent a very small percentage of cells, especially if these cells should be used without in vitro expansion as mentioned in some studies. A solution may be found in harvesting very large quantities of adipose tissue, which may be possible only in certain circumstances. Another alternative is the genetic modification of ASCs' total population to confer them a higher myogenic potential. We used the hMADS cells and postulated that overexpression of a key myogenic gene such as MyoD, which is a myogenesis master gene, may significantly increase their endogenous myogenic potential [26]. hMADS cells expressing MyoD were obtained by transduction with a recombinant MyoD lentiviral vector, encoding mouse MyoD under the control of phosphoglycerate kinase (PGK) promoter [12]. These cells undergone spectacular modifications and will be designated hereafter as MyoD-hMADS cells.

12.3.7 In Vitro Myogenic Differentiation Potential of MyoD-hMADS Cells

MyoD-hMADS cells cultured under myogenic conditions formed multinuclear myotubes and expressed early and late differentiation myogenic markers to the same extent than genuine myoblasts, unlike wild-type or LacZ-hMADS cells transduced with a PGK-nlsLacZ lentiviral vector and used as a negative control (Fig. 12.3).

In co-culture experiments, they largely fused with DMD myoblasts to form DMD-hMADS cell hybrid myotubes, exhibiting a restoration of expression of dystrophin, which is the DMD-lacking protein. This capacity of fusion with myoblasts is a basic feature of muscle cells and was therefore acquired by MyoD-hMADS cells. The presence of dystrophin necessarily indicated that the MyoD-hMADS cell genome was

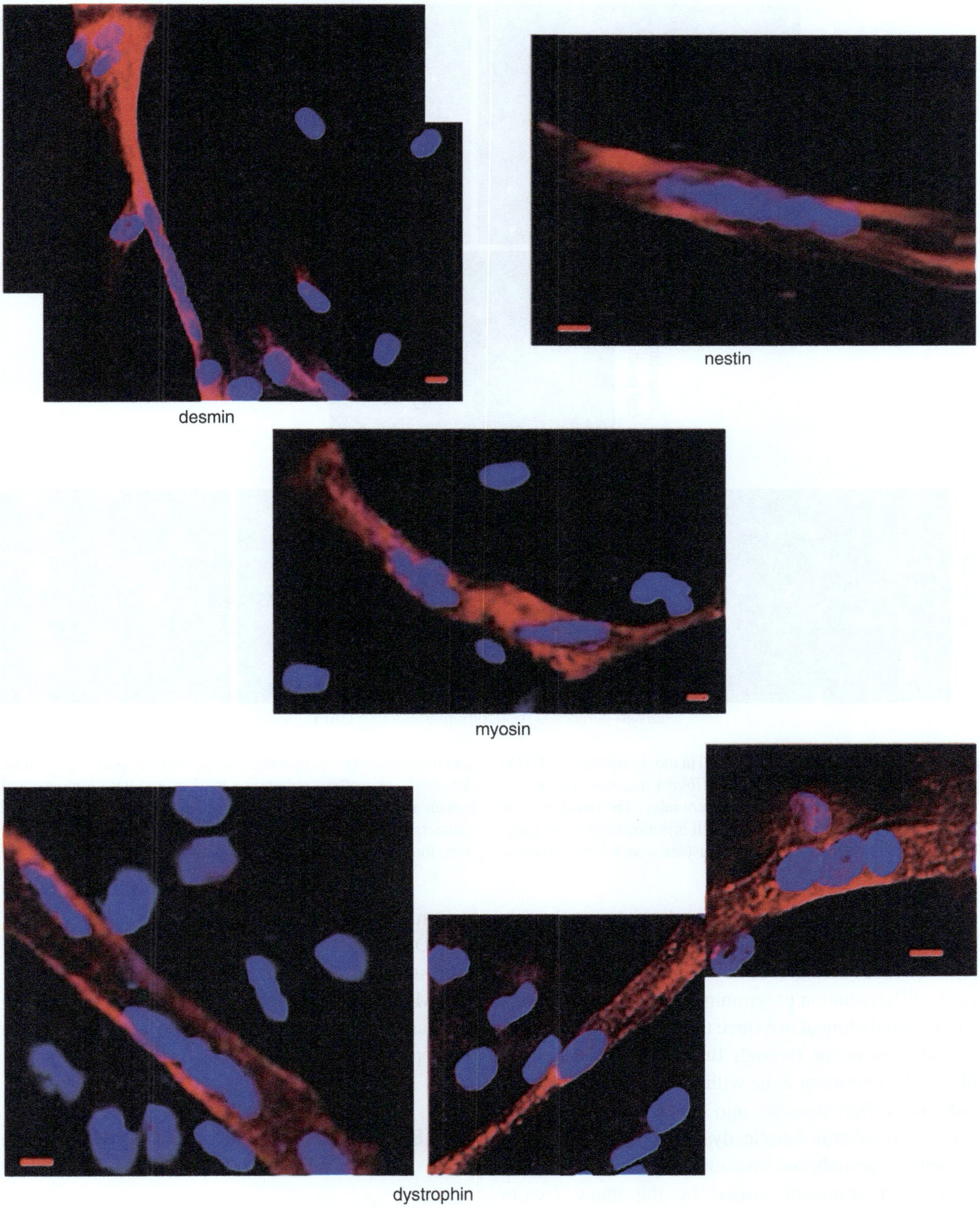

Fig. 12.3 In vitro myogenic differentiation of hMADS cells forced by MyoD expression. Unlike wild-type hMADS cells, MyoD-hMADS cells fuse very efficiently into myotubes when cultured under myogenic conditions. Muscle proteins desmin, nestin, myosin and dystrophin were detected with specific antibodies revealed with a fluorescent second antibody (*red*). Nuclei were counterstained with Hoechst 34580 dye (*blue*) to detect multinucleated myotubes. Bar = 20 μm

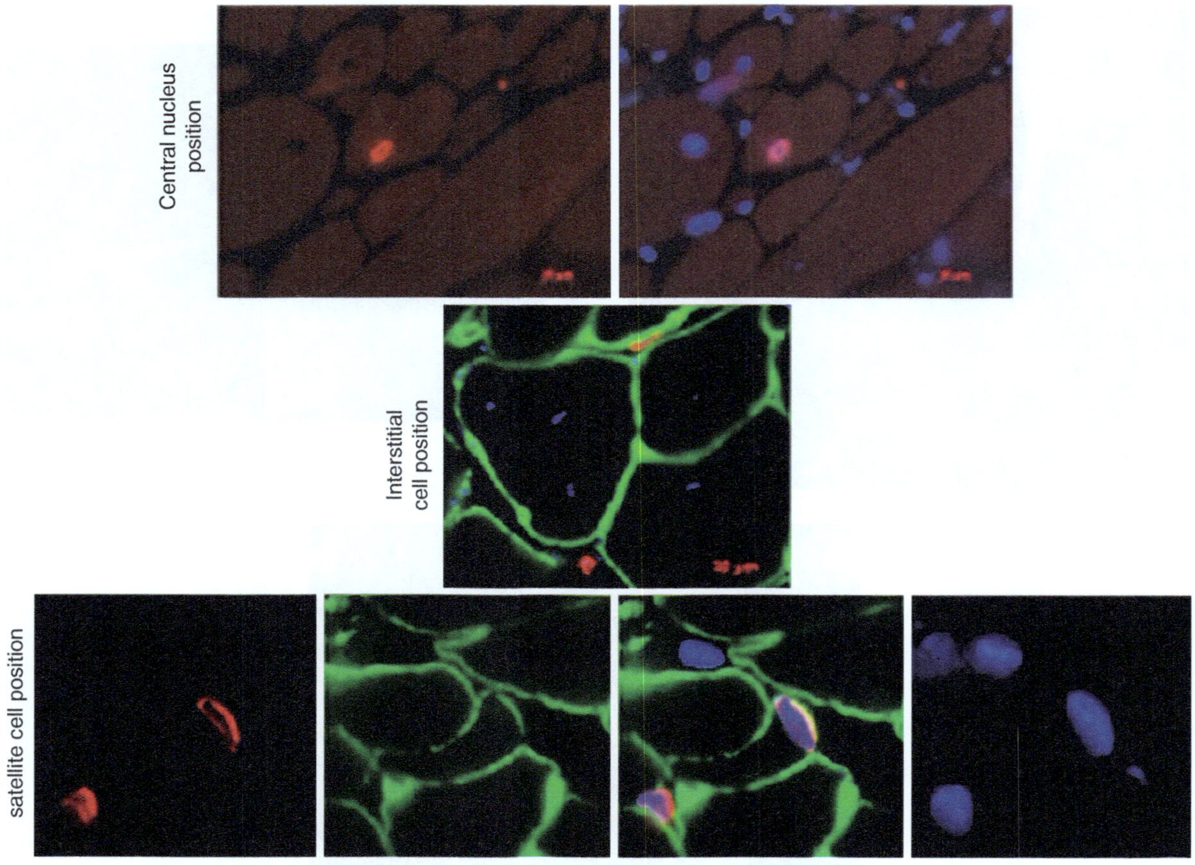

Fig. 12.4 hMADS cell-derived nuclei in mouse muscles. 500,000 hMADS cells were injected in the *Tibialis anterior* muscles of Rag2−/− gamma C−/− immunodeficient mice. The muscles were injured by cold lesion at the time of cell injection to promote muscle regeneration. One month later, muscles were taken and transversal cryo-sections were stained with a human-specific anti-lamin A/C (*red*), as nucleus membrane marker, and with a mouse and human anti-laminin (*green*), as basal lamina marker. Nuclei were counterstained with DAPI dye (*blue*). hMADS cell-derived nuclei were found in all muscle compartments

expressed in hybrid myotubes and consequently, this showed that the cellular fusion of DMD myoblasts with modified hMADS cells did not disturb the myogenic differentiation programme. This is a prerequisite for potential clinical use since the muscle repair can be expected to occur through fusion with host's myoblasts. An important issue with multipotent cells is the risk of differentiation into undesirable lineages. Moreover, several muscle dystrophies present intramuscular fibro-adipose invasion, which creates an adipogenic environment caused by the many factors secreted by fat cells. It was therefore decisive to investigate the adipogenic differentiation potential of MyoD-hMADS cells. Wild-type or LacZ-hMADS cells readily enter adipogenic differentiation when they are submitted to adipogenic differentiation conditions at confluency. This strong adipogenic potential of hMADS cells was clearly inhibited by the forced expression of MyoD. Together, these results showed that MyoD-hMADS cells have a promising in vitro myogenic potential that could be further exploited for dystrophic muscle repair.

12.3.8 In Vivo Contribution of MyoD-hMADS Cells to Muscle Repair

The impact of MyoD expression in hMADS cells on muscle repair was assessed by MyoD-hMADS cell transfer into regenerating muscles of Rag2−/− gC−/− immunodeficient mice, in comparison with wild-type and LacZ-hMADS cells. Muscle regeneration was

Fig. 12.5 In vivo muscle repair: detection in mouse muscles of muscle fibres derived from MyoD-hMADS cells. Approximately 500,000 MyoD-hMADS cells were injected in the cryo-injured *Tibialis anterior* muscle of immunodeficient Rag2–/– γC–/– mice. Contribution of injected cells to muscle regeneration was studied 1 month later on cryostat transversal sections with a human-specific anti-spectrin antibody. Spectrin is a sub-membrane protein expressed in muscle fibres. The figure represents assembled views of a muscle area with a high content in muscle fibres derived from MyoD-hMADS cells

routinely induced by cold lesion of the easily accessible *Tibialis anterior* muscles of the hind limbs. Four weeks later, the presence of hMADS cell-derived nuclei were examined with a human-specific anti-lamin A/C antibody. Positive nuclei were found either within the muscle fibres or scattered between fibres, sometimes gathered in clusters, or in satellite cell position, showing that hMADS cells survived in the mouse muscles in different locations (Fig. 12.4). Much more MyoD-hMADS cell-derived nuclei than wild-type or LacZ-hMADS cell-derived nuclei were found integrated in muscle fibres. Serial muscle sections showed that human nuclei were always detected among mouse nuclei within the same fibre, indicating that hMADS cells fused with mouse regenerating fibres, as could be expected from the in vitro results.

Then, the presence of hMADS cells-derived fibres was studied through the expression of human-specific muscle markers. Human spectrin, which is expressed at the sarcolemma of muscle fibres, was clearly detected at the membrane of many fibres that were always located in regenerated regions displaying human lamin A/C-positive nuclei (Fig. 12.5). The presence of hMADS cell-derived muscle fibres was

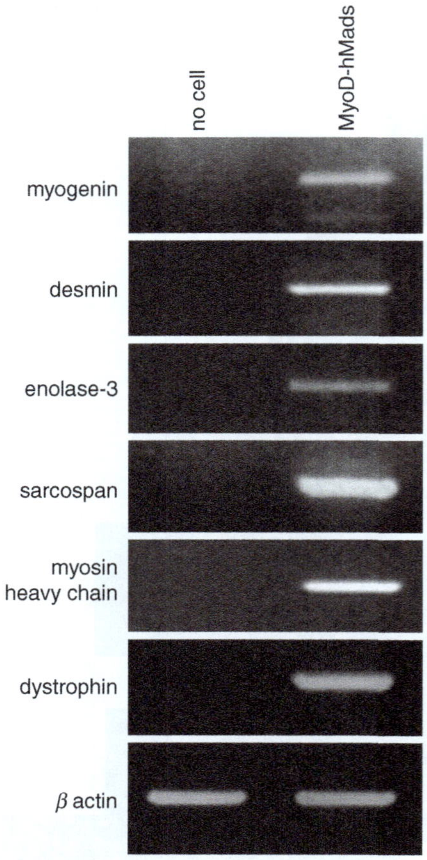

Fig. 12.6 In vivo muscle repair: detection in mouse muscles of human muscle markers derived from MyoD-hMADS cells. Approximately 500,000 MyoD-hMADS cells were injected in the cryo-injured *Tibialis anterior* muscle of immunodeficient Rag2−/− γC−/− mice. Cryostat sections not used for immunofluorescence studies were pooled and submitted to RNA extraction for RT-PCR analyses. Thirty cycles were used for PCR human-specific amplifications. Equilibrium of cDNA quantities was verified with mouse beta actin

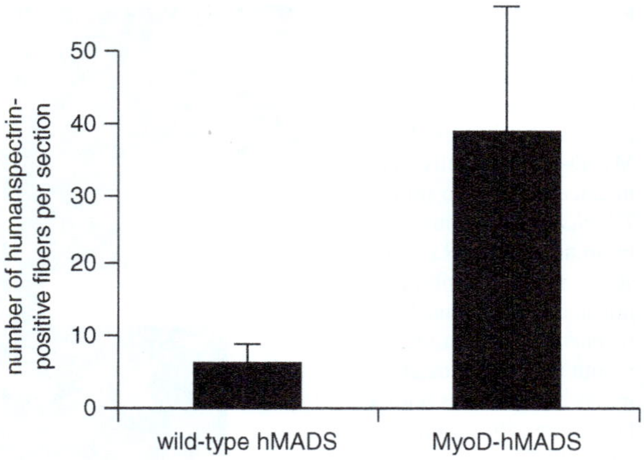

Fig. 12.7 Effect of MyoD expression in hMADS cells on mouse muscle repair. Approximately 500,000 wild-type or MyoD-hMADS cells were injected into the cryo-injured *Tibialis anterior* muscles of immunodeficient Rag2−/− γC−/− mice. One month later, human spectrin-positive fibres were counted on transversal cryostat sections after immunolabelling. For each mouse, 10 sections were counted, and the histograms represent the means ± SEM obtained from 18 muscles injected with wild-type hMADS cells and 6 muscles injected with MyoD-hMADS cells

confirmed by the expression of several other human muscle markers, such as delta-sarcoglycan and dystrophin at the membrane fibres, or by RT-PCR-specific amplifications (Fig. 12.6). The extent of human spectrin labelling was used to quantify the in vivo effect of modification of hMADS cells by MyoD. The total number of human spectrin-positive fibres was found to be about five times higher with MyoD-hMADS cells than with wild-type or LacZ-hMADS cells (Fig. 12.7). Of note, no tumours or other adverse side effects were observed within the duration of the experiments.

In summary, MyoD forced expression in hMADS cells enhances their engraftment in regenerating muscles and their contribution to muscle repair. In addition, it blocks the natural adipogenic potential of hMADS cells. The efficiency of this strategy has to be compared in animal models of dystrophic muscles with the most efficient cells used in this field, namely the vessel-derived mesoangioblasts [24] and the peripheral blood AC133+ cells [27].

12.4 Conclusions

ASCs are abundant and easily expandable adult stem cells which exhibit a myogenic differentiation potential lower than the adipogenic, osteogenic, or chondrogenic potential.

However, since they are very efficiently transduced with lentiviral vectors, they can be genetically modified with key myogenic genes. MyoD has been used but other genes such as Pax3 should be tested [11]. Genetically modified ASCs meet several requirements to be used as tools for cell-based therapy of muscle dystrophies. They can be harvested without major difficulty and trauma for the patients, they are easily expandable to obtain large quantities compatible with muscle therapy in humans. However, the genetic modification with myogenic factors should be adapted to allow rapid proliferation of the cells, since the expression of MyoD, for instance, leads to cell cycle withdrawal. This can be done by using an inducible promoter driving the transforming myogenic factors. Such promoters are available and could be pharmacologically induced only after the needed amount of cells are obtained. Finally, the level of contribution to regeneration of recipient's muscles remains to be assessed in dystrophic animal models to compare the efficiency of MyoD-hMADS cells with other cell-based therapies.

References

1. Ailhaud G (2006) Adipose tissue as a secretory organ: from adipogenesis to the metabolic syndrome. C R Biol 329(8): 570–577
2. Bacou F, el Andalousi RB, Daussin PA et al (2004) Transplantation of adipose tissue-derived stromal cells increases mass and functional capacity of damaged skeletal muscle. Cell Transplant 13(2):103–111
3. Bezaire V, Mairal A, Ribet C et al (2009) Contribution of adipose triglyceride lipase and hormone-sensitive lipase to lipolysis in hMADS adipocytes. J Biol Chem 284(27): 18282–18291
4. Darimont C, Zbinden I, Avanti O et al (2003) Reconstitution of telomerase activity combined with HPV-E7 expression allow human preadipocytes to preserve their differentiation capacity after immortalization. Cell Death Differ 10(9): 1025–1031
5. Di Rocco G, Iachininoto MG, Tritarelli A et al (2006) Myogenic potential of adipose-tissue-derived cells. J Cell Sci 119(Pt 14):2945–2952
6. Elabd C, Chiellini C, Massoudi A et al (2007) Human adipose tissue-derived multipotent stem cells differentiate in vitro and in vivo into osteocyte-like cells. Biochem Biophys Res Commun 361(2):342–348
7. Elabd C, Basillais A, Beaupied H et al (2008) Oxytocin controls differentiation of human mesenchymal stem cells and reverses osteoporosis. Stem Cells 26(9):2399–2407
8. Elabd C, Chiellini C, Carmona M et al (2009) Human multipotent adipose-derived stem cells differentiate into functional brown adipocytes. Stem Cells 27(11):2753–2760
9. Farini A, Razini P, Erratico S et al (2009) Cell based therapy for Duchenne muscular dystrophy. J Cell Physiol 221(3): 526–534
10. Galitzky J, Langin D, Verwaerde P et al (1997) Lipolytic effects of conventional beta 3-adrenoceptor agonists and of CGP 12,177 in rat and human fat cells: preliminary pharmacological evidence for a putative beta 4-adrenoceptor. Br J Pharmacol 122(6):1244–1250
11. Gang EJ, Bosnakovski D, Simsek T et al (2008) Pax3 activation promotes the differentiation of mesenchymal stem cells toward the myogenic lineage. Exp Cell Res 314(8): 1721–1733
12. Goudenege S, Pisani DF, Wdziekonski B et al (2009) Enhancement of myogenic and muscle repair capacities of human adipose-derived stem cells with forced expression of MyoD. Mol Ther 17(6):1064–1072
13. Jeninga EH, Bugge A, Nielsen R et al (2009) Peroxisome proliferator-activated receptor gamma regulates expression of the anti-lipolytic G-protein-coupled receptor 81 (GPR81/Gpr81). J Biol Chem 284(39):26385–26393
14. Lafontan M, Sengenes C, Galitzky J et al (2000) Recent developments on lipolysis regulation in humans and discovery of a new lipolytic pathway. Int J Obes Relat Metab Disord 24(suppl 4):S47–S52
15. Lee JH, Kemp DM (2006) Human adipose-derived stem cells display myogenic potential and perturbed function in hypoxic conditions. Biochem Biophys Res Commun 341(3):882–888
16. Mizuno H, Zuk PA, Zhu M et al (2002) Myogenic differentiation by human processed lipoaspirate cells. Plast Reconstr Surg 109(1):199–209; discussion 210–211
17. Nedergaard J, Bengtsson T, Cannon B (2007) Unexpected evidence for active brown adipose tissue in adult humans. Am J Physiol Endocrinol Metab 293(2):E444–E452
18. Plaisant M, Fontaine C, Cousin W et al (2009) Activation of hedgehog signaling inhibits osteoblast differentiation of human mesenchymal stem cells. Stem Cells 27(3): 703–713
19. Poitou C, Divoux A, Faty A et al (2009) Role of serum amyloid a in adipocyte-macrophage cross talk and adipocyte cholesterol efflux. J Clin Endocrinol Metab 94(5):1810–1817
20. Rodriguez A-M, Elabd C, Delteil F et al (2004) Adipocyte differentiation of multipotent cells established from human adipose tissue. Biochem Biophys Res Commun 315(2): 255–263
21. Rodriguez A-M, Pisani D, Dechesne CA et al (2005) Transplantation of a multipotent cell population from human adipose tissue induces dystrophin expression in the immunocompetent mdx mouse. J Exp Med 201(9):1397–1405
22. Rosen ED, Spiegelman BM (2000) Molecular regulation of adipogenesis. Annu Rev Cell Dev Biol 16:145–171
23. Saito T, Dennis JE, Lennon DP et al (1995) Myogenic expression of mesenchymal stem cells within myotubes of mdx mice in vitro and in vivo. Tissue Eng 1(4):327–343
24. Sampaolesi M, Blot S, D'Antona G et al (2006) Mesoangioblast stem cells ameliorate muscle function in dystrophic dogs. Nature 444(7119):574–579
25. Sennitt MV, Kaumann AJ, Molenaar P et al (1998) The contribution of classical (beta1/2-) and atypical beta-adrenoceptors to the stimulation of human white adipocyte lipolysis and right atrial appendage contraction by novel beta3-adrenoceptor agonists of differing selectivities. J Pharmacol Exp Ther 285(3):1084–1095
26. Tapscott SJ, Davis RL, Thayer MJ et al (1988) MyoD1: a nuclear phosphoprotein requiring a Myc homology region to convert fibroblasts to myoblasts. Science 242(4877): 405–411
27. Torrente Y, Belicchi M, Sampaolesi M et al (2004) Human circulating AC133(+) stem cells restore dystrophin expression and ameliorate function in dystrophic skeletal muscle. J Clin Invest 114(2):182–195
28. van Marken Lichtenbelt WD, Vanhommerig JW, Smulders NM et al (2009) Cold-activated brown adipose tissue in healthy men. N Engl J Med 360(15):1500–1508

29. Vernochet C, Azoulay S, Duval D et al (2003) Differential effect of HIV protease inhibitors on adipogenesis: intracellular ritonavir is not sufficient to inhibit differentiation. AIDS 17(15):2177–2180
30. Vernochet C, Azoulay S, Duval D et al (2004) Human immunodeficiency virus protease inhibitors accumulate into cultured human adipocytes and alter expression of adipocytokines. J Biol Chem 280(3):2238–2243
31. Vernochet C, Azoulay S, Duval D et al (2005) Human immunodeficiency virus protease inhibitors accumulate into cultured human adipocytes and alter expression of adipocytokines. J Biol Chem 280(3):2238–2243
32. Vieira NM, Bueno CR Jr, Brandalise V et al (2008) SJL dystrophic mice express a significant amount of human muscle proteins following systemic delivery of human adipose-derived stromal cells without immunosuppression. Stem Cells 26(9):2391–2398
33. Virtanen KA, Lidell ME, Orava J et al (2009) Functional brown adipose tissue in healthy adults. N Engl J Med 360(15):1518–1525
34. Wabitsch M, Brenner RE, Melzner I et al (2001) Characterization of a human preadipocyte cell strain with high capacity for adipose differentiation. Int J Obes Relat Metab Disord 25(1):8–15
35. Wabitsch M, Bruderlein S, Melzner I et al (2000) LiSa-2, a novel human liposarcoma cell line with a high capacity for terminal adipose differentiation. Int J Cancer 88(6): 889–894
36. Wakitani S, Saito T, Caplan AI (1995) Myogenic cells derived from rat bone marrow mesenchymal stem cells exposed to 5-azacytidine. Muscle Nerve 18(12):1417–1426
37. Zaragosi LE, Ailhaud G, Dani C (2006) Autocrine fibroblast growth factor 2 signaling is critical for self-renewal of human multipotent adipose-derived stem cells. Stem Cells 24(11):2412–2419
38. Zheng B, Cao B, Li G et al (2006) Mouse adipose-derived stem cells undergo multilineage differentiation in vitro but primarily osteogenic and chondrogenic differentiation in vivo. Tissue Eng 12(7):1891–1901
39. Zilberfarb V, Pietri-Rouxel F, Jockers R et al (1997) Human immortalized brown adipocytes express functional beta3-adrenoceptor coupled to lipolysis. J Cell Sci 110(Pt 7): 801–807
40. Zuk PA, Zhu M, Ashjian P et al (2002) Human adipose tissue is a source of multipotent stem cells. Mol Biol Cell 13(12):4279–4295
41. Zuk PA, Zhu M, Mizuno H et al (2001) Multilineage cells from human adipose tissue: implications for cell-based therapies. Tissue Eng 7(2):211–228

Role of Adipose Stem Cells Therapy in Obesity

13

Yves-Gérard Illouz, Aris Sterodimas, and Alexandra Condé Green

Contents

13.1 **Introduction** 133

13.2 **Adipose Tissue: An Endocrine Organ** 133

13.3 **Role of Adipose Tissue as a Secretory Organ in Obesity** 134

13.4 **Role of Adipose Stem Cells in Obesity** 135

13.5 **Adipose Stem Cells Use as a Treatment Strategy for Obesity** 135

13.5.1 Adipogenesis and Angiogenesis 135

13.5.2 Brown Adipose Tissue and Its Beneficial Metabolic Effects 136

13.5.3 Induced Pluripotent Stem Cells 137

13.6 **Future Research on Obesity** 137

13.7 **Conclusion** 138

References 138

Y.-G. Illouz (✉)
Plastic Surgery Department, Saint Louis Hospital, Avenue Claude-Vellefaux, Paris, France
e-mail: yillouz2000@hotmail.com

A. Sterodimas
Plastic and Reconstructive Surgery Department, IASO General Hospital, 264 Mesogeion Avenue, Athens, Greece
e-mail: aris@sterodimas.com

A.C. Green
Consultant Plastic Surgeon, 1391 Caserta Place, Ottawa, ON, Canada
e-mail: acondegreen@yahoo.com

13.1 Introduction

Rates of obesity have risen dramatically in recent decades and are expected to rise even more as a result of changing lifestyles and demography. More western-style diets, less exercise mean that corpulence is taking hold in adults as well as children, in developed and underdeveloped countries. As a consequence, metabolic diseases most prominently diabetes, hyperlipidemia, cardiovascular and kidney diseases are increasing dramatically in all nations, at all ages with a great risk of blindness, stroke, and amputations. This global epidemic of obesity and secondary sequelae can severely impact the quality of life of those who are afflicted and represent a public health issue.

Cell replacement therapy is one of many research avenues being pursued as potential treatment strategy for these conditions. With the increase interest in adipose-derived stem cells (ADSCs), researchers have begun to ask if a new treatment approach is on the horizon – can stem cells that are derived from adult adipose tissue be used to treat obesity?

13.2 Adipose Tissue: An Endocrine Organ

Humans and other mammals have three main adipose tissue depots: visceral white adipose tissue (WAT), subcutaneous white adipose tissue and brown adipose tissue (BAT); each of which possesses unique cell-autonomous properties. White fat cells are the "conventional" form of fat that persist in adulthood, particularly in humans. They are cells full of lipid

Y.-G. Illouz and A. Sterodimas (eds.), *Adipose Stem Cells and Regenerative Medicine*,
DOI: 10.1007/978-3-642-20012-0_13, © Springer-Verlag Berlin Heidelberg 2011

droplets that accumulate under the skin and around internal organs. Brown fat cells are cells containing small lipid droplets tucked between tiny energy factories called mitochondria. In rodents, brown fat cells are found throughout the body and are present during the entire life cycle. In humans, they are principally found in the neck area of newborns, helping their tiny bodies generate heat, also throughout infancy. Brown fat cells largely disappear by adulthood, but their precursors still remain in the body, lodged in white-fat depots. Depending on its localization, white adipose tissue (WAT) presents different metabolic and functional properties [33]. Visceral adipose tissue can induce detrimental metabolic effects in contrast to subcutaneous white adipose tissue and brown adipose tissue (BAT) that appear to be metabolically protective and have the potential to benefit metabolism by improving glucose homeostasis and increasing energy consumption [46].

Because of its simple macroscopic appearance, adipose tissue has long been considered an organ of storage, lipid synthesis, and lipid breakdown. The concept of adipose tissue as an endocrine organ is recent as the adipocyte is no longer considered a passive bystander, because it actively secretes many members of the cytokine family, such as leptin, tumor necrosis factor (TNF)-alpha, and interleukin-6 (IL6), among other cytokine signals, which influence peripheral fuel storage, mobilization, combustion, as well as energy homeostasis [18]. The properties of adipocytes in different adipose tissue depots could represent an intrinsic heterogeneity of adipocytes that might be regulated by fundamental developmental genes [3, 20].

Like bone marrow, adipose tissue is derived from embryonic mesoderm and generates a stromal fraction composed of preadipocytes, smooth muscle cells, endothelial cells, macrophages, and fibroblasts [9]. The preadipocytes progressively accumulate lipids and acquire the characteristics of mature adipocytes, exhibiting a high turnover rate, indicating that fat cell number is tightly regulated and constant in adulthood in both lean and obese individuals [39]. Because these mature adipocytes are terminally differentiated cells, they are considered incapable of division, therefore the apparent increase in adipocyte number is thought to originate from adipogenesis, the proliferation or differentiation of adipocyte progenitor cells [6].

13.3 Role of Adipose Tissue as a Secretory Organ in Obesity

The plasticity of adipose tissue is reflected by its remarkable ability to expand or to reduce in size throughout the adult lifespan. There is growing evidence of a causal link between what happens in adipose tissue and obesity [53]. Obesity apparently results from chronic energy surplus and excess lipid storage in WAT [49]. In this condition, adipocytes hypertrophy and lose their functionality [1]. This creates a dysequilibrium between lipogenesis and lipolysis, impaired transcriptional regulation of key factors that control adipogenesis, and a lack of sensitivity to external signals, as well as failure in the signal transduction process [48].

In vitro hypoxic adipocytes secrete inflammatory molecules such as tumor necrosis factor (TNF) α, interleukin (IL) 1, IL6, macrophage inflammatory protein (MIP), and plasminogen activator inhibitor-1 [52]. These chemokines presumably clear out necrotic cells and the macrophages cause inflammation of tissues, rendering adipose cells to become resistant to the effects of insulin, causing diverse metabolic diseases [31, 50].

Recently, Bone Morphogenic Proteins (BMP) has been reported to play a role in adipoctyte differentiation, providing instructive signals for adipose cell fate determination and regulating adipocyte function. Adenoviral-mediated expression of BMP7 in mice results in a significant increase in brown rather than white fat mass and leads to an increase in energy expenditure and a reduction in weight gain. These data reveal the important role of BMP7 in promoting brown adipocyte differentiation and thermogenesis in vivo and in vitro, and provide a potential new therapeutic approach for the treatment of obesity [38].

Additionally, apart from imbalance between food intake and caloric utilization, the development of obesity also depends on the balance between WAT, which is the primary site of energy storage, and BAT, which is specialized for energy expenditure [21]. Excess accumulation of WAT causes obesity, while BAT by the expression of the tissue-specific uncoupling protein 1 (UCP1) affects whole-body metabolism and may alter insulin sensitivity and modify susceptibility to weight gain [29, 51]. UCP1 is an inner mitochondrial membrane protein, unique marker for BAT, which short circuits the proton electrochemical gradient, so that oxygen consumption is no longer coupled to adenosine triphosphate synthesis, generating heat as a consequence [15].

BAT is therefore, primarily thermogenic, efficiently burning lipids and converting nutrients into heat through the abundant number of mitochondria that surrounds it; however, it is almost inexistent in adult humans. Over the past years, several independent research teams used a combination of positron-emission tomography and computed tomography (PET/CT) imaging, immunohistochemistry, and gene and protein expression assays to prove conclusively that adult humans do have functional BAT [12]. Substantial depots were identified in a region extending from the anterior neck to the thorax, in greater quantity in women, with a female to male ratio greater than 2:1. The probability of the detection of BAT was inversely correlated with years of age, and body-mass index especially in older people, suggesting a potential role of BAT in adult human metabolism [13]. Estimates suggest that if 50 g of maximally stimulated BAT were present, it could account for up to 20% of daily energy expenditure in adult human [35]. Although very scarce in adult humans, the thermogenic capacity of even small amounts of BAT makes it an attractive therapeutic target for inducing weight loss through energy expenditure.

13.4 Role of Adipose Stem Cells in Obesity

Adipose tissue contains adipose-derived stem cells (ADSCs) among other cells, which possess the ability to differentiate into multiple lineages such as adipocytes, osteocytes, and chondrocytes, a property of value for the repair and replacement of various cell types. These adipose precursor cells are a heterogeneous cells population, consisting of fibroblast-like multipotential stem cells [32]. They reside in a specialized environment called a niche, that controls many aspects of their behavior – quiescence, proliferation, and differentiation [27]. This location is well suited to receive environmental stimuli such as nutritional cues that can trigger the cells to leave their niche and mature into fat cells. Therefore, niche identification and characterization are significant and dynamic areas of stem cell biology, with many recent revelations. It is also important to underscore that to be defined as a niche, the microenvironment must provide regulatory inputs to the resident stem cell beyond simply representing a physical location. Immunohistochemical methods, combined with GFP marking, showed that adipose stem cells are found in the wall of blood vessels that supply adipose tissue depots but are absent from blood vessels that supply other tissues [43, 24].

The identification of the vasculature as an adipose stem cell niche also supports earlier studies indicating that adipocytes form in intimate juxtaposition with blood vessels and that adipogenesis and angiogenesis are tightly orchestrated. Human adipose tissue-derived stem cells can differentiate into endothelial cells and improve postnatal neovascularization. As a matter of fact, activated adipocytes produce multiple angiogenic factors including leptin, angiopoietins, HGF, GM-CSF, VEGF, FGF-2, TGF-β, which either alone or collectively stimulate neovascularization during fat mass expansion [7].

Adipocyte hyperplasia is associated with obesity and arises due to adipogenic differentiation of resident multipotent stem cells in the vascular stroma of adipose tissue and remote stem cells of other organs [34]. There also seems to be a positive balance of adipocyte turnover involving proliferation ADSCs and several transcriptional differences from adipose tissue enlargement in obesity [42]. These findings raise an interesting theory, in which expansion of fat mass seems to be dependent on angiogenesis, and suppression of angiogenesis might provide a novel therapeutic approach for the prevention and treatment of obesity [4].

13.5 Adipose Stem Cells Use as a Treatment Strategy for Obesity

Current pharmacotherapeutic options for treating obesity and related metabolic disorders remain limited and most of the time ineffective. Many options are being investigated in order to define potential targets and approaches for the treatment of these conditions [8].

13.5.1 Adipogenesis and Angiogenesis

Preadipocytes as previously mentioned, secrete high levels of a number of angiogenic factors that stimulate angiogenesis in adipose tissue. Recruitment of inflammatory cells, macrophages which increases with obesity also significantly contributes to adipose neovascularization which supports the notion that adipose tissue development requires continuous remodeling, maturation, and patterning of the vasculature. Also, this plasticity of the adipose vasculature may represent the outcome of a net balance between angiogenic factors and inhibitors, which determine growth or regression of adipose tissue. Coordinative communications between adipokines and other angiogenic factors are particularly important to understand the relationship of global control of adipose tissue expansion and local angiogenic responses. For example, high levels of serum leptin restrict fat mass expansion via endocrine stimulation of the central negative feedback loop, and paradoxically, leptin might promote adipogenesis through paracrine stimulation of angiogenesis. Inversely, high levels of serum adiponectin might restrain adipose tissue growth via inhibition of angiogenesis locally.

These observations have led scientists to believe that one of the therapeutic interventions of obesity might be achieved by targeting the vasculature [5]. There is a high probability that many of the angiogenic factors expressed in the adipose tissue might function as survival factors for the vasculature. For example, both VEGF and IGFs are important survival angiogenic factors for endothelial cells, and withdrawal of these factors may lead to endothelial cell apoptosis. Thus, it is possible that anti-angiogenic agents might block the functions of these survival factors, leading to regression of preexisting adipose vasculature and shrinkage of the fat mass. Furthermore, known angiogenesis inhibitors and endogenous protein inhibitors such as angiostatin and endostatin have been implicated in weight reduction as well as adipose tissue loss in mice. Some studies have shown that the angiogenesis inhibitor-treated adipose tissue undergoes remarkable vascular remodeling and contains an increased number of apoptotic cells and a decreased number of proliferating endothelial cells. These inhibitors have also shown the advantage of normalizing insulin sensitivity, preventing the development of type II diabetes in addition to body weight gain [36].

13.5.2 Brown Adipose Tissue and Its Beneficial Metabolic Effects

Because brown fat cells burn calories through adaptive thermogenesis, scientists think that finding ways to encourage the development of brown fat might be good for treating obesity. Therefore, researchers are actively investigating potential sources of new precursor cells that could give rise to mature brown cells, including different types of stem cells that could be candidates for use in this treatment. ADSCs isolated from the SVF by a series of processing methods following liposuction [10, 11] are now being explored as a possible tool to promote the beneficial metabolic effects of subcutaneous WAT and BAT. They are able, upon exposure to a specific PPARgamma or PPARalpha agonist, to switch to a brown phenotype by expressing both UCP1 and CIDEA mRNA [14]. A recent study in mouse models, showed that cyclooxygenase (COX)-2, a rate-limiting enzyme in prostaglandin (PG) synthesis, is a downstream effector of beta-adrenergic signaling in WAT and is required for the induction of BAT in WAT depots. PG shifted the differentiation of defined ADSCs toward a brown adipocyte phenotype. Over-expression of COX-2 in WAT induced de novo BAT recruitment in WAT, increasing systemic energy expenditure and protecting mice against high-fat diet-induced obesity. Thus, COX-2 appears integral to de novo BAT recruitment, which suggests that the PG pathway regulates systemic energy homeostasis [49].

In another study, using DNA chips to analyze how precursor cells give rise to mature brown fat cells, a strong genetic pattern was identified. Apparently, reducing the level of necdin, a protein which has been found to be over-expressed in obesity, is essential for precursor cells to give rise to brown fat cells. Researchers have also found that a transcription factor called CREB is involved in this reduction. "As we learn more about the genesis of brown fat cells and the genes governing them, we may be able to target those genes with drugs or other agents to create powerful tools to fight obesity" [47].

Therefore, transplantation of adipose tissue is being explored as a possible tool to promote the beneficial metabolic effects of BAT as well as ADSCs. When adipose tissue is transplanted between lean and obese mice, the cells in the grafts grow or shrink to those of the host, in size and fatty acid composition [2]. Thus, the transplants seem to respond similarly to normal adipose tissue pads, and the host environment is important in determining some aspects of adipose tissue cell fate [16]. In other terms, subcutaneous WAT could be harvested, collected, and molded to a brown-like, energy burning, adipocyte fat, that could be used to cure obesity and metabolic dysfunctions. Young adipose-derived stromal cells demonstrated significantly higher levels of VEGF production, proliferation, and tubulogenesis than those derived from aged, streptozotocin-induced, and db/db mice in both normoxia and hypoxia. Although aged and diabetic adipose-derived stromal cells retained the ability to up-regulate VEGF secretion, proliferation, and tubulogenesis in response to hypoxia, the response was blunted compared with young controls. ADSCs from elderly donors can lose their capacity to differentiate; optimal ADSCs should preferably be obtained from young, healthy donors and have a normal karyotype and a high potential for proliferation and differentiation in vivo [37].

Ultimately, the clinical applicability of adipose tissue transplantation for the treatment of obesity and metabolic disorders resides in the achievable level of safety, reliability, and efficacy compared with other treatments, and also the patients' conception and acceptance of the procedure [46, 54].

13.5.3 Induced Pluripotent Stem Cells

Some studies have proposed to induce pluripotent stem cells that are identical to natural pluripotent stem cells but have a forced expression of certain genes (cloning) in order to obtain a greater quantity of cells for treatment strategy of obesity. The study showed that ADSCs were the most amenable to reprogramming, and were a more clinically relevant cell type for this use; and that fat tissue is easily accessed, grown easily and rapidly in cultures [44]. Ex vivo engineering of SVF could involve the overexpression of beneficial adipokines, such as adiponectin and leptin, or involve the promotion of a beneficial brown cell fate. A recent report shows that, upon ex vivo engineering, human ADSCs are able to switch to a brown phenotype by expressing UCP1 after reprogramming by forced expression of UCP1, by retinoic acid and BMP7 treatment delivered intracellularly [30]. This switch is accompanied by an increase in oxygen consumption and uncoupling [25]. Furthermore, the expression of UCP1 protein is associated to stimulation of respiration by beta-AR agonists, including beta3-AR agonist, making ADSCs an invaluable cell model to screen for drugs stimulating the formation and/or the uncoupling capacity of human brown adipocytes that could help dissipate excess caloric intake of individuals [14].

An in vivo experiment has shown that implanted murine preadipocytes induced vigorous angiogenesis and formed fat pads in a mouse dorsal skin-fold chamber. The newly formed vessels subsequently remodeled into a mature vascular network, whereas the preadipocytes differentiated into adipocytes as confirmed by increased aP2 expression. Adipocyte differentiation into fat tissue formation was inhibited by transfection of preadipocytes with a peroxisome proliferator-activated receptor γ dominant-negative construct. Also, inhibition of angiogenesis by vascular endothelial growth factor receptor-2 (VEGFR2) blocking antibody, not only reduced angiogenesis and tissue growth but also inhibited preadipocyte differentiation suggesting that blockade of VEGF signaling can inhibit in vivo adipose tissue formation [19].

13.6 Future Research on Obesity

Obesity is a complex metabolic disorder influenced by a mixture of genetic and environmental factors, including control of appetite and energy expenditure, availability and nutritional content of food, and development of adipocyte cell mass [20]. It occurs with different degrees of fat accumulation in different depots, and is commonly associated with type II diabetes mellitus, hypertension, coronary heart disease, dyslipidemia, gallbladder disease, hepatic steatosis, sleep apnea, stroke, endometrial disorder, and cancer. Interestingly, most of these obesity-related disorders are closely associated with vascular dysfunctions. Adipose tissue is highly vascularized, and each adipocyte is nourished by an extensive capillary network. Recently, angiogenesis research has become one of the most important areas in biomedical research. Both genetic and high-calorie diet-induced obesity seems to require the switch of an angiogenic phenotype in adipose tissues to support adipogenesis. The fast expansion of this research field demands development of rigorous, reliable, stable, convenient, and clinically relevant assay systems for disease diagnosis, prognosis, therapeutic evaluation, and drug discovery [26]. These findings have paved avenues for possible therapeutic intervention of obesity and obesity-associated disorders by targeting the vascular compartment and its numerous factors as described above.

Leptin is one of the known adipocyte-derived hormone that regulates food intake and energy homeostasis. It is considered a potent angiogenic factor as endothelial cells express the functional long form of leptin receptor

(OB-Rb). The vasodilation function of leptin may be essential to comply with the demand of growing adipose tissues for an elevated blood flow rate. Growing adipose tissue contains fenestrated capillaries that are essential for vascular permeability [17]. Under hypoxia, the expression ratios of angiogenic factors in adipose tissue might be altered because expression levels of VEGF and leptin are elevated by low oxygen. Thus, hypoxia probably plays a critical and additional role in controlling the balance between vessel growth and remodeling [28].

Gene therapy has also been shown to reduce body weight after injection of different kinds of genetic material into mice brains. DNA delivery and successful tissue transfection was observed in the areas of the body where ultrasound was applied after intravascular administration of microbubbles and plasmid DNA [45]. These microbubbles appeared to enhance ultrasound energy deposition in tissues and served as cavitation nuclei, increasing intracellular drug delivery [22]. Another method of interest, is the nonviral transfer of nucleic acids (DNA and siRNA) into human ADSCs, exhibiting the potential of targeted modification of stem cells [23]. Delivery of these reprogrammed ADSCs can be done by direct injection in the subcutaneous tissue, which makes it more accessible and easy to use [40, 41]. Novel uses of growth factors, regulators of differentiation, and scaffolds should also be explored, in order to better purify, modulate, expand, maintain brown adipose tissue which could lead to treatment of obesity.

13.7 Conclusion

Some stem cell and antiaging centers are already advertising and offering patients who are undergoing esthetic procedures and present certain degenerative diseases, an opportunity to improve their health through cellular therapy, correcting their medical conditions from the source by regenerating the tissues and organs that are causing the ailments.

Since we are using autologous stem cells, this therapy has no risk of tissue rejection, and no secondary effects would be ideal for the treatment of obesity. Animal and in vitro studies already support the notion of transplantation of reprogrammed and genetically engineered ADSCs, and as researchers learn more about the mechanisms that govern stem cell programming, differentiation, and renewal, their ability to identify, isolate, and culture candidate stem cells will continue to improve. However, long-term studies are needed to assess the efficacy and safety of stem cell therapy to treat obesity. Additionally, the data for antiangiogenic agents offers an exciting new therapeutic option for the prevention and treatment of obesity. As more antiangiogenic agents become available for experimental and clinical applications, they provide an outstanding opportunity to test this possibility in the near future.

References

1. Armani A, Mammi C, Marzolla V et al (2010) Cellular models for understanding adipogenesis, adipose dysfunction, and obesity. J Cell Biochem 110(3):564–572
2. Ashwell M, Meade CJ (1981) Obesity: can some fat cells enlarge while others are shrinking? Lipids 16:475–478
3. Billon N et al (2007) The generation of adipocytes by the neural crest. Development 134:2283–2292
4. Brakenhielm E, Cao Y (2008) Angiogenesis in adipose tissue. Methods Mol Biol 456:65–81
5. Brakenhielm E et al (2004) Angiogenesis inhibitor, TNP-470, prevents diet-induced and genetic obesity in mice. Circ Res 94:1579–1588
6. Butterwith SC (1997) Regulators of adipocyte precursor cells. Poult Sci 76:118–123
7. Cao Y (2007) Angiogenesis modulates adipogenesis and obesity. J Clin Invest 117(9):2362–2368
8. Cao Y (2010) Adipose tissue angiogenesis as a therapeutic target for obesity and metabolic diseases. Nat Rev Drug Discov 9(2):107–115
9. Casteilla L, Charrière G, Laharrague P et al (2004) Tissus adipeux, chirurgie plastique et reconstructrice: le retour aux sources. Ann Chir Plast Esthét 49:409–418
10. Condé-Green A et al (2010) Effects of centrifugation on cell composition and viability of aspirated adipose tissue processed for transplantation. Aesthet Surg J 30(2):249–255
11. Condé-Green A et al (2010) Influence of decantation, washing and centrifugation on adipocyte and mesenchymal stem cell content of aspirated adipose tissue: a comparative study. J Plast Reconstr Aesthet Surg 63(8):1375–1381
12. Cypess AM, Kahn CR (2010) Brown fat as a therapy for obesity and diabetes. Curr Opin Endocrinol Diabetes Obes 17(2):143–149, Review
13. Cypess AM, Lehman S, Williams G et al (2009) Identification and importance of brown adipose tissue in adult humans. N Engl J Med 360(15):1509–1517
14. Elabd C, Chiellini C, Carmona M et al (2009) Human multipotent adipose-derived stem cells differentiate into functional brown adipocytes. Stem Cells 27(11):2753–2760
15. Enerback S (2010) Brown adipose tissue in humans. Int J Obes 34(Suppl 1):S43–S46
16. Enser M, Ashwell M (1983) Fatty acid composition of triglycerides from adipose tissue transplanted between obese and lean mice. Lipids 18:776–780

17. Eriksson A et al (2003) Small GTP-binding protein Rac is an essential mediator of vascular endothelial growth factor-induced endothelial fenestrations and vascular permeability. Circulation 107:1532–1538
18. Frühbeck G, Gómez-Ambrosi J, Muruzábal FJ, Burrell MA (2001) The adipocyte: a model for integration of endocrine and metabolic signaling in energy metabolism regulation. Am J Physiol Endocrinol Metab 280:E827–E847
19. Fukumura D, Ushiyama A, Duda DG et al (2003) Paracrine regulation of angiogenesis and adipocyte differentiation during in vivo adipogenesis. Circ Res 93(9):e88–e97
20. Gesta S et al (2006) Evidence for a role of developmental genes in the origin of obesity and body fat distribution. Proc Natl Acad Sci USA 103:6676–6681
21. Gesta S, Tseng YH, Kahn CR (2007) Developmental origin of fat: tracking obesity to its source. Cell 131(2):242–256
22. Hernot S, Klibanov AL (2008) Microbubbles in ultrasound-triggered drug and gene delivery. Adv Drug Deliv Rev 60(10):1153–1166
23. Hoelters J, Ciccarella M, Drechsel M et al (2005) Nonviral genetic modification mediates effective transgene expression and functional RNA interference in human mesenchymal stem cells. J Gene Med 7(6):718–728
24. Hooper AT, Butler JM, Nolan DJ et al (2009) Engraftment and reconstitution of hematopoiesis is dependent on VEGFR2-mediated regeneration of sinusoidal endothelial cells. Cell Stem Cell 4:263–274
25. Ishibashi J, Seale P (2010) Medicine. Beige can be slimming. Science 328(5982):1113–1114
26. Jensen LD, Cao R, Cao Y (2009) In vivo angiogenesis and lymphangiogenesis models. Curr Mol Med 9(8):982–991
27. Jones DL, Wagers AJ (2008) No place like home: anatomy and function of the stem cell niche. Nat Rev Mol Cell Biol 9:11–21
28. Kamba T et al (2006) VEGF-dependent plasticity of fenestrated capillaries in the normal adult microvasculature. Am J Physiol Heart Circ Physiol 290:H560–H576
29. Lowell BB, Susulic V, Hamann A et al (1993) Development of obesity in transgenic mice after genetic ablation of brown adipose tissue. Nature 366:740–742
30. Mercader J, Palou A, Luisa Bonet M (2010) Induction of uncoupling protein-1 in mouse embryonic fibroblast-derived adipocytes by retinoic acid. Obesity (Silver Spring) 18(4): 655–662
31. Murdoch C, Giannoudis A, Lewis CE (2004) Mechanisms regulating the recruitment of macrophages into hypoxic areas of tumors and other ischemic tissues. Blood 104:2224–2234
32. Perrini S, Cignarelli A, Ficarella R et al (2009) Human adipose tissue precursor cells: a new factor linking regulation of fat mass to obesity and type 2 diabetes? Arch Physiol Biochem 115(4):218–226
33. Prunet-Marcassus B, Cousin B, Caton D et al (2006) From heterogeneity to plasticity in adipose tissues: site-specific differences. Exp Cell Res 312:727–736
34. Qian SW, Li X, Zhang YY et al (2010) Characterization of adipocyte differentiation from human mesenchymal stem cells in bone marrow. BMC Dev Biol 10:47
35. Rothwell NJ, Stock JM (1983) Luxuskon-sumption, diet-induced thermogenesis and brown fat: the case in favor. Clin Sci (Lond) 64:19–23
36. Rupnick MA et al (2002) Adipose tissue mass can be regulated through the vasculature. Proc Natl Acad Sci USA 99:10730–10735
37. Schipper BM, Marra KG, Zhang W et al (2008) Regional anatomic and age effects on cell function of human adipose-derived stem cells. Ann Plast Surg 60:538–544
38. Schulz TJ, Tseng YH (2009) Emerging role of bone morphogenetic proteins in adipogenesis and energy metabolism. Cytokine Growth Factor Rev 20(5-6):523–531
39. Spalding KL, Arner E, Westermark PO et al (2008) Dynamics of fat cell turnover in humans. Nature 453:783–787
40. Sterodimas A, de Faria J, Nicaretta B, Papadopoulos O, Papalambros E, Illouz YG (2010) Cell-assisted lipotransfer. Aesthet Surg J 30(1):78–81
41. Sterodimas A, de Faria J, Nicaretta B, Pitanguy I (2010) Tissue engineering with adipose-derived stem cells (ADSCs): current and future applications. J Plast Reconstr Aesthet Surg 63(11):1886–1892
42. Suga H, Eto H, Inoue K et al (2009) Cellular and molecular features of lipoma tissue: comparison with normal adipose tissue. Br J Dermatol 161(4):819–825
43. Tang W, Zeve D, Suh JM et al (2008) White fat progenitor cells reside in the adipose vasculature. Science 322:583–586
44. Tat PA, Sumer H, Jones KL, Upton K, Verma PJ (2010) The efficient generation of induced pluripotent stem (iPS) cells from adult mouse adipose tissue-derived and neural stem cells. Cell Transplant 19(5):525–536
45. Tinkov S, Bekeredjian R, Winter G et al (2009) Microbubbles as ultrasound triggered drug carriers. J Pharm Sci 98(6): 1935–1961
46. Tran TT, Kahn CR (2010) Transplantation of adipose tissue and stem cells: role in metabolism and disease. Nat Rev Endocrinol 6(4):195–213
47. Tseng YH, Cypess AM, Kahn CR (2010) Cellular bioenergetics for obesity therapy. Nat Rev Drug Discov 9(6): 465–482
48. Vázquez-Vela ME, Torres N et al (2008) White adipose tissue as endocrine organ and its role in obesity. Arch Med Res 39(8):715–728
49. Vegiopoulos A, Müller-Decker K, Strzoda D et al (2010) Cyclooxygenase-2 controls energy homeostasis in mice by de novo recruitment of brown adipocytes. Science 328(5982): 1113–1114
50. Wentworth JM, Naselli G, Brown WA et al (2010) Pro-inflammatory CD11c+ CD206+ adipose tissue macrophages are associated with insulin resistance in human obesity. Diabetes 59(7):1648–1656
51. Yang X, Enerbäck S, Smith U (2003) Reduced expression of FOXC2 and brown adipogenic genes in human subjects with insulin resistance. Obes Res 11:1182–1191
52. Ye J, Gao Z, Yin J et al (2007) Hypoxia is a potential risk factor for chronic inflammation and adiponectin reduction in adipose tissue of ob/ob and dietary obese mice. Am J Physiol Endocrinol Metab 293:E1118–E1128
53. Yudkin JS (2003) Adipose tissue, insulin action and vascular disease: inflammatory signals. Int J Obes Relat Metab Disord 27(suppl 3):S25–S28
54. Zeve D, Tang W, Graff J (2009) Fighting fat with fat: the expanding field of adipose stem cells. Cell Stem Cell 5(5):472–481

Stem Cell Based Cardioregeneration and Adipose Tissue

14

Atta Behfar and Andre Terzic

Contents

14.1 **Introduction** 141

14.2 **The Stem Cell Paradigm** 142
14.2.1 Overview of Heart Disease 142
14.2.2 Stem Cell Plasticity 142
14.2.3 Direct Versus Indirect Repair Mechanism 143

14.3 **Cell-Based Myocardial Regeneration** 143
14.3.1 Hurdles in Translation 143

14.4 **Non-adipose Adult-Derived Stem Cells** 145
14.4.1 Myoblasts 145
14.4.2 Bone Marrow-Derived Cells 146
14.4.3 Mesenchymal Stem Cells (KDR$^-$CD14$^-$,34$^-$, 45$^-$;CD90$^+$,105$^+$,133$^+$,271$^+$) 146
14.4.4 Endothelial Progenitor Cells (CD45$^-$CD31$^-$CD34$^+$CD144$^-$KDR$^+$) 146
14.4.5 Resident Cardiac Stem Cells 147

14.5 **Adult-Derived Adipose Stem Cells** 147
14.5.1 Cell Phenotypes Isolated from Adipose Tissue 147
14.5.2 Pre-clinical Evaluation of ATSC in Myocardial Infarction 148
14.5.3 Mechanism of ATSC Benefit in the Myocardium 148
14.5.4 Clinical Translation 148

14.6 **Future Directions** 149

14.7 **Conclusions** 149

References 149

A. Behfar (✉) • A. Terzic
Marriott Heart Disease Research Program, Division of Cardiovascular Diseases, Department of Medicine, Mayo Clinic Center for Regenerative Medicine , Rochester, MN, 55905, USA
e-mail: behfar.atta@mayo.edu; terzic.andre@mayo.edu

14.1 Introduction

Stem cells are promising novel therapeutics candidates for organ regeneration, and have engendered a level of excitement within the medical and surgical community not seen since solid organ transplantation. Stem cells were initially hypothesized to reside within the confines of the bone marrow by Cohnheim in the nineteenth century [31]. They were not formally discovered until McCulloch and Till identified colony-forming cells from the bone marrow, as described in a series of papers in the 1960s. These investigators coined the term "stem cell" in 1969 to describe clonal hematopoietic cells [12, 151]. This work defined the stem cell phenotype as *plastic*, demonstrating the capacity for transdifferentiation into a phenotype outside of the original cellular lineage, and clonal, i.e., maintaining an aptitude for self-renewal [51, 149, 151].

Introduced into clinical practice with the advent of bone marrow transplantation, the first example of stem cell-based regenerative therapy was seen with the restoration of hematopoietic content for individuals suffering from hematologic disorders. The first attempt to utilize this resource to reverse myocardial damage was made in 1992 [85]. With the discovery of human embryonic stem cells in 1998 the notion of using stem cells as potential therapeutics for solid organ disease gained appeal [125, 141]. Cardiovascular medicine, and cardiac surgery in particular, embraced this novel platform, to soon thereafter implement multiple pre-clinical and clinical trials intended to identify a feasible and safe approach for cell-based heart repair.

Y.-G. Illouz and A. Sterodimas (eds.), *Adipose Stem Cells and Regenerative Medicine*, DOI: 10.1007/978-3-642-20012-0_14, © Springer-Verlag Berlin Heidelberg 2011

The advent of stem cell therapy thus has changed the focus of medical practice from palliative towards curative solutions [100]. Application of regenerative paradigms provides an approach to reverse the injury caused by ischemic and non-ischemic damage to the myocardium [10]. Today, both safety and feasibility of cell therapy have been established in over 3,000 patients following initial clinical trial testing, mostly through use of adult bone marrow stem cells derived and delivered in an unmodified state. Though this experience has provided significant advance in the delivery of recruited stem cells, fundamental clinical evaluation of stem cells has yet to identify an ideal cell source, mode of large-scale expansion, humanized culture, optimization of packaging, or a mode of action [28, 104]. The multiplicity of cell phenotypes transplanted to assess myocardial benefit has resulted in sub-optimal benefit and consequently impeded early adoption of this technology within the general practice of cardiology and cardiac surgery. To this end, definitive characterization of the cellular phenotype and its mode of action in the heart have been identified as a critical next step towards achieving success in regeneration [17, 42].

Human mesenchymal stem cells (hMSC) have been increasingly considered with their favorable regenerative potential and immunotolerant profiles [14]. The current gold standard for the derivation of this stem cell resource is the bone marrow [8]. However, translation of hMSC into clinical practice has been hampered by protracted in vitro culture durations and the zoonotic condition underlying cell processing. To this end, alternative tissue sources for hMSC derivation have been evaluated to expedite large-scale cell culture. Adipose tissue has recently been described as a favorable tissue alternative to bone marrow, giving rise to a significantly higher number of hMSC that in culture conditions appear to maintain a higher proliferative capacity.

The intent of this chapter is highlight different stem cell approaches, provide a rationale behind the use of specific stem cell technologies, assess stem cell utility based on current clinical trials, and discuss the current and projected use of adipose-derived stem cells for myocardial regeneration.

14.2 The Stem Cell Paradigm

14.2.1 Overview of Heart Disease

Congestive heart disease is the largest source of repeat hospitalization and mortality in the developed world [17, 64, 65]. Affecting more than 5 million Americans and nearly 20 million individuals world-wide, 550,000 annual cases of heart failure occur in the United States alone, with individuals suffering from a 5-year mortality of over 50% accounting for the greater than 50,000 deaths per year [76]. Overall, the cost of cardiovascular disease in the United States was 80 billion dollars in 2008, translating into greater than 160 billion dollars in productivity. By 2023, cost of cardiovascular disease is projected to exceed 150 billion dollars, impacting a major socio-economic burden on society [68, 76, 139].

Acute measures to revascularize the myocardium have dramatically blunted mortality from myocardial infarction [123], but in turn have resulted in an epidemic increase in the number of patients suffering from heart failure. Current heart failure management works to blunt disease progression though symptomatic palliation. However, this approach lacks the capacity to prevent organ failure [24]. Reaching the precipice of morbidity, patients are offered costly measures such as left ventricular assist devices or organ transplantation. But, as only a limited number of patients can access such therapies, development and implementation of regenerative approaches are needed.

14.2.2 Stem Cell Plasticity

The stem cell paradigm challenges the preeminent view that the myocardium is incapable of repair [21, 63]. This notion was challenged with evaluation of female hearts transplanted into male recipients, identifying exogenous (y-chromosome positive) stem cells from the recipient with capacity to home and integrate within the transplanted myocardium [108]. This hypothesis was furthermore validated through quantification of myocardial C-14 content, in radiation exposed individuals demonstrating cardiomyocyte turnover at a rate of ~1% per year at age 20 to 0.4% per year at age 75. As such, the myocardium is estimated on average to have 45% of its content regenerated by age 50 [22].

Stem cell plasticity provides an opportunity to boost the reparative capacity of the myocardium [43, 134]. Efforts to implement this new approach have been based on recruitment and delivery of adult stem cells [104]. Yet, clinical studies show limited benefit due to significant variability in outcome departing from the repair efficacy seen in pre-clinical studies [1, 41, 44, 116].

Stem cell therapies were initially hypothesized to directly replace lost or non-viable myocardium with newly generated myocytes from transplanted cells. After a decade of evaluation, it is increasingly evident that this intuitively logical premise is too simplistic to account for the complexity that is germane to myocardial regeneration [28, 138]. Infarction results in loss of approximately 40 g of human myocardium, with each gram containing approximately 20 million cardiomyocytes. Injury of greater than 70 g typically results in cardiogenic shock [27, 60, 96, 104, 147]. Regeneration requires restoration of at least 5 g of functional myocardial tissue [142]. When broadly evaluated, a small but detectable positive functional and perfusion advantage is noted with all forms of cell therapy, underscoring the potential for myocardial repair [54, 60, 96].

14.2.3 Direct Versus Indirect Repair Mechanism

Accordingly, novel iterations of the regenerative paradigm move beyond a hypothesis where transplanted stem cells act as new building blocks to rebuild the failing organ, to one that alters the myocardial molecular landscape to engender a healing microenvironment. In this way, paracrine signaling within the heart modulates inflammation, ischemia tolerance, endogenous repair, and inotropy, to promote a condition favorable for repair. Regenerative models have thus been amended to include the concept of indirect regeneration where enriched endogenous capacity for neoangiogenesis, cytoprotection of vulnerable myocardium, and activation of reparative resident cardiac stem cells collectively provide the putative basis for stem cell benefit (Fig. 14.1) [21, 43, 48, 56, 69, 104].

14.3 Cell-Based Myocardial Regeneration

A multitude of cellular phenotypes has been considered for use in the damaged heart. Pre-clinical benefit of cell transplantation has been documented irrespective of cellular origin, surface marker profile, phenotype, or capacity for differentiation. However, this repair efficacy is not translated into clinical practice. Here, hurdles impeding widespread translation, approaches for up-scaled manufacture and transplantation along with a brief overview of pre-clinical and clinical experience with non-adipose tissue-derived cytotypes will be provided.

14.3.1 Hurdles in Translation

Pre-clinical studies indicate a "universal" functional benefit following stem cell delivery, irrespective of the diverse repertoire utilized [96]. Indeed, indirect trophic mechanisms were hypothesized, after benefit was noted within 72 h of administration, absent any evidence of persistent engraftment. Initial pre-clinical trial testing focuses on feasibility of large-scale phenotype derivation and safety in vivo. Optimization of these steps prior to evaluation of efficacy and clinical trial testing is essential to maintain the integrity of potency during translation. Subtle changes occurring during large-scale culture, cryostorage, transport, and cell handing at the time of transplantation has a profound impact on the ultimate reparative potency of stem cells as exemplified in the REPAIR-AMI versus ASTAMI trials [121].

14.3.1.1 Up-Scaling Stem Cell Culture

Large-scale stem cell production to yield a biologic capable of withstanding regulatory scrutiny requires humanization of the culture condition and good manufacturing practice (GMP) compliance. Historically stem

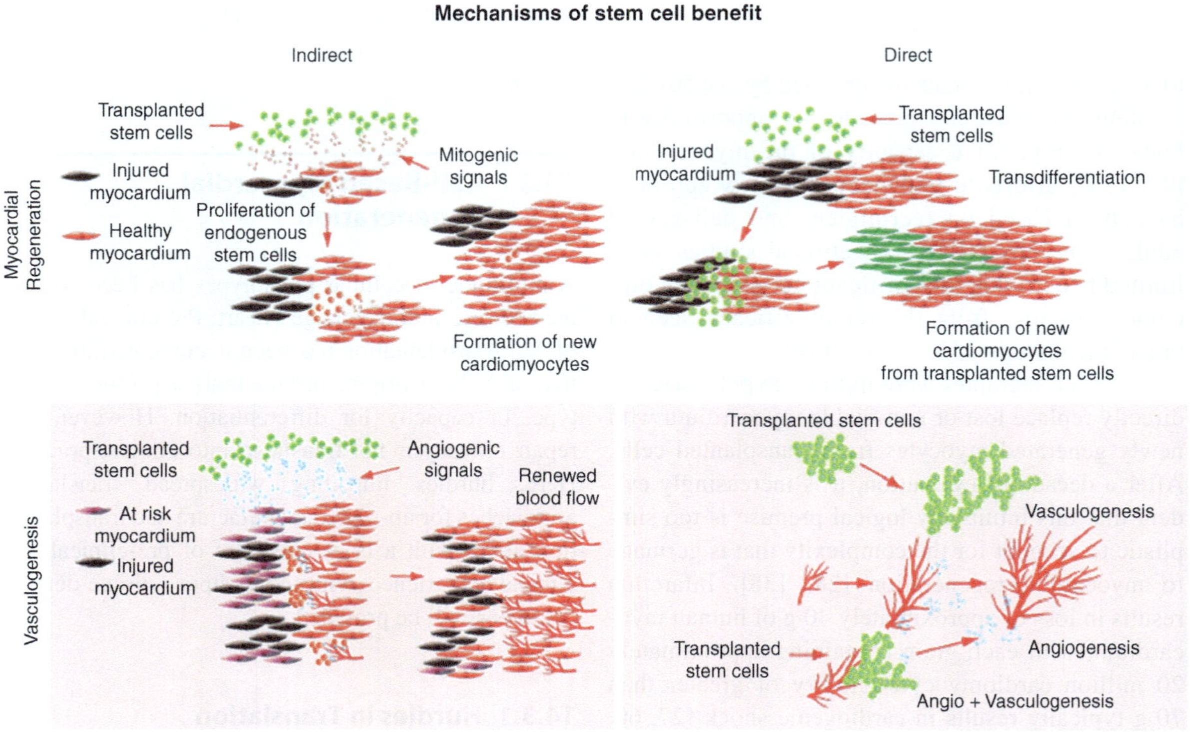

Fig. 14.1 Regenerative models have thus been amended to include the concept of indirect regeneration where enriched endogenous capacity for neoangiogenesis, cytoprotection of vulnerable myocardium, and activation of reparative resident cardiac stem cells collectively provide the putative basis for stem cell benefit

cell culture has relied on fetal bovine serum [33]. This standard has in large part been applied to cells produced for clinical trials, as the overall risk of zoonoses is thought to be small. However, greater scrutiny has been called for to ensure creation of pathogen-free master cell banks, elimination of culture variability, and avoidance of immune response to bovine protein following repeat therapy [29, 30, 45, 57, 62, 81, 97, 122, 133, 144]. Testing and commercialization of humanized substitutes has thus become a point of emphasis paving the way for expedited translation [23, 83, 115, 124].

14.3.1.2 Quality Control

Despite use of stem cell in numerous clinical studies, there continues to be lack of consensus with regard to tissue source and harvest, derivation, culture conditions along with packaging and transport [54]. In addition, no fixed standard linking cellular phenotype to a specific mode of reparative action has been established, resulting in heterogeneous quality control requirements in the release of stem cells for transplantation [42, 43, 150]. Conversely, with the implementation of bone marrow transplantation, vaccination and antibody-based therapeutics within medical practice, guidelines regulating development and delivery of biologics provide significant direction and oversight to ensure GMP compliance in the manufacture of clinical-grade stem cells [99]. Thus, regardless of phenotype, the manufacture, composition, and release of all post-Phase-II biologics delivered must conform to a high degree of quality as determined by full traceability, sterility, homogeneity, and purity [39, 132].

14.3.1.3 Modes of Delivery

One essential, but often overlooked, element in cell-based therapy is mode of delivery. Regardless of the mode of action, efficient stem cell engraftment within injured myocardium is required to achieve therapeutic benefit. This remains an elusive component of translation as no clinically viable method currently exists to track cell survival and homing within the heart following transplantation [105]. Currently, stem cells are delivered via four routes: peripheral intravenous injection (IV); intracoronary catheter-based delivery; catheter-based

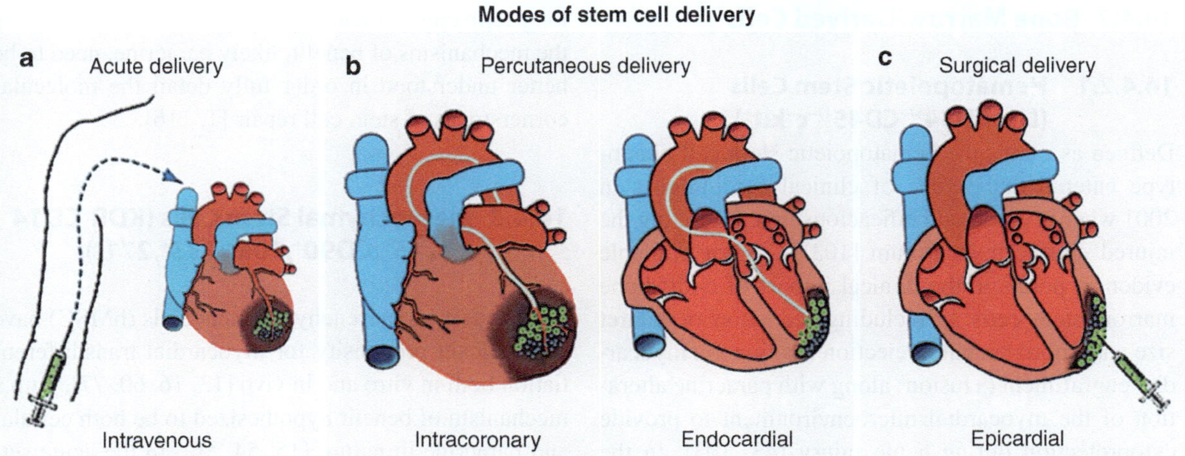

Fig. 14.2 Stem cells are delivered via four routes: peripheral intravenous injection (IV); intracoronary catheter-based delivery; catheter-based endocardial transplantation; and epicardial injection during cardiothoracic surgery

endocardial transplantation; and epicardial injection during cardiothoracic surgery (Fig. 14.2) [9].

Intravenous delivery provides the smallest degree of myocardial homing; however, if the mode of action solely relies upon paracrine secretion into the circulation, this approach would be an attractive option due to its broad clinical applicability. Pre-clinical studies provide proof-of-concept for this approach by demonstrating that, despite the pulmonary homing of IV-injected stem cells, the cardioprotective influence of this phenotype was preserved, in the context of myocardial infarction, due to the bioavailablity of secreted anti-inflammatory proteins [73]. Intracoronary catheter-based delivery has been utilized in multiple clinical trials [6, 25, 40, 67, 117]. Though limited to facilities with significant catheter-based skill, this approach provides an option for concentrated delivery to the site of myocardial injury [66]. Infusion of stem cells through the coronary venous system (coronary sinus) has also been tested, though anatomic variability limits its capacity for specific targeting [105, 129, 130, 140]. Percutaneous transendocardial transplantation is utilized in sub-acute and chronic infarction, to overcome the limited influx of homing and chemokine signaling typically observed during the acute phase of injury [4, 35, 153]. This approach is limited to centers of excellence capable of coupling intervention with techniques such as voltage mapping, computed tomography (CT), magnetic resonance imaging (MRI) or echocardiography to guide site-specific cell delivery [102, 127]. Epicardial cell transplantation is limited to individuals that have a primary indication for cardiac surgery. No approach to date has been singled out as the best technique. Yet, some approaches are more applicable to specific clinical scenarios or mode of action, cellular resource and patient population and will likely dictate the method of delivery.

14.4 Non-adipose Adult-Derived Stem Cells

14.4.1 Myoblasts

Myoblasts are a skeletal muscle-derived source with capacity for clonogenic propagation. These progenitors were the first to be tested in the clinical setting for myocardial repair [85, 90]. This phenotype is thought to provide therapeutic benefit by reserving or aborting the remodeling process through structural support of the post-infarction myocardial scar region [52, 126]. Indeed, a study by Dib and colleagues demonstrated this principle by tracking transplanted cells in hearts explanted from patients undergoing transplantation [37]. Although initially promising, the arrhythmogenic outcome noted in an initial patients, taken together with a limited evidence for long-term functional benefit, prompted a decline in enthusiasm for this approach [91]. However, genetic and catheter-based modifications are currently underway to improve the overall impact of this cell-based approach [38, 58, 89].

14.4.2 Bone Marrow-Derived Cells

14.4.2.1 Hematopoietic Stem Cells ($Lin^-CD34^{+/-}CD45^{+/-}c\text{-}kit^+$)

Defined as c-kit$^+$, the hematopoietic stem cell phenotype entered the psyche of clinical cardiologists in 2001 when it was found efficacious in regenerating the injured murine myocardium [103]. Current available evidence, points to the clinical impact of c-kit$^+$ bone marrow stem cells as including reduction of infarct size and improvement in ejection fraction via myocardial engraftment (±fusion) along with paracrine alteration of the myocardial microenvironment to provide cytoprotection during acute injury [43, 103]. In the acute setting, unsorted bone marrow mononuclear cells (BMMNC) consisting of endothelial progenitors, hematopoietic stem cells, monocytes, and stromal stem cells are typically utilized. A limited number of studies characterize identity prior to transplantation including TOPCARE-AMI and REPAIR-AMI [117–119, 135]. Here bone marrow stem cells were harvested using autologous serum and sorted for a CD34/CD45 positive state favoring a hematopoietic lineage. In addition, a study by Bartunek and colleagues utilized CD133 to isolated therapeutic stem cells but found that benefit came at the price of increased in-stent stenosis [6]. Overall, recent meta-analyses identify more than 80 studies evaluating cellular efficacy in repair of myocardial infarction. Though divergent outcomes are reported, the majority of studies show a marginal degree of ejection fraction and scar size benefit [1, 6, 53, 67, 70, 71, 78, 86, 87, 117, 119]. Thus far, studies have demonstrated a favorable safety profile with feasibility demonstrated for stem cell derivation and delivery [19, 119]. Specific studies such as the ASTAMI study revealed no benefit in the short term [80] and BOOST revealed that although post-infarction benefit is observed, it is lost upon long-term follow-up [93]. Cellular preparation and storage may account for such inter-trial discrepancy [121]. In particular, BOOST's utilization of undefined BMMNC in contrast to a well-defined CD34/45 positive population in REPAIR-AMI may have created a discrepancy in outcome. Overall the benefit of unfractionated or refined BMMNC in the treatment of acute myocardial infarction has been modest at best, with an ejection fraction benefit of $3.6 \pm 1.8\%$, infarct size decrease of $-5.5 \pm 3.6\%$, and left ventricular end systolic volume decrease of -4.8 ± 3.4 mL. As there is little evidence to indicate long-term engraftment following intravascular delivery the mechanisms of benefit, likely paracrine, need to be better understood in order fully detail the molecular cornerstones of stem cell repair [1, 116].

14.4.3 Mesenchymal Stem Cells ($KDR^-CD14^-, 34^-, 45^-; CD90^+, 105^+, 133^+, 271^+$)

Patient-derived mesenchymal stem cells (hMSC) have a significant propensity for myocardial transdifferentiation both in vitro and in vivo [13, 16, 60, 77], with a mechanism of benefit hypothesized to be both cellular and paracrine in nature [15, 54, 56]. In the acute setting hMSC demonstrate paracrine signaling to reduce apoptosis [36, 56, 73, 113]. In the chronic setting, benefit of hMSC is documented through demonstration of myocardial implantation with neovasculogenesis and de novo cardiogenesis [18, 60, 109]. Recently, Hare and colleagues demonstrated safety and benefit of allogenic hMSC therapy in patients with sub-acute anterior myocardial infarction [61]. At 12 months follow-up, hMSC-treated patients demonstrated a $5.2 \pm 1.9\%$ ejection fraction improvement compared to $1.8 \pm 1.5\%$ in placebo-treated counterparts. With ease of ex vivo propagation and immunoprivilege, this cell phenotype will be increasingly considered for allogenic and autologous use.

14.4.4 Endothelial Progenitor Cells ($CD45^-CD31^-CD34^+CD144^-KDR^+$)

Endothelial progenitors (EPC) have the capacity to clonally divide, form vascular networks in vitro, and participate in repair through neovasculogenesis [4, 74, 75, 110, 143, 153]. EPC mobilization in myocardial infarction was first described in 2001 [128]. The molecular signaling prompting cellular exodus from the bone marrow has been cataloged to include factors such as erythropoietin (EPO), vascular endothelial growth factor (VEGF), GCSF, and SDF-1 [35, 36, 46, 53, 128]. Few studies have assessed EPC role in myocardial infarction, the largest is the circulating progenitor cell (CPC, KDR$^+$/CD31$^+$/CD105$^+$) arm of TOPCARE-AMI [117]. Though a combination of BMMNC and EPC was utilized, significant benefit was documented. The majority of studies focus on cellular mobilization instead direct transplantation.

The HEBE-III trial documented the safety of erythropoietin following coronary reperfusion, though no benefit was observed [20]. GCSF at the time of reperfusion also has not provided a therapeutic yield [46]. While delivery of EPC may be useful in the neorevascularization of patients with intractable angina [79, 148], use of mobilizing agents at the time of myocardial infarction does not appear to provide benefit. Lack of factor-based mobilization efficacy may be due to an already upregulated endogenous response exhausting bone marrow stores, insufficient exogenous induction, or patient morbidity [41, 101, 112].

14.4.5 Resident Cardiac Stem Cells

The discovery of adult cardiac stem cells and their ex vivo capacity for expansion [21, 92], has created interest in the heart as a potential stem cell source. In adult hearts Sca-1 (murine) and c-Kit (human) positive populations have demonstrated the capacity to undergo myocardial transdifferentiation, while Islet-1 (Isl-1) positive cells have been identified only in embryonic and neonatal hearts as having ex vivo capacity for cardiogenesis [106, 114]. The most promising therapeutic is the c-kit positive population [21, 92], readily isolated patient heart specimens demonstrating clonal ex vivo proliferation [11, 131], significant myocardial homing capacity and potential for regeneration in pre-clinical models. Clinical evaluation of this cytotype is currently underway in the CADUCEUS (CArdiosphere-Derived aUtologous Stem CElls to reverse ventricUlar dySfunction) and SCIPO (Myocardial Regeneration Using Cardiac Stem Cells) trials [34, 69, 137, 145].

14.5 Adult-Derived Adipose Stem Cells

Adipose tissue-derived stem cells (ATSCs) have become the focus of many recent investigations due to practicality of their use. Derivation of stem cell from tissues such as heart, skeletal muscle, or bone marrow comes with the cost of painful and time-consuming procedures. This is in contrast to the isolation of adult stem cells from adipose tissue. In most patients, this resource is readily accessible via minimally invasive excision or microlipoaspiration procedures. In addition, adipose tissue has been shown to harbor significantly higher stem cell density [33, 49] compared to other tissues, resulting in enhanced yield despite smaller quantities of patient tissue [5, 50].

In the myocardium, several recent studies demonstrate the benefit of ATSCs to include direct contribution to neovasculogenesis and de novo cardiogenesis, in addition to indirect paracrine action for cardioprotection, neoangiogenesis, and neuron spindle formation [3, 11, 136]. Furthermore, in pre-clinical evaluation, treatment with ATSCs corresponded with a diminished propensity for arrhythmogenic risk without any evidence for uncontrolled growth, systemic complications, or tumorigenic change [5]. This section will comprehensively review the pre-clinical and translational work done to definitively evaluate the role of ATSCs in myocardial regeneration.

14.5.1 Cell Phenotypes Isolated from Adipose Tissue

Mononuclear cells are typically isolated from adipose tissue using collagenase-based digestion approaches [55], through recently commercialized tools [120]. Mononuclear cells freshly isolated from adipose tissue are typically heterogeneous [49] containing adipose tissue-derived mesenchymal stem cells (AT-hMSC; $CD34^-/45^-$;$CD44^+/CD90^+/105^+$) [33], endothelial cells ($CD34^+$/c-kit^+) [94], and hematopoietic stem cells ($CD11b^+/34^+/45^+$) [32]. In addition, smooth muscle cells (likely from disrupted microvascular walls) are also detected [156, 157]. Each stem cell population detected within adipose tissue has shown, in vitro, the capacity to undergo lineage-specific differentiation. Endothelial cells, when cultured on matrigel demonstrate the capacity to form capillary-like networks within a VEGF-rich medium. AT-hMSC following culture in either FBS or human pooled platelet lysate (PL) have the capacity for over 100 population doublings and demonstrate the capacity to uniformly maintain the hMSC surface marker expression profile [33, 154]. When placed in an induction medium, these multipotent stem cells demonstrate the capacity to differentiate into adipocytes, osteobalsts, chondrocytes, and cardioblasts [33, 47, 59, 157]. Furthermore, AT-hMSC have demonstrated the capacity to form tissue from all three germinal layer with the capacity, both in vitro and in vivo, for hepato-, neuro-, and cardiogenesis [1, 107, 136, 155].

14.5.2 Pre-clinical Evaluation of ATSC in Myocardial Infarction

Following discovery of ATSC, a number of studies were undertaken to evaluate the benefit of this cellular resource on the myocardium. Multiple groups have specifically studied the role of ATSC in pig and rodent models of acute myocardial infarction. Here, stem cells are delivered following ligation of the left anterior descending artery and demonstrated a cardioprotective influence preserving ejection fraction and reducing end systolic volumes [2, 26, 146]. In chronic heart disease the experience is less robust. The majority of studies have focused on either pig, rodent, or rabbit models of sub-acute myocardial infarction with delivery of AT-hMSC (derived either from brown or white fat) directly into the peri-infarction region of the myocardium with an overall improvement in myocardial pump function and remodeling [26, 95, 120, 146, 152].

14.5.3 Mechanism of ATSC Benefit in the Myocardium

With the potential for therapeutic benefit, tracking the fate of transplanted ATSCs within the myocardium has been implemented in order to elucidate their role and function in myocardial restoration. Several hypotheses have been proposed to explain the beneficial effect of ATSC in heart disease. Currently, ATSC influence is thought to be multifactorial and based on recent evidence includes direct differentiation into new cardiomyocytes along with direct participation of ATSC in neovasculogenesis [33, 94, 107]. Indirect mechanisms of benefit have also been suggested with secretion of pro-regenerative factors such as vascular endothelial growth factor (VEGF), hepatocyte growth factor (HGF), and Insulin-like Growth Factor (IGF) into the myocardial microenvironment serving an inductive role upon endogenous cardiac stem cells [98, 113]. Secretion of heat shock proteins and anti-oxidants has also been documented with AT-hMSC indicating a cardioprotective role at the time of acute injury [82]. Histological studies have been utilized to directly probe for the ultimate location and phenotype of implanted cells. However, these works are limited to evaluation and quantify cell survival or homing within the myocardium without an evaluation of cell fate. Typically, the studies focus on detection of human-specific markers within the recipient model's myocardium. Tool utilized to this end include use of human-specific surface markers, genome, or genetic manipulation of transplanted stem cells to allow whole animal stem cell tracking and detection [5, 26, 88].

14.5.4 Clinical Translation

With pre-clinical studies signaling a beneficial effect on the myocardium, in 2007 two clinical studies were initiated to assess the benefit of ATSC in patients. These trials have as code names "AdiPOse-derived Stem ceLLs in the treatment of patients with st-elevation myOcardial infarction" (APOLLO) and "a randomized clinical trial of adiPose-deRived stEm and regenerative Cells In the treatment of patients with non-revaScularizable ischEmic myocardium" (PRECISE). Both are prospective Phase I placebo-controlled, double-blinded studies designed to assess safety and feasibility of this approach. In regards to safety, major adverse cardiac and cerebral events (MACCE) will be assessed, while for feasibility the capacity to ascertain myocardial pump function improvement will be monitored. As documented in clinicaltrials.gov, the APOLLO study as of 2010 has completed the recruiting phase and is actively treating patients. Inclusion criteria for this study includes age >20 and <80 years, both genders, and an ability to undergo liposuction. Key pathological inclusion criteria is acute myocardial infarction for a minimum of 2 and maximum of 12 h requiring percutaneous coronary intervention and unresponsive to nitroglycerin, along with successful revascularization of the culprit lesion, area of hypo- or akinesis (as determined by ventriculogram) corresponding to the culprit lesion resulting in an LVEF of >30% and <50%. The APOLLO trial utilizes a mechanical device to process and generate the potentially therapeutic stem cells.

The PRECISE trial prospectively assesses the benefit on ATSC in individuals 20–75 years of age with coronary artery disease that is not amenable to revascularization. These individuals must demonstrate hemodynamic stability, ability to undergo liposuction, and have the capacity to walk on a treadmill. Currently, the study is ongoing but not actively recruiting. In addition to these pilot studies, larger studies will be required to demonstrate the long-term efficacy of ATSC therapy [7].

14.6 Future Directions

With the aging of population worldwide, the cardiovascular disease epidemic will remain a major cause of morbidity and mortality. Though advances in molecular diagnostics allow for early detection and prevention, the treatment of heart disease remains largely palliative in nature. Cell-based therapeutics provide an approach to utilize adult stem cells to rectify the distortion of pump function following myocardial injury [16, 134]. Both pre-clinical and clinical studies demonstrate a favorable trend, yet it still remains uncertain whether implanted somatic tissue-derived stem cells retain the capacity to reliably engage in regeneration. This in particular appears to be true of bone marrow-derived stem cells, recruited from patients suffering from heart disease. Transplantation of this resource has resulted in significant interpatient variability in efficacy culminating in a marginal degree of ejection fraction benefit [1, 116]. Adipose tissue-derived stem cells may provide a potential solution to this issue as these stem cells reside in a typically insulated microenvironment when compared to bone marrow [154]. However, change of tissue resource may not be sufficient to ensure regenerative outcome. To this end, preemptive differentiation of adult tissue-derived stem cells may be required. This approach has been demonstrated with direct injection of adult stem cells into blastocysts providing natural ectopic cues to guide lineage specification [77]. In addition, modification of the DNA using factors such as 5-azacytidine has also been attempted [72]. Though effective, these approaches provide a limited cellular yield that is non-viable for translation. Alternatively, harnessing embryonic cues to guide cardiogenesis in a recombinant fashion would provide a platform by which to scale-up lineage specification of adult stem cells [15, 16, 18]. Recombinant approaches to cardiogenically prime stem cells prior to transplantation have been demonstrated as efficacious in ensuring increased beneficial outcome [18].

Adipose tissue-derived stem cells obtained from healthy donors have been shown to demonstrate beneficial outcome following implantation into disease myocardium. However, as was seen with bone marrow, subpopulations of adipose-derived stem cells are not likely to be innately primed for cardiogenesis at the time of transplantation. Thus, utilization of this novel cellular resource paired with cardioinductive lineage specification, may provide a powerful therapeutic platform for the repair of injured myocardium [84]. Furthermore, as the cellular yield of adipose tissue is heterogeneous, a careful understanding of each cell's role in repair must be determined such that an optimized mixture of cells, pre-specified or not, is designed to provide maximal benefit upon transplantation. To this end, a standardized method to isolate and expand the desired cell phenotype from adipose tissue is first required with assessment of cell functionality. Next, a preemptive fate commitment is needed to ensure maximized cellular efficacy. Finally, combination of cell phenotypes to identify stem cell combination with the highest level of regenerative efficacy for acute, subacute, and chronic myocardial infarction repair needs to be determined [17].

14.7 Conclusions

Adipose tissue has demonstrated significant capacity to improve myocardial function with engrafted stem cells demonstrating the capacity for contribution both via direct differentiation and indirect regenerative paracrine signaling in the host myocardial microenvironment. Thus, this tissue resource holds great promise in providing the raw stem cell material to generate therapeutic products in the treatment of cardiovascular disease. However, due to the heterogenous nature of derived stem cells, further evaluation is needed to delineate the mechanism of stem cell benefit within the myocardium and implement next generation approaches such as linage specification and combinatorial stem cell therapy.

References

1. Abdel-Latif A, Bolli R, Tleyjeh IM et al (2007) Adult bone marrow-derived cells for cardiac repair: a systematic review and meta-analysis. Arch Intern Med 167(10):989–997
2. Alt E, Pinkernell K, Scharlau M et al (2010) Effect of freshly isolated autologous tissue resident stromal cells on cardiac function and perfusion following acute myocardial infarction. Int J Cardiol 144(1):26–35
3. Ashjian PH, Elbarbary AS, Edmonds B et al (2003) In vitro differentiation of human processed lipoaspirate cells into early neural progenitors. Plast Reconstr Surg 111(6):1922–1931
4. Askari AT, Unzek S, Popovic ZB et al (2003) Effect of stromal-cell-derived factor 1 on stem-cell homing and tissue

regeneration in ischaemic cardiomyopathy. Lancet 362 (9385):697–703
5. Bai X, Alt E (2010) Myocardial regeneration potential of adipose tissue-derived stem cells. Biochem Biophys Res Commun 401(3):321–326
6. Bartunek J, Vanderheyden M, Vandekerckhove B et al (2005) Intracoronary injection of CD133-positive enriched bone marrow progenitor cells promotes cardiac recovery after recent myocardial infarction: feasibility and safety. Circulation 112(9 Suppl):I178–I183
7. Bartunek J, Dimmeler S, Drexler H et al (2006) The consensus of the task force of the European Society of Cardiology concerning the clinical investigation of the use of autologous adult stem cells for repair of the heart. Eur Heart J 27(11):1338–1340
8. Bartunek J, Behfar A, Vanderheyden M, Wijns W, Terzic A (2008) Mesenchymal stem cells and cardiac repair: principles and practice. J Cardiovasc Transl Res 1(2):115–119
9. Bartunek J, Sherman W, Vanderheyden M, Fernandez-Aviles F, Wijns W, Terzic A (2009) Delivery of biologics in cardiovascular regenerative medicine. Clin Pharmacol Ther 85(5):548–552
10. Bartunek J, Vanderheyden M, Hill J, Terzic A (2010) Cells as biologics for cardiac repair in ischaemic heart failure. Heart 96(10):792–800
11. Bearzi C, Rota M, Hosoda T et al (2007) Human cardiac stem cells. Proc Natl Acad Sci USA 104(35):14068–14073
12. Becker AJ, McCulloch CE, Till JE (1963) Cytological demonstration of the clonal nature of spleen colonies derived from transplanted mouse marrow cells. Nature 197:452–454
13. Behfar A, Terzic A (2006) Derivation of a cardiopoietic population from human mesenchymal stem cells yields cardiac progeny. Nat Clin Pract Cardiovasc Med 3(Suppl 1): S78–S82
14. Behfar A, Terzic A (2008) Mesenchymal stem cells: engineering regeneration. Clin Transl Sci 1(1):34–35
15. Behfar A, Perez-Terzic C, Faustino RS et al (2007) Cardiopoietic programming of embryonic stem cells for tumor-free heart repair. J Exp Med 204(2):405–420
16. Behfar A, Faustino RS, Arrell DK, Dzeja PP, Perez-Terzic C, Terzic A (2008) Guided stem cell cardiopoiesis: discovery and translation. J Mol Cell Cardiol 45(4):523–529
17. Behfar A, Crespo-Diaz R, Nelson TJ, Terzic A, Gersh BJ (2010) Stem cells: clinical trials results the end of the beginning or the beginning of the end? Cardiovasc Hematol Disord Drug Targets 10(3):186–201
18. Behfar A, Yamada S, Crespo-Diaz R et al (2010) Guided cardiopoiesis enhances therapeutic benefit of bone marrow human mesenchymal stem cells in chronic myocardial infarction. J Am Coll Cardiol 56(9):721–34
19. Beitnes JO, Hopp E, Lunde K et al (2009) Long-term results after intracoronary injection of autologous mononuclear bone marrow cells in acute myocardial infarction: the ASTAMI randomised, controlled study. Heart 95(24): 1983–1989
20. Belonje AM, Voors AA, van Gilst WH et al (2008) Effects of erythropoietin after an acute myocardial infarction: rationale and study design of a prospective, randomized, clinical trial (HEBE III). Am Heart J 155(5):817–822
21. Beltrami AP, Barlucchi L, Torella D et al (2003) Adult cardiac stem cells are multipotent and support myocardial regeneration. Cell 114(6):763–776
22. Bergmann O, Bhardwaj RD, Bernard S et al (2009) Evidence for cardiomyocyte renewal in humans. Science 324(5923): 98–102
23. Bieback K, Hecker A, Kocaomer A et al (2009) Human alternatives to fetal bovine serum for the expansion of mesenchymal stromal cells from bone marrow. Stem Cells 27(9):2331–2341
24. Braunwald E (2008) Biomarkers in heart failure. N Engl J Med 358(20):2148–2159
25. Britten MB, Abolmaali ND, Assmus B et al (2003) Infarct remodeling after intracoronary progenitor cell treatment in patients with acute myocardial infarction (TOPCARE-AMI): mechanistic insights from serial contrast-enhanced magnetic resonance imaging. Circulation 108(18):2212–2218
26. Cai L, Johnstone BH, Cook TG et al (2009) IFATS collection: human adipose tissue-derived stem cells induce angiogenesis and nerve sprouting following myocardial infarction, in conjunction with potent preservation of cardiac function. Stem Cells 27(1):230–237
27. Caulfield JB, Leinbach R, Gold H (1976) The relationship of myocardial infarct size and prognosis. Circulation 53(3 Suppl):I141–I144
28. Chien KR, Domian IJ, Parker KK (2008) Cardiogenesis and the complex biology of regenerative cardiovascular medicine. Science 322(5907):1494–1497
29. Cobo F, Concha A (2007) Application of microarray technology for microbial diagnosis in stem cell cultures: a review. Cytotherapy 9(1):53–59
30. Cobo F, Cabrera C, Catalina P, Concha A (2006) General safety guidances in stem cell bank installations. Cytotherapy 8(1):47–56
31. Cohnheim J (1867) Inflammation and sepsis. Pathol Anat Physiol Klin Med 40:1–79
32. Cousin B, Andre M, Arnaud E, Penicaud L, Casteilla L (2003) Reconstitution of lethally irradiated mice by cells isolated from adipose tissue. Biochem Biophys Res Commun 301(4):1016–1022
33. Crespo-Diaz R, Behfar A, Perez-Terzic C, Bartunek J, Dietz AB, Terzic A (2011) Platelet lysate consisting of a natural repair proteome supports human mesenchymal stem cell proliferation and chromosomal stability.Cell Transplant. 2010 Nov 19. [Epub ahead of print]
34. Dawn B, Stein AB, Urbanek K et al (2005) Cardiac stem cells delivered intravascularly traverse the vessel barrier, regenerate infarcted myocardium, and improve cardiac function. Proc Natl Acad Sci USA 102(10):3766–3771
35. De Falco E, Porcelli D, Torella AR et al (2004) SDF-1 involvement in endothelial phenotype and ischemia-induced recruitment of bone marrow progenitor cells. Blood 104(12): 3472–3482
36. Deuse T, Peter C, Fedak PW et al (2009) Hepatocyte growth factor or vascular endothelial growth factor gene transfer maximizes mesenchymal stem cell-based myocardial salvage after acute myocardial infarction. Circulation 120(11 Suppl):S247–S254
37. Dib N, McCarthy P, Campbell A et al (2005) Feasibility and safety of autologous myoblast transplantation in patients with ischemic cardiomyopathy. Cell Transplant 14(1):11–19
38. Dib N, Dinsmore J, Lababidi Z et al (2009) One-year follow-up of feasibility and safety of the first U.S., randomized, controlled study using 3-dimensional guided catheter-based delivery of autologous skeletal myoblasts for ischemic

cardiomyopathy (CAuSMIC study). JACC Cardiovasc Interv 2(1):9–16
39. Dietz AB, Padley DJ, Gastineau DA (2007) Infrastructure development for human cell therapy translation. Clin Pharmacol Ther 82(3):320–324
40. Dill T, Schachinger V, Rolf A et al (2009) Intracoronary administration of bone marrow-derived progenitor cells improves left ventricular function in patients at risk for adverse remodeling after acute ST-segment elevation myocardial infarction: results of the Reinfusion of Enriched Progenitor cells And Infarct Remodeling in Acute Myocardial Infarction study (REPAIR-AMI) cardiac magnetic resonance imaging substudy. Am Heart J 157(3): 541–547
41. Dimmeler S, Leri A (2008) Aging and disease as modifiers of efficacy of cell therapy. Circ Res 102(11):1319–1330
42. Dimmeler S, Zeiher AM (2008) Cell therapy of acute myocardial infarction: open questions. Cardiology 113(3): 155–160
43. Dimmeler S, Zeiher AM, Schneider MD (2005) Unchain my heart: the scientific foundations of cardiac repair. J Clin Invest 115(3):572–583
44. Dimmeler S, Burchfield J, Zeiher AM (2008) Cell-based therapy of myocardial infarction. Arterioscler Thromb Vasc Biol 28(2):208–216
45. Doerr HW, Cinatl J, Sturmer M, Rabenau HF (2003) Prions and orthopedic surgery. Infection 31(3):163–171
46. Engelmann MG, Theiss HD, Hennig-Theiss C et al (2006) Autologous bone marrow stem cell mobilization induced by granulocyte colony-stimulating factor after subacute ST-segment elevation myocardial infarction undergoing late revascularization: final results from the G-CSF-STEMI (Granulocyte Colony-Stimulating Factor ST-Segment Elevation Myocardial Infarction) trial. J Am Coll Cardiol 48(8):1712–1721
47. Estes BT, Wu AW, Guilak F (2006) Potent induction of chondrocytic differentiation of human adipose-derived adult stem cells by bone morphogenetic protein 6. Arthritis Rheum 54(4):1222–1232
48. Forrester JS, White AJ, Matsushita S, Chakravarty T, Makkar RR (2009) New paradigms of myocardial regeneration post-infarction: tissue preservation, cell environment, and pluripotent cell sources. JACC Cardiovasc Interv 2(1):1–8
49. Fraser JK, Wulur I, Alfonso Z, Hedrick MH (2006) Fat tissue: an underappreciated source of stem cells for biotechnology. Trends Biotechnol 24(4):150–154
50. Fraser JK, Zhu M, Wulur I, Alfonso Z (2008) Adipose-derived stem cells. Methods Mol Biol 449:59–67
51. Fuchs E, Tumbar T, Guasch G (2004) Socializing with the neighbors: stem cells and their niche. Cell 116(6):769–778
52. Gavira JJ, Perez-Ilzarbe M, Abizanda G et al (2006) A comparison between percutaneous and surgical transplantation of autologous skeletal myoblasts in a swine model of chronic myocardial infarction. Cardiovasc Res 71(4):744–753
53. Ge J, Li Y, Qian J et al (2006) Efficacy of emergent transcatheter transplantation of stem cells for treatment of acute myocardial infarction (TCT-STAMI). Heart 92(12): 1764–1767
54. Gersh BJ, Simari RD, Behfar A, Terzic CM, Terzic A (2009) Cardiac cell repair therapy: a clinical perspective. Mayo Clin Proc 84(10):876–892
55. Gimble J, Guilak F (2003) Adipose-derived adult stem cells: isolation, characterization, and differentiation potential. Cytotherapy 5(5):362–369
56. Gnecchi M, He H, Liang OD et al (2005) Paracrine action accounts for marked protection of ischemic heart by Akt-modified mesenchymal stem cells. Nat Med 11(4):367–368
57. Gregory CA, Reyes E, Whitney MJ, Spees JL (2006) Enhanced engraftment of mesenchymal stem cells in a cutaneous wound model by culture in allogenic species-specific serum and administration in fibrin constructs. Stem Cells 24(10):2232–2243
58. Haider H, Lei Y, Ashraf M (2008) MyoCell, a cell-based, autologous skeletal myoblast therapy for the treatment of cardiovascular diseases. Curr Opin Mol Ther 10(6):611–621
59. Halvorsen YD, Franklin D, Bond AL et al (2001) Extracellular matrix mineralization and osteoblast gene expression by human adipose tissue-derived stromal cells. Tissue Eng 7(6):729–741
60. Hare JM, Chaparro SV (2008) Cardiac regeneration and stem cell therapy. Curr Opin Organ Transplant 13(5):536–542
61. Hare JM, Traverse JH, Henry TD et al (2009) A randomized, double-blind, placebo-controlled, dose-escalation study of intravenous adult human mesenchymal stem cells (prochymal) after acute myocardial infarction. J Am Coll Cardiol 54(24):2277–2286
62. Horwitz EM, Gordon PL, Koo WK et al (2002) Isolated allogeneic bone marrow-derived mesenchymal cells engraft and stimulate growth in children with osteogenesis imperfecta: Implications for cell therapy of bone. Proc Natl Acad Sci USA 99(13):8932–8937
63. Hsieh PC, Segers VF, Davis ME et al (2007) Evidence from a genetic fate-mapping study that stem cells refresh adult mammalian cardiomyocytes after injury. Nat Med 13(8): 970–974
64. Hunt SA, Baker DW, Chin MH et al (2001) ACC/AHA guidelines for the evaluation and management of chronic heart failure in the adult: executive summary a report of the American College of Cardiology/American Heart Association Task Force on Practice Guidelines (Committee to revise the 1995 guidelines for the evaluation and management of heart failure): developed in collaboration with the International Society for Heart and Lung Transplantation; endorsed by the Heart Failure Society of America. Circulation 104(24): 2996–3007
65. Hunt SA, Baker DW, Chin MH et al (2001) ACC/AHA guidelines for the evaluation and management of chronic heart failure in the adult: executive summary. A report of the American College of Cardiology/American Heart Association Task Force on Practice Guidelines (Committee to revise the 1995 guidelines for the evaluation and management of heart failure). J Am Coll Cardiol 38(7):2101–2113
66. Janssens S (2010) Stem cells in the treatment of heart disease. Annu Rev Med 61:287–300
67. Janssens S, Dubois C, Bogaert J et al (2006) Autologous bone marrow-derived stem-cell transfer in patients with ST-segment elevation myocardial infarction: double-blind, randomised controlled trial. Lancet 367(9505):113–121
68. Jennifer Manfrè MI (2007) An Unhealthy America: economic burden of chronic disease. The Milken Institute Review http://www.fightchronicdisease.org/media-center/releases/milken-institute-study-chronic-disease-costs-us-economy-more-1-trillion-annual

69. Johnston PV, Sasano T, Mills K et al (2009) Engraftment, differentiation, and functional benefits of autologous cardiosphere-derived cells in porcine ischemic cardiomyopathy. Circulation 120(12):1075–1083, 1077 p following 1083
70. Kang HJ, Lee HY, Na SH et al (2006) Differential effect of intracoronary infusion of mobilized peripheral blood stem cells by granulocyte colony-stimulating factor on left ventricular function and remodeling in patients with acute myocardial infarction versus old myocardial infarction: the MAGIC Cell-3-DES randomized, controlled trial. Circulation 114(1 Suppl):I145–I151
71. Kang S, Yang YJ, Li CJ, Gao RL (2008) Effects of intracoronary autologous bone marrow cells on left ventricular function in acute myocardial infarction: a systematic review and meta-analysis for randomized controlled trials. Coron Artery Dis 19(5):327–335
72. Labovsky V, Hofer EL, Feldman L et al (2010) Cardiomyogenic differentiation of human bone marrow mesenchymal cells: role of cardiac extract from neonatal rat cardiomyocytes. Differentiation 79(2):93–101
73. Lee RH, Pulin AA, Seo MJ et al (2009) Intravenous hMSCs improve myocardial infarction in mice because cells embolized in lung are activated to secrete the anti-inflammatory protein TSG-6. Cell Stem Cell 5(1):54–63
74. Leone AM, Rutella S, Bonanno G et al (2006) Endogenous G-CSF and CD34+ cell mobilization after acute myocardial infarction. Int J Cardiol 111(2):202–208
75. Leone AM, Valgimigli M, Giannico MB et al (2009) From bone marrow to the arterial wall: the ongoing tale of endothelial progenitor cells. Eur Heart J 30(8):890–899
76. Liao L, Anstrom KJ, Gottdiener JS et al (2007) Long-term costs and resource use in elderly participants with congestive heart failure in the Cardiovascular Health Study. Am Heart J 153(2):245–252
77. Liechty KW, MacKenzie TC, Shaaban AF et al (2000) Human mesenchymal stem cells engraft and demonstrate site-specific differentiation after in utero transplantation in sheep. Nat Med 6(11):1282–1286
78. Lipinski MJ, Biondi-Zoccai GG, Abbate A et al (2007) Impact of intracoronary cell therapy on left ventricular function in the setting of acute myocardial infarction: a collaborative systematic review and meta-analysis of controlled clinical trials. J Am Coll Cardiol 50(18):1761–1767
79. Losordo DW, Schatz RA, White CJ et al (2007) Intramyocardial transplantation of autologous CD34+ stem cells for intractable angina: a phase I/IIa double-blind, randomized controlled trial. Circulation 115(25):3165–3172
80. Lunde K, Solheim S, Aakhus S et al (2006) Intracoronary injection of mononuclear bone marrow cells in acute myocardial infarction. N Engl J Med 355(12):1199–1209
81. Mackensen A, Drager R, Schlesier M, Mertelsmann R, Lindemann A (2000) Presence of IgE antibodies to bovine serum albumin in a patient developing anaphylaxis after vaccination with human peptide-pulsed dendritic cells. Cancer Immunol Immunother 49(3):152–156
82. Madonna R, De Caterina R (2010) Adipose tissue: a new source for cardiovascular repair. J Cardiovasc Med (Hagerstown) 11(2):71–80
83. Mannello F, Tonti GA (2007) Concise review: no breakthroughs for human mesenchymal and embryonic stem cell culture: conditioned medium, feeder layer, or feeder-free; medium with fetal calf serum, human serum, or enriched plasma; serum-free, serum replacement nonconditioned medium, or ad hoc formula? All that glitters is not gold! Stem Cells 25(7):1603–1609
84. Marban E, Malliaras K (2010) Boot camp for mesenchymal stem cells. J Am Coll Cardiol 56(9):735–737
85. Marelli D, Desrosiers C, El-alfy M, Kao RL, Chiu RC (1992) Cell transplantation for myocardial repair: an experimental approach. Cell Transplant 1(6):383–390
86. Martin-Rendon E, Brunskill S, Doree C et al (2008) Stem cell treatment for acute myocardial infarction. Cochrane Database Syst Rev (4):CD006536
87. Martin-Rendon E, Brunskill SJ, Hyde CJ, Stanworth SJ, Mathur A, Watt SM (2008) Autologous bone marrow stem cells to treat acute myocardial infarction: a systematic review. Eur Heart J 29(15):1807–1818
88. Mazo M, Planat-Benard V, Abizanda G et al (2008) Transplantation of adipose derived stromal cells is associated with functional improvement in a rat model of chronic myocardial infarction. Eur J Heart Fail 10(5):454–462
89. Menasche P (2009) Cell-based therapy for heart disease: a clinically oriented perspective. Mol Ther 17(5):758–766
90. Menasche P, Hagege AA, Vilquin JT et al (2003) Autologous skeletal myoblast transplantation for severe postinfarction left ventricular dysfunction. J Am Coll Cardiol 41(7): 1078–1083
91. Menasche P, Alfieri O, Janssens S et al (2008) The Myoblast autologous grafting in ischemic cardiomyopathy (MAGIC) trial: first randomized placebo-controlled study of myoblast transplantation. Circulation 117(9):1189–1200
92. Messina E, De Angelis L, Frati G et al (2004) Isolation and expansion of adult cardiac stem cells from human and murine heart. Circ Res 95(9):911–921
93. Meyer GP, Wollert KC, Lotz J et al (2006) Intracoronary bone marrow cell transfer after myocardial infarction: eighteen months' follow-up data from the randomized, controlled BOOST (BOne marrOw transfer to enhance ST-elevation infarct regeneration) trial. Circulation 113(10):1287–1294
94. Miranville A, Heeschen C, Sengenes C, Curat CA, Busse R, Bouloumie A (2004) Improvement of postnatal neovascularization by human adipose tissue-derived stem cells. Circulation 110(3):349–355
95. Miyahara Y, Nagaya N, Kataoka M et al (2006) Monolayered mesenchymal stem cells repair scarred myocardium after myocardial infarction. Nat Med 12(4):459–465
96. Murry CE, Reinecke H, Pabon LM (2006) Regeneration gaps: observations on stem cells and cardiac repair. J Am Coll Cardiol 47(9):1777–1785
97. Muul LM, Tuschong LM, Soenen SL et al (2003) Persistence and expression of the adenosine deaminase gene for 12 years and immune reaction to gene transfer components: long-term results of the first clinical gene therapy trial. Blood 101(7):2563–2569
98. Nakagami H, Maeda K, Morishita R et al (2005) Novel autologous cell therapy in ischemic limb disease through growth factor secretion by cultured adipose tissue-derived stromal cells. Arterioscler Thromb Vasc Biol 25(12): 2542–2547
99. Nelson TJ, Faustino RS, Chiriac A, Crespo-Diaz R, Behfar A, Terzic A (2008) CXCR4+/FLK-1+ biomarkers select a

cardiopoietic lineage from embryonic stem cells. Stem Cells 26(6):1464–1473

100. Nelson TJ, Behfar A, Yamada S, Martinez-Fernandez A, Terzic A (2009) Stem cell platforms for regenerative medicine. Clin Transl Sci 2(3):222–227
101. Numaguchi Y, Sone T, Okumura K et al (2006) The impact of the capability of circulating progenitor cell to differentiate on myocardial salvage in patients with primary acute myocardial infarction. Circulation 114(1 Suppl): I114–I119
102. Opie SR, Dib N (2006) Surgical and catheter delivery of autologous myoblasts in patients with congestive heart failure. Nat Clin Pract Cardiovasc Med 3(Suppl 1): S42–S45
103. Orlic D, Kajstura J, Chimenti S et al (2001) Bone marrow cells regenerate infarcted myocardium. Nature 410(6829): 701–705
104. Passier R, van Laake LW, Mummery CL (2008) Stem-cell-based therapy and lessons from the heart. Nature 453 (7193):322–329
105. Perin EC, Lopez J (2006) Methods of stem cell delivery in cardiac diseases. Nat Clin Pract Cardiovasc Med 3(Suppl 1):S110–S113
106. Pfister O, Mouquet F, Jain M et al (2005) CD31- but Not CD31+ cardiac side population cells exhibit functional cardiomyogenic differentiation. Circ Res 97(1):52–61
107. Planat-Benard V, Menard C, Andre M et al (2004) Spontaneous cardiomyocyte differentiation from adipose tissue stroma cells. Circ Res 94(2):223–229
108. Quaini F, Urbanek K, Beltrami AP et al (2002) Chimerism of the transplanted heart. N Engl J Med 346(1):5–15
109. Quevedo HC, Hatzistergos KE, Oskouei BN et al (2009) Allogeneic mesenchymal stem cells restore cardiac function in chronic ischemic cardiomyopathy via trilineage differentiating capacity. Proc Natl Acad Sci USA 106(33): 14022–14027
110. Rafii S, Lyden D (2003) Therapeutic stem and progenitor cell transplantation for organ vascularization and regeneration. Nat Med 9(6):702–712
111. Rangappa S, Fen C, Lee EH, Bongso A, Sim EK (2003) Transformation of adult mesenchymal stem cells isolated from the fatty tissue into cardiomyocytes. Ann Thorac Surg 75(3):775–779
112. Rauscher FM, Goldschmidt-Clermont PJ, Davis BH et al (2003) Aging, progenitor cell exhaustion, and atherosclerosis. Circulation 108(4):457–463
113. Rehman J, Traktuev D, Li J et al (2004) Secretion of angiogenic and antiapoptotic factors by human adipose stromal cells. Circulation 109(10):1292–1298
114. Reinecke H, Minami E, Zhu WZ, Laflamme MA (2008) Cardiogenic differentiation and transdifferentiation of progenitor cells. Circ Res 103(10):1058–1071
115. Reinisch A, Hofmann NA, Obenauf AC et al (2009) Humanized large-scale expanded endothelial colony-forming cells function in vitro and in vivo. Blood 113(26): 6716–6725
116. Rosenzweig A (2006) Cardiac cell therapy – mixed results from mixed cells. N Engl J Med 355(12):1274–1277
117. Schachinger V, Assmus B, Britten MB et al (2004) Transplantation of progenitor cells and regeneration enhancement in acute myocardial infarction: final one-year results of the TOPCARE-AMI Trial. J Am Coll Cardiol 44(8):1690–1699
118. Schachinger V, Erbs S, Elsasser A et al (2006) Improved clinical outcome after intracoronary administration of bone-marrow-derived progenitor cells in acute myocardial infarction: final 1-year results of the REPAIR-AMI trial. Eur Heart J 27(23):2775–2783
119. Schachinger V, Erbs S, Elsasser A et al (2006) Intracoronary bone marrow-derived progenitor cells in acute myocardial infarction. N Engl J Med 355(12):1210–1221
120. Schenke-Layland K, Strem BM, Jordan MC et al (2009) Adipose tissue-derived cells improve cardiac function following myocardial infarction. J Surg Res 153(2):217–223
121. Seeger FH, Tonn T, Krzossok N, Zeiher AM, Dimmeler S (2007) Cell isolation procedures matter: a comparison of different isolation protocols of bone marrow mononuclear cells used for cell therapy in patients with acute myocardial infarction. Eur Heart J 28(6):766–772
122. Selvaggi TA, Walker RE, Fleisher TA (1997) Development of antibodies to fetal calf serum with arthus-like reactions in human immunodeficiency virus-infected patients given syngeneic lymphocyte infusions. Blood 89(3):776–779
123. Setoguchi S, Glynn RJ, Avorn J, Mittleman MA, Levin R, Winkelmayer WC (2008) Improvements in long-term mortality after myocardial infarction and increased use of cardiovascular drugs after discharge: a 10-year trend analysis. J Am Coll Cardiol 51(13):1247–1254
124. Shahdadfar A, Fronsdal K, Haug T, Reinholt FP, Brinchmann JE (2005) In vitro expansion of human mesenchymal stem cells: choice of serum is a determinant of cell proliferation, differentiation, gene expression, and transcriptome stability. Stem Cells 23(9):1357–1366
125. Shamblott MJ, Axelman J, Wang S et al (1998) Derivation of pluripotent stem cells from cultured human primordial germ cells. Proc Natl Acad Sci USA 95(23):13726–13731
126. Sherman W (2007) Myocyte replacement therapy: skeletal myoblasts. Cell Transplant 16(9):971–975
127. Sherman W, Martens TP, Viles-Gonzalez JF, Siminiak T (2006) Catheter-based delivery of cells to the heart. Nat Clin Pract Cardiovasc Med 3(Suppl 1):S57–S64
128. Shintani S, Murohara T, Ikeda H et al (2001) Mobilization of endothelial progenitor cells in patients with acute myocardial infarction. Circulation 103(23):2776–2779
129. Siminiak T, Kalawski R, Fiszer D et al (2004) Autologous skeletal myoblast transplantation for the treatment of postinfarction myocardial injury: phase I clinical study with 12 months of follow-up. Am Heart J 148(3):531–537
130. Siminiak T, Burchardt P, Kurpisz M (2006) Postinfarction heart failure: surgical and trans-coronary-venous transplantation of autologous myoblasts. Nat Clin Pract Cardiovasc Med 3(Suppl 1):S46–S51
131. Smith RR, Barile L, Cho HC et al (2007) Regenerative potential of cardiosphere-derived cells expanded from percutaneous endomyocardial biopsy specimens. Circulation 115(7):896–908
132. Sotiropoulou PA, Perez SA, Salagianni M, Baxevanis CN, Papamichail M (2006) Characterization of the optimal culture conditions for clinical scale production of human mesenchymal stem cells. Stem Cells 24(2):462–471
133. Spees JL, Gregory CA, Singh H et al (2004) Internalized antigens must be removed to prepare hypoimmunogenic

mesenchymal stem cells for cell and gene therapy. Mol Ther 9(5):747–756
134. Srivastava D, Ivey KN (2006) Potential of stem-cell-based therapies for heart disease. Nature 441(7097): 1097–1099
135. Strauer BE, Brehm M, Zeus T et al (2002) Repair of infarcted myocardium by autologous intracoronary mononuclear bone marrow cell transplantation in humans. Circulation 106(15):1913–1918
136. Talens-Visconti R, Bonora A, Jover R et al (2006) Hepatogenic differentiation of human mesenchymal stem cells from adipose tissue in comparison with bone marrow mesenchymal stem cells. World J Gastroenterol 12(36): 5834–5845
137. Tang XL, Rokosh G, Sanganalmath SK et al (2010) Intracoronary administration of cardiac progenitor cells alleviates left ventricular dysfunction in rats with a 30-day-old infarction. Circulation 121(2):293–305
138. Tendera M, Wojakowski W (2009) Cell therapy – success does not come easy. Eur Heart J 30(6):640–641
139. Thom T, Haase N, Rosamond W et al (2006) Heart disease and stroke statistics – 2006 update: a report from the American Heart Association Statistics Committee and Stroke Statistics Subcommittee. Circulation 113(6):e85–e151
140. Thompson CA, Nasseri BA, Makower J et al (2003) Percutaneous transvenous cellular cardiomyoplasty. A novel nonsurgical approach for myocardial cell transplantation. J Am Coll Cardiol 41(11):1964–1971
141. Thomson JA, Itskovitz-Eldor J, Shapiro SS et al (1998) Embryonic stem cell lines derived from human blastocysts. Science 282(5391):1145–1147
142. Trainini JC, Lago N, de Paz J et al (2004) Myoblast transplantation for myocardial repair: a clinical case. J Heart Lung Transplant 23(4):503–505
143. Traktuev DO, Merfeld-Clauss S, Li J et al (2008) A population of multipotent CD34-positive adipose stromal cells share pericyte and mesenchymal surface markers, reside in a periendothelial location, and stabilize endothelial networks. Circ Res 102(1):77–85
144. Tuschong L, Soenen SL, Blaese RM, Candotti F, Muul LM (2002) Immune response to fetal calf serum by two adenosine deaminase-deficient patients after T cell gene therapy. Hum Gene Ther 13(13):1605–1610
145. Urbanek K, Torella D, Sheikh F et al (2005) Myocardial regeneration by activation of multipotent cardiac stem cells in ischemic heart failure. Proc Natl Acad Sci USA 102(24):8692–8697
146. Valina C, Pinkernell K, Song YH et al (2007) Intracoronary administration of autologous adipose tissue-derived stem cells improves left ventricular function, perfusion, and remodelling after acute myocardial infarction. Eur Heart J 28(21):2667–2677
147. van Laake LW, Passier R, Doevendans PA, Mummery CL (2008) Human embryonic stem cell-derived cardiomyocytes and cardiac repair in rodents. Circ Res 102(9): 1008–1010
148. van Ramshorst J, Bax JJ, Beeres SL et al (2009) Intramyocardial bone marrow cell injection for chronic myocardial ischemia: a randomized controlled trial. JAMA 301(19):1997–2004
149. Wagers AJ, Weissman IL (2004) Plasticity of adult stem cells. Cell 116(5):639–648
150. Welt FG, Losordo DW (2006) Cell therapy for acute myocardial infarction: curb your enthusiasm? Circulation 113(10):1272–1274
151. Worton RG, McCulloch EA, Till JE (1969) Physical separation of hemopoietic stem cells differing in their capacity for self-renewal. J Exp Med 130(1):91–103
152. Yamada Y, Yokoyama S, Wang XD, Fukuda N, Takakura N (2007) Cardiac stem cells in brown adipose tissue express CD133 and induce bone marrow nonhematopoietic cells to differentiate into cardiomyocytes. Stem Cells 25(5): 1326–1333
153. Yamaguchi J, Kusano KF, Masuo O et al (2003) Stromal cell-derived factor-1 effects on ex vivo expanded endothelial progenitor cell recruitment for ischemic neovascularization. Circulation 107(9):1322–1328
154. Zhu Y, Liu T, Song K, Fan X, Ma X, Cui Z (2008) Adipose-derived stem cell: a better stem cell than BMSC. Cell Biochem Funct 26(6):664–675
155. Zhu XY, Zhang XZ, Xu L, Zhong XY, Ding Q, Chen YX (2009) Transplantation of adipose-derived stem cells overexpressing hHGF into cardiac tissue. Biochem Biophys Res Commun 379(4):1084–1090
156. Zuk PA, Zhu M, Mizuno H et al (2001) Multilineage cells from human adipose tissue: implications for cell-based therapies. Tissue Eng 7(2):211–228
157. Zuk PA, Zhu M, Ashjian P et al (2002) Human adipose tissue is a source of multipotent stem cells. Mol Biol Cell 13(12):4279–4295

Adipose Stem Cell Engineering: Characterization and Current Application in Neurological Tissue Repair

15

Guangfan Chi and Youngsook Son

Contents

15.1 Introduction 155
15.2 Accessible Alternative Sources for Neural Stem Cells 156
15.3 NCSCs of Adipose Tissue Origin 157
15.4 Neural Induction of ASCs 157
15.5 Induction of ASCs 158
15.6 Application of Induced SCs to Spinal Cord Injury and Brain Ischemic Stroke 159
15.7 Conclusion 160
References 160

15.1 Introduction

Adult adipose stem cells (ASCs) would have several advantages over multipotent bone marrow stromal cells (BMSCs) if their capacity for differentiation could be made comparable to that of BMSCs. For many years, bone marrow was considered the major source of adult stem cells for tissue engineering applications. However, it was recently noted that adipose tissue contains more adult stem cells than the bone marrow. According to previous reports, the BMSC frequency in bone marrow is between 1 in 25,000 and 1 in 100,000 mononuclear cells, while ASCs in adipose tissue represent approximately 2% of total lipoaspirate cells [43]. Additionally, adipose tissue is more broadly distributed in the body, and its presence just beneath the skin (subcutaneous fat) and in the abdomen (abdominal fat) renders it easily accessible. Therefore, ASCs are an attractive, readily available type of adult stem cell that has become increasingly popular for use in mesenchymal tissue repair as well as other therapeutic applications.

Adipose tissue comprises several different cell types whose populations differ according to harvesting and cell culture methods. ASCs can be harvested using a simple procedure: the harvested adipose tissue is digested using enzymes, filtered through a strainer, and collected by centrifugation. The spin-down pellet, which is called the stromal vascular fraction (SVF), contains ASCs as well as a variety of other cells, including adipocytes, pericytes, fibroblasts, vascular smooth muscle cells, endothelial (progenitor) cells, immune cells, and hematopoietic stem cells [24]. The SVF can be used directly as a source of stem cells or expanded by culturing through many passages for other investigations.

G. Chi • Y. Son (✉)
Department of Genetic Engineering, College of Life Science and Graduate School of Biotechnology, Kyung Hee University, Yongin, South Korea
e-mail: ysson@khu.ac.kr

Y.-G. Illouz and A. Sterodimas (eds.), *Adipose Stem Cells and Regenerative Medicine*,
DOI: 10.1007/978-3-642-20012-0_15, © Springer-Verlag Berlin Heidelberg 2011

Typical ASCs possess characteristics that are similar to BMSCs [17], including (a) adherence to plastic; (b) multipotent differentiation potential to osteoblasts, adipocytes, and chondroblasts; and (c) expression of specific surface antigens such as $CD34^-$, $CD45^-$, CD73+,CD90+, and $CD105^+$. Recently, Crisan et al. (2008) suggested that some mesenchymal stem cells (MSCs) from fat tissue are perivascular cells that are associated with blood vessel walls expressing $CD140b^+$(PDGFR-β), $CD146^+$, and $NG2^+$ [11]; these cells have multipotent differentiation potential similar to that of traditional BMSCs. Since adipose tissue includes many vascular structures, this finding may provide a reason why stem cells are more abundant in fat tissue than in bone marrow. Furthermore, numerous reports have suggested that ASCs have the ability to differentiate not only to cells of mesodermal lineages but also to neuronal and glial cells *in vitro* and *in vivo* [9, 31, 36, 43, 45, 96, 97, 100]. These reports imply that ASCs also harbor a subset of adult stem cells that can be transdifferentiated to neuroectodermal cells or a small number of neural crest stem cell (NCSC) types, which migrate during embryonic development and persist in the adipose tissue.

15.2 Accessible Alternative Sources for Neural Stem Cells

Adult neural stem/progenitor cells in the brain are localized in the subventricular zone of the lateral ventricle and the subgranular zone within the dentate gyrus of the hippocampus [16, 48]. In the spinal cord, adult neural stem/progenitor cells have been reported to be quiescent and located in the ependymal layer or in the subependymal zone of the central canal [34, 74, 86]. However, these neural progenitor cells are deeply localized in brain or spinal cord and, thus, are not easily accessible for autogenic cell therapeutics. Allogeneic fetal midbrain tissues have been used for the treatment of Parkinson's disease [21] or Huntington's disease [4, 66], and these fetal tissues have been found to alleviate the symptoms of such patients. However, ethical questions regarding the use of tissue derived from aborted human fetuses as well as issues of tissue availability, limited safety, and quality control have historically been raised. Moreover, transplantation of allogeneic fetal tissue or neural stem cells carries the risk of disease transmission to the recipient and immune rejection of the donor cells.

Embryonic stem (ES) cells and induced pluripotent stem (iPS) cells have similar characteristics of pluripotent differentiation potential and can give rise to all three germinal cell layers of the body *in vitro* and *in vivo*. These ES and iPS cells are therefore alternative cell sources for treating neural disease. According to previous reports, oligodendrocyte progenitor cells and progenitors of motor neurons can be induced from mice and human ES cells, and these cells participate in the functional repair of spinal cord injuries [14, 25, 40, 53, 64]. Nevertheless, ethical issues involving the use of ES cells, as well as problems related to tumorigenesis, will need to be overcome for future application in regenerative medicine. In 2006, Yamanaka and his team first documented that the *Klf4*, *Sox2*, *Oct4*, and *c-Myc* genes transformed somatic cells back into a pluripotent-like state [14]. Since then, numerous papers have reported on iPS cell generation [1, 19, 28, 42, 49, 58, 65, 75, 76, 78, 87, 99] and their subsequent application in tissue regeneration and disease treatment. Further, several papers have reported that dopaminergic neurons generated from iPS cells alleviated some deficits in an animal model of Parkinson disease [75, 87]. However, similar to ES cells, iPS cells are also involved in tumor formation. Therefore, clinical applications of genetically altered iPS cells in patients with neurological diseases will also need long-term investigation.

In recent years, numerous papers have reported that mesenchymal stem cells (MSCs) from bone marrow can be used to treat neurological disorders or diseases. Some reports have stated that murine MSCs differentiate to astrocytes, giving rise to neuronal phenotypes after transplantation into mouse brain [3, 6, 46, 56]. Other studies using a variety of protocols have attempted to differentiate MSCs to neural cells. According to the reagents and culture methods used, protocols can be divided into four categories: methods using chemical compounds [13, 35, 51, 60, 91], methods using growth factors [30, 32, 71, 80], neural sphere culture methods [7, 12, 26, 77, 88, 89], and combination methods [27, 73]. Among these methods, those using chemical compounds have been invalidated by other researchers who proved that after treatment with chemical reagents, neuron-like morphological changes were caused by cellular toxicity and a rapid disruption of the actin cytoskeleton [54, 62]. Although all other methods resulted in cells that exhibited neural cell phenotypes after induction *in vitro*, direct evidence is lacking as to whether these cells can functionally behave like neuronal or glial cells *in vivo*. A smaller number of reports has demonstrated that transplanted cells differentiate to neural cells *in vitro*, but their numbers were insufficient to influence tissue repair [57].

Recent investigations of neurogenesis-promoting functions of MSCs in central nervous system (CNS) injury models have focused mainly on other characteristics of MSCs; *i.e.*, their secretion of neurotrophic factors such as glial cell-derived neurotrophic factor (GDNF), brain-derived neurotrophic factor (BDNF), nerve growth factor (NGF), and other factors such as Angiopoietin-1 and vascular endothelial growth factor [10, 63, 95, 98]. Additionally, MSCs have also been shown to have an immunomodulating function *in vivo* [22, 38]. Through the actions of these secreted trophic factors and this immunomodulating function, MSCs stimulate angiogenesis and neural stem cell migration to the injury site, enhancing neural cell survival and differentiation, eventually improving neural function in CNS injury or disease.

15.3 NCSCs of Adipose Tissue Origin

NCSCs are pluripotent stem cells [50, 84] that arise during embryonic development; they are shed off and migrate from the neural tube through dorsolateral and ventral pathways to surrounding tissue. Under diverse environments, NCSCs can differentiate to several different cell types, including pigment cells (melanocytes), neurons and glial cells of the peripheral nervous system (PNS), and some classes of endocrine cells. In head and neck development, NCSCs also yield mesodermal cells, including connective tissue cells, vascular smooth muscle cells, tendons, dermis, odontoblasts, cartilages, and bone [18, 50]. Therefore, in the body, mesenchymal cells do not solely originate from the mesoderm. Billon et al. (2007) reported that the neuroectoderm/neural crest is a source of adipocytes in mouse embryonic stem cell-derived cultures [5]. In addition, using Cre-lox fate mapping in Sox10-Cre transgenic mice, the authors demonstrated that a subset of adipocytes originate from neural crest cells during normal development [5]. Furthermore, clonally cultured quail pigment cells were shown to differentiate into glia, myofibroblasts, and multipotent neural crest-like precursor cells when treated with endothelin-3 [68]. Based on these reports, there is some possibility for terminally differentiated adipocytes or adipocyte precursor cells to dedifferentiate into NCSC-like cells and progress to neural cells under neural-inducing conditions.

Nagoshi et al. (2008) and Morikawa et al. (2009) support the idea that some BMSCs originate from neural crest cells [59, 61]. Using adult transgenic mice encoding neural crest-specific P0-Cre/Floxed-EGFP and Wnt1-Cre/Floxed-EGFP transgenic mice, EGFP-positive BMSCs were isolated from bone marrow, and it was found that these cells could form spheres. Moreover, these spheroid-forming cells expressed NCSC genes and were able to differentiate into neurons, glial cells, and myofibroblasts. These results suggest that a subset of BMSCs originates from NCSCs [59, 61]. Furthermore, a series of articles demonstrated that rodent and human skin dermis comprise spheroid-forming cells – termed skin-derived precursors (SKPs) – that have multiple potentialities to differentiate to mesenchymal lineages and neuroectodermal lineage cells, including neurons, glia, smooth muscle cells, and adipocytes. SKPs are considered to be NCSC-like cells that persist in the dermis [20, 33, 55, 83, 90]. These NCSCs from different tissue origins all have common features: they form spheroids in serum-free media composed of N2 or B27 supplements, bFGF, and EGF; they consecutively self-renew in this culture condition for many passages; and they have multipotent differentiation capabilities. Recently, Chi et al. (2010) found that skin subcutaneous fat tissue also contains spheroid-forming cells and that these cells effectively differentiate into Schwann cells (SCs), osteocytes, and adipocytes under specific culture conditions. Moreover, in a comparative study using neonate and postnatal rats, the spheroids derived from either subcutaneous fat tissue or skin dermis both strongly expressed neural crest markers of Sox2, Twist, slug, nestin, and Sox 9 (Chi G and Son Y, unpublished data). Taken together, it is very likely that as in the skin dermis, NCSCs also reside in subcutaneous fat tissue.

15.4 Neural Induction of ASCs

Traditional ASCs are considered to be mesenchymal stem cells with multipotent differentiation capacity and similar cell surface antigen expression patterns [96, 100]. This kind of ASC is isolated from adipose tissue after enzymatic digestion followed by a series of monolayer cultures in 10% fetal bovine serum (FBS) – or fetal calf serum-containing basal medium. Therefore, after cell dissociation from adipose tissue, methods for induction of ASCs to neural-like cells are similar to the protocols used for induction of BMSCs to neural cells.

During the time period 2000–2004, the estimation of neural differentiation relied mainly on changes to a neuron-like cell morphology and expression of some immature neuron cell markers such as nestin, NeuN, and neuron-specific enolase (NSE). However, these cells did not reach the state of expression observed in mature neuron markers such as microtubule-associated protein-2 (MAP2), glial acidic fibrillary protein (GFAP), and galactocerebroside (GalC) [45]. In addition, there was no proof that the induced neuron-like cells revealed electrophysiologic or synaptic functions. Considering other studies in which undifferentiated ASCs were found to express markers characteristic of neural cells, such as NSE and NeuN [2], these induction methods could not be confirmed to result in successful neural differentiation. However, these data provide some evidence that a population of cells with early neuronal progenitor qualities or pluripotent stem cells resides in adipose tissue [2, 45, 69, 70, 81, 94, 100]. Several reports have suggested that ASCs comprise a heterogeneous population and that some subpopulation cells were diminished more quickly through serial passages in monolayer culture than others. These lost cell populations may have the ability to differentiate to multiple lineages; *i.e.*, osteogenic, adipogenic, chondrogenic, and neural [23, 100].

To improve neural induction from adipose tissue *in vitro*, an initial selection method needs to be developed to enrich the subpopulation of these neuronal progenitor-like cells from adipose tissue and new combinations of inducing agents also need to be developed. As an example, by adopting spheroid cultures at very early passages of Rhesus monkey ASC monolayer cultures in serum-free medium containing B27, bFGF, and EGF and through subsequent induction of neurogenesis in neurobasal medium containing only B27, Kang et al. (2004) obtained a markedly higher neurogenic potential with ASCs than with bone marrow MSCs. Additionally, the differentiated cells expressed MAP2ab and GFAP. Although the authors did not study the electrical activity of the differentiated neurons, this study implies that ASCs and BMSCs retain some differences in neural differentiation potential and that ASC culture methods may be very important for enhancing neural-inducing potential [36]. Furthermore, by using combined monolayer and spheroid methods, Kang et al. induced rat ASCs to oligodendrocyte precursor cells. Then, when those were intravenously injected in a spinal cord injury model, approximately 30–35% of cells migrated to and survived at the lesion site, differentiating to neurons and oligodendrocytes [37]. Jiang et al. (2008) reported that after simply culturing in conventional DMEM media containing 1% FBS, 100 ng/mL bFGF, and 10 m/mL forskolin for 7 days, human ASCs expressed immature and mature neural and glial markers such as nestin, Tuj1, MAP2, GFAP, and CNPase. Moreover, these induced cells displayed electrophysiological characteristic of voltage-dependent tetrodotoxin-sensitive sodium currents, outward potassium currents, and prominent negative resting membrane potentials under whole-cell patch clamp recordings [29]. This is an exciting result; however, to demonstrate functional equivalence with neuronal cells, a physical connection via a synapse to other endogenous neuronal cells and synaptic transmission of an electrical signal must be exhibited in an *in vivo* environment.

15.5 Induction of ASCs

Presently, numerous investigations using ASCs for neuron differentiation are underway but significant progress has not yet been made in this field. Nevertheless, approaches to the differentiation of bone marrow MSCs [8, 15, 39, 82] and ASCs into SCs have become a recent research topic, and several relevant articles have been published [9, 31, 43, 44, 67, 79, 92, 97]. During the induction process, most studies used ASCs obtained after several passages of monolayer culture utilizing a reagent combination of 10% FBS, retinoic acid, forskolin, bFGF, and heregulin beta-1 (or GGF-2) or 1% FBS, N2 supplement, forskolin, and heregulin beta-1 [97] (Fig. 15.1). Other studies performed spheroid culture from the primary culture stage and these spheroid-forming cells were separated and induced to SCs using the above-mentioned inducing agents [9, 97]. After induction, the cells displayed typical characteristics of SC morphology and expressed SC markers such as S100, GFAP, and p75. However, similar to the studies mentioned above, most studies have confined themselves to observing morphological changes and detecting the expression of a limited number of genes. According to reports, p75 is also expressed as a marker for ASC [93] and an S100 gene is already expressed in preinduced ASCs as well as postinduced cells [9]. Thus, more specific and comprehensive markers denoting induced SCs need to be adopted,

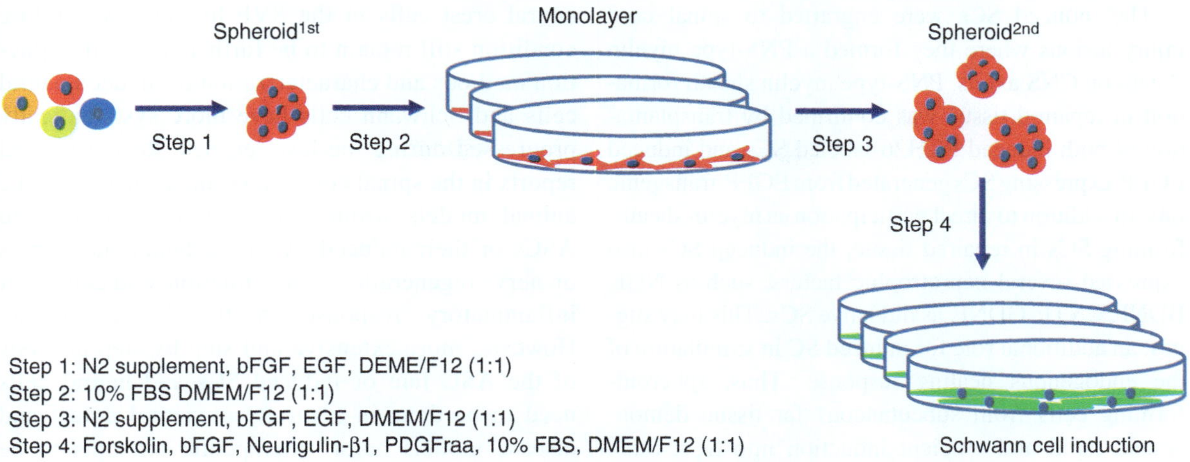

Fig. 15.1 Schematic diagram of Schwann cell induction from subcutaneous fat tissue. The cells immediately after dissociation from subcutaneous fat tissue were cultured in serum-free spheroid culture condition (*step 1*). After 10-day spheroid culture, the cells were dissociated into single cells and cultured in monolayer on the adherent surface with 10% FBS, DMEM/F12 medium up to passage five (*step 2*). The cells were induced to form the spheroid for 10 days (*step 3*) and induced to differentiate to Schwann cells (*step 4*)

depending on the origin and stages of differentiation. For example, Sox10 is only expressed in NCSCs but is downregulated after neural differentiation while remaining persistent in SCs and melanocytes [27, 47]. Therefore, Sox10 is a useful marker by which to identify induced SCs if the ASCs are truly differentiated to SC-like cells. More comprehensive sets of markers for SCs and glial cells, such as Sox10, p75, S100, Krox-20, L1, PLP/DM20, PMP22, ErbB2, PDGFr-aa, O4, A2B5, P0, and MBP, should be evaluated in comparison with ASCs, spheroids, and naïve SCs [9]. Radtke et al. (2009) reported that induced SCs stimulated dorsal root ganglion neurons after co-culture [67]. This is not surprising, given the data that ASCs and induced SCs all strongly expressed NGF, BDNF, CTNF, and GDNF genes.

15.6 Application of Induced SCs to Spinal Cord Injury and Brain Ischemic Stroke

The key trait of naïve SCs is the formation of the myelin sheath on axons, especially *in vivo*. Similar to neural induction of ASCs, most data presently available on induced SCs are limited to *in vitro* assessments; therefore, it is premature to assess their equivalence to naïve SCs in myelin sheath formation on nerve axons *in vivo*. More convincing data documenting myelin sheath formation *in vivo* and the technological advancement of ASC usage for treating neural tissue damage was recently reported by Chi et al. [9]. Unlike commonly used culture methods, spheroids, whose potential might be gradually diminished during traditional monolayer culture of ASCs, were selected out at the first stage of primary culture of adipose tissue. By expanding the spheroid-forming cells in monolayer culture and then reforming the spheroids, slower cell growth in the spheroid culture was overcome and SC-like cells were more efficiently induced (Fig. 15.1). This is one step further toward cell engineering of ASCs. By selecting the spheroid-forming cells at the first step of cell isolation, subpopulation cells expressing mRNAs of NGF, BDNF, CNTF, GDNF, p75, S100, PMP22, PLP/DM20, ErbB2 PDFGr-aa, and NES were enriched. Then, by further inducing the cells to SCs, SC-specific transcription factors, such as Krox-20 and L1, were expressed. Thus, induced SCs expressed NGF, BDNF, CNTF, GDNF, p75, S100, PMP22, PLP/DM20, ErbB2 PDFGr-aa, NES, Krox-20, and L1 genes identically to naïve SCs.

The induced SCs were engrafted to spinal cord injury lesions where they formed a PNS-type myelin sheath on CNS axons. PNS-type myelin sheath formation in repaired tissue was confirmed by transplantation of both induced PKH26-labeled SCs and induced EGFP-expressing SCs generated from EGFP-transgenic rats. In addition to direct participation as myelin sheath-forming SCs in repaired tissue, the induced SCs also expressed several neurotrophic factors, such as NGF, BDNF, CNTF, GDNF, as did naïve SCs. This may suggest an additional role for induced SC in stimulation of the endogenous healing response. Thus, spheroid-forming cells from subcutaneous fat tissue demonstrated rapid and efficient induction into SCs. Such cells show therapeutic promise for repair of damage to the CNS and PNS even though these data show a shortcoming in that not all the induced cells participated in myelin sheath formation at the lesion site.

Besides trying to induce ASCs to neural cells and SCs *in vitro*, other groups using normal ASCs directly performed *in vivo* experiments in ischemic stroke rat models [41, 52, 85] and sciatic nerve injury models [72]. In the stroke models, ASCs attenuated inflammation [41, 52], reduced neuron apoptosis [41, 85], reduced oxidative stress [52], and enhanced angiogenesis and neurogenesis [52]. Interestingly, grafted ASCs did not express neuronal or glial markers but did express endothelial markers [41, 52]. In the sciatic nerve injury model, nerve regeneration was promoted, but the transplanted cells did not differentiate into SCs.

The differentiation of ASCs to neurons was not completely successful. However, SC differentiation of ASCs *in vitro* and *in vivo* as well as the beneficial roles of ASCs in reduction of inflammation and apoptosis and enhancement of angiogenesis and nerve regeneration promises future therapeutic alternatives to naïve neural and glial cells for treatment of neural disease or disorders.

15.7 Conclusion

ASCs are readily available adult stem cells, which have become increasingly popular for the use in neural tissue repair in addition to our expectation for mesenchymal tissue repair. Even though the critical issues like the possibility of neural transdifferentiation of ASCs or selective expansion of residual neural crest cells in the SVF by selective culture condition still remain to be further resolved, induction methods and characterization of induced neural cells and Schwann cells have more systematically progressed during the last decades. Recent several reports in the spinal cord injury and ischemic stroke animal models strongly support that transplanted ASCs or their induced SCs have beneficial effects on nerve regeneration, remyelination, and control of inflammatory response in the injured tissue. However, more extensive and step by step analysis of the ASC fate or induced SCs and neural cells need to be done in a variety of neural defect and disease models before therapeutic attempts to be made.

Acknowledgements This work was supported by a Stem Cell Research Center Grant (SC3240) from the twenty-first century New Frontier Research Program and by the future based technology development program of the National Research Foundation (NRF) funded by the Korean government (MEST) (2010-0020405) awarded to Professor Youngsook Son.

References

1. Asgari S, Pournasr B, Salekdeh GH, Ghodsizadeh A, Ott M, Baharvand H (2010) Induced pluripotent stem cells: a new era for hepatology. J Hepatol 53:738–751
2. Ashjian PH, Elbarbary AS, Edmonds B, DeUgarte D, Zhu M, Zuk PA, Lorenz HP, Benhaim P, Hedrick MH (2003) *In vitro* differentiation of human processed lipoaspirate cells into early neural progenitors. Plast Reconstr Surg 111:1922–1931
3. Azizi SA, Stokes D, Augelli BJ, DiGirolamo C, Prockop DJ (1998) Engraftment and migration of human bone marrow stromal cells implanted in the brains of albino rats: similarities to astrocyte grafts. Proc Natl Acad Sci USA 95:3908–3913
4. Bachoud-Lévi AC, Gaura V, Brugières P, Lefaucheur JP, Boissé MF, Maison P, Baudic S, Ribeiro MJ, Bourdet C, Remy P, Cesaro P, Hantraye P, Peschanski M (2006) Effect of fetal neural transplants in patients with Huntington's disease 6 years after surgery: a long-term follow-up study. Lancet Neurol 5:303–309
5. Billon N, Iannarelli P, Monteiro MC, Glavieux-Pardanaud C, Richardson WD, Kessaris N, Dani C, Dupin E (2007) The generation of adipocytes by the neural crest. Development 134:2283–2292
6. Brazelton TR, Rossi FM, Keshet GI, Blau HM (2000) From marrow to brain: expression of neuronal phenotypes in adult mice. Science 290:1775–1779
7. Bunnell BA, Ylostalo J, Kang SK (2006) Common transcriptional gene profile in neurospheres-derived from pATSCs, pBMSCs, and pNSCs. Biochem Biophys Res Commun 343:762–771

8. Caddick J, Kingham PJ, Gardiner NJ, Wiberg M, Terenghi G (2006) Phenotypic and functional characteristics of mesenchymal stem cells differentiated along a Schwann cell lineage. Glia 54:840–849
9. Chi GF, Kim MR, Kim DW, Jiang MH, Son Y (2010) Schwann cells differentiated from spheroid-forming cells of rat subcutaneous fat tissue myelinate axons in the spinal cord injury. Exp Neurol 222:304–317
10. Crigler L, Robey RC, Asawachaicharn A, Gaupp D, Phinney DG (2006) Human mesenchymal stem cell subpopulations express a variety of neuro-regulatory molecules and promote neuronal cell survival and neuritogenesis. Exp Neurol 198:54–64
11. Crisan M, Yap S, Casteilla L, Chen CW, Corselli M, Park TS, Andriolo G, Sun B, Zheng B, Zhang L, Norotte C, Teng PN, Traas J, Schugar R, Deasy BM, Badylak S, Buhring HJ, Giacobino JP, Lazzari L, Huard J, Péault B (2008) A perivascular origin for mesenchymal stem cells in multiple human organs. Cell Stem Cell 3:301–313
12. Croft AP, Przyborski SA (2004) Generation of neuroprogenitor-like cells from adult mammalian bone marrow stromal cells *in vitro*. Stem Cells Dev 13:409–420
13. Deng W, Obrocka M, Fischer I, Prockop DJ (2001) *In vitro* differentiation of human marrow stromal cells into early progenitors of neural cells by conditions that increase intracellular cyclic AMP. Biochem Biophys Res Commun 282: 148–152
14. Deshpande DM, Kim YS, Martinez T, Carmen J, Dike S, Shats I, Rubin LL, Drummond J, Krishnan C, Hoke A, Maragakis N, Shefner J, Rothstein JD, Kerr DA (2006) Recovery from paralysis in adult rats using embryonic stem cells. Ann Neurol 60:32–44
15. Dezawa M, Takahashi I, Esaki M, Takano M, Sawada H (2001) Sciatic nerve regeneration in rats induced by transplantation of *in vitro* differentiated bone-marrow stromal cells. Eur J Neurosci 14:1771–1776
16. Doetsch F, Caiillé I, Lim DA, García-Verdugo JM, Alvarez-Buylla A (1999) Subventricular zone astrocytes are neural stem cells in the adult mammalian brain. Cell 97:703–716
17. Dominici M, Le Blanc K, Mueller I, Slaper-Cortenbach I, Marini F, Krause D, Deans R, Keating A, Prockop Dj, Horwitz E (2006) Minimal criteria for defining multipotent mesenchymal stromal cells. The International Society for Cellular Therapy position statement. Cytotherapy 8:315–317
18. Dupin E, Creuzet S, Le Douarin NM (2006) The contribution of the neural crest to the vertebrate body. Adv Exp Med Biol 589:96–119
19. Eggenschwiler R, Cantz T (2009) Induced pluripotent stem cells generated without viral integration. Hepatology 49:1048–1049
20. Fernandes KJ, McKenzie IA, Mill P, Smith KM, Akhavan M, Barnabé-Heider F, Biernaskie J, Junek A, Kobayashi NR, Toma JG, Kaplan DR, Labosky PA, Rafuse V, Hui CC, Miller FD (2004) A dermal niche for multipotent adult skin-derived precursor cells. Nat Cell Biol 6:1082–1093
21. Freed CR, Greene PE, Breeze RE, Tsai WY, DuMouchel W, Kao R, Dillon S, Winfield H, Culver S, Trojanowski JQ, Eidelberg D, Fahn S (2001) Transplantation of embryonic dopamine neurons for severe Parkinson's disease. N Engl J Med 344:710–719
22. Gordon D, Pavlovska G, Glover CP, Uney JB, Wraith D, Scolding NJ (2008) Human mesenchymal stem cells abrogate experimental allergic encephalomyelitis after intraperitoneal injection, and with sparse CNS infiltration. Neurosci Lett 448:71–73
23. Gronthos S, Franklin DM, Leddy HA, Robey PG, Storms RW, Gimble JM (2001) Surface protein characterization of human adipose tissue-derived stromal cells. J Cell Physiol 189:54–63
24. Han J, Koh YJ, Moon HR, Ryoo HG, Cho CH, Kim I, Koh GY (2010) Adipose tissue is an extramedullary reservoir for functional hematopoietic stem and progenitor cells. Blood 115:957–964
25. Harper JM, Krishnan C, Darman JS, Deshpande DM, Peck S, Shats I, Backovic S, Rothstein JD, Kerr DA (2004) Axonal growth of embryonic stem cell-derived motoneurons *in vitro* and in motoneuron-injured adult rats. Proc Natl Acad Sci USA 101:7123–7128
26. Hermann A, Gastl R, Liebau S, Popa MO, Fiedler J, Boehm BO, Maisel M, Lerche H, Schwarz J, Brenner R, Storch A (2004) Efficient generation of neural stem cell-like cells from adult human bone marrow stromal cells. J Cell Sci 117:4411–4422
27. Hermann A, Liebau S, Gastl R, Fickert S, Habisch HJ, Fiedler J, Schwarz J, Brenner R, Storch A (2006) Comparative analysis of neuroectodermal differentiation capacity of human bone marrow stromal cells using various conversion protocols. J Neurosci Res 83:1502–1514
28. Huangfu D, Osafune K, Maehr R, Guo W, Eijkelenboom A, Chen S, Muhlestein W, Melton DA (2008) Induction of pluripotent stem cells from primary human fibroblasts with only Oct4 and Sox2. Nat Biotechnol 26:1269–1275
29. Jang S, Cho HH, Cho YB, Park JS, Jeong HS (2010) Functional neural differentiation of human adipose tissue-derived stem cells using bFGF and forskolin. BMC Cell Biol 11:25
30. Jiang Y, Henderson D, Blackstad M, Chen A, Miller RF, Verfaillie CM (2003) Neuroectodermal differentiation from mouse multipotent adult progenitor cells. Proc Natl Acad Sci USA 100(Suppl 1):11854–11860
31. Jiang L, Zhu JK, Liu XL, Xiang P, Hu J, Yu WH (2008) Differentiation of rat adipose tissue-derived stem cells in to Schwann-like cells *in vitro*. Neuroreport 19:1015–1019
32. Jin K, Mao XO, Batteur S, Sun Y, Greenberg DA (2003) Induction of neuronal markers in bone marrow cells: differential effects of growth factors and patterns of intracellular expression. Exp Neurol 184:78–89
33. Joannides A, Gaughwin P, Schwiening C, Majed H, Sterling J, Compston A, Chandran S (2004) Efficient generation of neural precursors from adult human skin: astrocytes promote neurogenesis from skin-derived stem cells. Lancet 364:172–178
34. Johansson CB, Momma S, Clarke DL, Risling M, Lendhl U, Frisén J (1999) Identification of a neural stem cell in the adult mammalian central nervous system. Cell 96:25–34
35. Jori FP, Napolitano MA, Melone MA, Cipollaro M, Cascino A, Altucci L, Peluso G, Giordano A, Galderisi U (2005) Molecular pathways involved in neural *in vitro* differentiation of marrow stromal stem cells. J Cell Biochem 94: 645–655

36. Kang SK, Putnam LA, Ylostalo J, Popescu IR, Dufour J, Belousov A, Bunnell BA (2004) Neurogenesis of Rhesus adipose stromal cells. J Cell Sci 117:4289–4299
37. Kang SK, Shin MJ, Jung JS, Kim YG, Kim CH (2006) Autologous adipose tissue-derived stromal cells for treatment of spinal cord injury. Stem Cells Dev 15:583–594
38. Kassis I, Grigoriadis N, Gowda-Kurkalli B, Mizrachi-Kol R, Ben-Hur T, Slavin S, Abramsky O, Karussis D (2008) Neuroprotection and immunomodulation with mesenchymal stem cells in chronic experimental autoimmune encephalomyelitis. Arch Neurol 65:753–761
39. Keilhoff G, Goihi A, Langnäse K, Fansa H, Wolf G (2006) Transdifferentiation of mesenchymal stem cells into Schwann cell-like myelination cells. Eur J Cell Biol 85:11–24
40. Keirstead HS, Nistor G, Bernal G, Totoiu M, Cloutier F, Sharp K, Steward O (2005) Human embryonic stem cell-derived oligodendrocyte progenitor cell transplants remyelinate and restore locomotion after spinal cord injury. J Neurosci 25:4694–4705
41. Kim JM, Lee ST, Chu K, Jung KH, Song EC, Kim SJ, Sinn DI, Kim JH, Park DK, Kang KM, Hyung Hong N, Park HK, Won CH, Kim KH, Kim M, Kun Lee S, Roh JK (2007) Systemic transplantation of human adipose stem cells attenuated cerebral inflammation and degeneration in a hemorrhagic stroke model. Brain Res 1183:43–50
42. Kim D, Kim CH, Moon JI, Chung YG, Chang MY, Han BS, Ko S, Yang E, Cha KY, Lanza R, Kim KS (2009) Generation of human induced pluripotent stem cells by direct delivery of reprogramming proteins. Cell Stem Cell 4:472–476
43. Kingham PJ, Kalbernatten DF, Mahay D, Armstrong SJ, Wiberg M, Terenghi G (2007) Adipose-derived stem cells differentiate into a Schwann cell phenotype and promote neurite outgrowth *in vitro*. Exp Neurol 207:267–274
44. Kingham PJ, Mantovani C, Terenghi G (2009) Notch independent signalling mediates Schwann cell-like differentiation of adipose derived stem cells. Neurosci Lett 467: 164–168
45. Kokai LE, Rubin JP, Marra KG (2005) The potential of adipose-derived adult stem cells as a source of neuronal progenitor cells. Plast Reconstr Surg 116:1453–1460
46. Kopen GC, Prockop DJ, Phinney DG (1999) Marrow stromal cells migrate throughout forebrain and cerebellum, and they differentiate into astrocytes after injection into neonatal mouse brains. Proc Natl Acad Sci USA 96:10711–10716
47. Kuhlbrodt K, Herbarth B, Sock E, Hermans-Borgmeyer I, Wegner M (1998) Sox10, a novel transcriptional modulator in glial cells. J Neurosci 18:237–250
48. Kuhn HG, Dickinson-Anson H, Gage FH (1996) Neurogenesis in the dentate gyrus of the adult rat: age-related decrease of neuronal progenitor proliferation. J Neurosci 16:2027–2033
49. Laustriat D, Gide J, Peschanski M (2010) Human pluripotent stem cells in drug discovery and predictive toxicology. Biochem Soc Trans 38:1051–1057
50. Le Douarin NM, Creuzet S, Couly G, Dupin E (2004) Neural crest cell plasticity and its limits. Development 131: 4637–4650
51. Lee OK, Ko YC, Kuo TK, Chou SH, Li HJ, Chen WM, Chen TH, Su Y (2004) Fluvastatin and lovastatin but not pravastatin induce neuroglial differentiation in human mesenchymal stem cells. J Cell Biochem 93:917–928
52. Leu S, Lin YC, Yuen CM, Yen CH, Kao YH, Sun CK, Yip HK (2010) Adipose-derived mesenchymal stem cells markedly attenuate brain infarct size and improve neurological function in rats. J Transl Med 8:63
53. Liu S, Qu Y, Stewart TJ, Howard MJ, Chakrabortty S, Holekamp TF, McDonald JW (2000) Embryonic stem cells differentiate into oligodendrocytes and myelinate in culture and after spinal cord transplantation. Proc Natl Acad Sci USA 97:6126–6131
54. Lu P, Blesch A, Tuszynski M (2004) Induction of bone marrow stromal cells to neurons: differentiation, transdifferentiation, or artifact? J Neurosci Res 77:174–191
55. McKenzie IA, Biernaskie J, Toma JG, Midha R, Miller FD (2006) Skin-derived precursors generate myelinating Schwann cells for the injured and dysmyelinated nervous system. J Neurosci 26:6651–6660
56. Mezey E, Chandross KJ, Harta G, Maki RA, McKercher SR (2000) Turning blood into brain: cells bearing neuronal antigens generated *in vivo* from bone marrow. Science 290: 1779–1782
57. Montzka K, Lassonczyk N, Tschöke B, Neuss S, Führmann T, Franzen R, Smeets R, Brook GA, Wöltje M (2009) Neural differentiation potential of human bone marrow-derived mesenchymal stromal cells: misleading marker gene expression. BMC Neurosci 10:16
58. Moriguchi H, Chung RT, Mihara M, Sato C (2010) Generation of human induced pluripotent stem cells from liver progenitor cells by only small molecules. Hepatology 52:1169–1170
59. Morikawa S, Mabuchi Y, Niibe K, Suzuki S, Nagoshi N, Sunabori T, Shimmura S, Nagai Y, Nakagawa T, Okano H, Matsuzaki Y (2009) Development of mesenchymal stem cells partially originate from the neural crest. Biochem Biophys Res Commun 379:1114–1119
60. Muñoz-Elias G, Woodbury D, Black IB (2003) Marrow stromal cells, mitosis, and neuronal differentiation: stem cell and precursor functions. Stem Cells 21:437–448
61. Nagoshi N, Shibata S, Kubota Y, Nakamura M, Nagai Y, Satoh E, Morikawa S, Okada Y, Mabuchi Y, Katoh H, Okada S, Fukuda K, Suda T, Matsuzaki Y, Toyama Y, Okano H (2008) Ontogeny and multipotency of neural crest-derived stem cells in mouse bone marrow, dorsal root ganglia, and whisker pad. Cell Stem Cell 2:392–403
62. Neuhuber B, Gallo G, Howard L, Kostura L, Mackay A, Fischer I (2004) Reevaluation of *in vitro* differentiation protocols for bone marrow stromal cells: disruption of actin cytoskeleton induces rapid morphological changes and mimics neuronal phenotype. J Neurosci Res 77:192
63. Neuhuber B, Timothy Himes B, Shumsky JS, Gallo G, Fischer I (2005) Axon growth and recovery of function supported by human bone marrow stromal cells in the injured spinal cord exhibit donor variations. Brain Res 1035:73–85
64. Nistor GI, Totoiu MO, Haque N, Carpenter MK, Keirstead HS (2005) Human embryonic stem cells differentiate into oligodendrocytes in high purity and myelinate after spinal cord transplantation. Glia 49:385–396
65. Okita K, Ichisaka T, Yamanaka S (2007) Generation of germline-competent induced pluripotent stem cells. Nature 448:313–317
66. Olanow CW, Goetz CG, Kordower JH, Stoessl AJ, Sossi V, Brin MF, Shannon KM, Nauert GM, Perl DP, Godbold J,

Freeman TB (2003) A double-blind controlled trial of bilateral fetal nigral transplantation in Parkinson's disease. Ann Neurol 54:403–414

67. Radtke C, Schmitz B, Spies M, Kocsis JD, Vogt PM (2009) Peripheral glial cell differentiation from neurospheres derived from adipose mesenchymal stem cells. Int J Dev Neurosci 27:817–823
68. Real C, Glavieux-Pardanaud C, Le Douarin NM, Dupin E (2006) Clonally cultured differentiated pigment cells can dedifferentiate and generate multipotent progenitors with self-renewing potential. Dev Biol 300:656–669
69. Safford KM, Hicok KC, Safford SD, Halvorsen YD, Wilkison WO, Gimble JM, Rice HE (2002) Neurogenic differentiation of murine and human adipose-derived stromal cells. Biochem Biophys Res Commun 294:371–379
70. Safford KM, Safford SD, Gimble JM, Shetty AK, Rice HE (2004) Characterization of neuronal/glial differentiation of murine adipose-derived adult stromal cells. Exp Neurol 187:319–328
71. Sanchez-Ramos J, Song S, Cardozo-Pelaez F, Hazzi C, Stedeford T, Willing A, Freeman TB, Saporta S, Janssen W, Patel N, Cooper DR, Sanberg PR (2000) Adult bone marrow stromal cells differentiate into neural cells *in vitro*. Exp Neurol 164:247–256
72. Santiago LY, Clavijo-Alvarez J, Brayfield C, Rubin JP, Marra KG (2009) Delivery of adipose-derived precursor cells for peripheral nerve repair. Cell Transplant 18:145–158
73. Scintu F, Reali C, Pillai R, Badiali M, Sanna MA, Argiolu F, Ristaldi MS, Sogos V (2006) Differentiation of human bone marrow stem cells into cells with a neural phenotype: diverse effects of two specific treatments. BMC Neurosci 7:14
74. Shihabuddin LS (2008) Adult rodent spinal cord-derived neural stem cells: isolation and characterization. Methods Mol Biol 438:55–66
75. Soldner F, Hockemeyer D, Beard C, Gao Q, Bell GW, Cook EG, Hargus G, Blak A, Cooper O, Mitalipova M, Isacson O, Jaenisch R (2009) Parkinson's disease patient-derived induced pluripotent stem cells free of viral reprogramming factors. Cell 136:964–977
76. Stadtfeld M, Nagaya M, Utikal J, Weir G, Hochedlinger K (2008) Induced pluripotent stem cells generated without viral integration. Science 322:945–949
77. Suzuki H, Taguchi T, Tanaka H, Kataoka H, Li Z, Muramatsu K, Gondo T, Kawai S (2004) Neurospheres induced from bone marrow stromal cells are multipotent for differentiation into neuron, astrocyte, and oligodendrocyte phenotypes. Biochem Biophys Res Commun 322:918–922
78. Takahashi K, Yamanaka S (2006) Induction of pluripotent stem cells from mouse embryonic and adult fibroblast cultures by defined factors. Cell 126:663–676
79. Tang YJ, Zhang LH, Liu JM, Dong WR, Guo JS, Wang HH, Dai X, Chen YH, Xiao YQ (2009) Induced differentiation of rat adipose-derived stem cells into Schwann-like cells. Nan Fang Yi Ke Da Xue Xue Bao 29:680–684
80. Tao H, Rao R, Ma DD (2005) Cytokine-induced stable neuronal differentiation of human bone marrow mesenchymal stem cells in a serum/feeder cell-free condition. Dev Growth Differ 47:423–433
81. Tholpady SS, Katz AJ, Ogle RC (2003) Mesenchymal stem cells from rat visceral fat exhibit multipotential differentiation *in vitro*. Anat Rec A Discov Mol Cell Evol Biol 272:398–402
82. Tohill M, Mantovani C, Wiberg M, Terenghi G (2004) Rat bone marrow mesenchymal stem cells express glial markers and stimulate nerve regenratation. Neurosci Lett 362: 200–203
83. Toma JG, Akhavan M, Fernandes KJ, Barnabé-Heider F, Sadikot A, Kaplan DR, Miller FD (2001) Isolation of multipotent adult stem cells from the dermis of mammalian skin. Nat Cell Biol 3:778–784
84. Trainor PA, Melton KR, Manzanares M (2003) Origins and plasticity of neural crest cells and their roles in jaw and craniofacial evolution. Int J Dev Biol 47:541–553
85. Wei X, Zhao L, Zhong J, Gu H, Feng D, Johnstone BH, March KL, Farlow MR, Du Y (2009) Adipose stromal cells-secreted neuroprotective media against neuronal apoptosis. Neurosci Lett 462:76–79
86. Weiss S, Dunne C, Hewson J, Wohl C, Wheatley M, Peterson AC, Reynolds BA (1996) Multipotent CNS stem cells are present in the adult mammalian spinal cord and ventricular neuroaxis. J Neurosci 16:7599–7609
87. Wernig M, Zhao JP, Pruszak J, Hedlund E, Fu D, Soldner F, Broccoli V, Constantine-Paton M, Isacson O, Jaenisch R (2008) Neurons derived from reprogrammed fibroblasts functionally integrate into the fetal brain and improve symptoms of rats with Parkinson's disease. Proc Natl Acad Sci USA 105:5856–5861
88. Wislet-Gendebien S, Leprince P, Moonen G, Rogister B (2003) Regulation of neural markers nestin and GFAP expression by cultivated bone marrow stromal cells. J Cell Sci 116:3295–3302
89. Wislet-Gendebien S, Hans G, Leprince P, Rigo JM, Moonen G, Rogister B (2005) Plasticity of cultured mesenchymal stem cells: switch from nestin-positive to excitable neuron-like phenotype. Stem Cells 23:392–402
90. Wong CE, Paratore C, Dours-Zimmermann MT, Rochat A, Pietri T, Suter U, Zimmermann DR, Dufour S, Thiery JP, Meijer D, Beermann F, Barrandon Y, Sommer L (2006) Neural crest-derived cells with stem cell features can be traced back to multiple lineages in the adult skin. J Cell Biol 175:1005–1015
91. Woodbury D, Schwarz EJ, Prockop DJ, Black IB (2000) Adult rat and human bone marrow stromal cells differentiate into neurons. J Neurosci Res 61:364–370
92. Xu Y, Liu L, Li Y, Zhou C, Xiong F, Liu Z, Gu R, Hou X, Zhang C (2008) Myelin-forming ability of Schwann cell-like cells induced from rat adipose-derived stem cells *in vitro*. Brain Res 1239:49–55
93. Yamamoto N, Akamatsu H, Hasegawa S, Yamada T, Nakata S, Ohkuma M, Miyachi E, Marunouchi T, Matsunaga K (2007) Isolation of multipotent stem cells from mouse adipose tissue. J Dermatol Sci 48:43–52
94. Yang LY, Liu X, Sun B, Hui GZ, Fei J, Guo LH (2004) Adipose tissue-derived stromal cells express neuronal phenotypes. Chin Med J (Engl) 117:425–429
95. Yilmaz G, Alexander JS, Erkuran Yilmaz C, Granger DN (2010) Induction of neuro-protective/regenerative genes in stem cells infiltrating post-ischemic brain tissue. Exp Transl Stroke Med 2:11
96. Yoshimura H, Muneta T, Nimura A, Yokoyama A, Koga H, Sekiya I (2007) Comparison of rat mesenchymal stem cells derived from bone marrow, synovium, periosteum, adipose tissue, and muscle. Cell Tissue Res 327:449–462

97. Zavan B, Michelotto L, Lancerotto L, Della Puppa A, D'Avella D, Abatangelo G, Vindigni V, Cortivo R (2010) Neural potential of a stem cell population in the adipose and cutaneous tissues. Neurol Res 32:47–54
98. Zhang ZG, Zhang L, Croll SD, Chopp M (2002) Angiopoietin-1 reduces cerebral blood vessel leakage and ischemic lesion volume after focal cerebral embolic ischemia in mice. Neuroscience 113:683–687
99. Zhang J, Wilson GF, Soerens AG, Koonce CH, Yu J, Palecek SP, Thomson JA, Kamp TJ (2009) Functional cardiomyocytes derived from human induced pluripotent stem cells. Circ Res 104:e30–e41
100. Zuk PA, Zhu M, Ashjian P, De Ugarte DA, Huang JI, Mizuno H, Alfonso ZC, Fraser JK, Benhaim P, Hedrick MH (2002) Human adipose tissue is a source of multipotent stem cells. Mol Biol Cell 13:4279–4295

Adipose Stem Cell Engineering: Clinical Applications in Plastic and Reconstructive Surgery

16

Aris Sterodimas

Contents

16.1 **Introduction** 165

16.2 **Harvesting of ADSCs for Tissue Engineering** 166

16.3 **Isolation and Expansion of ADSCs** 166

16.4 **Current Human Applications of Adipose-Derived Stem Cells** 167

16.4.1 Stromal Enriched Lipograft Cases 169

16.5 **Current Animal Applications of Adipose-Derived Stem Cells** 169

16.6 **Adipose-Derived Stem Cells and Future Applications on Tissue Engineering** 173

16.7 **Conclusion** 176

References 177

A. Sterodimas
Plastic and Reconstructive Surgery Department, IASO General Hospital, 264 Mesogeion Avenue, Athens, Greece
e-mail: aris@sterodimas.com; www.sterodimas.com

16.1 Introduction

Standard approaches to soft-tissue reconstruction include autologous tissue flaps, autologous fat transplantation, and alloplastic implants. All of these approaches have disadvantages, including donor-site morbidity; implant migration, and foreign body reaction. Autologous fat transplantation, with a minimally invasive cannula harvest, has lower donor-site morbidity than tissue flaps do, but there is an unpredictable degree of resorption of the transplanted fat over time [20, 49, 65]. There is a major clinical need for strategies that adequately reconstruct the soft tissue defects after deep burns, tumor resection, or trauma. Human adipose tissue is an ideal source of autologous cells that is both plentiful and easily obtainable in large quantities through the simple surgical procedure of liposuction. Autologous fat transplantation is frequently used for a variety of cosmetic and reconstructive indications not limited to posttraumatic defects of the face and body, evolutional disorders such as hemifacial atrophy, sequelae of radiation therapy, and many esthetic uses such as lip and facial augmentation and wrinkle therapy [15, 19, 22, 39, 44, 50, 53, 54, 65]. In the past 20 years, the advancements in techniques and instrumentation have produced results that make fat grafting a viable option for soft tissue augmentation [21, 51, 52]. Fat is a living tissue that must be in close proximity to a nutritional and respiratory source to survive. Successful, three-dimensional sculpting requires meticulous planning, and optimizing the harvesting, storage, and transplantation of adipose tissue. There is though an unpredictable degree of resorption of the transplanted fat and repeated treatment sessions are usually needed in order to achieve the final result [11, 26].

Y.-G. Illouz and A. Sterodimas (eds.), *Adipose Stem Cells and Regenerative Medicine*,
DOI: 10.1007/978-3-642-20012-0_16, © Springer-Verlag Berlin Heidelberg 2011

Regenerative medicine is classified into cell therapy that does not require a scaffold and tissue engineering that requires a scaffold and bioactive substances such as growth factors, though both need adult stem cells. Recently, the role of adult stem cells in adipose tissue has gained much interest [64]. The isolation of a population of progenitor cells from adipose tissue was first described in 1964 by Rodbell, whose work was done in rodents [56]. In 1977, Van and Roncari reported that adipose tissue of adult rats contains cells with the potential to proliferate and acquire morphological characteristics similar to those of adipocytes [72]. After the introduction of liposuction, adipose tissue harvesting has become easier [18]. Since then, this tissue has been gaining an increased importance as a stem cell source for a wide range of potential applications in tissue engineering and regenerative medicine strategies, mainly due to its wide availability and easy access. [76].

Lipoaspirate, an otherwise disposable byproduct of cosmetic surgery, has been shown to contain a putative population of stem cells, termed adipose-derived stem cells (ADSCs) that share many similarities to marrow stromal cells (MSCs) from bone marrow [67]. ADSCs reside within the stromal-vascular fraction (SVF) in fat tissue which is thought to harbor cells that display extensive proliferative capacity and multilineage potential [14]. Recently, approaches based on "ADSCs enriched lipograft" have been published [40, 62, 66, 73].

16.2 Harvesting of ADSCs for Tissue Engineering

Adipose-derived stem cells isolated from harvested fat are able to better withstand the mechanical trauma from the suction cannula and may allow for improved cell survival and generation of new fat tissue after transfer to another anatomic site [57]. There is evidence that the ability of adipose precursor cells to grow and differentiate varies among fat depots and changes with age [60]. A significant difference in the apoptotic susceptibility of ADSCs has been noted, with the superficial abdominal depot (above Scarpa's layer) significantly more resistant to apoptosis when compared with the other subcutaneous depots. Younger patients have an increased induction of differentiation of adipose cells in all depots, whereas the older patients only in the arm and thigh subcutaneous depots [60]. These observations have been confirmed by another study and the abdomen seems to be preferable to the hip/thigh region for harvesting adipose tissue for ADSCs culture [24]. Human ADSCs not only function as progenitor cells for in vivo adipogenesis, but also induce de novo adipogenesis and are deemed more advantageous as a cell source than mature adipocytes because they are easily cultured, easily expanded, and easily obtained [34].

16.3 Isolation and Expansion of ADSCs

Current methods for isolating ADSCs rely on a collagenase digestion followed by centrifugal separation to isolate the stromal/vascular cells from primary adipocytes. Centrifugation plays a beneficial role in concentrating SVF and ADSCs although excessive centrifugation can destroy adipocytes and ADSCs. In a previous study, it was found that centrifugation of aspirated fat at 1,200 × *g* decreases the fat volume by 30%, damaging 12% of the adipocytes and 0% of the ADSCs [33]. This has been confirmed by a recent study stating that cell survival rates are significantly lower when centrifugation forces of 1,200 × *g* are used for more than 5 min and of 3,000 × *g* when used for more than 1 min [66]. The actual recommendation is 1,200 × *g* as an optimized centrifugal force among the tested centrifugal forces for obtaining good short- and long-term results in adipose transplantation. A recent study has showed that the ADSCs concentration is significantly higher in lipoaspirate that has been washed by saline compared to the decanted and centrifuged lipoaspirate samples. However, the pellet collected at the bottom of the centrifuged lipoaspirate sample showed the highest concentration of ADSCs [9]. Isolated ADSCs are typically expanded in monolayer on standard tissue culture plastic with a basal medium containing 10% fetal bovine serum [12, 67]. In an effort to eliminate the use of animal products in human adipose stem cell cultures, a very low human serum expansion medium and a completely serum-free medium have been recently presented. These newly developed culture conditions provide a unique environment within which the study of ADSCs without the interference of animal serum can be done. They also allow rapid expansion of autologous ADSCs in culture for use in human clinical trials [47]. Platelet-rich plasma can additionally enhance the proliferation of human ADSCs. Preliminary results support the clinical application of platelet-rich plasma for cell-based, soft-tissue engineering and wound healing [25]. A standard expansion method has not been

Table 16.1 The current human clinical studies of adipose-derived stem cells

Technique	Delivery of ADSCs	Purpose	Type of study	Follow up	Patients	Institution	Results published
Lipodystrophy (AADSCTPL)	Autologous ADSCs	Transplantation in patients with lipodystrophy	Phase I interventional single group	1 year	10	Irmandade Santa Casa de Misericordia de Porto Alegre, Brazil	N/A
Cell-assisted lipotransfer strategy (CAL)	Autologous ADSCs rich lipoaspirate transplantation	Cosmetic and functional results of facial remodeling	Phase III interventional single group	9–13 months	6	University of Tokyo, Japan	[74]
Stromal-Enriched Lipograft (SEL)	Transplantation of stromal vascular fraction rich lipograft	Cosmetic results of facial remodeling	Phase III randomized single group	3 years	20	IASO General Hospital, Athens, Greece	[62]
Cell-assisted lipotransfer strategy (CAL)	ADSCs rich lipoaspirate transplantation	Cosmetic and functional results of reconstructive breast surgery	Phase III interventional single group	6–42 months	400	University of Tokyo, Japan	[73]
Stromal Enriched Lipograft (SEL)	Transplantation of stromal vascular fraction rich lipoaspirate	Cosmetic results of gluteal remodeling	Phase III randomized single group	3 years	50	IASO General Hospital, Athens, Greece	In Press
Lumpectomy (RESTORE-2)	ADRC-enhanced autologous fat transplant	Reconstruct breast deformities after lumpectomy	Phase IV nonrandomized	1 year	70	Jules Bordet Institute, Brussels, Belgium	N/A
Adipocell	Drug: Adipocell autologous cultured adipocytes (ANTG-adip)	Cosmetic result of depressed scar	Phase II/III Open label uncontrolled single group	12 weeks	36	Samsung Medical Center, Seoul, Korea	N/A

yet established and new expansion methods accelerating proliferation of ADSCs while preserving their multipotent differentiation capacities are needed in order to proceed to clinical applications [37]. To avoid the potential risks of animal-derived products such as viral transmission and immunologic reactions, the necessity of human-derived products in the manipulation of cells for cell-based therapies must be investigated and clarified [32].

16.4 Current Human Applications of Adipose-Derived Stem Cells

Soft tissue repair is theoretically the simplest application for adipose-derived cell therapies since the isolated cells presumably do not need to display any transdifferentiation potential. The current human clinical studies of ADSCs are given in Table 16.1.

A recent report showed ADSCs capability to provide not only cellular elements, but also numerous cytokines [45]. Autologous ADSCs show great promise for applications in repair of skin, rejuvenation of aging skin, and aging-related skin lesions [23]. A recent study reported the combination of platelet rich plasma and autologous adipose stem cells in order to regenerate tissue and achieve epithelialization of the wound, with a significant healing-time reduction. Furthermore, the minimally invasive technique was well accepted by patients, with a noteworthy improvement of the quality of life along with cost reduction due to the fewer number of medications [7]. The clinical effectiveness of adipose-derived adult stem cells-lipoaspirate transplantation in the treatment of radiation side effects has been also investigated. Clinical outcomes have led to a systematic improvement or remission of symptoms in all evaluated patients, possibly due to restoration of tissue vascularization and organ function [30, 55]. ADSCs, together

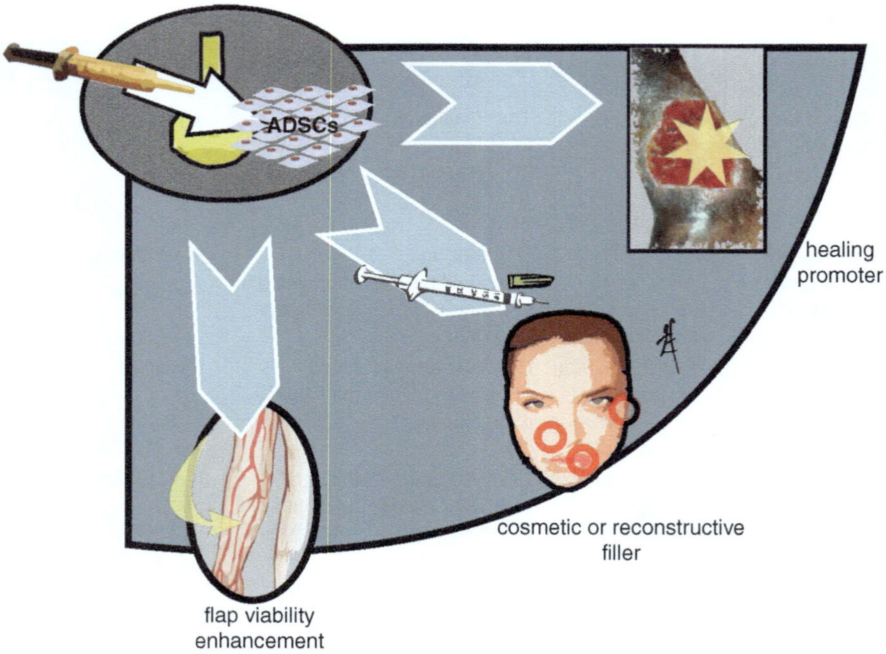

Fig. 16.1 A schematic representation of ADSCs applications in plastic and reconstructive surgery

with angiogenic and mitogenic factor of basic fibroblast growth factor and an artificial dermis, have been recently applied over excised irradiated skin defect and have been tested for differentiation and local stimulation effects in the radiation-exposed wounds. All the patients were uneventfully treated with minimal morbidities [2]. Recently, adipose stem cells have proved to selectively induce neovascularization and increase the viability of random pattern skin flaps. This mechanism might be both due to the direct differentiation of ADSCs into endothelial cells and the indirect effect of angiogenic growth factors released from ADSCs [35, 59, 71]. A microvascular custom-made ectopic bone flap has been created using ADSCs, beta-tricalcium phosphate and bone morphogenetic protein-2. The mature flap-bone structure was then transplanted in order to reconstruct the patients' defect caused after hemimaxillectomy [41]. Preclinical studies and early clinical series have also shown that ADSCs combined with scaffolds could possibly be used as fillers in plastic and reconstructive surgery [42]. Hyaluronic acid-based (HA) preadipocyte-seeded scaffolds have been evaluated for their adipoconductive potential and efficacy in humans. Preadipocytes were isolated from lipoaspirate material and seeded on HA scaffolds. After 8 weeks of implantation, all void spaces within the scaffolds were filled with cells with pronounced matrix deposition. This study proves that HA scaffolds are stable cell carriers and have the potential to generate volume-retaining tissue [68]. A schematic representation of ADSCs applications in plastic and reconstructive surgery is shown in Fig. 16.1. ADSCs have been approved and employed in clinical trials for restoring soft tissue defects. In adipose stem cell enriched lipograft, autologous ADSCs are used in combination with lipoinjection. A stromal vascular fraction (SVF) containing ADSCs is freshly isolated from two thirds of the aspirated fat and recombined with the other one third. This process converts relatively ADSC-poor aspirated fat to ADSC-rich fat (Fig. 16.2). Stromal enriched lipograft (SEL) has been used to treat body tissue defects caused by trauma or tumor removal [62]. ADSC-enriched lipograft has been used to treat craniofacial lipodystrophic and cosmetic defects with a high degree of success [62, 66, 67, 74]. Breast reconstruction and augmentation trials have enrolled the greatest number of patients for ADSC-assisted lipotransfer. Breast enhancement with artificial implants is one of the most frequently performed cosmetic surgeries but is associated with complications, such as capsular contracture, which lead to implant removal or replacement. Autologous transplantation of progenitor-supplemented adipose tissue has been performed [73, 75].

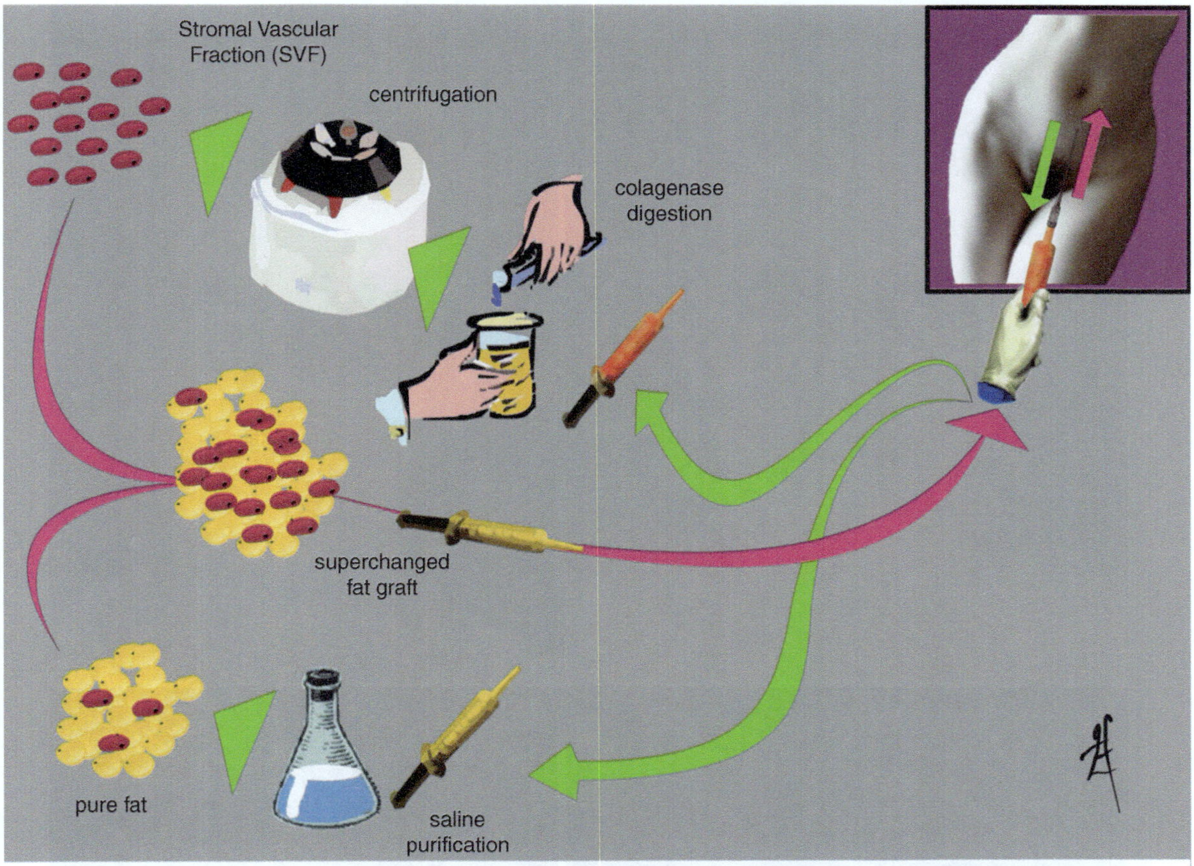

Fig. 16.2 In adipose stem cell enriched lipograft, autologous ADSCs are used in combination with lipoinjection. A stromal vascular fraction (*SVF*) containing ADSCs is freshly isolated from two thirds of the aspirated fat and recombined with the other one third. This process converts relatively ADSC-poor aspirated fat to ADSC-rich fat

16.4.1 Stromal Enriched Lipograft Cases

Case 1

A 57-year-old woman asked for rejuvenation of her face (Fig. 16.3a, b). She underwent a traditional round face lifting and they were transplanted 13 ml of stromal enriched lipograft. Postoperative photos after 2 years are shown (Fig. 16.3c, d).

Case 2

A 40-year-old woman asked for gluteal augmentation.(Fig. 16.4a, b). She underwent stromal enriched lipograft and received a total of 485 ml of graft in bilateral buttocks. After 2 years, the postoperative result is shown (Fig. 16.4c, d).

Case 3

A 32-year-old woman requested breast augmentation (Fig. 16.5a, b). She underwent stromal enriched lipograft and received 192 mL of graft on right breast and 177 mL on the left breast. After 1 year, the postoperative result is shown (Fig. 16.5c, d).

Case 4

A 65-year-old woman asked for body contour treatment of her left shoulder depression caused by trauma (Fig. 16.6a). She underwent stromal enriched lipograft and received 14 mL of graft. After 1 year, she was evaluated and her postoperative result is shown (Fig. 16.6b).

16.5 Current Animal Applications of Adipose-Derived Stem Cells

The current animal studies of ADSCs are given in Table 16.2.

A stimulus, such as appropriate growth factors [epidermal growth factor (EGF), transforming growth

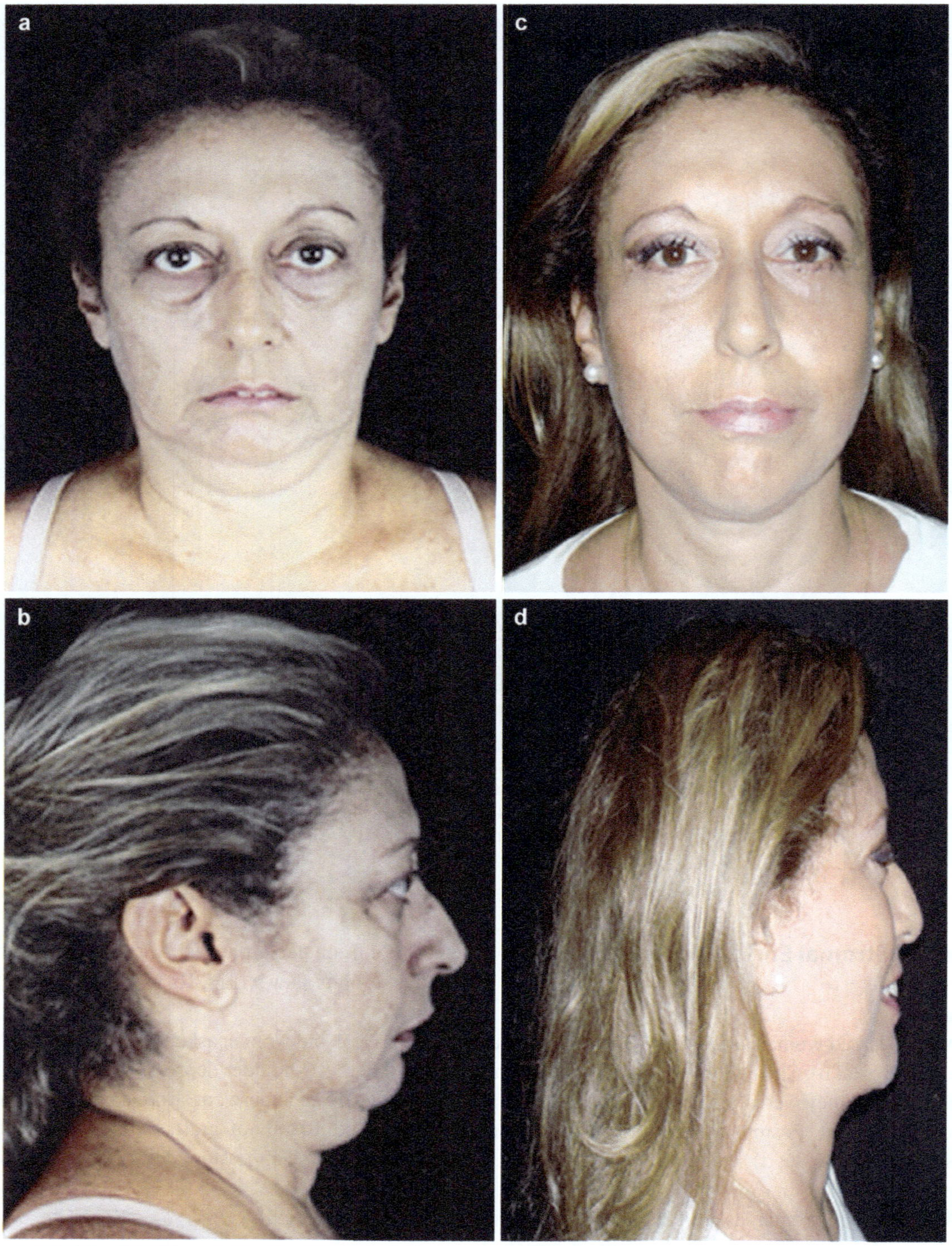

Fig. 16.3 (**a**, **b**) Preoperative photos of a 57-year-old woman requesting rejuvenation of her face. (**c**, **d**) Postoperative photos of a 57-year-old woman after undergoing traditional round face lifting and transplantation of 13 mL of stromal enriched lipograft

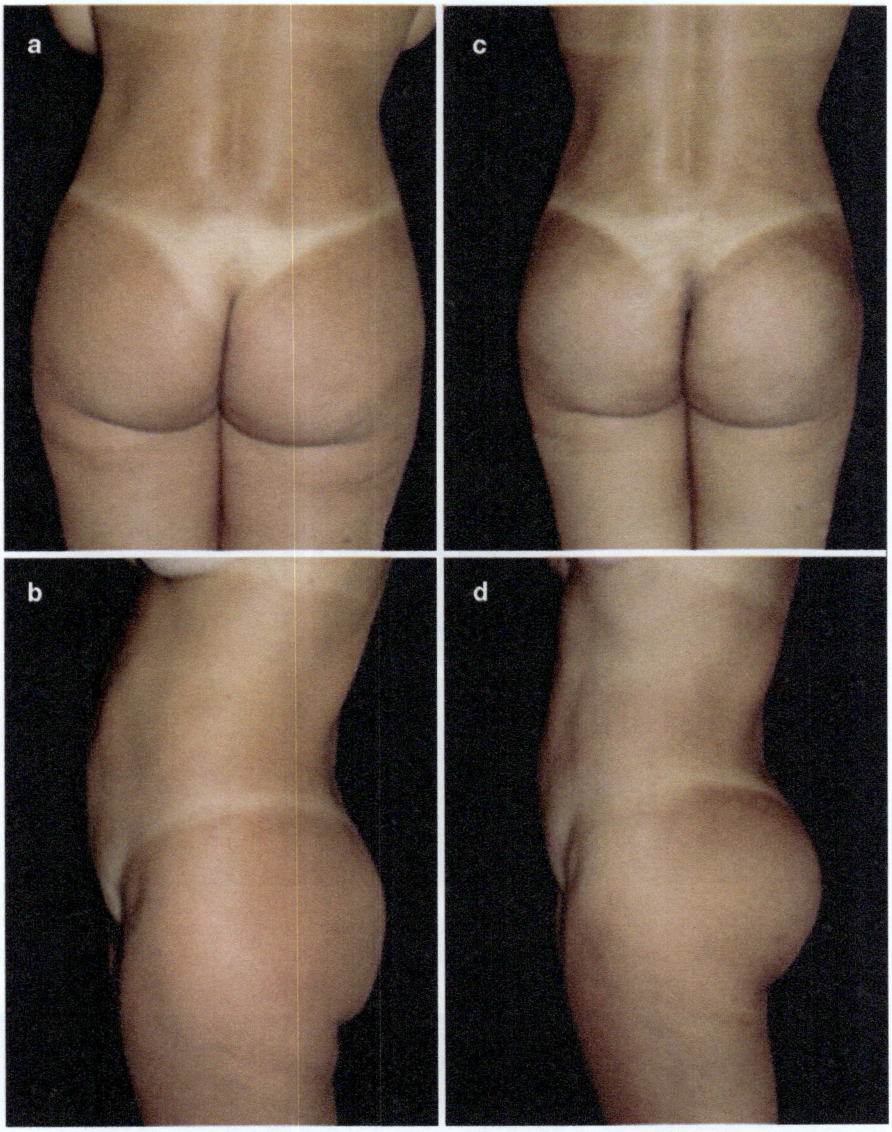

Fig. 16.4 (**a**, **b**) Preoperative photos of a 40-year-old woman asking for gluteal augmentation. (**c**, **d**) Postoperative photos of a 40-year-old woman after undergoing stromal enriched lipograft of 485 mL in the buttocks

factor-b (TGF-b), and platelet-derived growth factor (PDGF)], applied in vivo induces the migration of preadipocytes to the implant site. The cells subsequently proliferate and differentiate to form adipose tissue depot [27]. This de novo adiopogenesis has been demonstrated using subcutaneous injections consisting of Matrigel (a collagen-based gel derived from the basement membrane of a murine tumor) with basic fibroblast growth factor (bFGF) [31]. Within the implantation period, a visible fat pad was formed at the injection site, likely attributable to preadipocyte and endothelial cell migration to the injection site. The highly vascularized tissue of the omentum fragmented and combined with preadipocyte cells such that implantation in vivo results in a tissue mass consisting of high triacylglycerol content [38]. The omentum is highly vascularized and filled with adipose tissue, both ideal characteristics for engineering adipose tissue. Results following implantation indicated that the omentum implanted with preadipocytes had the ability to form a tissue mass with high triacylglycerol content, indicative of fat.

Injectable microcarrier beads combined with a hydrogel delivery medium to form a minimally invasive implant that will stimulate regeneration of host adipose cells and fill a soft-tissue void upon injection in vivo [6]. The system is comprised of cells seeded on biodegradable beads of an injectable size; these

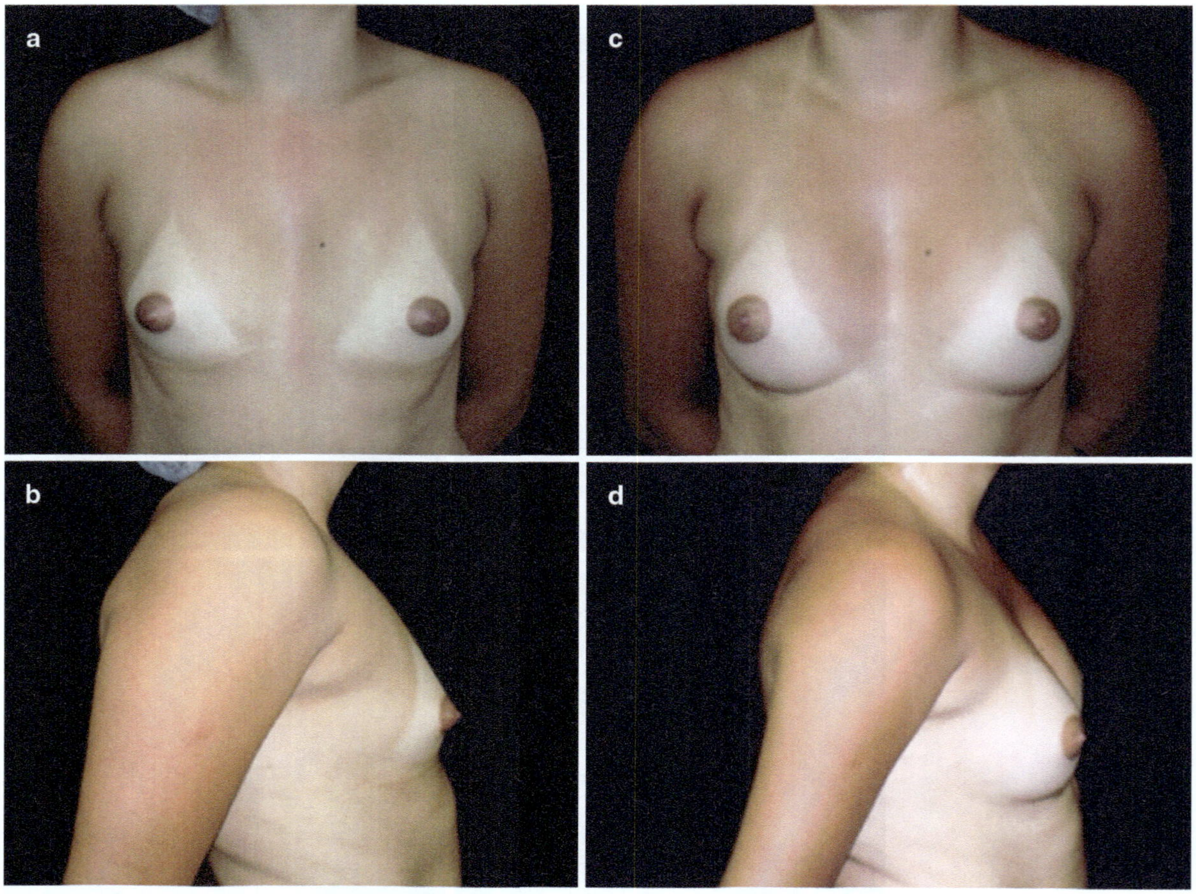

Fig. 16.5 (a, b) Preoperative photos of a 32-year-old woman requesting breast augmentation. (c, d) Postoperative photos of a 32-year-old woman after undergoing stromal enriched lipograft by receiving 192 mL of graft on right breast and 177 mL on the left breast

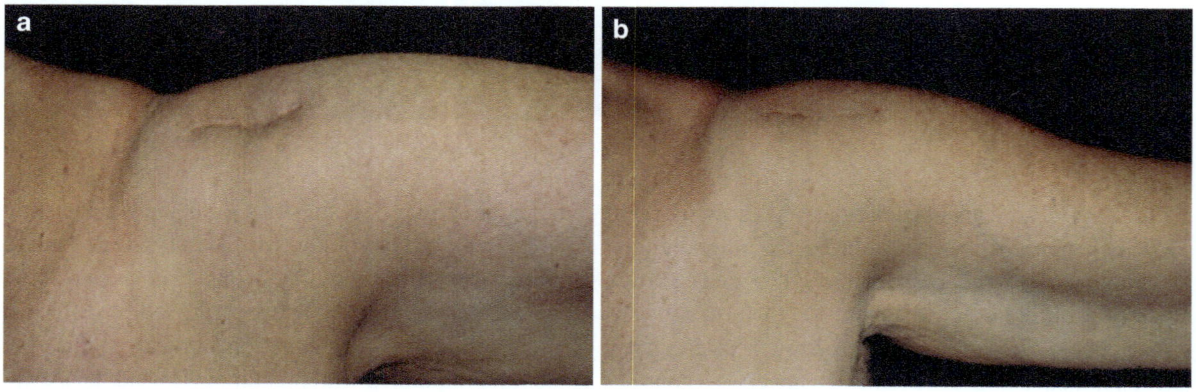

Fig. 16.6 (a) Preoperative photo of a 65-year-old woman asking for body contour treatment of her left shoulder depression. (b) Postoperative photo of a 65-year-old woman after undergoing stromal enriched lipograft of 14 mL

cellular constructs are then mixed with a hydrogel delivery medium, resulting in a composite that may be injected into a patient through a syringe at the defect site. Burg and coworkers have suggested the concurrent use of absorbable tissue expanders as temporary "space fillers" to allow the injection of composite cellular systems and thus facilitate the serial development of breast tissue in large defects.

Recently, preadipocyte cells cultured on absorbable polymeric scaffolds and implanted in vivo such

Table 16.2 The current animal studies of adipose-derived stem cells

Study	Scaffold used	Animal recipient	Result observed	Institution	Results published
Craniomaxillofacial tissue defect	Autologous ADSCs, bone morphogenetic protein-2, and periosteum	Pig	Revitalization of large-volume allograft mandible bone	Cincinnati Children's Hospital Medical Center, Cincinnati, Ohio, USA	[59]
Subcutaneous tissue defect	ADSCs transduced with vascular endothelial growth factor	Mice	Enhancement of the survival and quality of fat tissue	Southern Medical University, Guangzhou, China	[36]
Cartilage tissue engineering	ADSCs in silk-alginate Ear scaffold	Rabbit	Ear cartilage formation and volume maintenance	IASO General Hospital, Athens Greece	[61]
De novo adipogenesis	Laboratory made tissue engineered adipose tissue flap	Mice	Long-term Adipose fat volume maintenance	University of Melbourne, Melbourne, Australia	[13]
Injectable cell delivery vehicle for tissue defect	ADSCs cultured on porous collagenous microbeads	Lab	Ex vivo construction of injectable ADSCs seeded microbeads	University of Pittsburgh, Pittsburgh, PA 15261, USA	[57]
Full thickness wound defects	ADSCs and platelet-rich plasma therapy	Pig	Improvement of wound healing	Indiana University School of Medicine, Indiana, USA	[3]

that simultaneous cellular proliferation and scaffold resorption results in mature adipose tissue [4, 16, 48]. The concept is that the scaffold would then be implanted into the patient in order to fill the defect site. Scaffolds for adipose tissue engineering have typically been designed for restoration of tissue volume, as opposed to the restoration of tissue function. As a result, the scaffolds would ideally restore the esthetic function of the tissue by imparting a soft, smooth feel closely resembling that of natural tissue. Additionally, the surface topography of the pores within a scaffold may be altered to influence cellular behavior. Pore size and shape have been shown to effect cellular attachment by providing formed binding sites that cells may grow into. In the case of adipose tissue engineering, when seeding preadipocyte cells on porous scaffolds, the pores would ideally be large enough so as not to inhibit the proliferation and differentiation of preadipocytes. As preadipocytes mature, they significantly increase in size due to an increase in cell number and an increase in volume associated with lipid formation [10]. Adequate sizing of pores is essential in scaffold design for adipose tissue engineering.

In a recent study a novel, three-dimensional cell-copolymer construct resembling a human ear has been reported [61]. Schematic representation of auricular tissue engineering. (Fig. 16.7) The ADSCs are differentiated and cultured with the appropriate growth factors.(Fig. 16.8) The manufacture of 3D ear mold by rapid prototyping is shown in Fig. 16.9a. Adipose derived stem cells were obtained from inguinal fat pads of rabbits, were differentiated and expanded in vitro, and seeded onto 3D biodegradable alginate and silk polymer ear-shaped scaffolds (Fig. 16.9b). After implantation in the back of immunocompetent rabbits for 6 months, the cell/scaffold construct integrated is shown (Fig. 16.9c, d). This study demonstrates for the first time that it is possible to engineer an ear cartilage construct that resembles the human ear not only in shape but also in size and flexibility in a "real test" model.

16.6 Adipose-Derived Stem Cells and Future Applications on Tissue Engineering

Adipose tissue is believed to constitute an ideal source of uncultured stromal stem cells. The potential of using autologous adult stem cells derived from fat

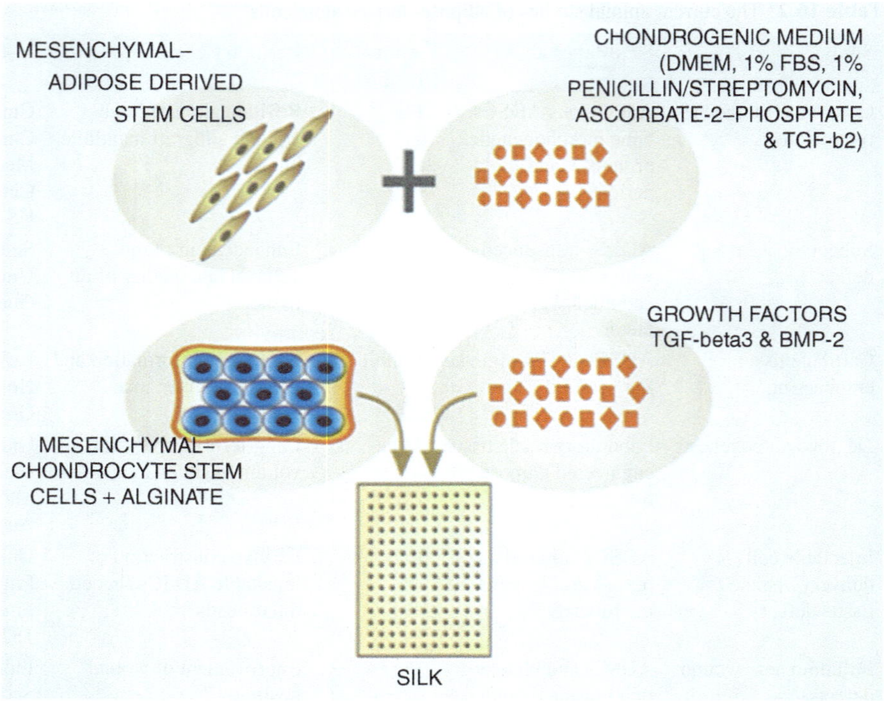

Fig. 16.7 Schematic representation of auricular tissue engineering

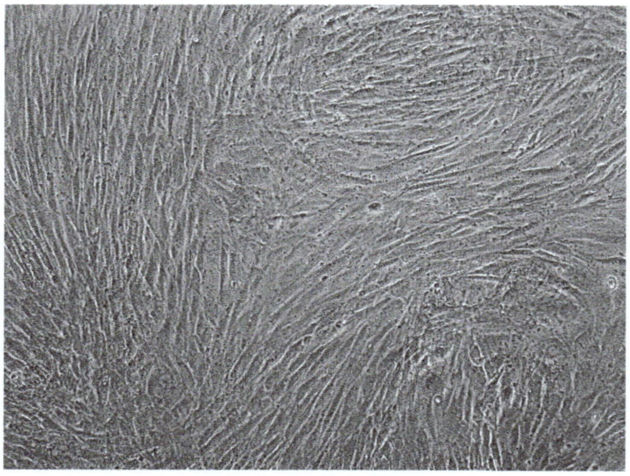

Fig. 16.8 ADSCs are differentiated in adipocytes and cultured with the appropriate growth factors after 14 days (EM ×10)

tissue is quickly becoming a clinical reality [62, 66, 67]. The ADSCs are isolated by enzymatic digestion, filtration, and centrifugation of the stromal vascular fraction that contains the adipose stem cells along with nonadherent cells. The SVF can be used directly as a source of stem cells. This process converts ADSCs-poor aspirated fat to ADSCs-rich fat. The number of functional ADSCs is likely to be important for tissue repair and remodeling and ADSCs differentiate into vascular endothelial cells which contribute to neoangiogenesis in the acute phase of fat transplantation. The presence of ADSCs has clinical implications for autologous fat transfer because ADSCs may contribute to neoangiogenesis in the acute phase by acting as endothelial progenitor cells or angiogenic-factor-releasing cells [17, 43]. Adipose-derived stem cells secrete angiogenic factors such as vascular endothelial growth factor and hepatocyte growth factor [5, 29]. ADSCs produce many antiapoptotic growth factors, and their secretion is significantly enhanced by hypoxia. In a recent study, hypoxia-treated ADSCs stimulated angiogenesis as well as maturation of the newly formed blood vessels in vivo [58]. ADSCs unregulated also their proneovascular activity in response to hypoxia, and may harbor the capacity to ischemic tissue and function cooperatively with existing vasculature to promote angiogenesis [69]. ADSCs can affect long-term survival of transplanted adipose by acting as preadipocytes [8]. The preliminary results suggest that adipose stem cell-enriched lipogaft is effective and safe for soft tissue augmentation and superior to conventional lipoinjection [62].

A recent study has demonstrated no statistical differences in adipocyte viability among abdominal fat, thigh fat, flank fat, or knee fat donor sites [70]. The

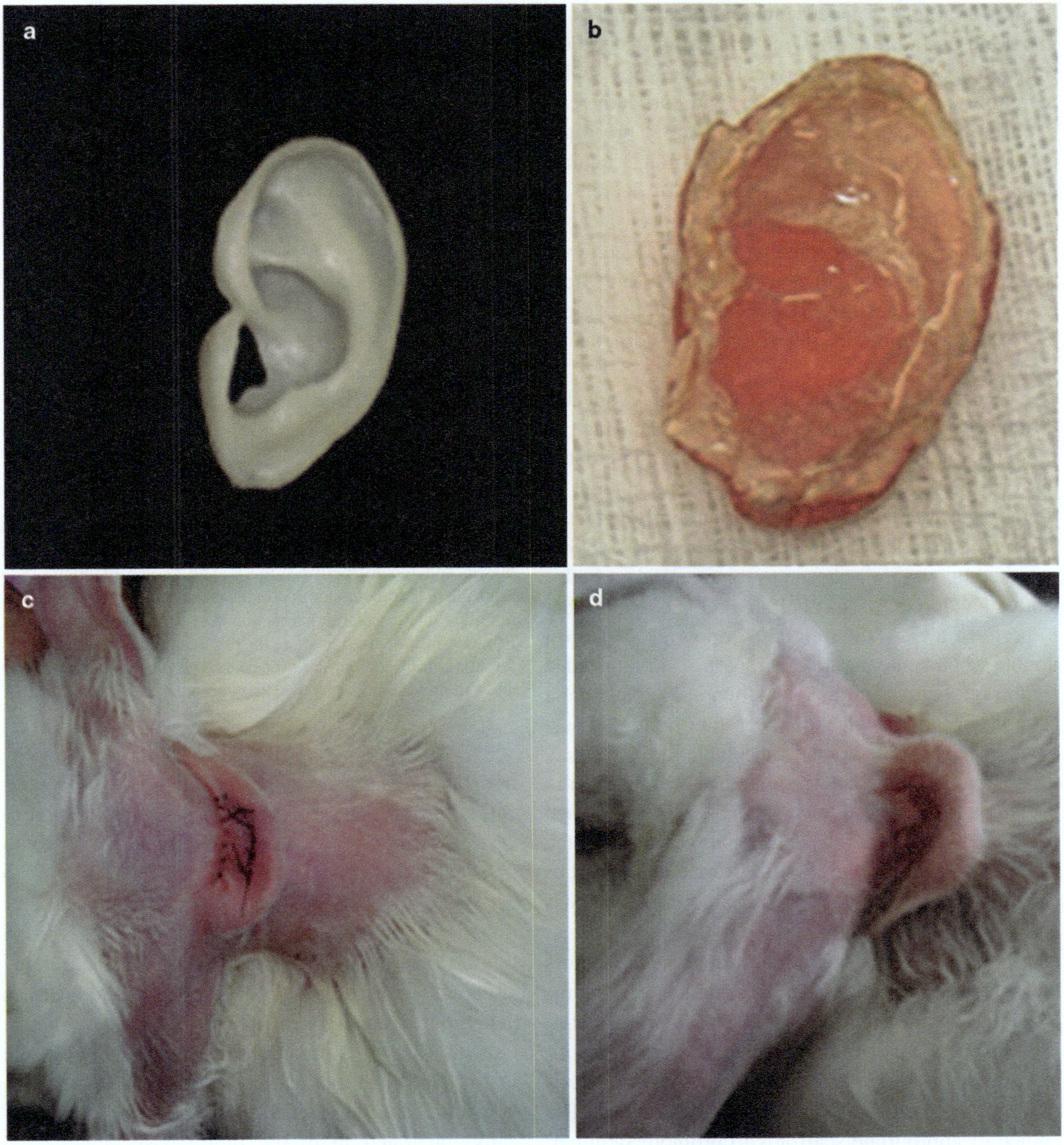

Fig. 16.9 (**a**) The manufacture of 3D ear mold by rapid prototyping. (**b**) Adipose-derived stem cells were obtained from inguinal fat pads of rabbits, were differentiated and expanded in vitro and seeded onto 3D biodegradable alginate and silk polymer ear-shaped scaffolds. (**c**) Ear scaffold implanted in the back of immunocompetent rabbit. (**d**) After implantation in the back of immunocompetent rabbits for 6 weeks, the cell/scaffold construct integrated is shown

abdomen though seems to be preferable to the hip/thigh region for harvesting adipose tissue when considering SVF cells for stem-cell-based therapies in one-step surgical procedures [24, 46].

In a recently published study, human preadipocytes were isolated from subcutaneous adipose tissue of patients and treated with bupivacaine, mepivacaine, ropivacaine, articaine/epinephrine, and lidocaine for

30 min. Viability was determined directly after treatment and during the following cultivation. While the immediate effects of mepivacaine and ropivacaine were only moderate, treatment with articaine/epinephrine and lidocaine strongly impaired preadipocyte viability. During long-term cultivation, articaine/epinephrine-treated cell viability markedly decreased, while other local anesthetics had no impact. Despite normal phenotypical appearance of cells treated with bupivacaine, mepivacaine, ropivacaine, and lidocaine, all local anesthetics markedly impaired adipocyte differentiation as determined by adiponectin expression [28].

The time needed in order to process the collected fat for SVF extraction is around 80–90 min. This is a limitation when compared to the time taken in order to perform the fat grafting to the face. The amount of donor fat available is also an important limiting factor. This is particularly important when the "ADSCs enriched lipograft" is chosen as treatment plan. In this study, the amount of lipoaspirated fat needed in order to perform the "ADSCs enriched lipograft" is almost double when compared to the lipoaspirated fat needed for the autologous fat transplantation in each treatment session. When though more than one treatment sessions of the autologous fat transplantation is needed, the total amount of lipoaspirated fat is comparable to the amount of lipoaspirated fat in "ADSCs enriched lipograft." There is an additional cost when the "ADSCs enriched lipograft" is performed. The cost of consumables needed for the preparation of "ADSCs enriched lipograft," the presence of tissue engineers, and the extra cost for operating theater use are factors that increase the final price of this surgical procedure.

Tissue engineering using ADSCs in animals continues to be a challenging problem. Long-term maintenance of the shape and dimension of the produced tissue engineered scaffolds remains a challenge. The choice of appropriate scaffolds to promote stem cell adhesion, proliferation, and differentiation is essential for successful tissue engineering (Fig. 16.10). Recent advances in nanotechnology may allow the development of nanostructured scaffolds with a cellular environment that maximally enhances not only cell expansion but also the neovascularization that is crucial for long-term maintenance of cell volume [1]. Bypassing ex vivo cell manipulation, the cell homing technique could eliminate the donor site morbidity and rejection, reducing the regulation issue in clinical translation. Biomaterials are used to guide the organization, growth, and differentiation of cells in the process of forming functional tissue and can provide both physical and chemical cues [63]. They need though to be tissue inductive in order to promote the proliferation of cells and tissue conductive in order to guide the migration of the cells. Biomolecules contribute in the structural integrity of tissue-engineered constructs and, at the same time, regulate their components. Growth factors, differentiation factors, angiogenic factors, and gene-modulated factors are the main components of the biomolecules and need to be strategically integrated in to the future tissue engineered constructs.

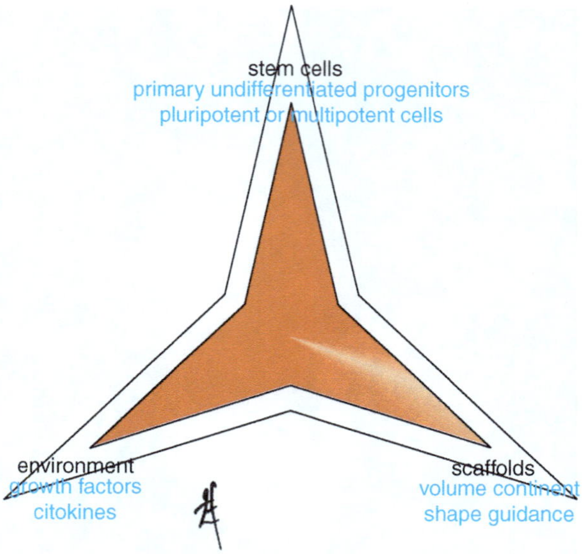

Fig. 16.10 Adipose Tissue Engineering: ADSCs have a great potential for use in tissue repair and regeneration in the field of plastic and reconstructive surgery. The choice of appropriate scaffolds to promote stem cell adhesion, proliferation, and differentiation is essential for successful tissue engineering. Biomaterials are used to guide the organization, growth, and differentiation of cells in the process of forming functional tissue and can provide both physical and chemical cues

16.7 Conclusion

ADSCs have a great potential for use in tissue repair and regeneration in the field of plastic and reconstructive surgery. The stromal enriched lipograft has already been used in clinical practice and the results have proved superior to traditional autologous fat grafting. Although the current animal and clinical applied strategies for adipose tissue engineering have advanced in the last years, complete understanding of the mechanisms of interactions among adipose stem cells, growth factors, and biomaterials in tissue engineering is needed in order to advance the end goal of developing "off the-shelf" tissue engineering products. The enthusiasm over what unquestionably represents a markedly innovative technique with huge therapeutic potential must be balanced though against stringent standards of scientific and clinical investigation.

References

1. Ahn JM, Mao JJ (2010) Adipose tissue engineering from adult human stem cells: a new concept in biosurgery. Facial Plast Surg 26(5):413–420, Epub 2010 Sep 17
2. Akita S, Akino K, Hirano A, Ohtsuru A, Yamashita S (2010) Mesenchymal stem cell therapy for cutaneous radiation syndrome. Health Phys 98(6):858–862
3. Blanton MW, Hadad I, Johnstone BH, Mund JA, Rogers PI, Eppley BL, March KL (2009) Adipose stromal cells and platelet-rich plasma therapies synergistically increase revascularization during wound healing. Plast Reconstr Surg 123(2 suppl):56S–64S
4. Brey EM, Patrick CW Jr (2000) Tissue engineering applied to reconstructive surgery. IEEE Eng Med Biol Mag 19(5):122–125
5. Bunnell BA, Flaat M, Gagliardi C, Patel B, Ripoll C (2008) Adipose-derived stem cells: isolation, expansion and differentiation. Methods 45(2):115–120, Epub 2008 May 29
6. Burg KJL, Austin CE, Culberson CR, Greene KG, Halberstadt CR, Holder Jr WD et al (2000) A novel approach to tissue engineering: injectable composites. In: Transactions of the 2000 world biomaterials congress. Kona, HI, 2000
7. Cervelli V, De Angelis B, Lucarini L, Spallone D, Balzani A, Palla L, Gentile P, Cerulli P (2010) Tissue regeneration in loss of substance on the lower limbs through use of platelet-rich plasma, stem cells from adipose tissue, and hyaluronic acid. Adv Skin Wound Care 23(6):262–272
8. Chung HM, Won CH, Sung JH (2009) Responses of adipose-derived stem cells during hypoxia: enhanced skin-regenerative potential. Expert Opin Biol Ther 9(12):1499–1508
9. Condé-Green A, de Amorim NF Gontijo, Pitanguy I (2010) Influence of decantation, washing and centrifugation on adipocyte and mesenchymal stem cell content of aspirated adipose tissue: a comparative study. J Plast Reconstr Aesthet Surg 63(8):1375–1381
10. Cornelius P, MacDougald OA, Lane MD (1994) Regulation of adipocyte development. Annu Rev Nutr 14:99–129
11. Donofrio LM (2008) Techniques in facial fat grafting. Aesthet Surg J 28(6):681–687
12. Estes BT, Diekman BO, Guilak F (2008) Monolayer cell expansion conditions affect the chondrogenic potential of adipose-derived stem cells. Biotechnol Bioeng 99(4):986–995
13. Findlay MW, Messina A, Thompson EW, Morrison WA (2009) Long-term persistence of tissue-engineered adipose flaps in a murine model to 1 year: an update. Plast Reconstr Surg 124(4):1077–1084
14. Grimes BR, Steiner CM, Merfeld-Clauss S, Traktuev DO, Smith D, Reese A, Breman AM, Thurston VC, Vance GH, Johnstone BH, Slee RB, March KL (2009) Interphase FISH demonstrates that human adipose stromal cells maintain a high level of genomic stability in long-term culture. Stem Cells Dev 18(5):717–724
15. Guerrerosantos J, Guerrerosantos F, Orozco J (2007) Classification and treatment of facial tissue atrophy in Parry-Romberg disease. Aesthetic Plast Surg 31(5):424–434
16. Halbleib M, Skurk T, de Luca C, von Heimburg D, Hauner H (2003) Tissue engineering of white adipose tissue using hyaluronic acid based scaffolds. I: in vitro differentiation of human adipocyte precursor cells on scaffolds. Biomaterials 24(18):3125–3132
17. Hamou C, Callaghan MJ, Thangarajah H, Chang E, Chang EI, Grogan RH, Paterno J, Vial IN, Jazayeri L, Gurtner GC (2009) Mesenchymal stem cells can participate in ischemic neovascularization. Plast Reconstr Surg 123(2 suppl): 45S–55S
18. Illouz YG (1983) Body contouring by lipolysis: a 5-year experience with over 3000 cases. Plast Reconstr Surg 72(5): 591–597
19. Illouz YG (1986) The fat cell "graft": a new technique to fill depressions. Plast Reconstr Surg 78(1):122–123
20. Illouz YG (1992) Adipoaspiration and "filling" in the face. Facial Plast Surg 8(1):59–71
21. Illouz YG (1996) History and current concepts of lipoplasty. Clin Plast Surg 23(4):721–730
22. Illouz YG, Sterodimas A (2009) Autologous fat transplantation to the breast: a personal technique with 25 years of experience. Aesthetic Plast Surg 33(5):706–715, Epub 2009 Jun 4
23. Jeong JH (2010) Adipose stem cells and skin repair. Curr Stem Cell Res Ther 5(2):137–140
24. Jurgens WJ, Oedayrajsingh-Varma MJ, Helder MN, Zandiehdoulabi B, Schouten TE, Kuik DJ, Ritt MJ, van Milligen FJ (2008) Effect of tissue-harvesting site on yield of stem cells derived from adipose tissue: implications for cell-based therapies. Cell Tissue Res 332(3):415–426, Epub 2008 Apr 1
25. Kakudo N, Minakata T, Mitsui T, Kushida S, Notodihardjo FZ, Kusumoto K (2008) Proliferation-promoting effect of platelet-rich plasma on human adipose-derived stem cells and human dermal fibroblasts. Plast Reconstr Surg 122(5): 1352–1360
26. Kaufman MR, Miller TA, Huang C, Roostaien J, Wasson KL, Ashley RK, Bradley JP (2007) Autologous fat transfer for facial recontouring: is there science behind the art? Plast Reconstr Surg 119(7):2287–2296
27. Kawaguchi N, Toriyama K, Nicodemou-Lena E, Inou K, Torii S, Kitagawa Y (1998) De novo adipogenesis in mice at the site of injection of basement membrane and basic fibroblast growth factor. Proc Natl Acad Sci USA 95(3):1062–1066
28. Keck M, Zeyda M, Gollinger K, Buriak S, Kamolz LP, Frey M, Stulnig TM (2010) Local anesthetics have a major impact on viability of preadipocytes and their differentiation to adipocytes. Plast Reconstr Surg 126(5):1500–1505

29. Kilroy GE, Foster SJ, Wu X et al (2007) Cytokine profile of human adipose-derived stem cells: expression of angiogenic, hematopoietic, and pro-inflammatory factors. J Cell Physiol 212:702–709
30. Kim WS, Park BS, Sung JH (2009) The wound-healing and antioxidant effects of adipose-derived stem cells. Expert Opin Biol Ther 9(7):879–887
31. Kimura Y, Ozeki M, Inamoto T, Tabata Y (2002) Time course of de novo adipogenesis in matrigel by gelatin microspheres incorporating basic fibroblast growth factor. Tissue Eng 8(4):603–613
32. Kurita M, Aiba-Kojima E, Shigeura T, Matsumoto D, Suga H, Inoue K, Eto H, Kato H, Aoi N, Yoshimura K (2008) Differential effects of three preparations of human serum on expansion of various types of human cells. Plast Reconstr Surg 122(2):438–448
33. Kurita M, Matsumoto D, Shigeura T, Sato K, Gonda K, Harii K, Yoshimura K (2008) Influences of centrifugation on cells and tissues in liposuction aspirates: optimized centrifugation for lipotransfer and cell isolation. Plast Reconstr Surg 121(3):1033–1041, discussion 1042–1043
34. Liu ZJ, Zhuge Y, Velazquez OC (2009) Trafficking and differentiation of mesenchymal stem cells. J Cell Biochem 106(6):984–991
35. Lu F, Mizuno H, Uysal CA, Cai X, Ogawa R, Hyakusoku H (2008) Improved viability of random pattern skin flaps through the use of adipose-derived stem cells. Plast Reconstr Surg 121(1):50–58
36. Lu F, Li J, Gao J, Ogawa R, Ou C, Yang B, Fu B (2009) Improvement of the survival of human autologous fat transplantation by using VEGF-transfected adipose-derived stem cells. Plast Reconstr Surg 124(5):1437–1446
37. Lund P, Pilgaard L, Duroux M, Fink T, Zachar V (2009) Effect of growth media and serum replacements on the proliferation and differentiation of adipose-derived stem cells. Cytotherapy 11(2):189–197
38. Masuda T, Furue M, Matsuda T (2004) Novel strategy for soft tissue augmentation based on transplantation of fragmented omentum and preadipocytes. Tissue Eng 10(11–12): 1672–1683
39. Matsudo PK, Toledo LS (1988) Experience of injected fat grafting. Aesthetic Plast Surg 12(1):35–38
40. Matsumoto D, Sato K, Gonda K, Takaki Y, Shigeura T, Sato T, Aiba-Kojima E, Iizuka F, Inoue K, Suga H, Yoshimura K (2006) Cell-assisted lipotransfer: supportive use of human adipose-derived cells for soft tissue augmentation with lipoinjection. Tissue Eng 12(12):3375–3382
41. Mesimäki K, Lindroos B, Törnwall J, Mauno J, Lindqvist C, Kontio R, Miettinen S, Suuronen R (2009) Novel maxillary reconstruction with ectopic bone formation by GMP adipose stem cells. Int J Oral Maxillofac Surg 38(3):201–209, Epub 2009 Jan 24
42. Moseley TA, Zhu M, Hedrick MH (2006) Adipose-derived stem and progenitor cells as fillers in plastic and reconstructive surgery (review). Plast Reconstr Surg 118(3 suppl): 121S–128S
43. Mylotte LA, Duffy AM, Murphy M, O'Brien T, Samali A, Barry F, Szegezdi E (2008) Metabolic flexibility permits mesenchymal stem cell survival in an ischemic environment. Stem Cells 26(5):1325–1336
44. Nicaretta B, Pereira LH, Sterodimas A, Illouz YG (2011) Autologous gluteal lipograft. Aesthetic Plast Surg 35(2): 216–224
45. Nie C, Yang D, Xu J, Si Z, Jin X, Zhang J (2011) Locally administered adipose-derived stem cells accelerate wound healing through differentiation and vasculogenesis. Cell Transplant 2011;20(2):205–216
46. Padoin AV, Braga-Silva J, Martins P, Rezende K, Rezende AR, Grechi B, Gehlen D, Machado DC (2008) Sources of processed lipoaspirate cells: influence of donor site on cell concentration. Plast Reconstr Surg 122(2):614–618
47. Parker AM, Shang H, Khurgel M, Katz AJ (2007) Low serum and serum-free culture of multipotential human adipose stem cells. Cytotherapy 9(7):637–646
48. Patrick CW Jr, Zheng B, Johnston C, Reece GP (2002) Long-term implantation of preadipocyte-seeded PLGA scaffolds. Tissue Eng 8(2):283–293
49. Pereira LH, Sterodimas A (2008) Free fat transplantation for the aesthetic correction of mild pectus excavatum. Aesthetic Plast Surg 32(2):393–396
50. Pereira LH, Sterodimas A (2009) Composite body contouring. Aesthetic Plast Surg 33(4):616–624, Epub 2009 May 12
51. Pereira LH, Sterodimas A (2009) Macroscopic and microscopic proof of long-term survival of gluteal fat transplantation. Plast Reconstr Surg 123(4):162e–163e
52. Pereira LH, Sterodimas A (2010) Long-term fate of transplanted autologous fat in the face. J Plast Reconstr Aesthet Surg 63(1):e68–e69, Epub 2009 Mar 12
53. Pereira LH, Sterodimas A (2010) Treatment of iatrogenic abdominal contour irregularities. Aesthetic Plast Surg 34(2): 129–135, Epub 2008 Aug 28
54. Pitanguy I (2007) Plastic surgery: personal recollections, contributions, and some thoughts: the Ohmori lecture: 18th I.S.A.P.S. Congress, Rio De Janeiro – 2006. Aesthetic Plast Surg 31(6):619–635
55. Rigotti G, Marchi A, Galiè M, Baroni G, Benati D, Krampera M, Pasini A, Sbarbati A (2007) Clinical treatment of radiotherapy tissue damage by lipoaspirate transplant: a healing process mediated by adipose-derived adult stem cells. Plast Reconstr Surg 119:1409–1422, discussion 1423–1424
56. Rodbell M (1964) Localization of lipoprotein lipase in fat cells of rat adipose tissue. J Biol Chem 239:753–755
57. Rubin JP, Bennett JM, Doctor JS, Tebbets BM, Marra KG (2007) Collagenous microbeads as a scaffold for tissue engineering with adipose-derived stem cells. Plast Reconstr Surg 120(2):414–424
58. Rubina K, Kalinina N, Efimenko A, Lopatina T, Melikhova V, Tsokolaeva Z, Sysoeva V, Tkachuk V, Parfyonova Y (2009) Adipose stromal cells stimulate angiogenesis via promoting progenitor cell differentiation, secretion of angiogenic factors, and enhancing vessel maturation. Tissue Eng A 15(8):2039–2050
59. Runyan CM, Jones DC, Bove KE, Maercks RA, Simpson DS, Taylor JA (2010) Porcine allograft mandible revitalization using autologous adipose-derived stem cells, bone morphogenetic protein-2, and periosteum. Plast Reconstr Surg 125(5):1372–1382
60. Schipper BM, Marra KG, Zhang W, Donnenberg AD, Rubin JP (2008) Regional anatomic and age effects on cell function

of human adipose-derived stem cells. Ann Plast Surg 60(5):538–544
61. Sterodimas A, Baptista LS, Pedrosa CS, da Silva KR, do Amaral R, Kochem M, Menezes H, de Faria J, de Oliveira M, Borojevic R (2008) Fibroin-based material from natural silk can be associated with alginate and mesenchymal progenitor cells. Key Eng Mater 396-398:437–440
62. Sterodimas A, de Faria J, Nicaretta B,Boriani F Autologous fat transplantation versus adipose derived stem cells enriched lipograft: a study. Aesthet Surg J [Epub ahead of print]
63. Sterodimas A, de Faria J, Correa WE, Pitanguy I (2009) Tissue engineering and auricular reconstruction: a review. J Plast Reconstr Aesthet Surg 62(4):447–452, Epub 2008 Dec 27
64. Sterodimas A, De Faria J, Correa WE, Pitanguy I (2009) Tissue engineering in plastic surgery: an up-to-date review of the current literature (review). Ann Plast Surg 62(1): 97–103
65. Sterodimas A, Huanquipaco JC, de Souza Filho S, Bornia FA, Pitanguy I (2009) Autologous fat transplantation for the treatment of Parry-Romberg syndrome. J Plast Reconstr Aesthet Surg 62(11):e424–e426
66. Sterodimas A, de Faria J, Nicaretta B, Papadopoulos O, Illouz YG (2010) Cell assisted lipotransfer. Aesthet Surg J 30(1):94–100
67. Sterodimas A, de Faria J, Nicaretta B, Pitanguy I (2010) Tissue engineering with adipose-derived stem cells (ADSCs): current and future applications. J Plast Reconstr Aesthet Surg 63(11):1886–1892
68. Stillaert FB, Di Bartolo C, Hunt JA, Rhodes NP, Tognana E, Monstrey S, Blondeel PN (2008) Human clinical experience with adipose precursor cells seeded on hyaluronic acid-based spongy scaffolds. Biomaterials 29(29):3953–3959, Epub 2008 Jul 17
69. Thangarajah H, Vial IN, Chang E, El-Ftesi S, Januszyk M, Chang EI, Paterno J, Neofytou E, Longaker MT, Gurtner GC (2009) IFATS series: adipose stromal cells adopt a proangiogenic phenotype under the influence of hypoxia. Stem Cells 27(1):266–274
70. Ullmann Y, Shoshani O, Fodor A, Ramon Y, Carmi N, Eldor L, Gilhar A (2005) Searching for the favorable donor site for fat injection: in vivo study using the nude mice model. Dermatol Surg 31(10):1304–1307
71. Uysal CA, Ogawa R, Lu F, Hyakusoku H, Mizuno H (2010) Effect of mesenchymal stem cells on skin graft to flap prefabrication: an experimental study. Ann Plast Surg 65(2):237–244
72. Van RL, Roncari DA (1977) Isolation of fat cell precursors from adult rat adipose tissue. Cell Tissue Res 181(2): 197–203
73. Yoshimura K, Sato K, Aoi N, Kurita M, Hirohi T, Harii K (2008) Cell-assisted lipotransfer for cosmetic breast augmentation: supportive use of adipose-derived stem/stromal cells. Aesthetic Plast Surg 32(1):48–55, discussion 56–57. Epub 2007 Sep 1
74. Yoshimura K, Sato K, Aoi N, Kurita M, Inoue K, Suga H, Eto H, Kato H, Hirohi T, Harii K (2008) Cell-assisted lipotransfer for facial lipoatrophy: efficacy of clinical use of adipose-derived stem cells. Dermatol Surg 34:1178–1185
75. Yoshimura K, Asano Y, Aoi N, Kurita M, Oshima Y, Sato K, Inoue K, Suga H, Eto H, Kato H, Harii K (2010) Progenitor-enriched adipose tissue transplantation as rescue for breast implant complications. Breast J 16(2):169–175, Epub 2009 Nov 12
76. Zuk PA, Zhu M, Mizuno H, Huang JI, Futrell WJ, Katz AJ, Benhaim P, Lorenz HP, Hedrick MH (2001) Multilineage cells from human adipose tissue: implications for cell-based therapies. Tissue Eng 7:211–226

Orthopedic Use of Adipose-Derived Stem Cells

17

Josh E. Schroeder, Shaul Beyth, and Meir (Iri) Liebergall

Contents

17.1 **Introduction** 181

17.2 **The Use of Adipose-Derived Stem Cells for Bone Regeneration** 182

17.3 **The Use of Adipose-Derived Stem Cells for Chondrogenesis** 184

17.4 **Spinal Applications of Adipose-Derived Stem Cells** 186

17.5 **The Use of Adipose-Derived Stem Cells for Tendon and Ligament Regeneration** 187

17.6 **Turning Fat into Muscle** 188

17.7 **Conclusions** 188

References 189

J.E. Schroeder • S. Beyth • M. Liebergall (✉)
Orthopedic Surgery Department, Hadassah Hebrew University Medical Center, P.O. Box, 1200, Kiryat Hadassah, Jerusalem, Israel
e-mail: liebergall@hadassah.org.il

17.1 Introduction

The musculoskeletal system is the largest system within the human body. The system includes: bone, cartilage, ligaments, tendons, and muscle [108]. It incorporates the axial and core parts of the body. In essence, the musculoskeletal system has two main tasks: it is in charge of motion, making sure that we can get from point A to point B smoothly and painlessly, and it protects internal vital organs. This system undergoes tremendous stress, with over three million steps taken per year [10], and withstands forces which can reach up to 10 times the body's weight [95]. The musculoskeletal system is designed to allow one to run an ultramarathon, jump from a cliff (as long as there is water below), and swim across the British Chanel without injuring any vital organ and without jeopardizing one's health.

However, there are limits to this system, and different components of the system have different thresholds and different healing and regeneration capacities. For example, when a bone is bent, the fine configuration of the bone fails and a fracture occurs. Bone has great healing capacities, about which we will elaborate later in the chapter, leading, in most cases, to new bone formation and replacement of the damaged bone with the same structure and characteristics as prior to the injury [69]. Cartilage, tendon, and ligaments are not so forgiving, although they will mostly heal following an acute injury. However, the injury is greater than a critical dimension, i.e., partial thickness injuries, scar tissue will replace the native tissue, with diminished functionality as compared with the native tissue [23, 40]. Even in bone, the most forgiving tissue of the

Y.-G. Illouz and A. Sterodimas (eds.), *Adipose Stem Cells and Regenerative Medicine*,
DOI: 10.1007/978-3-642-20012-0_17, © Springer-Verlag Berlin Heidelberg 2011

musculoskeletal system, when a critical size fracture [11] occurs or when a nonunion [69] is established, the system cannot heal itself and intervention is required.

The musculoskeletal system comprises specialized cells such as abundant extracellular matrix and progenitor cells, which are the reserve for future regeneration of the system. The system is dynamic with extracellular tissue constantly being deposited and absorbed [17]. With age, this turnover becomes less efficient marked by a decreasing number of progenitor cells, microvascular changes, and increased absorption and degeneration of extracellular proteins [16]. This process, along with other comorbidities, leads to osteoporosis [90], slower healing of fractures [42], osteoarthritis [32] and tendon failure, for example, rotator cuff tears of the shoulder [72].

Since the musculoskeletal system is required for every motion we take, these challenges have lead the scientific community to develop regeneration pathways for different components of the system.

At the base of this system, as in most other biological system, is the cell, a functional unit that is in charge of generating the extracellular components which together with other cooperators build the specialized tissue [79]. Although fat, bone, cartilage, and muscle seem very different, they are all generated from a single precursor: the mesenchymal stem cell (MSC) [79]. This multipotent cell, when placed in the right atmosphere or having received the appropriate signal, can convert to any one of the mesenchymal tissues. For many years, it has been known that MSCs are located in greater numbers in bone marrow [53] and when signaled proceed into the peripheral blood, homing in on the site of need and repairing the tissues as required [20]. When in the marrow, they are found in low concentrations of 1–2 cells per million, but increase in number when a fracture occurs [8]. For over 30 years, MSCs (as bone marrow aspirates) were added to nonunions in a nonpurified manner in an attempt to try to heal the fracture [22]. However, only years later with the development of different molecular assays [77], did exact isolation and characterization of MSCs occur. MSCs, using special probing signals roam the blood, settle in the fracture site, differentiate into bone cells, and start constructing the callus [29]. The number of stem cells differs from person to person and their number is negatively affected by increased age, female sex, diseases, different medications, and environmental factors. In particular, smoking reduces the number of stem cells [86]. These problems have led scientists to look for other sources for outer-skeletal MSCs [57, 86].

In today's world, there is a surplus of a specific tissue: fat [78]. With over half of the adult population overweight, looking at a positive product from fat is natural, especially since the rate of liposuctions is growing at a steady rate [5]. Since the start of the twenty-first century, it was demonstrated that there is an abundance of stem cells in fat tissue and that adipose-derived stem cells possess multiple differentiation capacities [38]. Additionally, adipose tissue is probably the most abundant and accessible source of adult stem cells in the adult human body. There are 300-fold more stem cells in fat tissue than in bone marrow aspirates [31]. Fat- and bone-derived MSCs are similar but are not identical [2]. There are differences between stem cells harvested from subcutaneous fat (these cells are the most similar to bone-derived stem cells) and stem cells harvested from visceral fat [45]. Brown fat, the fat abundant in early childhood, harbors very little stem cells while white fat, the fat which we tug around as we age, has large numbers of stem cells [80].

The entire scope of differences and similarities between the two types of cells is beyond the scope of this chapter; however, the general consensus is that freshly isolated stem cells from both bone marrow and adipose tissue are positive for multiple cellular markers: CD10, CD13, CD29, CD44, CD49e, CD90, CD105, CD166, and human leukocyte antigen (HLA)-ABC and negative for CD31, CD45, and HLA-DR [65]. These cells in the proper medium can be converted into cartilage, disc, and bone [38]. We will elaborate on each in the upcoming sections of this chapter.

17.2 The Use of Adipose-Derived Stem Cells for Bone Regeneration

When a fracture occurs, there is not only damage to the calcium-phosphate crystal structure of the bone but also to its organic matrix: the blood vessels and soft tissues surrounding the fracture. The energy that caused the fracture and the parts of bone which shifted propagate into the surrounding tissues, damaging blood vessels, fascia, muscles, and other connective tissue around the fracture [94]. As a consequence, a set of signals is triggered. These signals, local and systemic, are mediated by neuronal impulses [74], the hematoma at the site of the fracture [29], and trauma caused to the tissues surrounding the fracture. These signals which interact and are interchangeable can be classified into two categories: Inflammatory signals [36] (i.e., IL-1, IL-6, and TNF-α) and bone building and regenerating signals [82] (BMP growth factors and WNTs). These signals are in charge of the migration of phagocytotic cells to the area of the fracture, removing the necrotic tissue. Additionally, these signals trigger the ingrowth of new blood vessels at the site of the fracture, providing nutrients and recruiting cells leading to the proliferation of bone builder cells (osteoprogenitor cells), and ending with the formation of the new bone [82]. The process of cell recruitment differentiation is termed osteoinduction, while the entire cell activity to produce new bone is termed osteogenesis. One must have a scaffold on which the new bone will regenerate and this is termed osteoconduction [28]. This scaffold needs to be stable, with both ends of the fracture at close proximity; otherwise, the fracture will not heal.

Osteoprogenitor cells are needed to replace and create new bone tissue to replace the damaged bone. These cells can come via the periosteum at the fracture site via chemotaxis or through blood vessels entering the hematoma [29]. Specific mechanical and biological stimulates, i.e., osteoinduction, cause the MSCs to differentiate into osteoblasts, the bone forming cell. However, this migration does not suffice the needs for healing in all fractures. For example, in large or significant defects which in experimental studies are named "critical size bone defects," defined as fractures with a bone loss that is larger than twice the diameter of the bone, the natural migration of osteoprogenitor cells is not sufficient for healing [69]. In addition, as mentioned above, in elderly and diseased people the natural migration might not be enough for simple fractures to heal [13]. Thus, in these cases external augmentation is needed to assist the healing process. The optimal augmentation is a combination of a matrix as a scaffold for bone formation, cells which will build the bone and inductive growth factors, i.e., the combination of osteoinduction, osteoconduction, and osteogenesis. The gold standard for this augmentation is the use of bone graft taken most commonly from the iliac crest and placed in the fracture site. However, there is morbidity with such harvesting and only a limited amount of bone that can be harvested [6], therefore meriting a search for other sources of cells which will create bone, and developing artificial scaffolds on which these cells will create bone [17].

Adipose-derived stem cells have been shown to create bone tissue, both in vitro [79] and in animal models [46]. Cells need to grow for 2–4 weeks of culture, with the use of the same culture medium and environment as used for osteogenic differentiation of bone marrow harvested MSCs. When the culture medium is supplemented with dexamethasone and b-glycerolphosphate, the adipose-derived cells express gene characteristics of osteoblast cells, for example, BMP-2, BMP4, and PTH receptor 1 [51]. After several weeks these cells form a mineralized matrix. The addition of BMP-2 [26] and growth and differentiation factor 5 (GDF-5) [88] to the medium has been shown to increase the osteogenic potential of the cells. These cells can be tested for osteogenic traits by a selection of markers on their surface, by the creation of mineralized tissue (as tested by histological tests), or by the secretion of alkaline phosphatase [109]. When comparing the adipose-derived stem cells to bone marrow-derived cells, there are conflicting reports, some showing that the cells are equally potent, and others showing that adipose-derived MSCs are inferior to BM-derived MSCs in production of bone [51, 57, 64]. However, each study uses a different protocol, with slightly different mediums, so a final conclusion cannot be assessed.

In vivo, cells cannot be placed without a carrier, and therefore the scaffold is of critical importance to placing the cells at the fracture site and can provide additional signals via factors imbedded into the scaffold which increase transformation into bone producing cells: osteoblasts [44]. There are two experimental types of in vivo models to evaluate the potential of any substitute to induce bone formation: an orthotopic model in which the challenge is to repair the bone or fracture and a heterotopic one in which bone formation is induced at other tissues such as in subcutaneous tissue. In a rat femoral fracture model (orthotopic model), the placement of human adipose tissue-derived MSCs in the bone defect with only a collagen–ceramic carrier without additional genetic intervention did not bring about fracture healing [76], suggesting that more stimuli is needed to drive the cells into the bone linage. Adipose tissue-derived stem cells are different from bone marrow-derived stem cells where the default setting is to create bone. Hattori et al. [44] showed that when he cultured adipose-derived stem cells in a osteogenic medium for 2 weeks and then subcutaneously implanted them in mice (heterotopic model) and in atelocollagen honeycomb-shaped scaffolds with a membrane seal, the cells created bone that was equivalent to bone created by bone marrow-derived stem cells [44]. Yoon et al. [107] also showed that when adipose-derived stem cells are precultured in an osteogenic medium for 14 days, bone formation in vivo (when implanted in a critical size defect in a poly lactic-co-glycolic acid scaffold) is increased [107].

A different approach for the use of adipose-derived stem cells in bone formation from adipose cells is the manipulation of cells to produce osteogenic propagators via gene therapy pathways. In a study which manipulated adipose-derived stem cells to produce BMP-2, the fractures healed in 11 of 12 animals by 8 weeks after cell implantation [76]. The bone which was created was strong and was equivalent to the native bone of the rat. Unfortunately, the doses of BMP-2 produced by the stem cells in this study were high and are beyond the rage permitted in clinical use. In addition, in a clinical setting a more controlled release of BMP-2 will be needed. Genetic modification of cells is problematic and several safety issues need to be worked out (i.e., ruling out carcinogenic potential) prior to clinical use of these cell types.

There is very little clinical data regarding experiments in humans with bone defects and adipose-derived stem cells [20]. In a case report, a 7-year-old girl was successfully treated with adipose-derived stem cells for a bone defect in her calvaria [66]. In a second case report, that of a 68-year-old male who had lost part of the hard palate of the mouth following removal of a cyst resulting in an inability to chew or drink effectively. To address this defect, the surgical team harvested 200 g autologous subcutaneous adipose tissue, isolated and culture expanded the ASCs in autologous serum and then seeded them in the presence of bone morphogenetic protein 2 (BMP-2 onto a β-tricalcium phosphate scaffold fashioned into the shape of the defect). The construct was implanted into the patient's rectus abdominis muscle and was radiographically followed for evidence of mineralization over an 8-month period. At that time, the construct was resected and transplanted to the maxillofacial defect, where its intact epigastric artery blood supply was reanastomosed to the facial artery. The patient recovered full oral function and remained without complication [39, 71].

Since a two-stage procedure (with ex vivo proliferation of stem cells) has increased costs and morbidity, a one-stage method to induce osteogenesis was developed by Helder et al. [45].Within 100 min, adipose-derived stem cells can be harvested and returned to a surgery site within a osteoinductive scaffold [45]. This method was tested in a sheep spinal fusion model and was shown to be successful in creating bone. It would be interesting to see if this methodology is as successful in the axial skeleton as well.

There are still several steps that need to be worked out prior to the common use of adipose-derived stem cells in a wide based clinical setting. Ex vivo proliferation of cells presents many questions as the methylation patterns of these cells change when they are removed from their native atmosphere [106]. In addition, industrial "clean room" levels must be kept in order to prevent cell contamination. Furthermore, tests to prevent malignant transformation of the cells must be conducted before these cells can be introduced into the clinical setting. Finally, it will be necessary to work out accessory inputs which are needed to trigger cells need in order to create bone.

In conclusion, adipose-derived stem cells can be pushed ex vivo and in vivo into an osteogenic pathway. The amount of bone generated may not be as abundant as the bone created by bone marrow-derived stem cells; however, the amount of bone appears to be enough in order to overcome bone defects in both small and large mammals. In addition, the fact that adipose-derived stem cells show increased vascularity, with an increased trend towards angiogenesis [83], when compared with bone marrow-derived MSCs, might assist in tackling one of the major problems of large bone defects: how to bring blood vessels to the newly implanted graft [60]. Once safety and efficacy tests are preformed, adipose deriving kits will most probably find their way into the orthopedic armamentarium.

17.3 The Use of Adipose-Derived Stem Cells for Chondrogenesis

Cartilage is a tissue with a small number of cells known as chondrocytes, which are responsible for maintaining a large extracellular matrix. In normal human cartilage, a chondrocyte with a cell diameter of 13 μm controls averaging 104,040–160,707 μm [3] of matrix.

The remaining mass is broadly made up of water (60–85%) and two categories of molecules: collagenous and noncollagenous [91].

The main function of cartilage is to produce a low friction surface capable of withstanding load, thus enabling motion while protecting the underlying bone [21]. Articular cartilage is highly specialized. It is avascular, aneural, and alymphatic and, at maturity, of low metabolic activity [21]. This results in tissue which is highly suited to its task, but with limited capacity for repair. In cartilage tissue there is a finely regulated balance of degradation and synthesis, which even if mildly disrupted can lead to major long-term problems. It has been demonstrated that early in the process of tissue damage there is a rapid loss of glycosaminoglycans from the tissue [93]. Thus, most large defects fail to heal, leading to a long-term prognosis of osteoarthritis [33]. Current treatment options for cartilage injuries include joint lavage, tissue debridement, microfracture of the subchondral bone to induce regeneration, or transplantation of autologous or allogeneic osteochondral grafts [3]. Success rates with these treatments vary greatly. While some of these approaches show promise, many lead to the formation of fibrous tissue, apoptosis, and further cartilage degeneration. As a result many articular cartilage tissue-engineering techniques have been developed attempting to create cartilage ex vivo that can be implanted in vivo and regenerate the damaged joint.

However, ex vivo regeneration of cartilage is problematic. Problems not only include trying to recreate the complex architecture of hyaline cartilage with the correct ratios and correct structural arrangement of each matrix component, but also achieving successful integration of the newly formed tissue into the defect following implementation. This requires both cartilage–cartilage and cartilage–bone integration [58].

Newly developed ex vivo cartilage composite requires chondrocytes and extracellular matrix. The cells can be adult chondrocytes harvested from a different site of the patient's cartilage [14], at a cost of loss of additional cartilage from the donor site, or can be MSCs which underwent modification and differentiated into chondrocytes [67]. Various combinations of growth factors, primarily TGF-β, FGF-2, PDGF-BB, bone morphogenetic proteins (BMPs), IGF-1, and insulin formulations have also been applied to promote chondrocyte differentiation [52].

Growth factors are not the only factor to be taken into account when regenerating cartilage, the structure in which the cells are placed plays an important role as well. Chondrocytes placed in a monolayer might undergo dedifferentiation [15], however, when placed into a suitable three dimensional (3D) structure and properly cultured, these cells apparently revert back to normal chondrocytes [104].

Three-dimensional scaffold materials must be biocompatible and accommodate cell adhesion, proliferation, and matrix synthesis. Many materials have been used both in vivo and in vitro scaffolds for cartilage regeneration. These include hydrogels, sponge-like scaffolds, and hybrid systems incorporating a hydrogel and a stiffer macroporous sponge [91, 100]. The scaffold needs a mechanical stiffness capable of withstanding repeated compression and shear while at the same time maintaining the chondrocytic phenotype and retaining a greater proportion of newly synthesized matrix products.

Adipose-derived stem cells have been shown to convert into chondrocytes when in the proper medium (with addition of TGFβ, ascorbate-2-phosphate, dexamethasone, and FGF-2) [30]. As with bone tissue, in vitro studies have tested the quality of the matrix produced by these adipose-derived chondrocytes and the results are not in agreement with several studies showing that there are no differences between the two types of cells and others showing that bone marrow-derived stem cells result in superior matrix production [63, 103]. There are differences between bone marrow-derived stem cells and adipose-derived cells in the effect of the different BMPs regarding induction of chondrogenesis, with only BMP-6 inducing chondrogenesis in adipose-derived stem cells and only a very weak effect seen from BMP-7 [89]. Adipose-derived stem cells have been shown to create cartilage in pellet culture and in various in vitro [51] three-dimensional scaffolds. However, the delivery of these agents together with MSCs into cartilaginous lesions in vivo has not yet resulted in sustained regeneration of articular cartilage [43, 63]. This is not a problem specific to adipose-derived stem cells, but rather a general problem faced when trying to regenerate cartilage. As in bone regeneration, there should be an optimal composite of cells both inductive and conductive environment. In addition, there are two major problems when passing from in vitro to in vivo. The first is how to stop the cartilage from becoming hypertrophic [73] (in nonprofessional terms: how to turn off the cartilage proliferation), a step seen in chondrogenic ossification in fractures as the next step in bone creation. The second problem is the integration of cartilage into the defect site. Since cartilage is avascular it receives its nutrition from adjacent subchondral bone and from synovial fluid. If the lesion is not fresh and the subchondral bone is obliterated, a method to create a composite graft of bone and cartilage needs to be created, thereby allowing integration into the healthy subchondral bone [58].

When taking all that has been done so far into account, none of the cartilage regeneration work has taken on the major problem of cartilage degeneration–osteoarthritis. Osteoarthritis is the most common form of cartilage degradation, involving general degeneration of the articular cartilage [62]. The greatest risk factors for osteoarthritis include aging, obesity, joint trauma, and mutations in cartilage-specific matrix proteins [41]. Current estimates of treatment costs, both indirect and direct, of osteoarthritis in the United States are greater than $65 billion annually [81]. Joints with large defects and end-stage degeneration are far beyond the reach of present accomplishments in chondrogenesis. Early intervention in younger people with smaller defects might decrease the rate of end-stage joint degeneration seen in older populations. The role of adipose-derived stem cell is far from being applicable as a clinical solution for these challenging tissue engineering problems.

17.4 Spinal Applications of Adipose-Derived Stem Cells

The spine is a complex structure composed of bony vertebrae, intra-vertebral discs, multiple ligaments, and muscles both protecting the spinal cord and providing a solid axis for energy transfer from the lower limbs and pelvis to the thorax and upper limbs. A thorough background on spinal anatomy and pathology is beyond the scope of this chapter. This chapter will focus on degenerative processes of the spine.

Low back pain is one of the most common problems in medicine. It is the second most common reason for physician visits [19]. Nine out of ten adults experience back pain at some point in their life and five out of ten working adults have back pain every year [59]. After mechanical pain, caused by strain, lumbar spine degeneration is the next most common cause of low back pain and it is a target of diagnostic and surgical intervention [98]. This degeneration is mediated by degeneration of the intra-vertebral discs causing pathological movements between spinal segments, resulting in pain and, at times, narrowing of the spinal canal, compressing nerves, and causing lower limb claudication [35].

In contrast to other musculoskeletal systems, the intervertebral disc exhibits degenerative changes in the late second or early third decade. There is disc degeneration in at least one lumbar level in 35% of subjects between 20 and 39 years of age and close to every person 60–80 years of age has disc degeneration [9]. It is quite difficult to differentiate the normal aging process from pathological degeneration in patients showing symptoms.

An intervertebral disc consists of three components: an inner gelatinous nucleus pulposus, an outer annulus fibrosus (AF), and cartilage end plates located superiorly and inferiorly [50].

The outer annulus fibrosus, is abundant in type I collagen and elastin, resisting tensile pressures. The nucleus pulposus consists of a proteoglycan and water gel loosely held together by irregular networks of fine type II collagen and elastin fibers [50]. The intervertebral disc is mainly avascular deriving its nutrition from vessels in the subchondral bone adjacent to the hyaline cartilage of the end plate. The disc has different cells maintaining the unique structure of each of the elements of the disc. Chondrocyte-like cells, present in the nucleus pulposus (NP cells), produce proteoglycan and collagen, which is the basic skeletal structure of a disc. Annulus fibrosus cells are more fibroblast like in nature producing collagen and elastin and maintaining AF structure [4].

In the normal disc, the extracellular proteins are responsible for retaining water within the matrix resulting in a high viscoelastic tissue with water retained within the nucleus pulposus, inflating the collagen framework. The disc is avascular, and the cells in the center of an adult lumbar disc are approximately 8 mm away from the nearest blood supply, depending on a complicated pathway extending from a capillary network that penetrates the subchondral plate. The nutrients diffuse from these capillaries across the cartilaginous end plates and through the dense disc matrix to the cells. Over the years, this network becomes occluded and transport is hampered causing ischemia of the disc, cell death, and disc degeneration [1].

In this processes, the strongly hydrophilic chondroitin-4-sulfate and chondroitin-6-sulfate decrease in number and keratin sulfate increase in number. This process leads to decreased water within the discs, decreasing the height of the intervertebral disc, as well as the resistance of the disc to axial loading [1].

In addition, the loss of disc height alters the facet joint mechanics resulting in the formation of osteophytes at the disc margins. Coincidentally, a decrease in intervertebral height causes buckling of the ligament flavum and facet overriding. Degenerative changes occur in a parallel manner in all three components, which may cause narrowing of the neuroforaminal and spinal canals [4].

The decrease in cell numbers in the AF leads to decreased production of collagen type I, leading to circumferential annular tears progressing to radial tears. Herniation can occur when nuclear materials protrude or extrude into the perineural space through these radial tears [105].

The common treatment of such a condition is devised of physical therapy, pain management, and if all else fails, surgery, to decompress the neural elements and fuse the two vertebrae. However, there is known long-term detrimental effects on the juxtafusion segments in the form of accelerated degeneration [12]. In past decades, nonfusion strategies have been attempted aimed at preserving the stability and mobility of the spinal segment, while relieving the clinical symptoms of the neural compression [12].

There are two general approaches to motion preservation in degenerative discs: replacement and regeneration. Disc replacement is based on a bimodular implant attached to each side of the vertebrae from which the disc was removed. However, the results of these implants, at first promising, are hampered by secondary spontaneous fusion between vertebra and loss of motion. In addition, degeneration of these implants is an ongoing question requiring further long-term studies [75].

Regeneration or repair of the disc using growth factors, gene therapy, and cell therapy are actively being researched [4]. Most of the attention is focused on the nucleus pulposus and how to rejuvenate the disc through increasing cell numbers, thus restoring matrix production [85]. The major limitation of this approach is that the nucleus pulposus is the sole degenerative part of the disc. The problems in the annulus and the bony end plates are equally important. If the blood supply to the disc is not rejuvenated, any cells implanted will die out prior to regenerating the disc [87].

The source of cells for disc regeneration is a relevant question as well. There is little known about the molecular profile of the AF and nucleus pulposus cells. It has therefore not been determined if stem cells will be able to convert to AF or NP cells [47, 85]. Co-culturing MSCs with disc cells appears to speed up the conversion to a chondrogenic/discogenic phenotype [56].

There have been only a few studies testing the effect of implantation of adipose-derived stem cells within disc tissue in vitro and in vivo in animal models. Adipose-derived stem cells were shown to differentiate into a nucleus pulposus-like phenotype when exposed to environmental factors similar to disc [71]. In a canine trial of adipose-derived regenerative cell transplantation, adipose-derived cells were transplanted under fluoroscopic guidance directly into a surgically damaged disc [34, 84]. Three interventions were used in this study: adipose-derived cells placed in hyaluronic acid (HA) carrier, HA alone, or no intervention, all deliveries were guided by fluoroscopy. Statistical differences were found between the control and each intervention. No significant difference could be shown between HA alone and no intervention. These results suggest that cells harvested from adipose tissue might offer a reliable source of regenerative potential capable of bio-restitution [34]; however, no control of adipose-derived stem cells injected alone was provided. In addition, the long-term outcome of these discs was not determined.

When spinal fusion is needed, adipose-derived stem cells were shown to propagate bone fusion, as with axial bone [68]. These principles were described earlier in the bone subchapter.

17.5 The Use of Adipose-Derived Stem Cells for Tendon and Ligament Regeneration

Ligaments and tendons are important structures designed to transmit forces and facilitate joint articulation, providing motion and stability in the musculoskeletal system. As such, these tissues are frequently injured during sports and work related activities. The healing responses following injuries to different ligaments and the consequences of the healing can vary greatly. For example, the anterior cruciate ligament of knee has limited healing capabilities as compared with other fully disrupted ligaments [102].

There have been many approaches to fully disrupted ligaments, including: direct repair; autograft; and allograft, and synthetic grafts aimed at replacing the damaged tissue [102].

The use of stem cells as a natural source for cellular replacement in the damaged matrix has been tried and MSCs were identified that when implanted in the injured rat's MCL, differentiated into fibroblasts [101]. Further, when MSCs are seeded in an Achilles' tendon with a critical gap tendon injury, the MSCs assisted in healing particularly as collagen fibers appeared to be better aligned than those in the controls [55]. In vivo studies testing an intra-articular injection of MSCs in a rat model with partially transacted ACL showed healing [18]. Few studies have been performed concerning adipose-derived stem cells, but in vitro and small animal in vivo studies show similar results, thus indicating that the ADSC can be utilized for tendon regeneration as well [97].

17.6 Turning Fat into Muscle

Skeletal muscle cells under normal conditions remain at a constant number, proliferating under strain and increasing fibers, synapses, and improving oxygen consumption as a result of mechanical stimuli. The ability of adult skeletal muscle to generate new fibers resides in a reserve population of mononucleotide precursors termed satellite cells [92]. They reside in grooves or depressions between the basal lamina and sarcolemma of mature fibers and make up 2–10% of sublaminar muscle nuclei, depending on the type of fibers with which they associate. In response to environmental cues such as injury, satellite cells are activated and proliferate and convert into myoblasts (with signals as MyoD and HGF) [37, 99]. These cells fuse to each other to form new myotomes or become incorporated into preexisting muscle fibers. The regeneration capacity of satellite cells is limited with the cells undergoing senescence after several divisions [92]. In addition, in certain muscular diseases this cell population is not suitable to regenerate the muscle since the cells are also diseased, e.g., Duchene muscular dystrophy.

In vitro work has shown that MSCs can turn into myoblasts [61]. There have been several trials to incorporate different bone marrow-derived stem cells and embryonic stem cells into muscle, specifically in degenerative muscle diseases; however, the frequency of stem cell incorporation into skeletal muscle was low and could not reverse the amelioration of the dystrophic phenotype [70].

Fat and muscle are interlinked. When a muscle is denervated, the fatty content of the muscle increases [27, 96], indicating either that there are fat fibers in the muscle or that muscle progenitor cells convert to fat cells under specific conditions [7]. Adipose-derived stem cells seem to have several subpopulations which react in a different manner when converted to muscle cell [25]. One subpopulation was stimulated in vitro towards muscle cells when cultured with muscle cells in the same cultures and a second group could be directly converted to muscle cells even without contact to muscle fibers. Studies have shown that when muscles are injured, they set off signals which convert adipose-derived stem cells into muscle tissue [54].

Most of the work being done with stem cells and muscle fibers is directed towards cardiac muscle, hoping to assist in regeneration of damaged heart tissue following an ischemic insult [48]. Trials have been directed at direct regeneration of muscle or regeneration of blood vessels and increasing oxygen supply to cardiac muscle, thus increasing endogenous healing of this tissue [49].

There is very limited work in vivo with adipose-derived stem cells and skeletal muscle regeneration and then it is only in small animals [24]. There is still a long way to go before adipose-derived stem cells can be converted to skeletal muscle in the clinical setting.

17.7 Conclusions

Adipose-derived stem cells have immense potential in orthopedic practice. Firstly, in many degenerative diseases the patients are obese, and harvesting fat to regenerate their arthritic joints, spine, tendons, or muscles seems like "killing two birds with one stone."

Secondly, adipose-derived stem cells are pro vascular, regenerating blood vessels at an increased rate as compared with bone marrow-derived stem cells. This property is advantageous in the healing of critical size bone defects and disc degeneration.

And lastly, the increased number of stem cells in fat tissue, when compared to bone marrow, will require less harvesting and will achieve the same result.

Although promising, the clinical work is still very limited, mainly focusing on bone and small defect cartilage regeneration. There is still much to do before this will become a common clinical solution to everyday orthopedic problems.

References

1. Adams MA, Roughley PJ (2006) What is intervertebral disc degeneration, and what causes it? Spine (Phila Pa 1976) 31:2151–2161
2. Aksu AE, Rubin JP, Dudas JR, Marra KG (2008) Role of gender and anatomical region on induction of osteogenic differentiation of human adipose-derived stem cells. Ann Plast Surg 60:306–322
3. Alford JW, Cole BJ (2005) Cartilage restoration, part 1: basic science, historical perspective, patient evaluation, and treatment options. Am J Sports Med 33:295–306
4. Alini M, Roughley PJ, Antoniou J, Stoll T, Aebi M (2002) A biological approach to treating disc degeneration: not for today, but maybe for tomorrow. Eur Spine J 11(Suppl 2): S215–S220
5. Aly AA (2010) Preface. The traditional, the new, and the future of abdominoplasty. Clin Plast Surg 37:ix
6. Arrington ED, Smith WJ, Chambers HG, Bucknell AL, Davino NA (1996) Complications of iliac crest bone graft harvesting. Clin Orthop Relat Res 329:300–309
7. Asakura A, Komaki M, Rudnicki M (2001) Muscle satellite cells are multipotential stem cells that exhibit myogenic, osteogenic, and adipogenic differentiation. Differentiation 68:245–253
8. Barry FP, Murphy JM (2004) Mesenchymal stem cells: clinical applications and biological characterization. Int J Biochem Cell Biol 36:568–584
9. Boden SD, Davis DO, Dina TS, Patronas NJ, Wiesel SW (1990) Abnormal magnetic-resonance scans of the lumbar spine in asymptomatic subjects: a prospective investigation. J Bone Joint Surg Am 72:403–408
10. Bohannon RW (2007) Number of pedometer-assessed steps taken per day by adults: a descriptive meta-analysis. Phys Ther 87:1642–1650
11. Bosch C, Melsen B, Vargervik K (1998) Importance of the critical-size bone defect in testing bone-regenerating materials. J Craniofac Surg 9:310–316
12. Botelho RV, Moraes OJ, Fernandes GA, Buscariolli YS, Bernardo WM (2010) A systematic review of randomized trials on the effect of cervical disc arthroplasty on reducing adjacent-level degeneration. Neurosurg Focus 28:E5
13. Boyd HB (1964) Nonunion of the shafts of long bones. Postgrad Med 36:315–320
14. Brittberg M, Lindahl A, Nilsson A, Ohlsson C, Isaksson O, Peterson L (1994) Treatment of deep cartilage defects in the knee with autologous chondrocyte transplantation. N Engl J Med 331:889–895
15. Brodkin KR, Garcia AJ, Levenston ME (2004) Chondrocyte phenotypes on different extracellular matrix monolayers. Biomaterials 25:5929–5938
16. Burkhardt R, Kettner G, Bohm W et al (1987) Changes in trabecular bone, hematopoiesis and bone marrow vessels in aplastic anemia, primary osteoporosis, and old age: a comparative histomorphometric study. Bone 8:157–164
17. Cancedda R, Dozin B, Giannoni P, Quarto R (2003) Tissue engineering and cell therapy of cartilage and bone. Matrix Biol 22:81–91
18. Caplan AI, Dennis JE (2006) Mesenchymal stem cells as trophic mediators. J Cell Biochem 98:1076–1084
19. Carey TS, Evans A, Hadler N, Kalsbeek W, McLaughlin C, Fryer J (1995) Care-seeking among individuals with chronic low back pain. Spine (Phila Pa 1976) 20:312–317
20. Chamberlain G, Fox J, Ashton B, Middleton J (2007) Concise review: mesenchymal stem cells: their phenotype, differentiation capacity, immunological features, and potential for homing. Stem Cells 25:2739–2749
21. Cohen NP, Foster RJ, Mow VC (1998) Composition and dynamics of articular cartilage: structure, function, and maintaining healthy state. J Orthop Sports Phys Ther 28:203–215
22. Connolly JF, Guse R, Tiedeman J, Dehne R (1989) Autologous marrow injection for delayed unions of the tibia: a preliminary report. J Orthop Trauma 3:276–282
23. Curl WW, Krome J, Gordon ES, Rushing J, Smith BP, Poehling GG (1997) Cartilage injuries: a review of 31,516 knee arthroscopies. Arthroscopy 13:456–460
24. De Matteis R, Zingaretti MC, Murano I et al (2009) In vivo physiological transdifferentiation of adult adipose cells. Stem Cells 27:2761–2768
25. Di Rocco G, Lachininoto MG, Tritarelli A et al (2006) Myogenic potential of adipose-tissue-derived cells. J Cell Sci 119:2945–2952
26. Dragoo JL, Choi JY, Lieberman JR et al (2003) Bone induction by BMP-2 transduced stem cells derived from human fat. J Orthop Res 21:622–629
27. Dulor JP, Cambon B, Vigneron P et al (1998) Expression of specific white adipose tissue genes in denervation-induced skeletal muscle fatty degeneration. FEBS Lett 439:89–92
28. Einhorn TA (1996) Enhancement of fracture healing. Instr Course Lect 45:401–416
29. Einhorn TA (1998) The cell and molecular biology of fracture healing. Clin Orthop Relat Res 355:S7–S21
30. Erickson GR, Gimble JM, Franklin DM, Rice HE, Awad H, Guilak F (2002) Chondrogenic potential of adipose tissue-derived stromal cells in vitro and in vivo. Biochem Biophys Res Commun 290:763–769
31. Estes BT, Diekman BO, Gimble JM, Guilak F (2010) Isolation of adipose-derived stem cells and their induction to a chondrogenic phenotype. Nat Protoc 5:1294–1311
32. Felson DT, Lawrence RC, Dieppe PA et al (2000) Osteoarthritis: new insights. Part 1: the disease and its risk factors. Ann Intern Med 133(8):635–646
33. Fife RS, Brandt KD, Braunstein EM et al (1991) Relationship between arthroscopic evidence of cartilage damage and radiographic evidence of joint space narrowing in early osteoarthritis of the knee. Arthritis Rheum 34:377–382
34. Ganey T, Hutton WC, Moseley T, Hedrick M, Meisel HJ (2009) Intervertebral disc repair using adipose tissue-derived stem and regenerative cells: experiments in a canine model. Spine (Phila Pa 1976) 34:2297–2304
35. Genevay S, Atlas SJ (2010) Lumbar spinal stenosis. Best Pract Res Clin Rheumatol 24:253–265

36. Gerstenfeld LC, Cullinane DM, Barnes GL, Graves DT, Einhorn TA (2003) Fracture healing as a post-natal developmental process: molecular, spatial, and temporal aspects of its regulation. J Cell Biochem 88:873–884
37. Gill R, Hitchins L, Fletcher F, Dhoot GK (2010) Sulf1A and HGF regulate satellite-cell growth. J Cell Sci 123:1873–1883
38. Gimble J, Guilak F (2003) Adipose-derived adult stem cells: isolation, characterization, and differentiation potential. Cytotherapy 5:362–369
39. Gimble JM, Guilak F, Bunnell BA (2010) Clinical and preclinical translation of cell-based therapies using adipose tissue-derived cells. Stem Cell Res Ther 1:19
40. Goldberg BA, Nowinski RJ, Matsen FA III (2001) Outcome of nonoperative management of full-thickness rotator cuff tears. Clin Orthop Relat Res 382:99–107
41. Goldring MB, Goldring SR (2010) Articular cartilage and subchondral bone in the pathogenesis of osteoarthritis. Ann NY Acad Sci 1192:230–237
42. Gruber R, Koch H, Doll BA, Tegtmeier F, Einhorn TA, Hollinger JO (2006) Fracture healing in the elderly patient. Exp Gerontol 41:1080–1093
43. Guilak F, Estes BT, Diekman BO, Moutos FT, Gimble JM (2010) 2010 Nicolas Andry Award: multipotent adult stem cells from adipose tissue for musculoskeletal tissue engineering. Clin Orthop Relat Res 468:2530–2540
44. Hattori H, Masuoka K, Sato M et al (2006) Bone formation using human adipose tissue-derived stromal cells and a biodegradable scaffold. J Biomed Mater Res B Appl Biomater 76(1):230–239
45. Helder MN, Knippenberg M, Klein-Nulend J, Wuisman PI (2007) Stem cells from adipose tissue allow challenging new concepts for regenerative medicine. Tissue Eng 13:1799–1808
46. Hicok KC, Du Laney TV, Zhou YS et al (2004) Human adipose-derived adult stem cells produce osteoid in vivo. Tissue Eng 10:371–380
47. Hiyama A, Mochida J, Sakai D (2008) Stem cell applications in intervertebral disc repair. Cell Mol Biol 54:24–32
48. Hoke NN, Salloum FN, Loesser-Casey KE, Kukreja RC (2009) Cardiac regenerative potential of adipose tissue-derived stem cells. Acta Physiol Hung 96:251–265
49. Hong SJ, Traktuev DO, March KL (2010) Therapeutic potential of adipose-derived stem cells in vascular growth and tissue repair. Curr Opin Organ Transplant 15:86–91
50. Humzah MD, Soames RW (1988) Human intervertebral disc: structure and function. Anat Rec 220:337–356
51. Im GI, Shin YW, Lee KB (2005) Do adipose tissue-derived mesenchymal stem cells have the same osteogenic and chondrogenic potential as bone marrow-derived cells? Osteoarthr Cartilage 13:845–853
52. Indrawattana N, Chen G, Tadokoro M et al (2004) Growth factor combination for chondrogenic induction from human mesenchymal stem cell. Biochem Biophys Res Commun 320:914–919
53. Jiang Y, Jahagirdar BN, Reinhardt RL et al (2002) Pluripotency of mesenchymal stem cells derived from adult marrow. Nature 418:41–49
54. Joe AW, Yi L, Natarajan A et al (2010) Muscle injury activates resident fibro/adipogenic progenitors that facilitate myogenesis. Nat Cell Biol 12:153–163
55. Juncosa-Melvin N, Boivin GP, Galloway MT, Gooch C, West JR, Butler DL (2006) Effects of cell-to-collagen ratio in stem cell-seeded constructs for Achilles tendon repair. Tissue Eng 12:681–689
56. Kandel R, Roberts S, Urban JP (2008) Tissue engineering and the intervertebral disc: the challenges. Eur Spine J 17(Suppl 4):480–491
57. Kern S, Eichler H, Stoeve J, Kluter H, Bieback K (2006) Comparative analysis of mesenchymal stem cells from bone marrow, umbilical cord blood, or adipose tissue. Stem Cells 24:1294–1301
58. Khan IM, Gilbert SJ, Singhrao SK, Duance VC, Archer CW (2008) Cartilage integration: evaluation of the reasons for failure of integration during cartilage repair: a review. Eur Cell Mater 16:26–39
59. Knauer SR, Freburger JK, Carey TS (2010) Chronic low back pain among older adults: a population-based perspective. J Aging Health 22(8):1213–1234
60. Kneser U, Schaefer DJ, Polykandriotis E, Horch RE (2006) Tissue engineering of bone: the reconstructive surgeon's point of view. J Cell Mol Med 10:7–19
61. Kocaefe C, Balci D, Balci HB, Can A (2010) Reprogramming of human umbilical cord stromal mesenchymal stem cells for myogenic differentiation and muscle repair. Stem Cell Rev 6(4):512–522
62. Koelling S, Miosge N (2009) Stem cell therapy for cartilage regeneration in osteoarthritis. Expert Opin Biol Ther 9: 1399–1405
63. Koga H, Muneta T, Nagase T et al (2008) Comparison of mesenchymal tissues-derived stem cells for in vivo chondrogenesis: suitable conditions for cell therapy of cartilage defects in rabbit. Cell Tissue Res 333:207–215
64. Lee JA, Parrett BM, Conejero JA et al (2003) Biological alchemy: engineering bone and fat from fat-derived stem cells. Ann Plast Surg 50:610–617
65. Lee RH, Kim B, Choi I et al (2004) Characterization and expression analysis of mesenchymal stem cells from human bone marrow and adipose tissue. Cell Physiol Biochem 14:311–324
66. Lendeckel S, Jodicke A, Christophis P et al (2004) Autologous stem cells (adipose) and fibrin glue used to treat widespread traumatic calvarial defects: case report. J Craniomaxillofac Surg 32:370–373
67. Li WJ, Tuli R, Huang X, Laquerriere P, Tuan RS (2005) Multilineage differentiation of human mesenchymal stem cells in a three-dimensional nanofibrous scaffold. Biomaterials 26:5158–5166
68. Lopez MJ, McIntosh KR, Spencer ND et al (2009) Acceleration of spinal fusion using syngeneic and allogeneic adult adipose derived stem cells in a rat model. J Orthop Res 27:366–373
69. McKibbin B (1978) The biology of fracture healing in long bones. J Bone Joint Surg Br 60-B:150–162
70. Meregalli M, Farini A, Parolini D, Maciotta S, Torrente Y (2010) Stem cell therapies to treat muscular dystrophy: progress to date. BioDrugs 24:237–247
71. Mesimaki K, Lindroos B, Tornwall J et al (2009) Novel maxillary reconstruction with ectopic bone formation by GMP adipose stem cells. Int J Oral Maxillofac Surg 38:201–209
72. Milgrom C, Schaffler M, Gilbert S, van Holsbeeck M (1995) Rotator-cuff changes in asymptomatic adults: the effect of age, hand dominance and gender. J Bone Joint Surg Br 77:296–298

73. Mueller MB, Tuan RS (2008) Functional characterization of hypertrophy in chondrogenesis of human mesenchymal stem cells. Arthritis Rheum 58:1377–1388
74. Nordsletten L, Madsen JE, Almaas R et al (1994) The neuronal regulation of fracture healing: effects of sciatic nerve resection in rat tibia. Acta Orthop Scand 65:299–304
75. Parkinson JF, Sekhon LH (2005) Cervical arthroplasty complicated by delayed spontaneous fusion: case report. J Neurosurg Spine 2:377–380
76. Peterson B, Zhang J, Iglesias R et al (2005) Healing of critically sized femoral defects, using genetically modified mesenchymal stem cells from human adipose tissue. Tissue Eng 11:120–129
77. Petite H, Viateau V, Bensaid W et al (2000) Tissue-engineered bone regeneration. Nat Biotechnol 18:959–963
78. Philipson T (2001) The world-wide growth in obesity: an economic research agenda. Health Econ 10:1–7
79. Pittenger MF, Mackay AM, Beck SC et al (1999) Multilineage potential of adult human mesenchymal stem cells. Science 284:143–147
80. Prunet-Marcassus B, Cousin B, Caton D, Andre M, Penicaud L, Casteilla L (2006) From heterogeneity to plasticity in adipose tissues: site-specific differences. Exp Cell Res 312:727–736
81. Rana AJ, Iorio R, Healy WL (2010) Hospital economics of primary THA decreasing reimbursement and increasing cost, 1990 to 2008. Clin Orthop Relat Res 469(2):355–361
82. Reddi AH (1998) Initiation of fracture repair by bone morphogenetic proteins. Clin Orthop Relat Res 335:S66–S72
83. Rehman J, Traktuev D, Li J et al (2004) Secretion of angiogenic and antiapoptotic factors by human adipose stromal cells. Circulation 109:1292–1298
84. Risbud MV, Albert TJ, Guttapalli A et al (2004) Differentiation of mesenchymal stem cells towards a nucleus pulposus-like phenotype in vitro: implications for cell-based transplantation therapy. Spine (Phila Pa 1976) 29:2627–2632
85. Risbud MV, Shapiro IM, Vaccaro AR, Albert TJ (2004) Stem cell regeneration of the nucleus pulposus. Spine J 4:348S–353S
86. Romanov YA, Svintsitskaya VA, Smirnov VN (2003) Searching for alternative sources of postnatal human mesenchymal stem cells: candidate MSC-like cells from umbilical cord. Stem Cells 21:105–110
87. Sebastine IM, Williams DJ (2007) Current developments in tissue engineering of nucleus pulposus for the treatment of intervertebral disc degeneration. Conf Proc IEEE Eng Med Biol Soc 2007:6401–6406
88. Shen FH, Zeng Q, Lv Q et al (2006) Osteogenic differentiation of adipose-derived stromal cells treated with GDF-5 cultured on a novel three-dimensional sintered microsphere matrix. Spine J 6:615–623
89. Shirasawa S, Sekiya I, Sakaguchi Y, Yagishita K, Ichinose S, Muneta T (2006) In vitro chondrogenesis of human synovium-derived mesenchymal stem cells: optimal condition and comparison with bone marrow-derived cells. J Cell Biochem 97:84–97
90. Smith R (1987) Osteoporosis: cause and management. Br Med J 294:329–332
91. Stoddart MJ, Grad S, Eglin D, Alini M (2009) Cells and biomaterials in cartilage tissue engineering. Regen Med 4:81–98
92. Tedesco FS, Dellavalle A, Diaz-Manera J, Messina G, Cossu G (2010) Repairing skeletal muscle: regenerative potential of skeletal muscle stem cells. J Clin Invest 120:11–19
93. Tiderius CJ, Olsson LE, Nyquist F, Dahlberg L (2005) Cartilage glycosaminoglycan loss in the acute phase after an anterior cruciate ligament injury: delayed gadolinium-enhanced magnetic resonance imaging of cartilage and synovial fluid analysis. Arthritis Rheum 52:120–127
94. Tull F, Borrelli J Jr (2003) Soft-tissue injury associated with closed fractures: evaluation and management. J Am Acad Orthop Surg 11:431–438
95. Turner CH, Robling AG (2003) Designing exercise regimens to increase bone strength. Exerc Sport Sci Rev 31:45–50
96. Uezumi A, Fukada S, Yamamoto N, Takeda S, Tsuchida K (2010) Mesenchymal progenitors distinct from satellite cells contribute to ectopic fat cell formation in skeletal muscle. Nat Cell Biol 12:143–152
97. Uysal AC, Mizuno H (2010) Tendon regeneration and repair with adipose derived stem cells. Curr Stem Cell Res Ther 5:161–167
98. van Tulder M, Koes B, Bombardier C (2002) Low back pain. Best Pract Res Clin Rheumatol 16:761–775
99. Wagner Alves DS, Goncalves W, Garrido Cavalcante WL, Pai-Silva MD, Gallacci M (2010) Nandrolone stimulates MyoD expression during muscle regeneration in the condition of myonecrosis induced by Bothrops jararacussu venom poisoning. J Toxicol Environ Health A 73:934–943
100. Warnke PH (2010) In-vivo tissue engineering of biological joint replacements. Lancet 376:394–396
101. Watanabe N, Woo SL, Papageorgiou C, Celechovsky C, Takai S (2002) Fate of donor bone marrow cells in medial collateral ligament after simulated autologous transplantation. Microsc Res Tech 58:39–44
102. Wilk KE, Andrews JR (1992) Current concepts in the treatment of anterior cruciate ligament disruption. J Orthop Sports Phys Ther 15:279–293
103. Winter A, Breit S, Parsch D et al (2003) Cartilage-like gene expression in differentiated human stem cell spheroids: a comparison of bone marrow-derived and adipose tissue-derived stromal cells. Arthritis Rheum 48:418–429
104. Yang KG, Saris DB, Geuze RE et al (2006) Impact of expansion and redifferentiation conditions on chondrogenic capacity of cultured chondrocytes. Tissue Eng 12:2435–2447
105. Yasuma T, Makino E, Saito S, Inui M (1986) Histological development of intervertebral disc herniation. J Bone Joint Surg Am 68:1066–1072
106. Yeo S, Jeong S, Kim J, Han JS, Han YM, Kang YK (2007) Characterization of DNA methylation change in stem cell marker genes during differentiation of human embryonic stem cells. Biochem Biophys Res Commun 359:536–542
107. Yoon E, Dhar S, Chun DE, Gharibjanian NA, Evans GR (2007) In vivo osteogenic potential of human adipose-derived stem cells/poly lactide-co-glycolic acid constructs for bone regeneration in a rat critical-sized calvarial defect model. Tissue Eng 13:619–627
108. Zaretsky HH, Richter EF, Eisenberg MG (2004) Medical aspects of disability: a handbook for the rehabilitation professional, 3rd edn. Springer, New York, NY, pp 31–55
109. Zuk PA, Zhu M, Ashjian P et al (2002) Human adipose tissue is a source of multipotent stem cells. Mol Biol Cell 13:4279–4295

18 Adipose-Derived Stem Cells: Characterization and Application in Urology

Ching-Shwun Lin and Tom F. Lue

Contents

18.1 **Introduction** 193
18.1.1 Vascular Versus Fat Stem Cells 193
18.1.2 ADSC Versus BMSC 194
18.1.3 Stromal Vascular Fraction Versus Adherent Cells 194
18.1.4 Adventitial Cells Versus Pericytes 195
18.1.5 Paracrine Action Versus Differentiation 197
18.1.6 Immunosuppression Versus Proinflammation 197
18.1.7 Procancer Versus Anticancer 197
18.1.8 Local Versus Systemic Administration 198

18.2 **ADSC Application in Urology** 198
18.2.1 Kidneys 198
18.2.2 Ureters 199
18.2.3 Bladder 199
18.2.4 Urethra 200
18.2.5 Penis 201
18.2.6 Testes 201

18.3 **Conclusions** 202

References 203

C.-S. Lin (✉) • T.F. Lue
Department of Urology, Knuppe Molecular Urology Laboratory, School of Medicine, University of California, San Francisco, CA 94143-0738, USA
e-mail: clin@urology.ucsf.edu; tlue@urology.ucsf.edu

18.1 Introduction

Adipose-derived stem cells (ADSC) originate from the stromal vascular fraction (SVF) of adipose tissue. Freshly isolated SVF cells are a heterogeneous mixture of endothelial cells, smooth muscle cells, pericytes, fibroblasts, mast cells, and preadipocytes [82]. Culturing of these cells under standard conditions eventually (within the first few passages) results in the appearance of a relatively homogenous population of mesodermal or mesenchymal cells [118]. However, many factors can influence the cellular composition of an ADSC culture, for example, species, age, tissue depot (e.g., subcutaneous vs. omentum), isolation procedure, culture condition, passage number, and cell storage. Additionally, the choice of experimental methods and reagents may also affect the outcome of any given study concerning ADSC's expression profile, differentiation potential, and therapeutic efficacy. Thus, it is not surprising that conflicting evidence and inconsistencies exist in the literature. For detailed information on the characterization of ADSC, the readers are advised to consult recent review articles such as these 2010 contributions [41, 46, 63, 117]. In this chapter, we will only consider some contrasting viewpoints relating to the clinical application of ADSC. Note that, depending on the situation, "ADSC" is used as a singular or a plural. Other abbreviations follow the same rule.

18.1.1 Vascular Versus Fat Stem Cells

At least a dozen different names have been given to the same cell type isolated from the SVF of adipose tissue.

Y.-G. Illouz and A. Sterodimas (eds.), *Adipose Stem Cells and Regenerative Medicine*,
DOI: 10.1007/978-3-642-20012-0_18, © Springer-Verlag Berlin Heidelberg 2011

One of such names, adipose tissue-derived stem cells (ADSC), was most commonly used in the early stage of this research field. However, in a 2007 review article Gimble et al. [40] advocated the use of "adipose-derived stem cells" (ASC), citing the adoption of this term by the International Fat Applied Technology Society. While "adipose-derived stem cells" is an improvement from "adipose tissue-derived stem cells," the abbreviation "ASC" has already been used for "adult stem cells" and thus may further confuse laypersons or even dedicated researchers. For example, in a recently submitted manuscript to Stem Cells and Development the authors cited a study on ASC (for adult stem cells) to support their viewpoints on ASC (for adipose-derived stem cells). Regardless, it is surprising that, despite its well-known connection with the vasculature, ADSC has never been given a name that would hint at such a connection. Even more troubling is that the two frequently and casually used terms "adipose stem cells" and "fat stem cells" almost always register in the mind of a layperson as something evil and whose mission is to make us fat and fatter. Only when explained that these cells actually come from blood vessels then the layperson becomes receptive to the idea of using ADSC as a therapeutic agent. However, strictly speaking, definitive proof that ADSC are vascular cells is still lacking, and the identity of ADSC within adipose tissue is a subject of ongoing debate. Still, there appears to be a consensus among ADSC researchers as far as its vascular origin is concerned; therefore, a name like "adipose vascular stem cells (AVSC)" would seem more fitting than any of the existing ones.

18.1.2 ADSC Versus BMSC

The vascular connection appears to apply to other types of mesenchymal stem cells (MSC) as well, including bone marrow stem cells (BMSC) [2, 9, 20, 22, 92]. However, due to the lack of a definitive MSC marker, the exact location and the cellular identity of MSC remain elusive. Nevertheless, the generally accepted perivascular localization implies a close relationship between MSC and blood vessels, and thus, in terms of availability as a therapeutic cell source, the more abundant the blood vessels are within a donor tissue site, the higher yield of MSC can be expected of that tissue site. When MSC yield is sufficiently high, it provides the option of transplanting the freshly isolated cells into patients without the need of cell culturing – a process that can be problematic due to the introduction of undesirable animal products and the possibility of microbial contamination, cell type alteration, and human errors.

The adipose tissue is endowed with an abundance of blood vessels, with each adipocyte surrounded by an extensive capillary network [21]. Furthermore, a recent study shows that the MSC frequency correlates with blood vessel density in the adipose tissue [23]. While reinforcing the concept of a vascular connection for MSC, this study also suggests that, being highly vascular, the adipose tissue could be an ideal source for harvesting large numbers of MSC. Indeed, it has been shown that, while the number of BMSC in bone marrow is approximately 1 in 25,000 to 1 in 100,000 nucleated cells, the average frequency of ADSC in processed lipoaspirate is approximately 2% of nucleated cells [95]. Thus, the yield of ADSC from 1 g of fat is approximately 5,000 cells, whereas the yield of BMSC is 100–1,000 cells per milliliter of morrow. Furthermore, while bone marrow can only be obtained in limited quantity, the adipose tissue is usually obtainable in abundance, especially in our increasingly obese society. The safety of the tissue isolation procedure is another advantage of ADSC over BMSC, as it has been shown that between 1994 and 2000 there were zero deaths on 66,570 liposuction procedures and a serious adverse event rate of only 0.068% [48]. Thus, while ADSC and BMSC are virtually identical in their therapeutic potential, the difference in their applicability is obvious.

18.1.3 Stromal Vascular Fraction Versus Adherent Cells

The isolation of ADSC consists of mincing adipose tissue, washing with saline to remove blood cells, incubation with collagenase to separate cells, and centrifugation at 1,200 g to obtain a pelleted SVF, which contains pericytes, endothelial cells, preadipocytes, ADSC, and other cells (see below). Alternatively, SVF can be obtained from aspirated adipose tissue although such preparations have a much lower ADSC count [27, 90]. Plating of SVF cells in plastic culture dishes permits the selection of adherent cells, which are mostly ADSC. For clinical application, the adherent cells, i.e., ADSC, are more homogeneous than SVF cells and thus may better conform to regulatory requirement. However, SVF has been shown to contain T regulatory

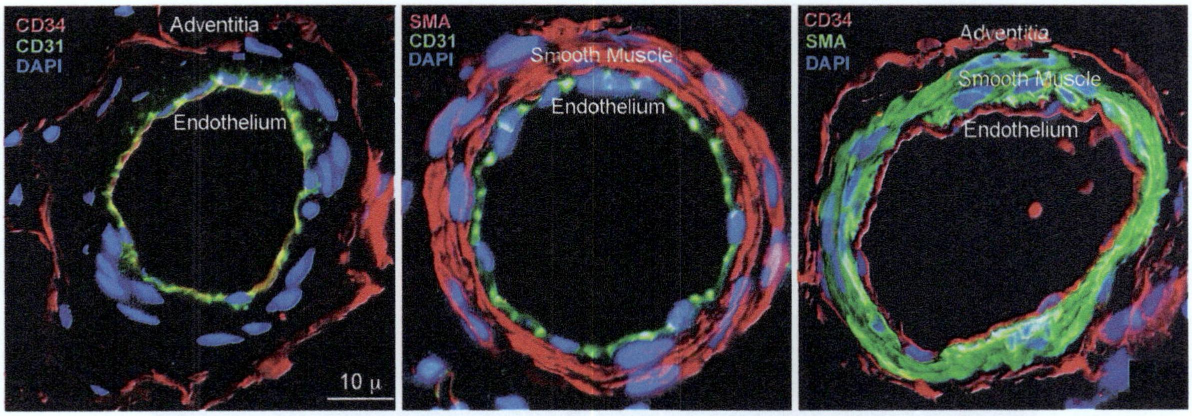

Fig. 18.1 Immunofluorescence localization of vascular and stem cell markers in arteries of adipose tissue. Human adipose tissues were immunostained for the indicated markers; a small artery (~75 μm in diameter) from each stained specimen was chosen for display. DAPI stain (*blue*) indicates cell nuclei. Note the lack of CD31 (or CD34) and SMA co-localization

cells [34] and hematopoietic stem cells [47] that may offer additional therapeutic benefits [53]. Moreover, by skipping the additional culturing procedure, the preparation of SVF cells is less likely to be contaminated. More importantly, when used in the SVF form, ADSC is the only stem cell type that can be isolated and autologously injected back to the patients on a same-day basis [104, 109]. Thus, for autologous application, SVF cells are preferred to ADSC.

18.1.4 Adventitial Cells Versus Pericytes

Although it has been well documented that CD34 expression in ADSC decreases gradually during culture, CD34 remains the most trusted marker when it comes to finding ADSC in tissue [62, 63, 96, 115]. In the arteries, arterioles, veins, and venules, CD34 is localized to the tunica intima and tunica adventitia; thus, the cross section of these CD34-stained blood vessels has the appearance of two concentric circles (endothelium and adventitia) sandwiching the unstained smooth muscle layer (Fig. 18.1). In the tunica intima, CD34 staining is easily distinguishable from that of CD31 because CD34 staining was homogeneous and contiguous while that of CD31 was discrete and more intense at intercellular junctions (Fig. 18.1). In the tunica adventitia, which is devoid of CD31 expression (Fig. 18.1), CD34 is expressed most likely in cells that have been proposed to be progenitor/stem cells [49, 97, 103, 111]. Together, these data suggest that the CD34+ cells in the adventitia (or the "vasculogenic zone") of blood vessels in the adipose tissue are the de facto ADSC.

There has been increasing interest in the possibility that MSC, including ADSC, are pericytes [2, 22, 86, 99]. In particular, Traktuev et al. [99] reported that ADSC could be defined by co-expression of CD34, CD140b (pericyte marker), and SMA (marker for both smooth muscle and pericyte). However, we have shown that in human adipose tissue neither CD140b nor SMA was co-expressed with CD34 [62], and these findings have been confirmed by two later studies [96, 115]. In addition, Suga et al. [96] showed that expression of pericyte markers was higher in CD34- than in CD34+ ADSC. Thus, because ADSC are known to lose CD34 expression during culture, it is possible that studies in favor of equating ADSC with pericytes were looking at cells whose CD34 expression was on the decline and whose pericyte marker expression on the rise as a result of adaptation to an artificial cell culture environment.

In their native tissue, CD34+ ADSC do not express pericyte marker CD140b or SMA. This conclusion, however, does not exclude the possibility that ADSC are pericytes because neither CD140b nor SMA can fully define pericytes [6]. In addition, it remains possible that there may exist a population of CD34- ADSC in adipose tissue. Still, it should be pointed out that equating ADSC with pericytes does not help improving our understanding of ADSC because the identity of pericytes is not any better defined than that of ADSC. In fact, in a recent review Corselli et al. [18] cautioned

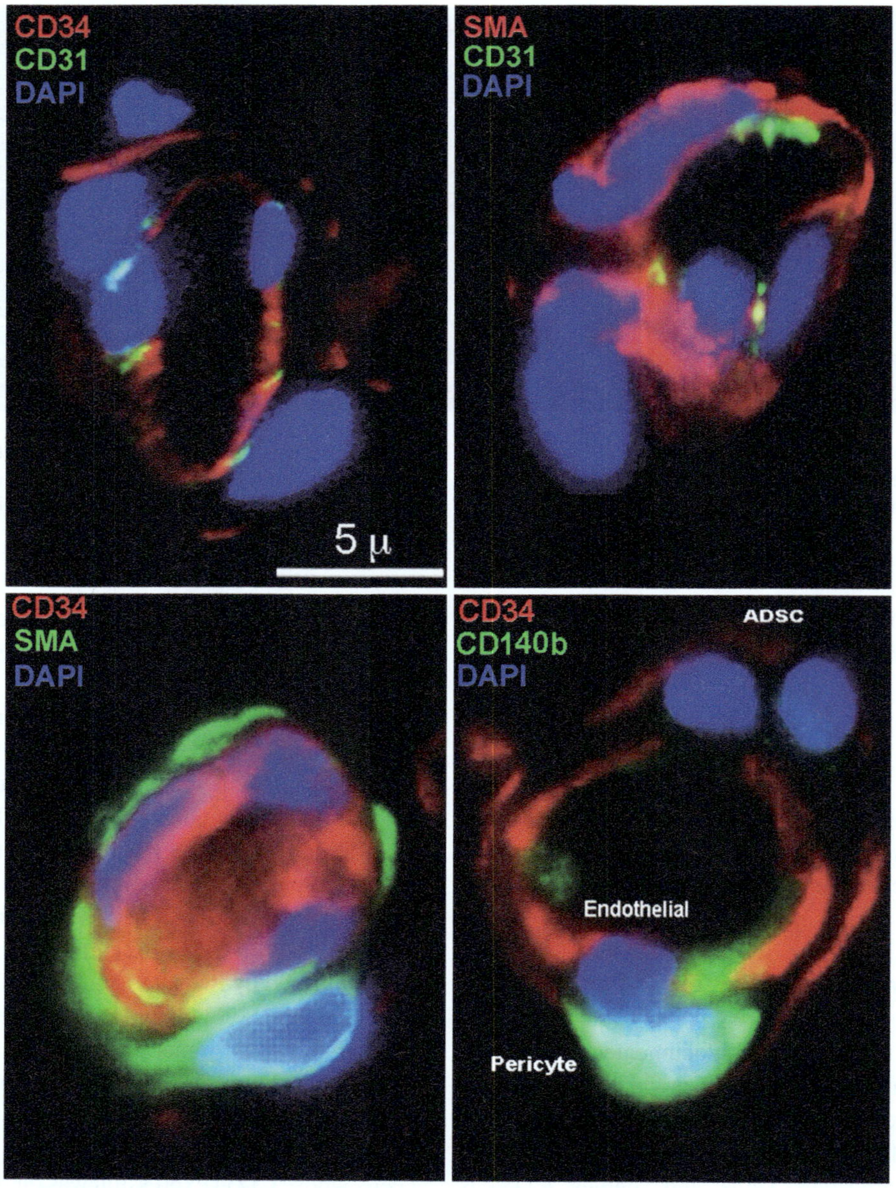

Fig. 18.2 Immunofluorescence localization of vascular and stem cell markers in capillaries of adipose tissue. Human adipose tissues were immunostained for the indicated markers; a capillary from each stained specimen was chosen for display. DAPI stain (*blue*) indicates cell nuclei. Materials in the lumen are blood clots (*lower left panel*). Note the lack of CD31 (or CD34) and SMA co-localization. CD34 and CD140b were not co-localized either. The putative endothelial cell, pericyte, and ADSC are indicated in the *lower right panel*

that pericytes have only been referred to in the strictest etymologic sense as periendothelial cells; that is, "the term *pericyte* has been used in its anatomic literal sense, without any functional connotation."

So, if ADSC reside in the adventitia, then how about capillaries, which do not have adventitia. In our 2008 paper [62], we showed that capillaries in adipose tissue contained three cell types: CD31+CD34+CD140b-SMA-endothelial cells, CD31-CD34-CD140b+SMA+pericytes, and CD31-CD34+CD140b-SMA-ADSC (Fig. 18.2). And, despite their localization to a very thin wall of the capillary, these three cell types are arranged with the endothelial being innermost, pericytes in the middle, and ADSC outermost (Fig. 18.2). Thus, similar to the situation with larger blood vessels, ADSC reside in the capillaries in a location that is equivalent to the adventitia. In this sense, capillary-bound ADSC are still adventitial cells.

18.1.5 Paracrine Action Versus Differentiation

While the original hypothesis underlying stem cell therapy was based on functional recovery as a consequence of stem cell differentiation, it is now clear that mechanisms other than cell differentiation are just as important if not more so. In particular, trophic and/or immunomodulatory factors secreted by stem cells have been shown to mediate functional improvements in several preclinical studies [85]. In the field of ADSC research, suppression of hepatocyte growth factor synthesis has been shown to markedly reduce the ability of ADSC to promote ischemic tissue revascularization [12]. Downregulation of cytokine (granulocyte chemotactic protein-2 or monocyte chemoattractant protein-1) or growth factor (VEGF) also reduced the proangiogenic action of ADSC [16]. Recently, we have also shown that both ADSC and lysates of ADSC were able to promote neuroregeneration [1]. Furthermore, we have found that ADSC secrete CXCL5 cytokine and which has neurotrophic properties [113]. In several other studies, functional improvements with little evidence of stem cell differentiation in the target tissue have also been interpreted as possible paracrine actions of ADSC although the specific secretory factors have yet to be identified. For example, Banas et al. [5] have demonstrated the therapeutic effects of ADSC on liver injury and identified several growth factors that were secreted by ADSC at high levels; however, no connection between these two lines of observation has been established.

18.1.6 Immunosuppression Versus Proinflammation

The above-mentioned paracrine actions are mostly concerned with their trophic effects. However, in a broader sense, paracrine actions also include their immunomodulatory effects. In fact, many studies have demonstrated immunomodulation as a key mechanism through which MSC exert therapeutic efficacy [39]. Still, while immunomodulation is generally equated with immunosuppression, it can also mean the opposite, i.e., proinflammation. In fact, a recent study contends that BMSC can be primed into either an immunosuppressive or a proinflammatory phenotype, depending on the downstream Toll-like receptor signaling [101]. While it remains to be seen how widespread this dual phenotype is among different MSC, it is clear that all reported ADSC's immunomodulatory effects are related to the immunosuppressive phenotype [53]. These include therapeutic effects on multiple sclerosis [87], rheumatoid arthritis [45], colitis [42], allergic rhinitis [17], and asthma [15, 81]. In addition, ADSC's immunomodulatory property has also been exploited as an effective method to attenuate graft-versus-host reaction [30, 105], and several studies have also demonstrated the feasibility of using ADSC for xenotransplantation [3, 17, 24, 42–45, 56, 77, 83].

18.1.7 Procancer Versus Anticancer

There are three different mechanisms through which cell-based therapy with ADSC and other types of MSC can possibly increase cancer risks. First, an ADSC preparation may contain cancer cells that preexisted in the donor site. This possibility was reported in our 2009 paper in which an angiosarcoma-like cell line was discovered among several human ADSC isolates [78]. Second, ADSC may become cancerous during culture. This possibility was reported in a 2005 paper in which ADSC was found to spontaneously transform into tumor cells after 4–5 months of culturing [89]. However, since it is highly unlikely that ADSC will be cultured for such an extended period before being used for therapy, the clinical relevance of this study is minimal. More importantly, this paper has been retracted on August 15, 2010 by five of the seven authors because the tumor cell line in question was found to be a result of contamination by HT1080 fibrosarcoma cell line [37]. Finally, the third possible mechanism is that ADSC transplantation may promote the progression of a preexisting cancer in the patient; for example, cancers of the brain [110], lung [55, 110], breast [73, 100, 116], prostate [65, 84], and other tissues [113]. However, it should be pointed out that all of these studies acquired their evidence by co-transplanting ADSC and a human cancer cell line (or metastatic pleural effusion cells [116]) in an immunocompromised animal. Up to this date, there has been no direct evidence that ADSC transplantation increases the progression of a preexisting cancer in humans or animals.

On the anticancer side, several studies have demonstrated the utilization of ADSC for cancer therapy. However, the majority of these studies simply utilize ADSC as a vehicle, such as carrying a prodrug or gene to the tumor site. Truly studying ADSC's potential as an anticancer agent is found in two studies. One, Cousin et al. [19] reported that ADSC provoked pancreatic cancer cell death both in vitro and in vivo. Second, Lee et al. [59] found that ADSC migrated toward brainstem glioma and caused reduction of tumor size, possibly through apoptosis. However, similar to the situation with the above-mentioned procancer studies, these two anticancer studies also relied on co-transplanting ADSC and the human cancer cell line in an immunocompromised animal. Thus, their clinical translation is also uncertain.

18.1.8 Local Versus Systemic Administration

In most published preclinical and clinical trials of stem cell therapy, the treatment cells were administered locally to the target organ or tissue. However, numerous studies have shown that MSC are capable of homing to sites of inflammation following tissue injury when injected intravenously. Thus, increasingly the intravenous route has been investigated for the treatment of both systemic and locally confined diseases. In ADSC field, feasibility of intravenous injection has been demonstrated in animal models of hindlimb ischemia [70], stroke [57], muscular dystrophy [66], rheumatoid arthritis [43], myocardial injury [58], liver injury [61], acute kidney injury [60], overactive bladder [52], and stress urinary incontinence [64]. Furthermore, case studies with intravenously injected ADSC for graft-versus host disease, refractory pure red cell aplasia, refractory chronic autoimmune thrombocytopenic purpura, rheumatoid arthritis, and multiple sclerosis have also shown promises [29, 31–33, 53, 87].

18.2 ADSC Application in Urology

The urinary system encompasses the kidneys and the lower urinary tract (LUT). While the kidneys filter blood and produce urine, the LUT is concerned with the temporary storage and the eventual elimination of urine. Thus, as far as this urine storage/elimination function is concerned, the LUT is defined as comprising the ureters, urinary bladder, and urethra. However, in men the urethra is anatomically a component of the prostate and the penis; therefore, both of these two reproductive organs are often considered part of the LUT and their diseases are traditionally studied within the urology discipline. The prostate, whose diseases are most often concerned with enlargement and cancer, is not a target for regenerative stem cell therapy; thus it will not be discussed further in this chapter. On the other hand, another male-specific organ, the testes, is also traditionally studied within the urology discipline, and it will be included for discussion in this chapter. Thus, in this chapter, with the exception of the prostate, all urinary (kidneys, ureters, bladder, and urethra) and male reproductive (penis and testes) organs are included for discussion.

18.2.1 Kidneys

Acute and chronic kidney disease is a major health problem worldwide with an overall mortality rate of 50–80% [14]. Current treatments are met with an acute shortage of compatible organs and limited adaptability of dialysis techniques. As such, alternative treatments such as stem cell-based therapy are currently being investigated. However, due to its anatomic complexity, the kidney has proven to be a very difficult organ for such applications [107]. In ADSC field, the first study on possible kidney cell differentiation was reported in 2005 by Brzoska et al. [10]. These authors showed that all-trans retinoic acid was able to induce cytokeratin-18 expression and nearly abolished vimentin expression in human ADSC. FACS analysis indicated that more than 80% of ADSC underwent such an epithelial differentiation. In 2009, the same group of authors reported a different strategy to derive kidney epithelial cells from ADSC [4]. By culturing human ADSC in conditioned media of human proximal tubular epithelial cells, the authors observed morphological changes toward an epithelial-like monolayer and phosphorylation of extracellular signal-regulated kinase 1 and 2. They also detected cytokeratin-18 expression and reduced expression of alpha-smooth muscle actin, CD49a and CD90. More recently, Li et al. [60] investigated possible renal epithelial differentiation of ADSC in vivo. These authors transplanted human ADSC into the kidney of an ischemia–reperfusion mouse model. The results showed that ADSC differentiated toward renal tubular epithelium at an early stage of renal injuries, and the differentiated donor cells replaced the vacant space left over by the dead cells. Moreover, the differentiated donor cells appeared to contribute to promoting host cell proliferation as well. Thus, both cell differentiation and paracrine actions appeared to play a role in ADSC's ability to repair injured kidney.

18.2.2 Ureters

Partial ureterectomy is the most common treatment in patients with ureter cancer. When the length of the dissected ureter is large, especially proximally, ureteral replacement is required to avoid urinary diversion. In 2006, Matsunuma et al. [69] reported an attempt to construct tissue-engineered ureters by using decellularized ureteral matrix seeded with cultured bladder uroepithelial cells (UEC). After transplantation of the seeded matrix into the subcutaneous space of nude mice or the omentum of nude rats, the seeded UEC began showing vacuolar degeneration in 3 days and gradually disappeared thereafter. Co-seeding the matrix with bone marrow-derived mononuclear cells (BM-MNC) before transplantation increased the survival of UEC, which was accompanied with increased microvessel density in the transplanted tissue. However, the increased number of capillaries was not due to differentiation of BM-MNC. Thus, it appears that BM-MNC enhanced neovascularization through paracrine actions. In 2010, Shen et al. [91] published a very similar study except that the decellularized ureteral matrix was replaced with a scaffold made of polycaprolactone-lecithin electrospun fibers and the UEC were obtained from rabbits instead of dogs. One significant difference is that the co-seeded cells were BMSC instead of BM-MNC, but the outcomes were the same, i.e., BMSC enhanced neovascularization without evidence of differentiation. In regard to ADSC, there has been no report of its application for the treatment of ureteral diseases.

18.2.3 Bladder

Stem cell research has been conducted in two situations relating to the bladder. One is the need to replace the entire or part of the bladder or to increase the existing bladder's size. The other is treatment of detrusor overactivity (DO) or overactive bladder (OAB), which is defined as urgency and often associated with frequency, nocturia, and urinary incontinence. Many risk factors are associated with DO/OAB; one of which is hyperlipidemia. In a rat model of hyperlipidemia-associated DO/OAB, we have recently shown that ADSC was able to reduce DO/OAB as demonstrated by conscious cystometry, and administration of ADSC through intra-bladder injection or tail vein injection was equally effective [52]. Moreover, functional improvement was accompanied with tissue improvement, as treated subjects showed enhanced vascular and nerve contents compared to control.

Bladder restoration or augmentation requires tissue engineering to recreate the native bladder milieu. Currently, the favored approach involves seeding a scaffold, usually an acellular matrix, with autologous bladder smooth muscle and epithelial cells. However, a significant drawback of this approach is the risk of reintroducing the pathologic condition (e.g., cancer) to the engineered tissue. For this and several other reasons, stem cells are considered ideal alternatives to the use of autologous bladder cells. To this end, differentiation of ADSC into smooth muscle cells in the bladder has been reported [88], and remarkably, in vitro urothelial differentiation of ADSC has also been demonstrated [67]. The importance of the latter study is two-fold. First, similar to the situation with kidney tubular cells, the bladder epithelial cells are structurally and functionally highly specialized cells; therefore, their transdifferentiation from other cell types is itself a remarkable feat. Second, ADSC are of mesenchymal lineage; therefore, their differentiation into epithelial cells is expectedly more difficult than differentiation into smooth muscle cells. In any event, this recent study showed that ADSC were able to express certain urothelial markers when co-cultured with preexisting urothelial cells. Interestingly, this probable urothelial differentiation of ADSC required direct cell–cell contact with the preexisting urothelial cells. Thus, it appears that conventional strategies such as growth factors and gene transfer will not be able to direct the differentiation of ADSC and perhaps other stem cells into functional urothelial cells.

Using ADSC for the purpose of bladder augmentation has been demonstrated in two recent studies. In one study [54], human ADSC were cultured in smooth muscle inductive media and seeded into 85:15 polylactic-glycolic acid bladder dome composites that were cast using an electropulled microfiber luminal surface combined with an outer porous sponge. The seeded and unseeded composites were then grafted in nude rats that underwent partial cystectomy. A third group of control animals had their partially cystectomized bladders sutured without grafting. The results showed that bladder capacity and compliance were maintained in the ADSC-seeded group throughout the 12-week course of monitoring. In contrast, these functional parameters deteriorated gradually with time in the unseeded group. However, control animals that received no grafts regained their baseline bladder capacities by week 12, thus demonstrating a long-term limitation of this model.

In another study [114], rabbit ADSC were seeded onto bladder acellular matrices that were prepared from allogenic rabbits. The seeded and unseeded matrices were then grafted in rabbits that underwent partial cystectomy. The results showed that at 24 weeks post-grafting in the ADSC-seeded group, the reconstructed bladders reached a mean volume of $94.68 \pm 3.31\%$ of the precystectomy bladder capacity. On the other hand, in the unseeded group, the reconstructed bladders reached a mean volume of $69.33 \pm 5.05\%$ of the pre-cystectomy bladder capacity. Immunohistochemical examination demonstrated regeneration of epithelial, smooth muscle, and nerve tissues in the seeded group. In contrast, the unseeded group displayed multilayered urothelium without evidence of organized muscle or nerve tissue.

From the above discussion, it can be concluded that ADSC has the potential to differentiate into bladder cells (smooth muscle and epithelium) and can be seeded onto scaffolds to improve the scaffold's success rate for bladder augmentation. ADSC can also improve urinary continence possibly through paracrine actions that promote tissue regeneration.

18.2.4 Urethra

The urethra is the most studied urological organ as far as stem cell therapy is concerned. This is perhaps due to the assumption that restoration of the urethral musculature alone would be sufficient to correct the most frequently encountered urethral problem, sphincter deficiency, which manifests symptomatically as stress urinary incontinence (SUI). While primarily a female concern – because of pregnancy and parturition-associated injuries to the urethra, SUI can also occur in men due to prostate surgeries. Initial cell-based experimental therapy for SUI involved the injection of autologous skeletal myoblasts into the vicinity of the urethral sphincter [108]. It then progressed to substituting myoblasts with skeletal muscle stem cells (SkMSC), and eventually several clinical trials with SKMSC were conducted, resulting in three publications from an Austrian group [71, 72, 94] and one from an American team [13]. Although clinical outcomes of these studies are generally favorable, a clear disadvantage of SkMSC is the requirement for complicated isolation procedure (several preplating and selection steps) and long-term culturing (more than 7 weeks), as skeletal muscle cannot be practically obtained in quantity from the patient and SkMSC exist in a small fraction relative to the tissue mass. Application of other types of SC, including BMSC, may also face the same problem if they are to be employed autologously. The only exception is ADSC because, in our increasingly obese society, adipose tissue is often considered dispensable, and the commonly performed liposuction procedure is capable of safely isolating large quantities of adipose tissue. Furthermore, it has been shown that ADSC can be isolated and injected back into the same patients for successful breast augmentation in approximately 4 h [109]; therefore, it is reasonable to expect that ADSC can be used to treat the much smaller urethra on a same-day basis – and without the need for culturing. Thus, as a first step toward this goal, we recently demonstrated the efficacy of ADSC in preventing SUI in an animal model [64]. We showed that tail vein injection of ADSC was equally effective as intra-urethral injection, thus pointing to the possibility of using the convenient intravenous route for administering ADSC clinically. We also showed that ADSC treatment restored not only the cellular (SMC) but also the extracellular (elastin) component in the experimentally injured rat urethra. Thus, it appears that ADSC has the potential to prevent SUI by correcting the underlying cellular and extracellular defects in the injured urethral sphincter.

Shortly after the above-mentioned publication of ours, two additional ADSC-SUI-related publications appeared. In one study, Fu et al. [35] used 5-azacytidine to induce ADSC to differentiate into myoblasts, which were then injected into the urethra of SUI rats.

At 1 and 3 months postinjection, both maximal bladder capacity and leak point pressure of the treated animals increased significantly when compared to control animals. Increased thickness of inferior muscularis in urethral mucosa and a greater number of large longitudinal muscle bundles were observed. In another study, Yamamoto et al. [104] performed periurethral injection of ADSC to treat two patients who developed SUI after radical prostatectomy. The results show that urinary incontinence improved 2 weeks after ADSC injection and continued to improve for up to 12 weeks. Ultrasonography detected the presence of injected "adipose tissue" (direct quote from the article) and a gradual increase in the blood flow to the injected area during the entire 12-week follow-up period. In addition, magnetic resonance imaging indicated a bulking effect at the injection site that persisted throughout the entire 12-week course. No significant adverse event was noted during the liposuction or ADSC injection procedure. No severe side effects such as pelvic pain, inflammation, or de novo urgency were observed after the treatment. Thus, periurethral injection of autologous ADSC appears to be a safe and feasible treatment modality for SUI.

18.2.5 Penis

Phosphodiesterase type-5 (PDE5) inhibitors are currently the first-line treatment of choice for men with erectile dysfunction (ED). However, PDE5 inhibitors are strictly contraindicated in men taking nitrate therapy and are known to cause a variety of adverse side effects that may reduce their suitability for some patients. More importantly, PDE5 inhibitors are less effective in treating certain types of ED including those associated with diabetes, hyperlipidemia, and surgery-induced cavernous nerve injuries. As such, alternative treatments, particularly those that can treat the underlying disease process of ED, would be preferable to current interventions. In this regard, one of the strategies currently being evaluated is stem cell therapy. In 2003, Deng et al. [25] showed that BMSC transduced with endothelial nitric oxide synthase (eNOS) were able to improve erectile function in aged rats. In 2004, Bochinski et al. [8] showed that ESC transduced with brain-derived neurotrophic factor improved erectile function in a rat model of postprostatectomy ED. In 2007, Bivalacqua et al. [7] demonstrated that BMSC alone or transduced with eNOS were able to reverse age-associated ED. Also in 2007, Song et al. [93] showed that immortalized human BMSC (by v-myc transfection) transplanted into rat corpus cavernosum could differentiate into endothelial cells and SMC. In 2008, Nolazco et al. [80] indicated that intracavernous injection of SkMSC could restore cavernous SMC and erectile function in aged rats. In 2009, Fall et al. [28] reported that intracavernous injection of BMSC improved erectile function in a rat model of postprostatectomy ED.

We first reported the therapeutic potential of ADSC for ED in 2009 [79]. Specifically, we showed that in vitro endothelial differentiation of ADSC was mediated by FGF2 and ADSC injected into rat penis appeared to have differentiated into endothelial cells. Since the cavernous endothelium plays a key role in penile erection and is often damaged in disease processes such as diabetes and hyperlipidemia, it is important that we provide further evidence that ADSC can restore endothelial function in the erectile tissue. To this end, we recently demonstrated the efficacy of ADSC to treat diabetes and hyperlipidemia-associated ED, respectively [36, 51]. In both studies, functional improvements were accompanied with restoration of the crucial endothelial and neural components in the penis, suggesting the curative prospect of ADSC treatment. In another study of postprostatectomy ED, ADSC was also effective in restoring the damaged nerves and improving erectile function [1]. In a recently submitted manuscript, we also described the application of ADSC for the construction of a nerve graft. Specifically, we prepared a decellularized adipose-derived matrix and seeded it with ADSC. We then used this seeded matrix to bridge the ends of transected cavernous nerves in rats. The results show that ADSC seeding encouraged axonal growth into the matrix, and which was accompanied by an improvement of erectile function. Thus, it appears that ADSC has the potential to treat various forms of ED.

18.2.6 Testes

The testes are both a reproductive and an endocrine organ. Specifically, they produce sperms for reproduction and secrete male sex hormones for regulating a wide array of developmental processes and physiological functions. Spermatogenesis, which is nurtured by

Sertoli cells within in the seminiferous tubules, requires the production of androgens from Leydig cells in the interstitia between tubules. Thus, defects in any of the three cell types (germ, Sertoli, and Leydig) can contribute to male infertility, and which can possibly be corrected by stem cell transplantation. In addition, because Leydig cells are capable of testosterone production, their derivation from stem cells may permit the treatment of male hypogonadism, which is currently treated by life-long testosterone supplementation.

In 2003, Toyooka et al. [98] showed that ESC could form male germ cells in vitro. Specifically they showed that the differentiation of ESC into male germ cells depended on embryoid body formation and was greatly enhanced by the inductive effects of bone morphogenic protein 4-producing cells. They further showed that the induced germ cells could participate in spermatogenesis when transplanted into reconstituted testicular tubules, demonstrating that ESC can produce functional germ cells in vitro. In 2004, another paper by Geijsen et al. [38] showed that ESC-derived embryoid bodies supported maturation of primordial germ cells into haploid male gametes, which, when injected into oocytes, restored the somatic diploid chromosome complement and developed into blastocysts. Also in 2004, Nayernia et al. [74] reported the in vitro generation of a germ cell line (SSC1) from the pluripotent teratocarcinoma cells and showed that the SSC1 cell line formed mature seminiferous tubule structures and supported spermatogenesis after transplantation into recipient testes. In 2006, West et al. [102] published a detailed protocol for the in vitro generation of germ cells from murine ESC. Most significantly, Nayernia et al. [76] demonstrated for the first time that ESC-derived germ cells were able to generate offspring mice. Finally, in 2009 Bucay et al. [11] showed that human ESC could differentiate into not only primordial germ cells but also Sertoli cells.

The above-mentioned studies are all related to ESC, which are expected to be able to differentiate into all cell types except the fertilized eggs; thus, their demonstration of germ cell differentiation is not completely a surprise. On the other hand, the work by Nayernia et al. [75] in 2006 was indeed remarkable: Differentiation of male germ cells from BMSC! And, while the study was done with murine BMSC, 1 year later another study using human BMSC was published [26], and this was followed in 2009 by another study also using human BMSC [50]. Thus, it appears that ASCs, at least MSC from bone marrow, can transdifferentiate into male germ cells.

In regard to Leydig cell differentiation, Yazawa et al. [106] demonstrated in 2006 that, when transplanted into immature rat testes, BMSC were able to engraft and differentiate into steroidogenic cells that were indistinguishable from Leydig cells. In 2007, Lue et al. [68] showed that BMSC transplanted into the testis of a busulfan-treated infertility mouse model differentiated into germ cells, Sertoli cells, and Leydig cells. Thus, it appears that BMSC transplantation represents a promising avenue for the treatment of male infertility and testosterone deficiency. Whether ADSC can be used for the same purposes remains to be seen.

18.3 Conclusions

The adipose tissue is one of the rare tissues that can be partially removed from a living person without causing harm. Its superficial location makes it more accessible than most other tissues, for example, bone marrow. In addition, the removal of adipose tissue is an intervention desired by many patients. Thus, being abundantly residing in an abundant, convenient, and dispensable tissue source, ADSC is the most ideal type of MSC for the application as a cell-based regenerative medicine. Due to their abundance, ADSC can be administered to patients either as freshly isolated SVF cells or as the more homogeneous cultured, adherent cells. When applied as SVF cells, both the isolation and transplantation of cells can be done on a same-day basis, making ADSC the only MSC type that can be autologously applied in such a fashion.

Similar to other types of MSC, ADSC are closely associated with blood vessels and may in fact be VSC. Due to the vasculature's dynamic capacity for growth and multipotent nature for diversification, VSC in tissues are individually at various stages and on different paths of differentiation. Therefore, when isolated and put in culture, these cells are expected to be heterogeneous in marker expression, renewal capacity, and differentiation potential. Although this heterogeneity of VSC does impose difficulties and cause confusions in basic science studies, its impact on the development of VSC as a therapeutic cell source has not been as apparent, as many preclinical and clinical trials have reported favorable outcomes.

Despite the growing appreciation of MSC's vascular origin, it remains unsettled where exactly MSC are located in the blood vessels. In regard to ADSC's localization, there appear to be two possible sites, namely, the intima where pericytes may reside and the adventitia where adventitial progenitor cells reside. The pericyte connection is mainly based on studies with cultured ADSC that may express both CD34 (ADSC marker) and CD140b (pericyte marker). However, in adipose tissue these two markers are mutually exclusively expressed. Therefore, while it cannot be ruled out that CD34-CD140b+cells (pericytes) may possess stem cell properties, these cells cannot be ADSC, which are classified as CD34+. However, since both SVF cells and freshly cultured ADSC do contain a mixture of CD34+ and CD34- cells and both CD34+ and CD34- cells are multipotent [96], it is possible that both pericytes and adventitial progenitor cells contribute to what constitutes an experimentally obtainable ADSC culture.

In addition to their differentiation capacity, ADSC secrete many trophic factors and immunosuppressive molecules. Thus, their therapeutic capacity is likely the result of these properties alone or in combination. For renal diseases, transplantation of human ADSC into the kidney of an ischemia–reperfusion mouse model showed that ADSC differentiated toward renal tubular epithelium at an early stage of renal injuries. Moreover, the differentiated donor cells appeared to contribute to promoting host cell proliferation as well. Thus, both cell differentiation and paracrine actions appeared to play a role in ADSC's ability to repair injured kidney. For ureteral diseases, two studies have demonstrated the angiogenic potential of bone marrow-derived cells to increase the survival of urothelial cells seeded on decellularized or synthetic ureteral grafts. No ADSC study has been reported in this regard.

Using ADSC for the purpose of bladder augmentation has been demonstrated in two recent studies with encouraging results. Transplantation of ADSC into the bladder or through intravenous injection has also shown promises in treating hyperlipidemia-associated overactive bladder. Similarly, transplantation of ADSC into the urethra or through intravenous injection has also shown effectiveness in preventing birth trauma-induced SUI. A small-scale clinical trial on the use of ADSC to treat postprostatectomy SUI has also obtained favorable outcomes. Intracavernous injection of ADSC has also demonstrated efficacy in treating various types of ED, including hyperlipidemia-associated, diabetic, and cavernous nerve injury. Finally, while ESC and BMSC have been shown to differentiate into male germ cells, Sertoli cells, and Leydig cells, no similar studies in ADSC field have been reported.

Despite these encouraging advances, the application of ADSC for treating urological diseases is still in its infancy and is expected to have go through many growing-up difficulties. For example, since most of ADSC treatment studies have been done in preclinical settings, it is quite possible that some of them may not translate into clinical successes due to the fact that a particular animal model is often designed to mimic a specific human disease entity. These animals are otherwise perfectly healthy, which of course is in sharp contrast to clinical situations where patients who would benefit most from regenerative interventions often have comorbidity and are taking multiple medications. Still, urologists in various countries where governmental regulations are more relaxed than in the USA have obtained interesting clinical results with the application of ADSC (personal communications). Thus, the prognosis for ADSC application in urology is cautiously optimistic.

References

1. Albersen M, Fandel TM, Lin G, Wang G, Banie L, Lin CS, Lue TF (2010) Injections of adipose tissue-derived stem cells and stem cell lysate improve recovery of erectile function in a rat model of cavernous nerve injury. J Sex Med 7:3331–3340
2. Amos PJ, Shang H, Bailey AM, Taylor A, Katz AJ, Peirce SM (2008) Ifats collection: the role of human adipose-derived stromal cells in inflammatory microvascular remodeling and evidence of a perivascular phenotype. Stem Cells 26:2682–2690
3. Arnalich-Montiel F, Pastor S, Blazquez-Martinez A, Fernandez-Delgado J, Nistal M, De Alio JL, Miguel MP (2008) Adipose-derived stem cells are a source for cell therapy of the corneal stroma. Stem Cells 26:570–579
4. Baer PC, Bereiter-Hahn J, Missler C, Brzoska M, Schubert R, Gauer S, Geiger H (2009) Conditioned medium from renal tubular epithelial cells initiates differentiation of human mesenchymal stem cells. Cell Prolif 42:29–37
5. Banas A, Teratani T, Yamamoto Y, Tokuhara M, Takeshita F, Osaki M, Kawamata M, Kato T, Okochi H, Ochiya T (2008) Ifats collection: in vivo therapeutic potential of human adipose tissue mesenchymal stem cells after transplantation into mice with liver injury. Stem Cells 26:2705–2712

6. Bergers G, Song S (2005) The role of pericytes in blood-vessel formation and maintenance. Neuro Oncol 7:452–464
7. Bivalacqua TJ, Deng W, Kendirci M, Usta MF, Robinson C, Taylor BK, Murthy SN, Champion HC, Hellstrom WJ, Kadowitz PJ (2007) Mesenchymal stem cells alone or ex vivo gene modified with endothelial nitric oxide synthase reverse age-associated erectile dysfunction. Am J Physiol Heart Circ Physiol 292:H1278–H1290
8. Bochinski D, Lin GT, Nunes L, Carrion R, Rahman N, Lin CS, Lue TF (2004) The effect of neural embryonic stem cell therapy in a rat model of cavernosal nerve injury. BJU Int 94:904–909
9. Brachvogel B, Moch H, Pausch F, Schlotzer-Schrehardt U, Hofmann C, Hallmann R, von der Mark K, Winkler T, Poschl E (2005) Perivascular cells expressing annexin a5 define a novel mesenchymal stem cell-like population with the capacity to differentiate into multiple mesenchymal lineages. Development 132:2657–2668
10. Brzoska M, Geiger H, Gauer S, Baer P (2005) Epithelial differentiation of human adipose tissue-derived adult stem cells. Biochem Biophys Res Commun 330:142–150
11. Bucay N, Yebra M, Cirulli V, Afrikanova I, Kaido T, Hayek A, Montgomery AM (2009) A novel approach for the derivation of putative primordial germ cells and sertoli cells from human embryonic stem cells. Stem Cells 27:68–77
12. Cai L, Johnstone BH, Cook TG, Liang Z, Traktuev D, Cornetta K, Ingram DA, Rosen ED, March KL (2007) Suppression of hepatocyte growth factor production impairs the ability of adipose-derived stem cells to promote ischemic tissue revascularization. Stem Cells 25:3234–3243
13. Carr LK, Steele D, Steele S, Wagner D, Pruchnic R, Jankowski R, Erickson J, Huard J, Chancellor MB (2008) 1-Year follow-up of autologous muscle-derived stem cell injection pilot study to treat stress urinary incontinence. Int Urogynecol J Pelvic Floor Dysfunct 19:881–883
14. Chhabra P, Brayman KL (2009) The use of stem cells in kidney disease. Curr Opin Organ Transplant 14:72–78
15. Cho KS, Roh HJ (2010) Immunomodulatory effects of adipose-derived stem cells in airway allergic diseases. Curr Stem Cell Res Ther 5:111–115
16. Cho HH, Kim YJ, Kim JT, Song JS, Shin KK, Bae YC, Jung JS (2009) The role of chemokines in proangiogenic action induced by human adipose tissue-derived mesenchymal stem cells in the murine model of hindlimb ischemia. Cell Physiol Biochem 24:511–518
17. Cho KS, Park HK, Park HY, Jung JS, Jeon SG, Kim YK, Roh HJ (2009) Ifats collection: immunomodulatory effects of adipose tissue-derived stem cells in an allergic rhinitis mouse model. Stem Cells 27:259–265
18. Corselli M, Chen CW, Crisan M, Lazzari L, Peault B (2010) Perivascular ancestors of adult multipotent stem cells. Arterioscler Thromb Vasc Biol 30:1104–1109
19. Cousin B, Ravet E, Poglio S, De Toni F, Bertuzzi M, Lulka H, Touil I, Andre M, Grolleau JL, Peron JM, Chavoin JP, Bourin P, Penicaud L, Casteilla L, Buscail L, Cordelier P (2009) Adult stromal cells derived from human adipose tissue provoke pancreatic cancer cell death both in vitro and in vivo. PLoS ONE 4:e6278
20. Covas DT, Panepucci RA, Fontes AM, Silva WA Jr, Orellana MD, Freitas MC, Neder L, Santos AR, Peres LC, Jamur MC, Zago MA (2008) Multipotent mesenchymal stromal cells obtained from diverse human tissues share functional properties and gene-expression profile with cd146+ perivascular cells and fibroblasts. Exp Hematol 36:642–654
21. Crandall DL, Hausman GJ, Kral JG (1997) A review of the microcirculation of adipose tissue: anatomic, metabolic, and angiogenic perspectives. Microcirculation 4:211–232
22. Crisan M, Yap S, Casteilla L, Chen CW, Corselli M, Park TS, Andriolo G, Sun B, Zheng B, Zhang L, Norotte C, Teng PN, Traas J, Schugar R, Deasy BM, Badylak S, Buhring HJ, Giacobino JP, Lazzari L, Huard J, Peault B (2008) A perivascular origin for mesenchymal stem cells in multiple human organs. Cell Stem Cell 3:301–313
23. da Silva ML, Sand TT, Harman RJ, Lennon DP, Caplan AI (2009) Msc frequency correlates with blood vessel density in equine adipose tissue. Tissue Eng A 15:221–229
24. DelaRosa O, Lombardo E, Beraza A, Mancheno-Corvo P, Ramirez C, Menta R, Rico L, Camarillo E, Garcia L, Abad JL, Trigueros C, Delgado M, Buscher D (2009) Requirement of ifn-gamma-mediated indoleamine 2,3-dioxygenase expression in the modulation of lymphocyte proliferation by human adipose-derived stem cells. Tissue Eng A 15:2795–2806
25. Deng W, Bivalacqua TJ, Chattergoon NN, Hyman AL, Jeter JR Jr, Kadowitz PJ (2003) Adenoviral gene transfer of enos: high-level expression in ex vivo expanded marrow stromal cells. Am J Physiol Cell Physiol 285:C1322–C1329
26. Drusenheimer N, Wulf G, Nolte J, Lee JH, Dev A, Dressel R, Gromoll J, Schmidtke J, Engel W, Nayernia K (2007) Putative human male germ cells from bone marrow stem cells. Soc Reprod Fertil Suppl 63:69–76
27. Eto H, Suga H, Matsumoto D, Inoue K, Aoi N, Kato H, Araki J, Yoshimura K (2009) Characterization of structure and cellular components of aspirated and excised adipose tissue. Plast Reconstr Surg 124:1087–1097
28. Fall PA, Izikki M, Tu L, Swieb S, Giuliano F, Bernabe J, Souktani R, Abbou C, Adnot S, Eddahibi S, Yiou R (2009) Apoptosis and effects of intracavernous bone marrow cell injection in a rat model of postprostatectomy erectile dysfunction. Eur Urol 56:716–725
29. Fang B, Song Y, Liao L, Zhang Y, Zhao RC (2007) Favorable response to human adipose tissue-derived mesenchymal stem cells in steroid-refractory acute graft-versus-host disease. Transplant Proc 39:3358–3362
30. Fang B, Song Y, Lin Q, Zhang Y, Cao Y, Zhao RC, Ma Y (2007) Human adipose tissue-derived mesenchymal stromal cells as salvage therapy for treatment of severe refractory acute graft-vs.-host disease in two children. Pediatr Transplant 11:814–817
31. Fang B, Song Y, Zhao RC, Han Q, Lin Q (2007) Using human adipose tissue-derived mesenchymal stem cells as salvage therapy for hepatic graft-versus-host disease resembling acute hepatitis. Transplant Proc 39:1710–1713
32. Fang B, Song Y, Li N, Li J, Han Q, Zhao RC (2009) Mesenchymal stem cells for the treatment of refractory pure red cell aplasia after major abo-incompatible hematopoietic stem cell transplantation. Ann Hematol 88:261–266
33. Fang B, Song YP, Li N, Li J, Han Q, Zhao RC (2009) Resolution of refractory chronic autoimmune thrombocytopenic purpura following mesenchymal stem cell transplantation: a case report. Transplant Proc 41: 1827–1830

34. Feuerer M, Herrero L, Cipolletta D, Naaz A, Wong J, Nayer A, Lee J, Goldfine AB, Benoist C, Shoelson S, Mathis D (2009) Lean, but not obese, fat is enriched for a unique population of regulatory t cells that affect metabolic parameters. Nat Med 15:930–939
35. Fu Q, Song XF, Liao GL, Deng CL, Cui L (2010) Myoblasts differentiated from adipose-derived stem cells to treat stress urinary incontinence. Urology 75:718–723
36. Garcia MM, Fandel TM, Lin G, Shindel AW, Banie L, Lin CS, Lue TF (2010) Treatment of erectile dysfunction in the obese type 2 diabetic zdf rat with adipose tissue-derived stem cells. J Sex Med 7:89–98
37. Garcia S, Bernad A, Martin MC, Cigudosa JC, de la Garcia-Castro J, Fuente R (2010) Pitfalls in spontaneous in vitro transformation of human mesenchymal stem cells. Exp Cell Res 316:1648–1650
38. Geijsen N, Horoschak M, Kim K, Gribnau J, Eggan K, Daley GQ (2004) Derivation of embryonic germ cells and male gametes from embryonic stem cells. Nature 427:148–154
39. Ghannam S, Bouffi C, Djouad F, Jorgensen C, Noel D (2010) Immunosuppression by mesenchymal stem cells: mechanisms and clinical applications. Stem Cell Res Ther 1:2
40. Gimble JM, Katz AJ, Bunnell BA (2007) Adipose-derived stem cells for regenerative medicine. Circ Res 100: 1249–1260
41. Gimble JM, Guilak F, Bunnell BA (2010) Clinical and preclinical translation of cell-based therapies using adipose tissue-derived cells. Stem Cell Res Ther 1:19
42. Gonzalez MA, Gonzalez-Rey E, Rico L, Buscher D, Delgado M (2009) Adipose-derived mesenchymal stem cells alleviate experimental colitis by inhibiting inflammatory and autoimmune responses. Gastroenterology 136:978–989
43. Gonzalez MA, Gonzalez-Rey E, Rico L, Buscher D, Delgado M (2009) Treatment of experimental arthritis by inducing immune tolerance with human adipose-derived mesenchymal stem cells. Arthritis Rheum 60:1006–1019
44. Gonzalez-Rey E, Anderson P, Gonzalez MA, Rico L, Buscher D, Delgado M (2009) Human adult stem cells derived from adipose tissue protect against experimental colitis and sepsis. Gut 58:929–939
45. Gonzalez-Rey E, Gonzalez MA, Varela N, O'Valle F, Hernandez-Cortes P, Rico L, Buscher D, Delgado M (2010) Human adipose-derived mesenchymal stem cells reduce inflammatory and t cell responses and induce regulatory t cells in vitro in rheumatoid arthritis. Ann Rheum Dis 69:241–248
46. Guilak F, Estes BT, Diekman BO, Moutos FT, Gimble JM (2010) 2010 Nicolas Andry award: multipotent adult stem cells from adipose tissue for musculoskeletal tissue engineering. Clin Orthop Relat Res 468:2530–2540
47. Han J, Koh YJ, Moon HR, Ryoo HG, Cho CH, Kim I, Koh GY (2010) Adipose tissue is an extramedullary reservoir for functional hematopoietic stem and progenitor cells. Blood 115:957–964
48. Housman TS, Lawrence N, Mellen BG, George MN, Filippo JS, Cerveny KA, DeMarco M, Feldman SR, Fleischer AB (2002) The safety of liposuction: results of a national survey. Dermatol Surg 28:971–978
49. Hu Y, Zhang Z, Torsney E, Afzal AR, Davison F, Metzler B, Xu Q (2004) Abundant progenitor cells in the adventitia contribute to atherosclerosis of vein grafts in apoe-deficient mice. J Clin Invest 113:1258–1265
50. Hua J, Pan S, Yang C, Dong W, Dou Z, Sidhu KS (2009) Derivation of male germ cell-like lineage from human fetal bone marrow stem cells. Reprod Biomed Online 19:99–105
51. Huang YC, Ning H, Shindel AW, Fandel TM, Lin G, Harraz AM, Lue TF, Lin CS (2010) The effect of intracavernous injection of adipose tissue-derived stem cells on hyperlipidemia-associated erectile dysfunction in a rat model. J Sex Med 7:1391–1400
52. Huang YC, Shindel AW, Ning H, Lin G, Harraz AM, Wang G, Garcia M, Lue TF, Lin CS (2010) Adipose derived stem cells ameliorate hyperlipidemia associated detrusor overactivity in a rat model. J Urol 183:1232–1240
53. Ichim TE, Harman RJ, Min WP, Minev B, Solano F, Rodriguez JP, Alexandrescu DT, De Necochea-Campion R, Hu X, Marleau AM, Riordan NH (2010) Autologous stromal vascular fraction cells: a tool for facilitating tolerance in rheumatic disease. Cell Immunol 264:7–17
54. Jack GS, Zhang R, Lee M, Xu Y, Wu BM, Rodriguez LV (2009) Urinary bladder smooth muscle engineered from adipose stem cells and a three dimensional synthetic composite. Biomaterials 30:3259–3270
55. Jeon ES, Lee IH, Heo SC, Shin SH, Choi YJ, Park JH, Park DY, Kim JH (2010) Mesenchymal stem cells stimulate angiogenesis in a murine xenograft model of a549 human adenocarcinoma through an lpa1 receptor-dependent mechanism. Biochim Biophys Acta 1801:1205–1213
56. Kang JW, Kang KS, Koo HC, Park JR, Choi EW, Park YH (2008) Soluble factors-mediated immunomodulatory effects of canine adipose tissue-derived mesenchymal stem cells. Stem Cells Dev 17:681–693
57. Kim JM, Lee ST, Chu K, Jung KH, Song EC, Kim SJ, Sinn DI, Kim JH, Park DK, Kang KM, Hyung Hong N, Park HK, Won CH, Kim KH, Kim M, Kun Lee S, Roh JK (2007) Systemic transplantation of human adipose stem cells attenuated cerebral inflammation and degeneration in a hemorrhagic stroke model. Brain Res 1183:43–50
58. Kim U, Shin DG, Park JS, Kim YJ, Park SI, Moon YM, Jeong KS (2009) Homing of adipose-derived stem cells to radiofrequency catheter ablated canine atrium and differentiation into cardiomyocyte-like cells. Int J Cardiol 146:371–378
59. Lee DH, Ahn Y, Kim SU, Wang KC, Cho BK, Phi JH, Park IH, Black PM, Carroll RS, Lee J, Kim SK (2009) Targeting rat brainstem glioma using human neural stem cells and human mesenchymal stem cells. Clin Cancer Res 15:4925–4934
60. Li K, Han Q, Yan X, Liao L, Zhao RC (2010) Not a process of simple vicariousness, the differentiation of human adipose-derived mesenchymal stem cells to renal tubular epithelial cells plays an important role in acute kidney injury repairing. Stem Cells Dev 19:1267–1275
61. Liang L, Ma T, Chen W, Hu J, Bai X, Li J, Liang T (2009) Therapeutic potential and related signal pathway of adipose-derived stem cell transplantation for rat liver injury. Hepatol Res 39:822–832
62. Lin G, Garcia M, Ning H, Banie L, Guo YL, Lue TF, Lin CS (2008) Defining stem and progenitor cells within adipose tissue. Stem Cells Dev 17:1053–1063

63. Lin CS, Xin ZC, Deng CH, Ning H, Lin G, Lue TF (2010) Defining adipose tissue-derived stem cells in tissue and in culture. Histol Histopathol 25:807–815
64. Lin G, Wang G, Banie L, Ning H, Shindel AW, Fandel TM, Lue TF, Lin CS (2010) Treatment of stress urinary incontinence with adipose tissue-derived stem cells. Cytotherapy 12:88–95
65. Lin G, Yang R, Banie L, Wang G, Ning H, Li LC, Lue TF, Lin CS (2010) Effects of transplantation of adipose tissue-derived stem cells on prostate tumor. Prostate 70:1066–1073
66. Liu Y, Yan X, Sun Z, Chen B, Han Q, Li J, Zhao RC (2007) Flk-1+ adipose-derived mesenchymal stem cells differentiate into skeletal muscle satellite cells and ameliorate muscular dystrophy in mdx mice. Stem Cells Dev 16:695–706
67. Liu J, Huang J, Lin T, Zhang C, Yin X (2009) Cell-to-cell contact induces human adipose tissue-derived stromal cells to differentiate into urothelium-like cells in vitro. Biochem Biophys Res Commun 390:931–936
68. Lue Y, Erkkila K, Liu PY, Ma K, Wang C, Hikim AS, Swerdloff RS (2007) Fate of bone marrow stem cells transplanted into the testis: potential implication for men with testicular failure. Am J Pathol 170:899–908
69. Matsunuma H, Kagami H, Narita Y, Hata K, Ono Y, Ohshima S, Ueda M (2006) Constructing a tissue-engineered ureter using a decellularized matrix with cultured uroepithelial cells and bone marrow-derived mononuclear cells. Tissue Eng 12:509–518
70. Miranville A, Heeschen C, Sengenes C, Curat CA, Busse R, Bouloumie A (2004) Improvement of postnatal neovascularization by human adipose tissue-derived stem cells. Circulation 110:349–355
71. Mitterberger M, Marksteiner R, Margreiter E, Pinggera GM, Colleselli D, Frauscher F, Ulmer H, Fussenegger M, Bartsch G, Strasser H (2007) Autologous myoblasts and fibroblasts for female stress incontinence: a 1-year follow-up in 123 patients. BJU Int 100:1081–1085
72. Mitterberger M, Pinggera GM, Marksteiner R, Margreiter E, Fussenegger M, Frauscher F, Ulmer H, Hering S, Bartsch G, Strasser H (2008) Adult stem cell therapy of female stress urinary incontinence. Eur Urol 53:169–175
73. Muehlberg FL, Song YH, Krohn A, Pinilla SP, Droll LH, Leng X, Seidensticker M, Ricke J, Altman AM, Devarajan E, Liu W, Arlinghaus RB, Alt EU (2009) Tissue-resident stem cells promote breast cancer growth and metastasis. Carcinogenesis 30:589–597
74. Nayernia K, Li M, Jaroszynski L, Khusainov R, Wulf G, Schwandt I, Korabiowska M, Michelmann HW, Meinhardt A, Engel W (2004) Stem cell based therapeutical approach of male infertility by teratocarcinoma derived germ cells. Hum Mol Genet 13:1451–1460
75. Nayernia K, Lee JH, Drusenheimer N, Nolte J, Wulf G, Dressel R, Gromoll J, Engel W (2006) Derivation of male germ cells from bone marrow stem cells. Lab Invest 86:654–663
76. Nayernia K, Nolte J, Michelmann HW, Lee JH, Rathsack K, Drusenheimer N, Dev A, Wulf G, Ehrmann IE, Elliott DJ, Okpanyi V, Zechner U, Haaf T, Meinhardt A, Engel W (2006) In vitro-differentiated embryonic stem cells give rise to male gametes that can generate offspring mice. Dev Cell 11:125–132
77. Niemeyer P, Vohrer J, Schmal H, Kasten P, Fellenberg J, Suedkamp NP, Mehlhorn AT (2008) Survival of human mesenchymal stromal cells from bone marrow and adipose tissue after xenogenic transplantation in immunocompetent mice. Cytotherapy 10:784–795
78. Ning H, Liu G, Lin G, Garcia M, Li LC, Lue TF, Lin CS (2009) Identification of an aberrant cell line among human adipose tissue-derived stem cell isolates. Differentiation 77:172–180
79. Ning H, Liu G, Lin G, Yang R, Lue TF, Lin CS (2009) Fibroblast growth factor 2 promotes endothelial differentiation of adipose tissue-derived stem cells. J Sex Med 6:967–979
80. Nolazco G, Kovanecz I, Vernet D, Gelfand RA, Tsao J, Ferrini MG, Magee T, Rajfer J, Gonzalez-Cadavid NF (2008) Effect of muscle-derived stem cells on the restoration of corpora cavernosa smooth muscle and erectile function in the aged rat. BJU Int 101:1156–1164
81. Park HK, Cho KS, Park HY, Shin DH, Kim YK, Jung JS, Park SK, Roh HJ (2010) Adipose-derived stromal cells inhibit allergic airway inflammation in mice. Stem Cells Dev 19:1811–1818
82. Pettersson P, Cigolini M, Sjostrom L, Smith U, Bjorntorp P (1984) Cells in human adipose tissue developing into adipocytes. Acta Med Scand 215:447–451
83. Plaschke K (2009) Human adult mesenchymal stem cells improve rat spatial cognitive function after systemic hemorrhagic shock. Behav Brain Res 201:332–337
84. Prantl L, Muehlberg F, Navone NM, Song YH, Vykoukal J, Logothetis CJ, Alt EU (2010) Adipose tissue-derived stem cells promote prostate tumor growth. Prostate 70:1709–1715
85. Prockop DJ (2009) Repair of tissues by adult stem/progenitor cells (mscs): controversies, myths, and changing paradigms. Mol Ther 17:939–946
86. Rajashekhar G, Traktuev DO, Roell WC, Johnstone BH, Merfeld-Clauss S, Van Natta B, Rosen ED, March KL, Clauss M (2008) Ifats collection: adipose stromal cell differentiation is reduced by endothelial cell contact and paracrine communication: role of canonical wnt signaling. Stem Cells 26:2674–2681
87. Riordan NH, Ichim TE, Min WP, Wang H, Solano F, Lara F, Alfaro M, Rodriguez JP, Harman RJ, Patel AN, Murphy MP, Lee RR, Minev B (2009) Non-expanded adipose stromal vascular fraction cell therapy for multiple sclerosis. J Transl Med 7:29
88. Rodriguez LV, Alfonso Z, Zhang R, Leung J, Wu B, Ignarro LJ (2006) Clonogenic multipotent stem cells in human adipose tissue differentiate into functional smooth muscle cells. Proc Natl Acad Sci USA 103:12167–12172
89. Rubio D, Garcia-Castro J, Martin MC, de la Fuente R, Cigudosa JC, Lloyd AC, Bernad A (2005) Spontaneous human adult stem cell transformation. Cancer Res 65:3035–3039
90. Schreml S, Babilas P, Fruth S, Orso E, Schmitz G, Mueller MB, Nerlich M, Prantl L (2009) Harvesting human adipose tissue-derived adult stem cells: resection versus liposuction. Cytotherapy 11:947–957
91. Shen J, Fu X, Ou L, Zhang M, Guan Y, Wang K, Che Y, Kong D, Steinhof G, Li W, Yu Y, Ma N (2010) Construction of ureteral grafts by seeding urothelial cells and bone marrow mesenchymal stem cells into polycaprolactone-lecithin electrospun fibers. Int J Artif Organs 33:161–170
92. Shi S, Gronthos S (2003) Perivascular niche of postnatal mesenchymal stem cells in human bone marrow and dental pulp. J Bone Miner Res 18:696–704

93. Song YS, Lee HJ, Park IH, Kim WK, Ku JH, Kim SU (2007) Potential differentiation of human mesenchymal stem cell transplanted in rat corpus cavernosum toward endothelial or smooth muscle cells. Int J Impot Res 19:378–385
94. Strasser H, Marksteiner R, Margreiter E, Mitterberger M, Pinggera GM, Frauscher F, Fussenegger M, Kofler K, Bartsch G (2007) Transurethral ultrasonography-guided injection of adult autologous stem cells versus transurethral endoscopic injection of collagen in treatment of urinary incontinence. World J Urol 25:385–392
95. Strem BM, Hicok KC, Zhu M, Wulur I, Alfonso Z, Schreiber RE, Fraser JK, Hedrick MH (2005) Multipotential differentiation of adipose tissue-derived stem cells. Keio J Med 54:132–141
96. Suga H, Matsumoto D, Eto H, Inoue K, Aoi N, Kato H, Araki J, Yoshimura K (2009) Functional implications of cd34 expression in human adipose-derived stem/progenitor cells. Stem Cells Dev 18:1201–1210
97. Torsney E, Mandal K, Halliday A, Jahangiri M, Xu Q (2007) Characterisation of progenitor cells in human atherosclerotic vessels. Atherosclerosis 191:259–264
98. Toyooka Y, Tsunekawa N, Akasu R, Noce T (2003) Embryonic stem cells can form germ cells in vitro. Proc Natl Acad Sci USA 100:11457–11462
99. Traktuev DO, Merfeld-Clauss S, Li J, Kolonin M, Arap W, Pasqualini R, Johnstone BH, March KL (2008) A population of multipotent cd34-positive adipose stromal cells share pericyte and mesenchymal surface markers, reside in a periendothelial location, and stabilize endothelial networks. Circ Res 102:77–85
100. Walter M, Liang S, Ghosh S, Hornsby PJ, Li R (2009) Interleukin 6 secreted from adipose stromal cells promotes migration and invasion of breast cancer cells. Oncogene 28:2745–2755
101. Waterman RS, Tomchuck SL, Henkle SL, Betancourt AM (2010) A new mesenchymal stem cell (msc) paradigm: polarization into a pro-inflammatory msc1 or an immunosuppressive msc2 phenotype. PLoS ONE 5:e10088
102. West JA, Park IH, Daley GQ, Geijsen N (2006) In vitro generation of germ cells from murine embryonic stem cells. Nat Protoc 1:2026–2036
103. Xu Q (2008) Stem cells and transplant arteriosclerosis. Circ Res 102:1011–1024
104. Yamamoto T, Gotoh M, Hattori R, Toriyama K, Kamei Y, Iwaguro H, Matsukawa Y, Funahashi Y (2010) Periurethral injection of autologous adipose-derived stem cells for the treatment of stress urinary incontinence in patients undergoing radical prostatectomy: report of two initial cases. Int J Urol 17:75–82
105. Yanez R, Lamana ML, Garcia-Castro J, Colmenero I, Ramirez M, Bueren JA (2006) Adipose tissue-derived mesenchymal stem cells have in vivo immunosuppressive properties applicable for the control of the graft-versus-host disease. Stem Cells 24:2582–2591
106. Yazawa T, Mizutani T, Yamada K, Kawata H, Sekiguchi T, Yoshino M, Kajitani T, Shou Z, Umezawa A, Miyamoto K (2006) Differentiation of adult stem cells derived from bone marrow stroma into leydig or adrenocortical cells. Endocrinology 147:4104–4111
107. Yokoo T, Kawamura T, Kobayashi E (2008) Kidney organogenesis and regeneration: a new era in the treatment of chronic renal failure? Clin Exp Nephrol 12:326–331
108. Yokoyama T, Huard J, Chancellor MB (2000) Myoblast therapy for stress urinary incontinence and bladder dysfunction. World J Urol 18:56–61
109. Yoshimura K, Sato K, Aoi N, Kurita M, Hirohi T, Harii K (2008) Cell-assisted lipotransfer for cosmetic breast augmentation: supportive use of adipose-derived stem/stromal cells. Aesthetic Plast Surg 32:48–55, discussion 56–47
110. Yu JM, Jun ES, Bae YC, Jung JS (2008) Mesenchymal stem cells derived from human adipose tissues favor tumor cell growth in vivo. Stem Cells Dev 17:463–473
111. Zengin E, Chalajour F, Gehling UM, Ito WD, Treede H, Lauke H, Weil J, Reichenspurner H, Kilic N, Ergun S (2006) Vascular wall resident progenitor cells: a source for postnatal vasculogenesis. Development 133:1543–1551
112. Zhang Y, Daquinag A, Traktuev DO, Amaya-Manzanares F, Simmons PJ, March KL, Pasqualini R, Arap W, Kolonin MG (2009) White adipose tissue cells are recruited by experimental tumors and promote cancer progression in mouse models. Cancer Res 69:5259–5266
113. Zhang H, Yang R, Wang Z, Lin G, Lue TF, Lin CS (2011) Adipose tissue-derived stem cells secrete CXCL5 cytokine with neurotrophic effects on cavernous nerve regeneration. J Sex Med 8:437–446
114. Zhu WD, Xu YM, Feng C, Fu Q, Song LJ, Cui L (2010) Bladder reconstruction with adipose-derived stem cell-seeded bladder acellular matrix grafts improve morphology composition. World J Urol 28:493–498
115. Zimmerlin L, Donnenberg VS, Pfeifer ME, Meyer EM, Peault B, Rubin JP, Donnenberg AD (2010) Stromal vascular progenitors in adult human adipose tissue. Cytom A 77:22–30
116. Zimmerlin L, Donnenberg AD, Rubin JP, Landreneau RJ, Basse P, Donnenberg VS (2011) Regenerative therapy and cancer: in vitro and in vivo studies of the interaction between adipose-derived stem cells and breast cancer cells from clinical isolates. Tissue Eng A 17:93–106
117. Zuk PA (2010) The adipose-derived stem cell: looking back and looking ahead. Mol Biol Cell 21:1783–1787
118. Zuk PA, Zhu M, Mizuno H, Huang J, Futrell JW, Katz AJ, Benhaim P, Lorenz HP, Hedrick MH (2001) Multilineage cells from human adipose tissue: implications for cell-based therapies. Tissue Eng 7:211–228

Adipose Stem Cell Engineering: Characterization and Current Application in Otolaryngology

19

Suzanne N. King and Susan Thibeault

Contents

19.1	**Introduction**	209
19.2	**Isolation of Adipose-Derived Stem Cells**	209
19.2.1	BM MSC Versus ASC	210
19.2.2	Multilineage Differentiation	210
19.3	**Tissue Engineering**	213
19.3.1	Cellular Delivery	213
19.3.2	Biomaterial Scaffolds	213
19.3.3	Organ Transplant	214
19.3.4	Gene Therapy	215
19.4	**Conclusions**	218
References		218

S.N. King • S. Thibeault (✉)
Division of Otolaryngology Head and Neck Surgery,
Department of Surgery,
University of Wisconsin - Madison,
K4/709 600 Highland Ave,
Madison, WI 53792-7375, USA
e-mail: kings@surgery.wisc.edu; thibeault@surgery.wisc.edu

19.1 Introduction

Adipose-derived stem cells (ASCs) provide plentiful possibilities for clinical application in Otolaryngology – Head and Neck Surgery, because of their abundant quantity in the body, their ability to self-renew with high proliferative capacity and their capabilities of multilineage differentiation [7]. ASCs act in a paracrine manner in injured tissue, stimulating the environment through secreted cytokines and growth factors. They can be delivered into the body directly, with fabricated scaffolds, or they can be regenerated in the body for organ replacement. Subsequently, diverse therapeutic applications are possible for plastic surgery, tissue repair and revascularization of ischemic tissues.

19.2 Isolation of Adipose-Derived Stem Cells

ASCs have gained therapeutic prominence as an accessible and abundant cell source able to differentiate to multiple lineages for tissue regeneration [4]. Many factors can influence the viability and yield of ASCs in fat depots, such as type of adipose tissue, surgical procedure, harvested anatomical region and culturing conditions (i.e. media, cell density, polystyrene dish). Previous studies comparing resection and two common methods of liposuction found that samples obtained from tumescent liposuction presented a greater yield of ASC and doubling time compared to ultrasound-assisted or resection [26, 35]. Further studies have investigated the yield and functionality of

Y.-G. Illouz and A. Sterodimas (eds.), *Adipose Stem Cells and Regenerative Medicine*,
DOI: 10.1007/978-3-642-20012-0_19,

ASC harvested from different fat depots using the tumescent liposuction procedure. Results demonstrated that a richer source of ASCs can be obtained from the abdomen when compared to the thigh; however, the thigh can still yield more ASCs than bone marrow stem cells (BM MSC) [12].

In adipose tissue a special cell population termed stromal vascular fraction (SVF) exists. SVF consists of undefined stem cells as well as immune, endothelial (blood vessel lining), progenitor (undifferentiated or premature precursor cells), and stromal (connective tissue) cells. Standard methods for harvesting stem cells from SVF involve treatment with collagenase for enzymatic digestion, centrifuge cycles and filtering to remove debris [39]. Three criteria have been defined to characterize human mesenchymal stem cells (MSC) – adherence to plastic, specific surface antigen expression and multipotent differentiation [3]. Homogeneous populations of ASCs can be expanded on tissue culture flasks and characterized by surface markers (CD29, CD44, CD73, CD90, and CD105) through flow cytometry [3, 5].

Fundamental challenges exist with current strategies for cell culture expansion and storage, focusing attention toward developing good manufacturing practices for future clinical applications. Current in vitro culturing methods call for a cocktail of cytokines and growth factors from variable species to support proliferation and differentiation of stem cells. The potential risks of infection that these supplements and media might have in a patient threaten the possibility of transplant rejection. Autologous human serum with platelet-factor rich supernatant has been suggested for clinical use as a potential substitute for fetal calf serum (FCS) [15]. Prolonged in vitro culturing has also been found to significantly alter the phenotype of ASCs, causing the formation of tumors in immunodeficient mice [33]. Cryopreservation techniques are being investigated to improve the long-term storage of freshly isolated or frozen ASCs within their fat depots. Such outcomes could possibly provide greater populations of viable ASCs for in vitro expansion and later patient use.

19.2.1 BM MSC Versus ASC

A major challenge for cell therapy and tissue engineering has been finding a suitable cell source that can fulfill the function of replacing or remodeling specific tissues. BM MSCs and ASCs are viable candidates because of their capacity for multilineage differentiation, immune suppression and angiogenic phenotypes. BM MSCs have been studied extensively in vitro and in vivo as a source for regenerating bone, cartilage, muscle tissue, and hematopoietic cells. However, procedures for harvesting BM MSCs generate limited amounts of cells and as donors age the time needed for in vitro growth lengthens. Adipose tissues have sufficient amounts of stem cells that can be derived from a single sample and harvested with minimally invasive procedures [37]. In addition, comparative studies have shown that ASCs exhibit similar surface markers and upregulate over 25 similar genes as BM MSCs and umbilical cord blood stem cells [37]. Recent studies by Hanson et al. have found that ASCs, BM MSCs and vocal fold fibroblasts (VFF) exhibit similar surface markers and capacity for multiple lineage differentiation (Fig. 19.1). Results indicate that MSCs are resident in the vocal fold lamina propria [6]. Lastly, previous in vitro and in vivo studies have shown that ASCs have similar immunosuppressive qualities with mismatched allogeneic tissue as BM MSCs, preventing the incidence of graft versus host disease [30, 38].

Single surgical procedures for the removal, isolation, differentiation, and injection of autologous stem cells back into the patient hold several challenges. ASCs provide a significant advantage to BM MSCs in that tissues harvested from the abdomen have been shown to yield enough ASCs for immediate use with cell-based therapies [12]. However, isolation of ASCs from SVF can be time consuming, entailing in vitro culturing or magnetic separation using flow cytometry. Successful single surgical procedures have been conducted with SVF and biomaterial scaffolds to differentiate the cells [8]. There is debate over the immunogenicity of SVF because studies have shown that they can elicit a T cell-proliferative response [22]. Further investigation is needed comparing the therapeutic differences of isolated ASCs to SVF.

19.2.2 Multilineage Differentiation

The clinical potential for ASCs is abundant and their capacity to differentiate into multiple lineages demonstrates the impact they can provide to regenerative therapies in Otolaryngology – Head and Neck Surgery. Isolated human adipocytic progenitor cells

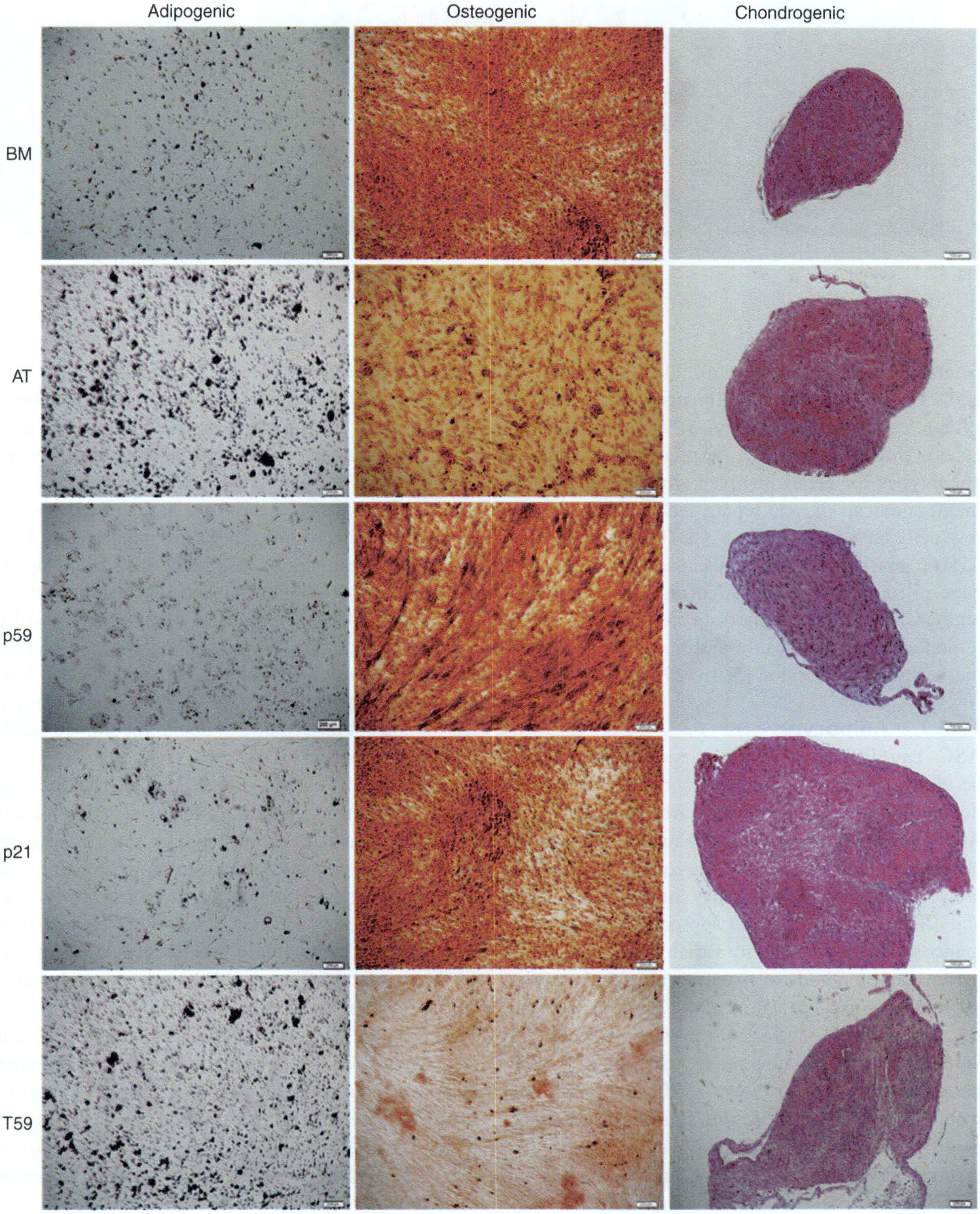

Fig. 19.1 Differentiation potential of ASC. Representative microscopic image views of the differentiation of human vocal fold fibroblasts (hVFF) cell lines p59, p21, T59, and T21 compared to bone marrow (BM)- and adipose tissue (AT)-derived mesenchymal stem cells into adipogenic (*left column*), osteogenic (*middle column*), and chondrogenic lineages (*right column*). Oil red O staining shows lipid vacuoles stained red, alizarin red S staining shows deposits of calcium crystals stained orange to brown, and safranin O staining shows cartilage-specific glycosaminoglycans. (Reprinted from Hanson et al. [6] with permission from Elsevier)

T21

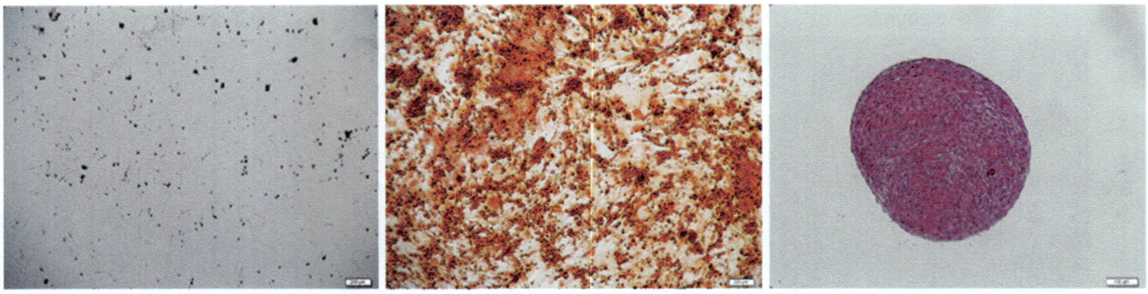

Fig. 19.1 (continued)

from lipoaspiration exhibit fibroblastic morphology and can differentiate into lineages of fat (adipocytes), cartilage (chondrocytes) and bone (osteocytes) (Fig. 19.1). They express transcriptional phenotypes for genes involved with tissue development, homeostasis, and repair, including the following cell surface receptors: CD105, STRO-1 and CD166 markers identifying capacity for multilineage differentiation; CD117 a marker for pluripotency; CD29 marker supporting angiogenesis; CD44 a marker for cell adhesion and migration; CD49e a marker for cell adhesion [5, 13, 29].

19.2.2.1 Adipocytes

ASCs can differentiate into adipocytes in vitro and be used as a filler to help contour facial regions or deficits in the vocal folds [25]. Although the use of ASCs can render many resemblances with current clinical applications with autologous fat grafts there are several differences that may affect the reliability and effectiveness of current approaches. The performance of fat grafts is highly variable, possible due to inconsistent methods of isolating fat, fat absorption or limited vascularization [24, 25]. In addition, because adipose tissue is a heterogeneous cell population, it is unknown how many ASCs are in the harvested tissue or the ratio of ASCs to matrix that would be needed for successful transplant. ASCs can be differentiated into mature adipocytes in vitro with insulin, IBMX, dexamethasone and indomethacine [37]. Characteristics of differentiated cells are similar to fat cells, expressing triglycerides for lipolysis, energy storage, and leptin, a fat specific protein [2]. In addition, studies have shown no apparent effects with donor age or body mass index with the differentiation or function of ASCs [2].

19.2.2.2 Chondrocytes

Injuries to cartilage caused by blunt trauma, prolonged inflammation, joint immobilization or surgical disruptions, can damage the matrix without gross tissue disruption. Cartilage tissue supports areas of the ear, nose and throat and consists of a matrix of collagen type II, proteoglycans and chondrocytes. The lack of lymphatic or blood vessels in the cartilage tissue prevents cells from migrating to the site of the injury to induce synthetic activity.

Currently, the only standard method approved by the FDA for cartilage reconstruction surgery involves in vitro expansion of autologous chondrocytes harvested from a patient's ribs, nose or ear. This technique has significant disadvantages due to the limited availability of tissue and morbidity to the region.

ASCs are an easily assessable strategy for replacing cells that are missing within functional tissues. Their proliferation and differentiation into chondrocytes in vitro requires the addition transforming growth factor (TGF-beta) and bone morphogenetic protein (BMP-6) to standard culture conditions to improve the expression of collagen II and cartilage specific glycoproteins [9]. Studies have focused on establishing specific cell markers for isolating homogeneous populations of ASCs, improving the effectiveness of differentiation. A recent study with ASCs reported a specific marker for isolating a homogeneous population of CD105+ cells with stronger chondrogenic potential in vitro [11]. The presence of CD105, a cell membrane glycoprotein, on ASCs provides a receptor complex for TGF-beta which enhances their growth, migration and differentiation.

19.2.2.3 Osteocytes

Defects to bony regions in the face or skull can be associated with inadequate wound healing due to the ingrowth of surrounding connective tissue and limited supply of growth factors and proteins to the defect. Bony portions of nose and face are composed of a matrix of collagen type I, glycoproteins and proteoglycans. After the creation of a wound the reformation of bone requires a balance of osteoinduction and osteoconduction development, initiated by the migration of fibroblasts, chonodrocytes and osteocytes into the deficit. Recruited immature cells secrete fibrous proteins to fill the voided region. During the final stages of healing cartilage drives endochondral ossification, mineralizing the extracellular matrix (ECM).

Reconstruction of bone using cell seeded scaffolds require materials that can osseointegrate anchoring biomaterials with internal stability that influence further in growth on the surface of bone. Several methods have been suggested for osteogenic differentiation of ASCs. Previous studies have demonstrated that insulin-like growth factor (IGF) and platelet-derived growth factor (PDGF) as key signaling pathways for enhancing osteogenic differentiation of human ASCs in vitro [18]. Osteogenesis of ASCs can exhibit osteoblast phenotypes, ALP and osteocalcin, as well as increased calcium deposition, indicating the ability to mineralize ECM [19].

19.3 Tissue Engineering

Current tissue engineering strategies to replace and remodel tissue functions of the head and neck include cell delivery with adipocytes, chondrocytes, and osteocytes, cell seeded biomaterial scaffolds, organ transplantation and gene therapy advances utilizing ASCs in head and neck cancer management.

19.3.1 Cellular Delivery

ASCs injected into an injured tissue can deliver natural growth factors and cytokines, promoting the regeneration of tissue. Their capacity to modulate the local environment and stimulate wound healing has made them a viable alternative approach for treatment of vocal fold scar and inflammation. Previous studies have shown that fibroblasts derived from a wounded or scarred tissue have phenotypic differences compared to normal fibroblasts [31]. Kumai et al. have suggested ASC delivery as a means of modulating the cellular environment in an injured vocal fold. To test this interaction, they co-cultured autologous ASCs with scar VFF from ferrets and results showed a decrease in collagen, minimal effects in hyaluronic acid (HA) and increases in proliferation [16]. This indicates that ASCs interaction with scar VFF could prevent stiffening in the vocal fold due to increased collagen production and maintain the viscoelastic properties through production of HA. To identify the ASC signal involved in changing the ECM composition, a further study was conducted analyzing ASCs secretion of hepatocyte growth factor (HGF) [17], a known inhibitor of fibrosis in the vocal folds [14]. Results indicated that ASCs increase their expression of HGF by 2.5-folds in co-culture with scar VFF. In addition, when HGF was inhibited in ASCs scar VFF expression of collagen did not decrease and HA had minimal effects. Results from these studies indicate that ASCs can help modulate collagen production in scar VFF by increasing their production of HGF, thereby, suggesting the use of ASCs as a model of cell delivery to help stabilize the level of HGF in the vocal folds. This method may provide a greater treatment advantage than HGF alone.

19.3.2 Biomaterial Scaffolds

The application of cell-scaffold tissue engineering approaches could revolutionize the management of a variety of disorders and diseases encountered by specialists in the ear, nose, larynx and airway. Biomaterial scaffolds can be designed to model diverse mechanical and chemical microenvironments, advancing our understanding of the pathophysiology of the head and neck anatomy. The inclusion of ASCs into a biomaterial scaffold can support remodeling of the ECM and contribute to long-term viability. The implementation also implies a greater potential for clinical application, because ASCs can be easily obtained for autonomous transplant.

Scaffold biomaterials can be composed of natural or synthetic-derived ECMs with advantages dependent on the structure and function needed to replicate the tissue of interest. There have been several FDA approved natural ECMs used in bone grafts to provide structural support and biological signaling. Synthetic materials are also commonly used because they inhibit the host response and their components have greater uniformity.

Tissue-engineered approaches for the repair and remodel of the vocal folds need to account for the unique physiologic components of this tissue. Researchers suggest that fabricated scaffolds for the repair of the vocal fold lamina propria should have the following criteria: viscoelastic properties similar to vocal fold tissue – biomechanical stability accounting for appropriate vibrational stress [32]. In a study by Park et al., comparisons between five biomaterial scaffolds – collagen, HA, fibrin, cogel of fibrin–collagen, and cogel of fibrin–HA – were made for proliferation, differentiation, and ECM gene expression of ASCs [28]. Results concluded higher growth rates and elastin production in fibrin-based hydrogels compared to collagen and HA. Expression of CD44 and CD105 was the highest in HA gels, indicating inhibition for differentiation. Gels containing collagen had increased decorin expression. The data suggest that fibrin and collagen scaffolds can promote differentiation of ASCs to elongated fibroblast-like morphology.

Another novel strategy for treatment of severe vocal fold scarring involves excising scar layers of the epithelial and lamina propria, and then restoring the segment with a Fibrin-ASC construct [21]. This in vitro design is a multifaceted approach, differentiating epithelial cells with EGF and an air interface on the surface and mesenchymal lineages in the deeper portions of the fibrin gel. A follow-up study assessed the vibratory properties of this model using a forced transducer. Differences in collagen concentration between the layers, which were similar to human vocal fold lamina propria were reported [20]. The present studies are in there infancy; however, it suggests a novel method of fabricating multiple layers for potential organ transplant.

19.3.3 Organ Transplant

Complications can occur with the complete or partial replacement of an organ, resulting in tissue rejection. The use of ASCs has not been approved by the FDA for clinical applications in the USA; however, studies with human ASCs are currently being conducted with individual cases in various countries.

Airway disorders can occur from acquired and congenital anomalies due to the development of infection, cancer, autoimmune deficiencies or trauma. Invasive tracheal malignancies or stenosis can obstruct a patient's airway, requiring immediate surgical intervention. Previous studies using tissue-engineered models for trachea reconstruction have demonstrated problems associated with host rejection or slow epithelial proliferation. Omori et al. performed the first successful in situ trachea transplant in 2003 with a tissue-engineered scaffold composed of marlex mesh tube covered by a collagen sponge (Fig. 19.2). Slow rate of epithelial regeneration on the surface of the scaffold created an environment at risk for infection [27]. In an attempt to accelerate growth, researchers investigated the use of autologous ASCs embedded in scaffolds in trachea defected rats (Fig. 19.3) [36]. Increased proliferation and migration of subepithelial cells by Day 7 was reported compared to scaffolds without ASCs. Neovascularization was also observed around the implanted ASCs, indicating enhanced angiogenesis. The use of ASCs embedded in the scaffold increased the rate of epithelialization on the surface, improving the functionality of this method for clinical applications.

Cleft of the palate is a malformation to the soft and hard tissues that make up the maxilla and palate, affecting a patient's speech. Palatal bone reconstruction of the alveolar portion of the palate can improve normal function and prevent the development of further deficits to the nasopharynx and maxilla. The current gold standard for reconstruction of the maxilla uses autologous cancellous bone grafts to stabilize the defect, which require numerous procedures to correct the growing alveolar portion of the mandible. Advancements in tissue engineering scaffolds combined with ASCs can provide an alternative strategy for repair. Adipose tissue is commonly used to fill large deficits with buccal fat pad grafts, providing a vascular supply and filling space between the cavities. This demonstrates their potential use in bone allografts vascularization, preventing a host response. The reconstruction of the large bony deficit has several mechanical challenges due to high strain and stress of this area. Further in vitro studies are needed to find a strong enough biomaterial with capabilities of growth.

Large head and neck tumor resections can leave devastating cosmetic and functional impairments, altering a patient's airway and swallow. Extensive surgical reconstruction may be needed to restore deformities that occur in the face, nose, mouth and throat. Current microvascular techniques harvest tissues from other areas of the body to fill deficits; however, these methods are invasive and risk the transfer of infection, leading to rejection. ASCs embedded in a scaffold has been utilized for partial maxilla regeneration in a case study for a patient with a hemimaxillectomy from a large recurrent keratocyst [23]. The craniofacial bone defect

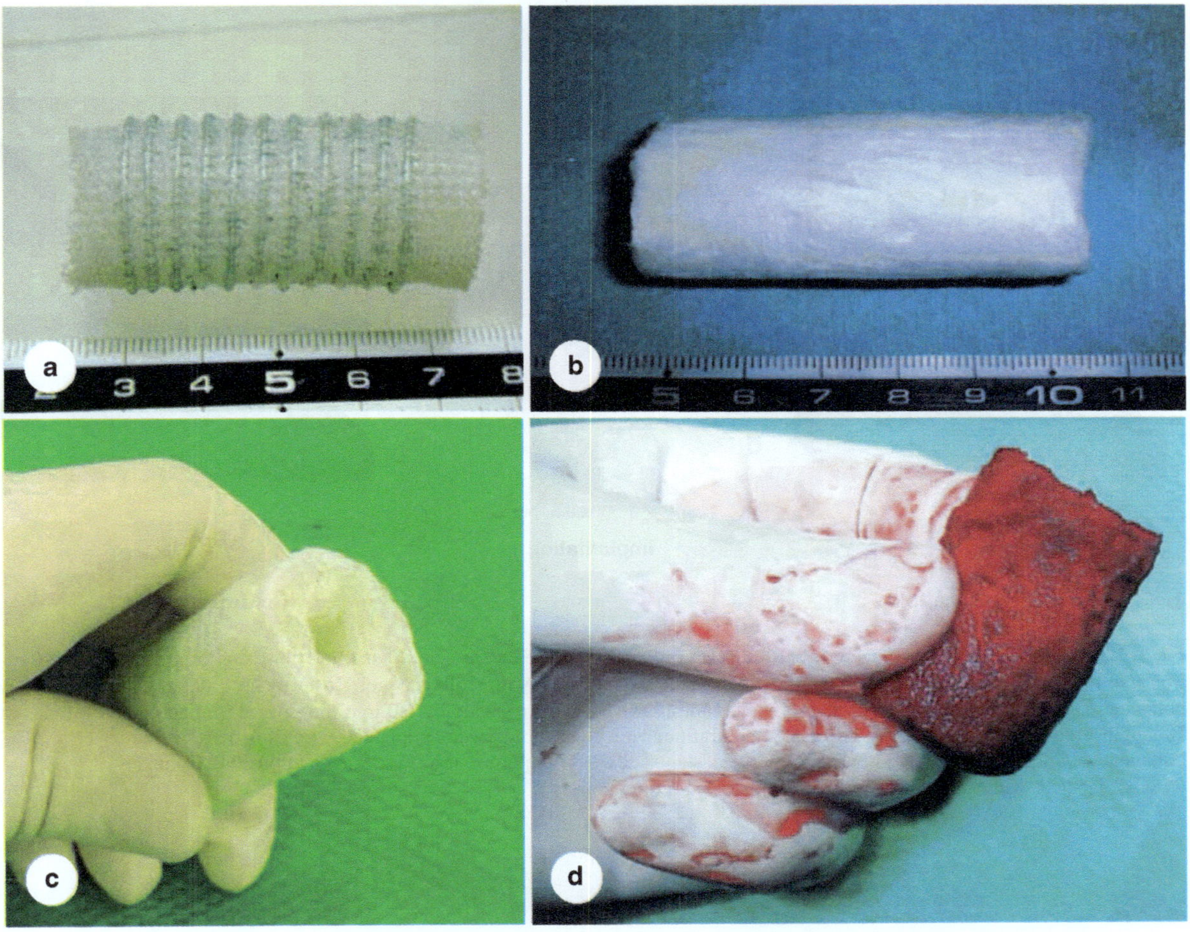

Fig. 19.2 Artificial material. (**a**) Marlex mesh tube with spiral ring. (**b**) Marlex mesh tube with spiral ring coated with collagen sponge. (**c**) Artificial material trimmed to size of defect. (**d**) Injection of autologous blood into collagen sponge. (Reprinted from Omori et al. [27] with permission from Annals Publishing Company)

was reconstructed without animal-derived reagents, using three surgical procedures (Fig. 19.4). The first surgery harvested autologous ASCs for in vitro expansion in basal media with 15% autologous serum. Cells were then cultured in rhBMP for osteogenesis and seeded in a betaTCP scaffold. This was then implanted in a preformed titanium cage with ASCs in betaTCP for in vivo culturing in the left rectus abdominis muscle to form the bony structure. Lastly, bone neotissue was implanted in the maxillary defect with anastomosis of the facial artery and vein. Radiological outcomes presented a mature maxillary bone after 4 months at which time temporary acrylic bridges were placed (Fig. 19.5). ASCs were able to function in low vascularized areas of the body, relying on endogenous signaling to form the ectopic bone in vivo.

19.3.4 Gene Therapy

Virally transduced genes can be used for long-term delivery to patients with a single genetic deficiency [34]. ASCs can provide an effective platform to deliver specific agents toward a target tumor. Properties that support the use of ASCs as delivery approaches for cancer therapy include the following: less invasive harvesting with high proliferation rate in vitro, high migration in vivo, low immunogenicity, no neurotoxicity or tumorigenicity and able to genetically modify the expression of proteins [10]. In head and neck cancer patients solid tumors are targeted, increasing the visibility of the tumor cells to the immune system in order to modify effector cells [34]. OncoVex is an oncolytic virus that generates granulocyte–macrophage colony-stimulating factor (GM-CSF)

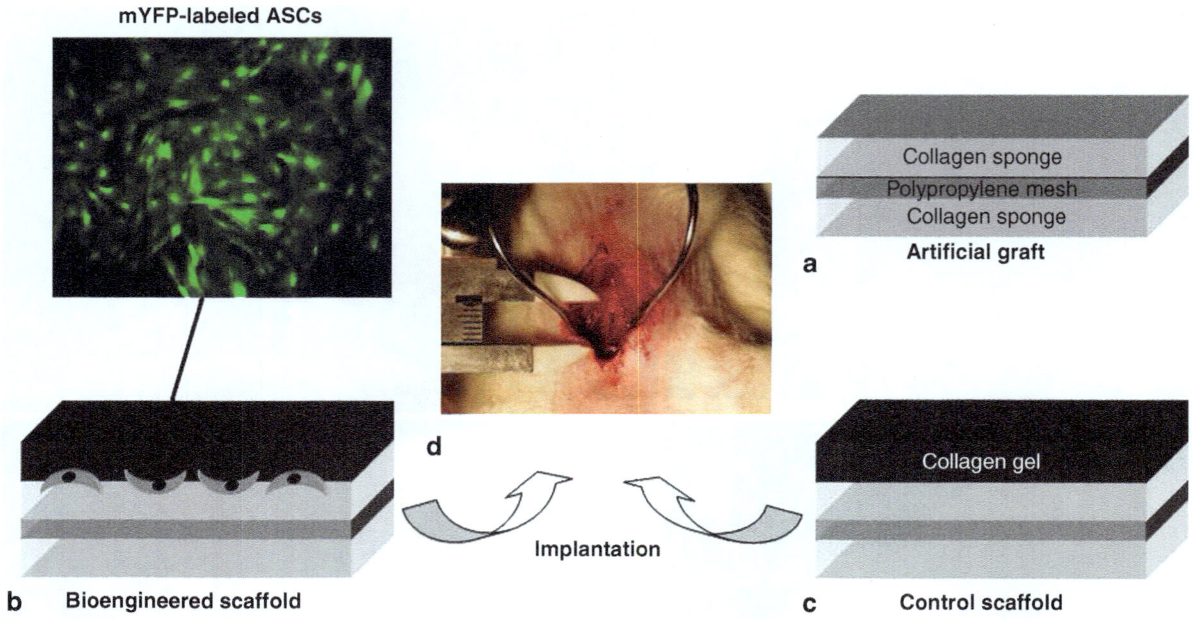

Fig. 19.3 Fabrication of bioengineered scaffold and implantation into rat trachea defects. (**a**) Artificial graft was made from collagen sponge scaffold with polypropylene mesh. (**b**) Bioengineered scaffold consisted of artificial graft stratified with collagen gel including autologous adipose-derived stem cells (ASCs) labeled with monomeric yellow fluorescent protein (mYFP). (**c**) Control scaffold consisted of artificial graft stratified with collagen gel. (**d**) Trachéal defects, approximately 3.0 mm wide by 6.0 mm long, were formed, and two types of scaffolds were implanted. (Reprinted from Suzuki et al. [36] with permission from Annals Publishing Company)

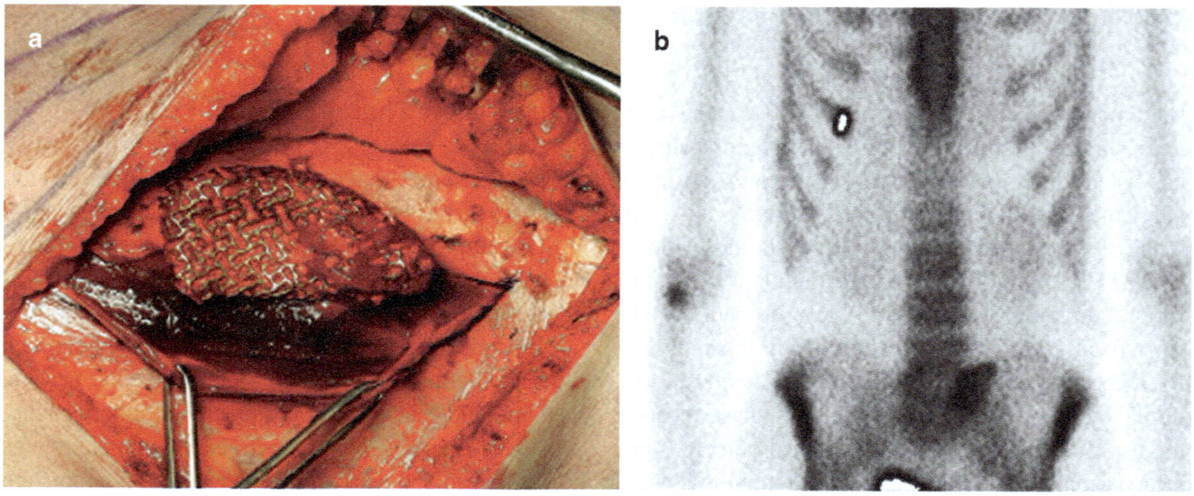

Fig. 19.4 Second and third procedures. (**a**) The titanium cage filled with beta-tricalcium phosphate and adipose stem cells, before insertion in the prepared rectus abdominis muscle pouch. (**b**) The bony regenerate in the rectus abdominis muscle. Skeletal scintigraphy was performed to confirm bone activity. (**c**) The rectus abdominis free-flap raised, muscle pouch and titanium cage opened. Note bleeding from the bone. The tissue-engineered bone was clinically confirmed to be rigid. (**d**) Histological section of the biopsy from the tissue-engineered bone showing normal mature bone structures. (Reprinted from Mesimäki et al. [23] with permission from John Wiley and Sons)

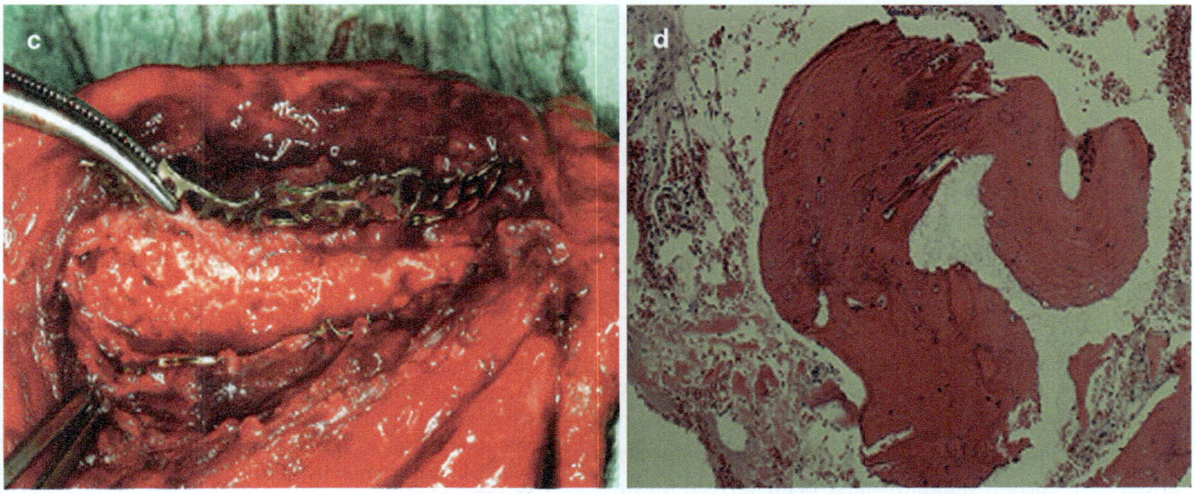

Fig. 19.4 (continued)

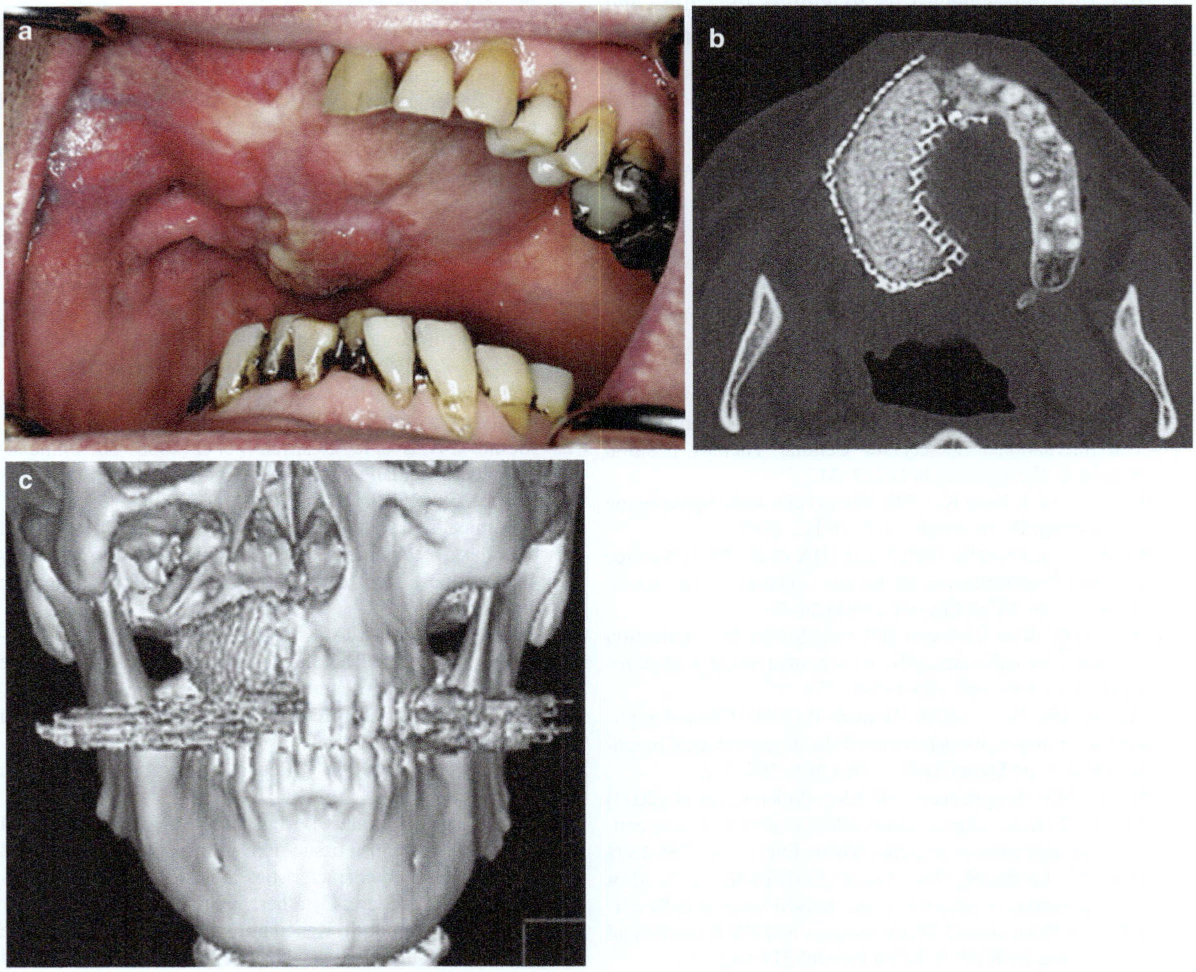

Fig. 19.5 Two months postoperatively. (**a**) The rectus abdominis muscle has atrophied nearly totally and epithelialized almost completely. Only a small area in the molar region was non-epithelialized. Note the well-formed buccal sulcus. Axial (**b**) and 3D CT scans (**c**) show the shape and normal bone density of the new maxilla. (Reprinted from Mesimäki et al. [23] with permission from John Wiley and Sons)

in tumor cells to kill and activate the immune response [1]. The current approach also includes chemoradiation and is in phase III clinical trials, studying patients with squamous cell carcinoma of the head and neck.

19.4 Conclusions

The advantages that adipose stem cell therapy provide to the treatment regimen of patients with head and neck disorders, does not come without controversy. Many of the current research breakthroughs introduced in this chapter have not been tested on humans and there is limited understanding about their long-term effects. No current standards have been derived for lineage differentiation and it is unknown if isolation or expansion procedures can present other medical issues. Further research is necessary to address the safety and effectiveness of using adipose-derived stem cells before the use of these unique cells can become a standard of care in clinical practice.

References

1. BioVex (2010) Available on Website: http://www.biovex.com/oncovex.html
2. Dicker A, Le Blanc K, Aström G et al (2005) Functional studies of mesenchymal stem cells derived from adult human adipose tissue. Exp Cell Res 308(2):283–290
3. Dominici M, Le Blanc K, Mueller I et al (2006) Minimal criteria for defining multipotent mesenchymal stromal cells. The International Society for Cellular Therapy position statement. Cytotherapy 8(4):315–317
4. Gomillion CT, Burg KJ (2006) Stem cells and adipose tissue engineering. Biomaterials 27(36):6052–6063
5. Gronthos S, Franklin DM, Leddy HA et al (2001) Surface protein characterization of human adipose tissue-derived stromal cells. J Cell Physiol 189(1):54–63
6. Hanson SE, Kim J, Johnson BH et al (2010) Characterization of mesenchymal stem cells from human vocal fold fibroblasts. Laryngoscope 120(3):546–551
7. Hanson SE, Thibeault SL, Hematti P (2010) Clinical applications of mesenchymal stem cells in laryngotracheal reconstruction. Curr Stem Cell Res Ther 5(3):268–272
8. Helder MN, Knippenberg M, Klein-Nulend J et al (2007) Stem cells from adipose tissue allow challenging new concepts for regenerative medicine. Tissue Eng 13(8):1799–1808
9. Hennig T, Lorenz H, Thiel A et al (2007) Reduced chondrogenic potential of adipose tissue derived stromal cells correlates with an altered TGFb receptor and BMP profile and is overcome by BMP-6. J Cell Physiol 211:682–691
10. Hu YL, Fu YH, Tabata Y et al (2010) Mesenchymal stem cells: a promising targeted-delivery vehicle in cancer gene therapy. J Control Release 147:154–162
11. Jiang T, Liu W, Lv X, Sun H et al (2010) Potent in vitro chondrogenesis of CD105 enriched human adipose-derived stem cells. Biomaterials 31(13):3564–3571
12. Jurgens WJ, Oedayrajsingh-Varma MJ, Helder MN (2008) Effect of tissue-harvesting site on yield of stem cells derived from adipose tissue: implications for cell-based therapies. Cell Tissue Res 332(3):415–426
13. Katz AJ, Tholpady A, Tholpady SS et al (2005) Cell surface and transcriptional characterization of human adipose-derived adherent stromal (hADAS) cells. Stem Cells 23:412–423
14. Kishimoto Y, Hirano S, Suehiro A et al (2009) Effect of exogenous hepatocyte growth factor on vocal fold fibroblasts. Ann Otol Rhinol Laryngol 118(8):606–611
15. Kocaoemer A, Kern S, Klüter H et al (2007) Human AB serum and thrombin-activated platelet-rich plasma are suitable alternatives to fetal calf serum for the expansion of mesenchymal stem cells from adipose tissue. Stem Cells 25:1270–1278
16. Kumai Y, Kobler JB, Park H et al (2009) Crosstalk between adipose-derived stem/stromal cells and vocal fold fibroblasts in vitro. Laryngoscope 119(4):799–805
17. Kumai Y, Kobler JB, Park H et al (2010) Modulation of vocal fold scar fibroblasts by adipose-derived stem/stromal cells. Laryngoscope 120(2):330–337
18. Levi B, James AW, Wan DC et al (2010) Regulation of human adipose-derived stromal cell osteogenic differentiation by insulin-like growth factor-1 and platelet-derived growth factor-alpha. Plast Reconstr 26(1):41–52
19. Liu G, Zhou H, Li Y et al (2008) Evaluation of the viability and osteogenic differentiation of cryopreserved human adipose-derived stem cells. Cryobiology 57(1):18–24
20. Long JL, Neubauer J, Zhang Z et al (2010) Functional testing of a tissue-engineered vocal fold cover replacement. Otolaryngol Head Neck Surg 142:438–440
21. Long JL, Zuk P, Berke GS et al (2010) Epithelial differentiation of adipose-derived stem cells for laryngeal tissue engineering. Laryngoscope 120:125–131
22. McIntosh K, Zvonic S, Garrett S et al (2006) The immunogenicity of human adipose-derived cells: temporal changes in vitro. Stem Cells 24(5):1246–1253
23. Mesimäki K, Lindroos B, Törnwall J et al (2009) Novel maxillary reconstruction with ectopic bone formation by GMP adipose stem cells. Int J Oral Maxillofac Surg 38:201–209
24. Mikus JL, Koufman JA, Kilpatrik SE (1995) Fate of liposuctioned and purified autologous fat injection in the canine vocal fold. Laryngoscope 105:17–27
25. Moseley TA, Zhu M, Hedrick MH (2006) Adipose-derived stem and progenitor cells as fillers in plastic and reconstructive surgery. Plast Reconstr Surg 118(3 Suppl):121S–128S
26. Oedayrajsingh-Varma MJ, van Ham SM, Knippenberg M et al (2006) Adipose tissue-derived mesenchymal stem cell yield and growth characteristics are affected by the tissue-harvesting procedure. Cytotherapy 8(2):166–177
27. Omori K, Nakamura T, Kanemaru S et al (2005) Regenerative medicine of the trachea: the first human case. Ann Otol Rhinol Laryngol 114(6):429–433
28. Park H, Karajanagi S, Wolak K et al (2010) Three-dimensional hydrogel model using adipose-derived stem

cells for vocal fold augmentation. Tissue Eng A 16(2): 535–543
29. Peroni D, Scambi I, Pasini A et al (2008) Stem molecular signature of adipose-derived stromal cells. Exp Cell Res 314(3):603–615
30. Puissant B, Barreau C, Bourin P et al (2005) Immunomodulatory effect of human adipose tissue-derived adult stem cells: comparison with bone marrow mesenchymal stem cells. Br J Hematol 129:118–129
31. Regan MC, Kirk SJ, Wasserkrug HL et al (1991) The wound environment as a regulator of fibroblast phenotype. J Surg Res 50(5):442–448
32. RoyChowdhury P, Klemuk S, Titze I et al (2009) Effects of fabrication parameters on viscoelastic shear modulus of 2,3-dialdehydecellulose membranes—potential scaffolds for vocal fold lamina propria tissue engineering. J Biomed Mater Res A 88A:680–688
33. Rubio D, Garcia-Castro J, Martin MC et al (2005) Spontaneous human adult stem cell transformation. Cancer Res 65:3035–3039
34. Scheller EL, Krebsbach PH (2009) Gene therapy: design and prospects for craniofacial regeneration. J Dent Res 88(7):585–596
35. Schreml S, Babilas P, Fruth S et al (2009) Harvesting human adipose tissue-derived adult stem cells: resection versus liposuction. Cytotherapy 11(7):947–957
36. Suzuki T, Kobayashi K, Tada Y et al (2008) Regeneration of the trachea using a bioengineered scaffold with adipose-derived stem cells. Ann Otol Rhinol Laryngol 117(6):453–463
37. Wagner W, Wein F, Seckinger A et al (2003) Comparative characteristics of mesenchymal stem cells from human bone marrow, adipose tissue, and umbilical cord blood. Exp Hematol 33(11):1402–1416
38. Yañez R, Lamana ML, García-Castro J et al (2006) Adipose tissue-derived mesenchymal stem cells have in vivo immunosuppressive properties applicable for the control of the graft-versus-host disease. Stem Cells 24:2582–2591
39. Zuk PA, Zhu M, Ashjian P et al (2002) Human adipose tissue is a source of multipotent stem cells. Mol Biol Cell 13:4279–4295

Adipose Stem Cell Technologies for Tissue Regeneration in Dentistry

20

Andrea Cochis and Lia Rimondini

Contents

20.1 **Introduction** 221

20.2 **Stem Cell from Different Sources** 222
20.2.1 Stem Cells from Adipose Tissue 222
20.2.2 Stem Cells from Dental Pulp, Human Exfoliated Deciduous Teeth 223

20.3 **Stem Cells from Periodontal Ligament and Precursors Cells (PCs) from Human Dental Follicle** 225

20.4 **Regeneration Strategies in Dentistry: Tooth and Tissue Regeneration** 226

20.5 **Other Potential Applications** 227

20.6 **Conclusions** 227

References 227

A. Cochis (✉) • L. Rimondini
Department of Medical Sciences, Università del Piemonte Orientale "Amedeo Avogadro", Via Solaroli 17, 20068 Novara, Italy
e-mail: andrea.cochis@med.unipmn.it; lia.rimondini@med.unipmn.it

20.1 Introduction

Although human body has a remarkable capacity of regeneration, some tissues have less capability to heal or to face the needs of regeneration. By definition, a stem cell is characterized by the ability of self-renewing and the capability to differentiate along multiple lineage pathways.

Embryonic stem cells are responsible for tissue development during histogenesis and they are theoretically able to differentiate into any type of tissue. However, some limitations are still on their use in practice including cell regulations and ethical considerations. In contrast, postnatal adult stem cells are immunocompatible and no ethical issues are related to their use. Cells obtained from bone marrow, adipose tissue, myocardial tissue, dermis, fibroblasts, periodontal ligament (PDL), and pulp are the representative of this type of adult stem cells. They generally show adipogenic, osteogenic, chondrogenic, myogenic and neurogenic potential differentiation in vitro when stimulate by suitable differentiation factors.

Based on these considerations, many technologies, including applications of stem cells alone or in combination with biomaterials, have been developed for regenerative medicine.

Cell therapies may be theoretically applied in all fields of regenerative medicine, including dentistry. In fact, caries, trauma, erosion and periodontal disease are pathologies characterized by the damage and loss of dental tissues and sometimes loss of the whole tooth. These groups of diseases are very common and affect millions of people worldwide in both developing and industrialized countries, representing the 10% of total public healthcare budget.

Y.-G. Illouz and A. Sterodimas (eds.), *Adipose Stem Cells and Regenerative Medicine*,
DOI: 10.1007/978-3-642-20012-0_20, © Springer-Verlag Berlin Heidelberg 2011

Restorations of damaged tooth tissues and substitution of missing teeth with artificial prostheses represent the traditional therapeutic solutions. Interest in dental-tissue regeneration applications continues to increase as clinically relevant methods alternative to traditional treatments with a potential high impact on socio-economical environment.

In this chapter the potential role of adipose tissue, dental pulp, periodontal ligament, and deciduous teeth as source for adult stem cells isolation will be discussed. Mesenchymal stem cells (MSCs) so obtained may be applied alone or in combination with suitable scaffolds to obtain regeneration of oral tissue.

20.2 Stem Cell from Different Sources

20.2.1 Stem Cells from Adipose Tissue

Tissue engineering is now contributing to new developments in several clinic fields and MSCs derived from tissues such as adipose tissue, dental pulp, or PDL may provide a novel opportunity to replace, repair and promote the regeneration of diseased or damaged tissues.

Several studies have shown that adipose-derived stem cells (hASCs) are able to differentiate in vitro into bone, cartilage, fat, muscle, endothelial, and neuron-like cells [4, 11, 14, 15, 26, 41]. Moreover, hASCs can be easily isolated from lipoaspirated tissue or with minimally invasive procedure of liposuction; with a few millimeters of raw tissue it is possible to isolate a significant number of hASCs [10]. Furthermore, some data from different laboratories confirm that independently from the age of donor, hASCs proliferate quite constantly over an extended period of time while showing clonogenic properties [3, 9, 19].

An interesting field of research is in hASCs phenotype profile characterization. Backsh has defined a phenotype characterized by a CD45-, CD14-, and CD34-profile [1], while Sengenes and De Girolamo have shown a very interesting population of hACSs expressing CD34 [3, 31]. Actually, further studies on the characterization of CD34+ hASCs sub-populations are in progress to evaluate if these cells are directed toward a prominent osteospecificity in vitro, likely permitting more rapid and efficient future applications [38]. hASCs have shown the ability to differentiate toward chondrocytes and adipocyte-like cells; in 2007, De Girolamo et al. [3] have demonstrate that the enhancement of sulfate glycosaminoglycans deposition was caused only by the chondrogenic differentiation process of differentiate samples of hASCs. These data suggest the potential involvement of hASCs as adjunctive tool in maxillofacial surgery for the treatment of bone defects and regeneration of cartilage in temporomandibular joint disease.

Another important consideration is that hASCs are also involved in the revascularization of new formed tissue. Rigotti et al. [30] have damaged a tissue with radiotherapy. Afterward, the same tissue has been treated with hASCs derived from lipoaspirate transplant and they have obtained signs of neovascularization of the target tissue which initially exhibited microvascular alterations similar to several chronic ischemic diseases.

Table 20.1 reports some of differentiation factors useful to address osteo-, condro-, and adipogenic phenotypes. In vitro differentiation is checked by staining (Fig. 20.1). In contrast, in vivo differentiation is induced by signals from the surrounding environment and, in this view, the scaffolds used to delivery cells acquire more importance in respect with the supplementation of differentiation factors.

Table 20.1 Media, serum supplementation and differentiating factors to obtain adipogenic, osteogenic and condrogenic differentiation of human adipose-derived stem cells

Adipogenic
Induction DMEM + 10% FBS 1 μM dexamethasone, 10 μg/mL insulin, 500 μM IBMX (3-isobutyl-1-methyl-xanthine), 200 μM indomethacin
Maintenance DMEM 10% FBS 10 μg/mL insulin
Osteogenic
DMEM + 10% FBS, 10 nM dexamethasone, 10 mM glicerol-2-phosphate, 150 μM L-ascorbic acid-2-phosphate, 10 nM cholecalciferol
Chondrogenic
DMEM + 1% FBS 100 nM dexamethasone, 110 mg/L sodium pyruvate, 150 μM L-ascorbic acid-2-phosphate, 1X ITS, 10 ng/mL TGF-β11

Source: Adapted from De Girolamo et al. [3]

The use of scaffolds represents an important clinical strategy while some data have reported a spontaneous differentiation into osteoblast-like cells of hACSs seeded in hydroxyapatite or titanium scaffolds (Fig. 20.2). These cells have shown a calcified extracellular matrix deposition after 3 weeks of cultures

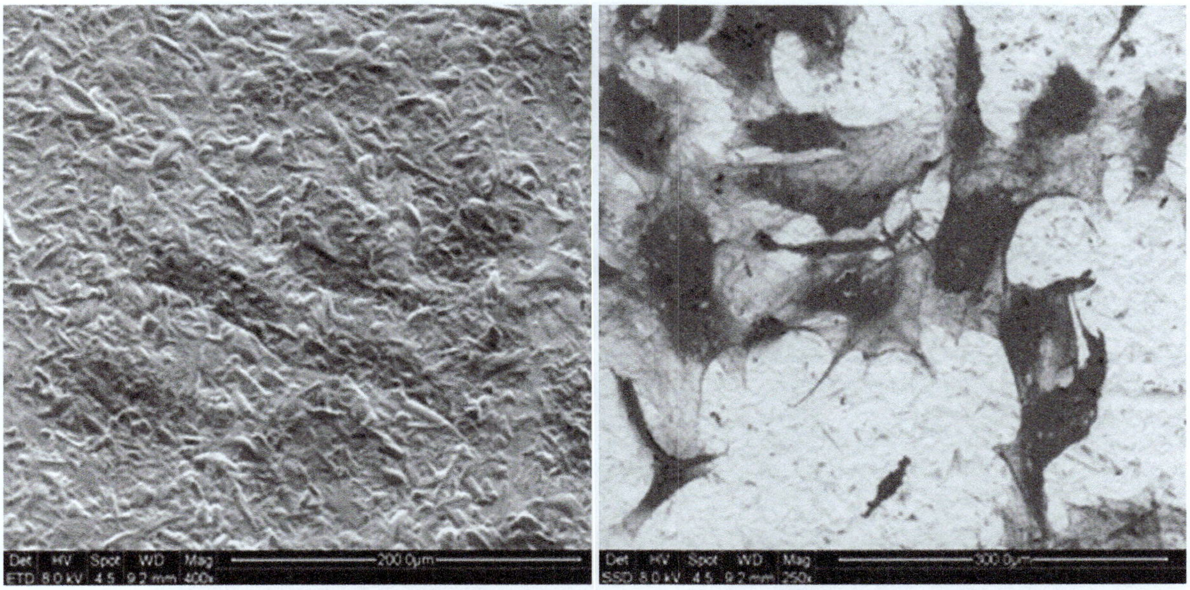

Fig. 20.1 Common staining to check differentiation of stem cells

compared to the same cells cultured in polystyrene. The effect is emphasized in case of supplementation with differentiating factors (Fig. 20.3). Thus, loading undifferentiated hASCs directly on scaffolds could shorten the time between the adipose tissue removal and the use of differentiated cells, allowing faster clinical applications. However, it remains to investigate if the employment of metabolic activity can be related with the chemical features of the scaffolds or more to the 3D structure which may allow the hASCs to activate specific intracellular pathways that cause extracelluar matrix deposition. That is why in the future it will be important to test different materials suitable for scaffolds and understand the behavior of hASCs and eventually study interactions with additional osteogenic stimulants or other growth factors.

20.2.2 Stem Cells from Dental Pulp, Human Exfoliated Deciduous Teeth

Several studies have demonstrated the presence of progenitor cells in dental pulp, defined as dental pulp stem cells (DPSCs), and their ability to activate upon injury undergoing proliferation and odontoblast differentiation to provide for dental repair [13, 33, 35]. A hierarchy of progenitor cells have been postulated in dental pulp with diverse differentiations potential; DPSCs niches are localized specially in the perivascular area of pulp from where cells can migrate to the injury site [33, 35]. Dental pulp stem cells exhibited a high proliferative rate in vitro, but actually some studies have shown also a good proliferative and regenerative activity in vivo of DPSCs when used alone or combined with medical scaffolds or proper biochemical stimuli [40]. About this theme, an important consideration is that adult DPSCs may express embryonic stem cell specific antigens: SSEA-4 and OCT-4 expression has been detected in dental pulp stem cell. Gronthos et al. have shown that STRO-1 + cells extracted from dental pulp of adult rat can differentiate toward adipogenic, neurogenic, myogenic and chondrogenic lineages in vitro. In vivo, Gronthos has demonstrated that DPSCs transplanted in conjunction with HA/TPC powder into immunocompromised mice can generate a pulp/dentine-like collagenous structure [13]. El-Backly et al. [8] have observed the presence of an osteo-dentin-like structure after 2 weeks from the subcutaneous implantation of DPSCs grafted with poly (lactic-co-glycolic acid) polymeric porous scaffolds in rabbits.

Another interesting class of dental stem cells is represented by stem cells isolated from dental pulp of human exfoliated deciduous teeth, also known as SHED. Cordeiro et al. seeded SHED onto dentine slices and implanted them subcutaneously in immunodeficient mice. After 14–28 days they observed a new

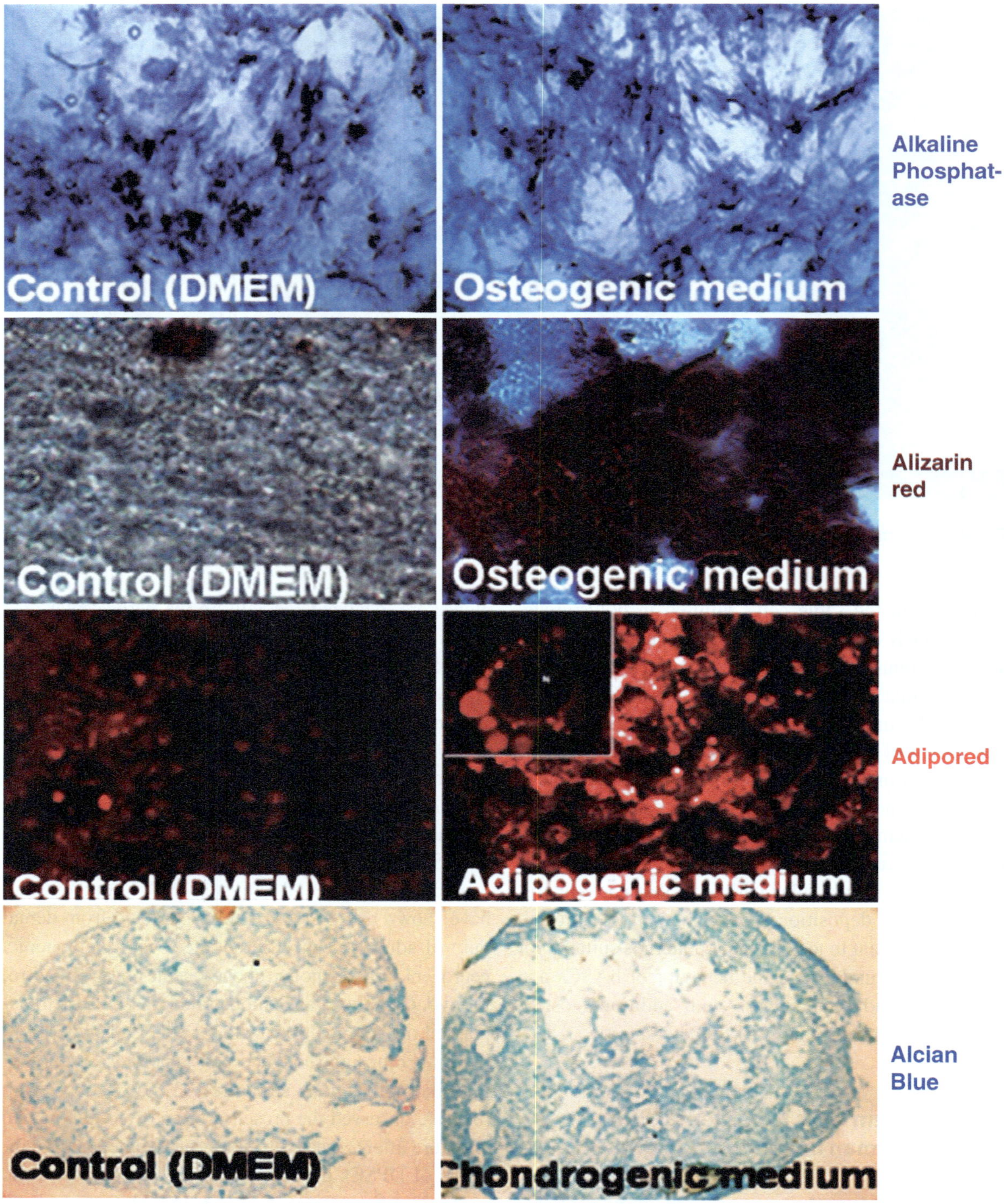

Fig. 20.2 Scanning electron microscopic graphs of adipose-derived stem cells seeded onto sandblasted and acid-etched titanium surfaces in differentiation toward osteogenic lineages. The cells appear spreaded and with polygonal features. These aspects may be observed with both secondary (*left*) and backscattered (*right*) electrons detection

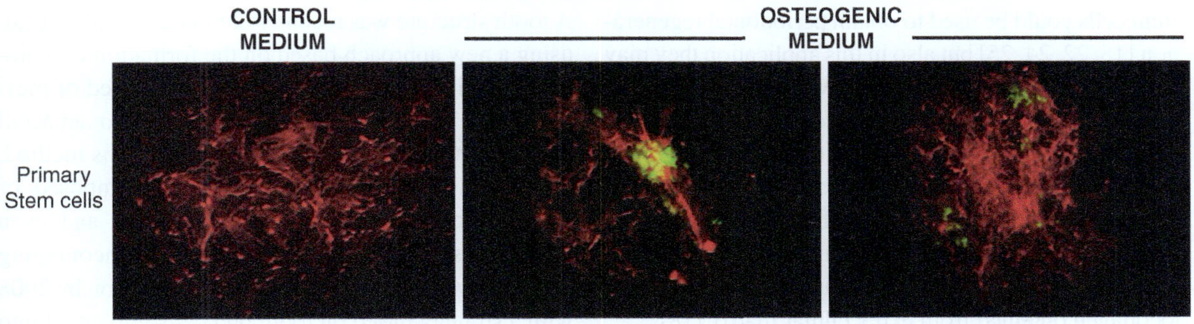

Fig. 20.3 Adipose-derived stem cells seeded onto titanium scaffolds and cultivated with basal (control) and osteogenic media. After 3 weeks in cultures, mineralized nodules stained by calcein (green fluorescence) were detected

dental pulp-like tissue with great similarity with physiological pulp. Moreover, SHED seeded on tooth scaffolds were capable of differentiating into odontoblast-like cells and showed characteristic of dentin-secreting odontoblast [2]. SHED have shown the ability to improve the new formed tissue organization, microvascular network and oxygen and nutrient influx. Recent evidences suggest that DPSCs cultures also contain multipotent neural crest stem cells demonstrated by their ability to differentiate into a number of neural crest-derived cell lineages including melanocytes [34]. Moreover, a sub-population of DPSCs which expressed Nanog, Oct-4 and Rex-1 was grafted into hippocampus of immune suppressed mice: stem cells from dental pulp were able to stimulate proliferation of endogenous neural cells and resulted in the recruitment of pre-existing neural progenitor cells and mature neurons to the site of the graft [16].

20.3 Stem Cells from Periodontal Ligament and Precursors Cells (PCs) from Human Dental Follicle

Periodontal ligament (PDL) is a connective tissue that connects the root cementum to alveolar bone. Its main function is to maintain the root into the alveolar socket under function.

Even PDL contains progenitor cells that can migrate and differentiate into cementoblasts and osteoblasts in response to lesions [12]. This had lead to suppose that stem cell population could exist in PDL. A first demonstration of this fact was provided by the identification of expression of stemness marker such as STRO-1 and CD146. Furthermore, these cells harvested from PDL had the potential to differentiate into oil red O-positive lipid laden fat cells and to form alizarin red positive nodules and an array of cementoblastic/osteoblastic markers (such as alkaline phosphatase and bone sialoprotein) if influenced with osteogenic medium [32].

Recently, these cells are indicated as periodontal ligament stem cells (PDLSCs). The use of PDLSCs seems to be very interesting both if used as single stem cells or specially if combined with carrier. PDLSCs have shown the capacity to differentiate into cementoblasts if transplanted subcutaneously into the dorsal surfaces of immunocompromised mice when hydroxyapatite tricalcium phosphate (HA/TCP) was used as a carrier [12]. Seo et al. [32] have shown the similarity between collagen fibers connecting to the cementum-like tissues and Sharpey's fibers, suggesting the potential to regenerate PDL attachment.

More recently, Li et al. performed a similar experiment in vivo: they have implanted autologous PDLSCs using HA/TCP as carrier in miniature pigs with created periodontal defects obtaining bone, cementum and PDLs [21].

Also precursor cells (PCs) from human dental follicles of wisdom teeth could represent an important source for periodontal regeneration. In fact, these cells are able to reproduce in vitro a membranous structure that could be compared with PDL; however, these encouraging data from in vitro experiments are not supported by in vivo results. Morsczeck et al. [27] have tried to implant PCs into immunocompromised mice using hydroxyapatite as scaffold but no sign of cementum or bone was found in histological sections of transplants.

Recently, some authors such as Mizuno and Kawaguchi have shown that not only dental-derived stem cells could be used to obtain periodontal regeneration [18, 22, 24, 25] but also in this application they may represent an advantage in respect with ASCs, even if in a recent study has been demonstrated that isolated ASCs together with platelet-rich plasma obtained from inbred rats were able to regenerate alveolar bone, cementum and a PDL-like tissue in rats with created periodontal tissue defect after 8 weeks [36] similar to what obtained with MSCs derived from bone marrow mixed with atelocollagen obtained from extracellular matrix [18].

20.4 Regeneration Strategies in Dentistry: Tooth and Tissue Regeneration

Humans are genetically programmed to replace their teeth once during childhood. The transition from deciduous teeth to adult permanent teeth is a dynamic process that combines the resorption of the roots of deciduous teeth and the eruption of the permanent teeth. When adult teeth are lost there are no ways to generate new teeth or to repair them. A clinical approach able to regenerate teeth could represent a new clinical way. However, even if it is possible to think about strategies based on direct transplantation or on biomaterials acting as template into the defected tissue in order to obtain regeneration, teeth structure complexity represent a strong limit for regenerative therapy. In fact, teeth are composed of both hard tissue (dentin, enamel, cementum, alveolar bone) and soft tissue (dental pulp, PDL, gingival) and so tooth development is regulated by the interactions between dental epithelium and dental mesenchyme. For this reason, it is necessary to consider two families of stem cells: epithelial (EpSC) and mesenchymal stem cells (MSCs). Therefore, a stem cell-based strategy for regeneration needs to be considered as a coupled strategy which involves both EpSCs and MSCs. In 2004, Ozahama et al. have combined in vitro a "novel" mesenchyme with an embryonic oral epithelium. This artificial and composite tissue has been implanted into adult renal capsules the formation of tooth and bone structures was observed as the results [29]. Again in 2004 Duailibi et al. have coupled cells isolated from rat tooth buds with scaffolds composed of polyglycolide/poly-L-lactide (PGA/PLLA): they have obtained the formation of an ectopic bioengineered tooth with pulp, dentine, and enamel tissues even with anomalous arrangement [6]. A tooth structure was recently obtained by Nakao et al. using a new approach based on the formation of a bioengineered tooth in mice mandible composed of mesenchymal and epithelial cells seeded into adjacent regions within a collagen gel drop. With this method, they have obtained a tooth structure composed of enamel, dentine, root, dental pulp, bone, and even blood vessels and nerve fibers [28]. Encouraging results have been obtained also by Duailibi in 2008 with a strategy based on tooth bud cells implanted into the jaw of adult rats for 12 weeks [7]. In a recent study, Yamada et al. [37] have underlined the potential role of bone marrow as a source of stem cells able to differentiate in ameloblast-like cells. However, it is actually still considered DPSCs as the cells with the major ability to regenerate a complete tooth structure.

The regeneration of PDL is an ambitious aim in periodontology and implantology because these cells are capable to differentiate into periodontal structure and promote the formation of both cementum-like mineralized tissue and bone. Actually, it is possible to confirm that MSCs are present in the PDLs and probably represent the most important source for regenerative therapy. However, it is important to underline that most of the studies have suggested the use of a mix of stem cells. Some cells may act as a motor which is able to guide stem cells toward a precise differentiation, whereas other cells represent the fuel for generation of new tissues. A first evidence of this idea has been proved by Inanç et al. [17] who co-cultured embryonic stem cells with human PDL fibroblastic cells for 21 days and after this period stem cells were able to differentiate toward the PDL fibroblastic progenitors. Yang et al. have performed two experiments, the first one in vitro and the second in vivo, with monodispersed human PDLSCs. Firstly, in vitro they have conditioned cells with ascorbic acid and selective medium in the presence of extracellular matrix. In this way they obtained cells with several phenotypic characteristics of cementoblast lineage such as upregulated activity of alkaline phosphatase, accelerated mineralization and the expression of bone sialoprotein and osteocalcin genes. Secondly, they transplanted these PDLSCs pellets into immunocompromised mice obtaining a regular aligned cementum/PDL-like complex [39].

Dentine seems to have a very important role on MSC differentiation toward periodontal tissues. Li et al. [20] have shown that PDL cells together with dentine implanted in nude mice are able to form PDL and root cementum with physiological architecture. These data prove that in the future the whole tooth could be engineered but, at the moment, it seems more realistic to consider only the regeneration of a single component of the tooth.

Cordeiro et al. [2] have demonstrated that both DPSCs and SHED are able to generate pulp and dentine under the induction by proper biochemical stimuli. Particularly, it is very interesting that SHED grafted onto dentine slices and implanted in immunocompromised mice are able to form a tissue with features very similar to those observed in physiological pulp. That is the proof that the morphogenetic signaling has been provided directly and only by the signals of the native dentine matrix confirming the signaling role of dentine previously described. Another important feature of DPSCs is that they are able to induce the vascularization of the new formed tissues. As previously discussed, El Backly et al. [8] have obtained an osteodentine new formed tissue after subcutaneous implantation of DPSCs grafted with poly polymeric porous scaffolds in rabbit that was characterized by typical tubular features.

20.5 Other Potential Applications

Recent data have demonstrated that MSCs have potent immune regulation capacity in vitro, enhancing their therapeutic appeal in the management of many immunological diseases such as acute graft-versus-host disease [20]. Consequently their use is also suggested for the treatment of some immunological disease in oral cavity such as oral lichen planus [5]. This effect observed also in other applications such as the treatment of rheumatoid arthritis [23] and opens new perspectives of application of stem cells in oral medicine.

20.6 Conclusions

The therapeutic potential of adult stem cells for regenerative purposes in dentistry is well accepted but further studies are necessary. A better understanding of molecular events that control tissue development, repair and regeneration is necessary. Interdisciplinary approaches combining cell biology and biomaterials are needed in order to develop dental engineered tissues which represent a great opportunity for future dental treatment strategies.

References

1. Backsh D, Song L, Tuan RS (2004) Adult mesenchymal stem cells: characterization, differentiation and applications in cell and gene therapy. J Cell Mol Med 8:301–316
2. Cordeiro MM, Dong Z, Kaneko T et al (2008) Dental pulp tissue engineering with stem cells from exfoliated deciduous teeth. J Endod 34:962–969
3. De Girolamo L, Sartori MF, Albisetti W et al (2007) Osteogenic differentiation of human adipose-derived stem cells: comparison of two different inductive media. J Tissue Eng Regen Med 1:154–157
4. De Ugarte DA, Morizono K, Elbarbary A et al (2003) Comparison of multi-lineage cells from human adipose tissue and bone marrow. Cells Tissues Organs 174:101–109
5. Ding G, Wang W, Liu Y, Zhang C, Wang S (2011) Mesenchymal stem cell transplantation: a potential therapy for oral lichen planus. Med Hypotheses 76(3):322–324
6. Duailibi MT, Duailibi SE, Young CS et al (2004) Bioengineered teeth from cultured rat tooth bud cells. J Dent Res 83:523–528
7. Dualibi SE, Dualibi MT, Zhang W et al (2008) Bioengineered dental tissues grown in the rat jaw. J Dent Res 87:745–750
8. El-Backly RM, Massoud AG, El-Badry AM et al (2008) Regeneration of dentine/pulp-like tissue using a dental pulp stem cell/poly(lactic-co-glycolic)acid scaffold construct in New Zealand white rabbits. Aust Endod J 34:52–67
9. Fraser JK, Wulur I, Alfonso Z et al (2006) Fat tissue: an underappreciated source of stem cell for biotechnology. Trends Biotechnol 24:150–154
10. Gimble JM, Guilak F (2003) Adipose-derived adult stem cells: isolation, characterization and differentiation potential. Cytotherapy 5:362–369
11. Gimble JM, Katz AJ, Bunnel BA (2007) Adipose-derived stem cells for regenerative medicine. Circ Res 100:1249–1260
12. Gould TR, Melcher AH, Brunette DM et al (1980) Migration and division of progenitor cell populations in periodontal ligament after wounding. J Periodontal Res 15:20–42
13. Gronthos S, Mankani M, Brahim J et al (2000) Postnatal human dental pulp stem cells (DPSCs) in vitro and in vivo. Proc Natl Acad Sci USA 97:13625–13630
14. Guilak F, Lott KE, Awad HA et al (2006) Clonal analysis of the differentiation potential of human adipose-derived adult stem cells. J Cell Physiol 206:229–237
15. Helder MN, Knippenberg M, Klein-Nulend J et al (2007) Stem cells from adipose tissue allow challenging new concepts for regenerative medicine. Tissue Eng 13:1799–1808
16. Huang AH, Snyder BR, Cheng PH et al (2008) Putative dental pulp-derived stem/stromal cells promote proliferation and differentiation of endogenous neural cells in the hippocampus of mice. Stem Cells 26:2654–2663
17. Inanç B, Elçin AE, Unsal E et al (2008) Differentiation of human embryonic stem cells on periodontal ligament fibroblasts in vitro. Artif Organs 22:100–109

18. Kawaguchi H, Hayashi H, Mizuno N et al (2005) Cell transplantation for periodontal diseases. A novel periodontal tissue regenerative therapy using bone marrow mesenchymal stem cells. Clin Calcium 15:99–104
19. Lee RH, Kim B, Choi I et al (2004) Characterization and expression analysis of mesenchymal stem cells from human bone marrow and adipose tissue. Cell Physiol Biochem 14:311–324
20. Li H, Guo Z, Zhu H, Li XS, Jiang X, Yao H, Wang X, Liao L, Liu Y, Wu Y, Zhang Y, Mao N (2010) Transplanted mesenchymal stem cells can inhibit the three developmental stages of murine acute graft-versus-host disease. In Vivo 24(5):659–666
21. Liu Y, Zheng Y, Ding G et al (2008) Periodontal ligament stem cells-mediated treatment for periodontitis in miniature swine. Stem Cells 26:1065–1073
22. Li Y, Zheng W, Liu JS et al (2011) Expression of osteoclastogenesis inducers in a tissue model of periodontal ligament under compression. J Dent Res 90(1):115–120
23. Liu J, Mao J, Chen L (2011) Epithelial-mesenchymal interactions as a working concept for oral mucosa regeneration. Tissue Eng B Rev 17(1):25–31
24. Liu Y, Mu R, Wang S, Long L, Liu X, Li R, Sun J, Guo J, Zhang X, Guo J, Yu P, Li C, Liu X, Huang Z, Wang D, Li H, Gu Z, Liu B, Li Z (2011) Therapeutic potential of human umbilical cord mesenchymal stem cells in the treatment of rheumatoid arthritis. Arthritis Res Ther 12(6):R210
25. Mizuno H (2010) Adipose-derived stem and stromal cells for cell-based therapy: current status of preclinical studies and clinical trials. Curr Opin Mol Ther 12:442–449
26. Mizuno H, Zuk PA, Zhu MH et al (2002) Myogenic differentiation by human processed lipoaspirate cells. Plast Reconstr Surg 109:199–211
27. Morsczeck C, Gotz W, Schierholz J et al (2005) Isolation of precursor cells (PCs) from human dental follicle of wisdom teeth. Matrix Biol 24:155–165
28. Nakao K, Morita R, Saji Y et al (2007) The development of a bioengineered organ germ method. Nat Methods 4:227–230
29. Ozahama A, Modino SA, Miletich A et al (2004) Stem-cell-based tissue engineering of murine teeth. J Dent Res 83:518–522
30. Rigotti G, Marchi A, Galiè M et al (2007) Clinical treatment of radiotherapy tissue damage by lipoaspirate transplant: a healing process mediated by adipose-derived adult stem cells. Plast Reconstr Surg 119:1409–1424
31. Sengenes C, Miranville A, Maumus M et al (2007) Chemotaxis and differentiation of human adipose tissue CD34+/CD31- progenitor cells: role of SDF-1 released by adipose tissue capillary endothelial cells. Stem Cells 25:2269–2276
32. Seo BM, Miura M, Gronthos S et al (2004) Investigation of multipotent postnatal stem cells from human periodontal ligament. Lancet 364:149–155
33. Shi S, Gronthos S (2003) Perivascular niche of postnatal mesenchymal stem cells in human bone marrow and dental pulp. J Bone Miner Res 18:696–704
34. Stevens A, Zuliani T, Olejnik C et al (2008) Human dental pulp stem cells differentiate into neural crest derived melonocytes and have label-retaining and sphere-forming abilities. Stem Cells Dev 17:1175–1184
35. Téclès O, Laurent P, Zygouritsas S et al (2005) Activation of human dental pulp progenitor stem cells in response to odontoblast injury. Arch Oral Biol 50:103–108
36. Tobita M, Uysal AC, Ogawa R et al (2008) Periodontal tissue regeneration with adipose-derived stem cells. Tissue Eng A 14:945–953
37. Yamada Y, Ito K, Nakamura S et al (2010) Promising cell-based therapy for bone regeneration using stem cells from deciduous teeth, dental pulp, and bone marrow. Cell Transplant. [Epub ahead of print]
38. Yananmoto N, Akamatsu H, Hasegawa S et al (2007) Isolation of multipotent stem cells from mouse adipose tissue. J Dermatol Sci 48:43–52
39. Yang Z, Jin F, Xhang X et al (2009) Tissue engineering of cementum/periodontal-ligament complex using a novel three-dimensional pellet cultivation system for human periodontal ligament stem cells. Tissue Eng Part C Method 15:571–581
40. Zhang W, Walboomers XF, Van Kuppevelt TH et al (2008) In vivo evaluation of human dental pulp stem cells differentiated towards multiple lineages. J Tissue Eng Regen Med 2:117–125
41. Zuk PA, Zhu M, Mizuno H et al (2001) Multilineage cells from human adipose tissue: implications for cell-based therapies. Tissue Eng 7:211–228

Adipose Stem Cells, Tissue Engineering, and Solid Organ Transplantation and Regeneration

21

Benoit Labbé, Valérie Trottier, Maryse Proulx, Caroline Vincent, and Julie Fradette

Contents

21.1	**Introduction**	229
21.2	**ASCs for Adipose Tissue Reconstruction and Transplantation**	230
21.2.1	Engineering Adipose Tissue In Vitro: Strategies to Achieve Volume Enhancement	230
21.2.2	The Self-Assembly Approach or Cell Sheet Technology in Tissue Engineering	231
21.3	**ASCs for the Production of Human Skin Substitutes and Their Transplantation**	232
21.3.1	Classical and New Cell Sources for Skin Reconstruction	232
21.3.2	Skin Reconstruction with Human ASCs and the Self-Assembly Approach	233
21.3.3	Reconstruction of a More Complete Trilayered Skin	237
21.4	**ASCs as Vehicles for the Stimulation of Skin Wound Repair**	238
21.5	**Conclusions**	239
References		240

B. Labbé • V. Trottier • M. Proulx • C. Vincent
Centre LOEX de l'Université Laval, Génie tissulaire et régénération: LOEX Aile-R, Centre de recherche FRSQ du Centre hospitalier affilié universitaire de Québec (CHA), 1401, 18 rue, Québec, QC, Canada G1J 1Z4 and Département de Chirurgie, Faculté de Médecine, Université Laval, Québec, QC, Canada

J. Fradette (✉)
LOEX Aile-R, Centre hospitalier *affilié* universitaire de Québec (CHA), 1401, 18e rue, Québec, QC, Canada G1J 1Z4
e-mail: julie.fradette@chg.ulaval.ca

21.1 Introduction

Regenerative medicine and tissue engineering both harness the potency of human cells to repair, regenerate, and even recreate tissues and organs with the goal of restoring their architecture and functionality. Mesenchymal cell populations are particularly well endowed for such purposes since they harbor subpopulations of multipotent cells and secrete many important bioactive molecules stimulating repair. Various connective tissues and stroma have now been used as sources of multipotent stem cells including bone marrow [17, 47, 80, 82], dermis [37, 98], umbilical cord Wharton's jelly [87, 105], and adipose tissue [110]. Subcutaneous lipoaspirated adipose tissue has recently become the focus of much attention given its abundancy, its accessibility through minimally invasive procedures, and its rather high stem cell content. The use of adipose-derived stem/stromal cells (ASCs) has rapidly expanded over the last years and their potential for regenerative therapies is described in an increasing number of studies (reviewed in [70, 93]).

The potential of ASCs is already investigated in a few clinical trials including fistula's repair (Spain, Germany, United Kingdom) [31] and cardiovascular repair with trials such as the PRECISE study for chronic myocardial ischemia (Spain) and the APOLLO trial for treatment of heart attacks (Netherlands, Spain) (www.clinicaltrials.gov). The secretion of potent growth factors from ASCs is associated with increased angiogenic activity that would stimulate endogenous repair mechanisms [16, 60, 68, 84]. Infused ASCs can be recruited at the site of damage, or stimulate the recruitment of endogenous stem cells, while specific cues from the

Y.-G. Illouz and A. Sterodimas (eds.), *Adipose Stem Cells and Regenerative Medicine*,
DOI: 10.1007/978-3-642-20012-0_21, © Springer-Verlag Berlin Heidelberg 2011

microenvironment might promote their differentiation into the appropriate lineage. These proposed mechanisms of action, consistently supported by the work of many investigators, are discussed in [35, 43].

While cell therapy strategies might be relatively straightforward and could ultimately be used as point of care services, substantial tissue loss requires an alternative to injectable delivery of therapeutic cells. Solid graftable tissues or substitutes are then necessary to repair more extended tissue damage. Substitutes can be engineered from natural or synthetic biomaterials. When enriched with progenitor cells, they become tissue equivalents of autologous or allogenic origin. Living biological substitutes might be the best option for the treatment of larger defects. ASCs are currently used for the engineering of substitutes that will ultimately be utilized in diversified fields of reconstructive surgery such as urology [46, 109], ophthalmology [4], orthopedics [40, 83, 101], to name a few. Still, to date, clinical applications of tissues engineered from ASCs are scarce [66, 94]. Autologous bone reconstruction using ASCs and a composite scaffold comprising beta tricalcium phosphate granules enriched with recombinant human BMP-2 has been one of the first application described in human [66]. The patient acted as its own "bioreactor," promoting the maturation of an ectopic bone flap previously created in vitro according to Good Manufacturing Practices (GMP) standards using autologous ASCs.

It is evident that the development of tissue-engineered substitutes that will both contribute to increase our knowledge of biological processes and be applicable in the clinical realm is of the utmost importance. We will next describe recent advances supporting the potential benefits of using ASCs to achieve adipose and skin regeneration through in vitro engineering of living substitutes.

21.2 ASCs for Adipose Tissue Reconstruction and Transplantation

21.2.1 Engineering Adipose Tissue In Vitro: Strategies to Achieve Volume Enhancement

Soft-tissue loss can be caused by trauma, tumor resection, extensive deep burns, as well as congenital or acquired anomalies. Despite great recent improvements, most of the current fat autograft techniques do not provide reproducible or satisfactory long-term results [76]. This is due in part to the fragility of adipose tissue and the disruption of vascularization due to the injection technique [12, 86], reviewed in [91]. Therefore, the demand for soft tissue substitutes in reconstructive and plastic surgery is continually increasing.

Different approaches have been used to reconstruct adipose tissue in vitro. Modeling such a specialized connective tissue represents a real challenge. Adipose substitutes should be easy to manipulate, soft and flexible in order to avoid discomfort once implanted. Human stem cells remain at the forefront of tissue engineering since they represent a valuable source of cells for in vitro reconstruction. ASCs have been combined with a variety of natural or synthetic materials in order to favor cell attachment, proliferation, and adipogenic differentiation within these biomaterials. For example, electrospun silk, hydrogels, as well as collagen beads, gels, or sponges have all been optimized to support and promote adipose tissue reconstruction (reviewed in [9, 33, 38]).

These various substitutes represent highly specialized tridimensional (3D) models for in vitro experimentation in addition to their expected use as tissue fillers. As such, in vivo evaluation of the constructs is necessary. Apart from volume retention imparted by lipid-filled adipocytes, the development of cysts or fibrosis has to be monitored. Moreover, vascularization of the grafted tissue is usually an excellent predictor of tissue survival and functionality. Different animal models can help establish the biocompatibility and performance of the various adipose substitutes, as reviewed by Dr Patrick [77]. However, long-term volume retention studies as well as assessment of the substitute's functionality are not routinely reported yet. They will likely become part of standardized criteria for assessment and comparison of diverse substitutes as the field of adipose tissue engineering matures.

Table 21.1 relates the most recent models developed for adipose tissue engineering using human ASCs. Interestingly, among these new strategies, a trend is seen towards the use of scaffolding elements of more natural origin such as decellularized placental membrane [29] or even processed human adipose tissue itself [21, 28, 104, 108]. One model is based on the self-assembly approach of tissue engineering. It results in the production of highly physiological

Table 21.1 Tissue engineering for adipose tissue reconstruction using human ASCs

Tissue engineering approach	Year	Authors
3D porous silk fibroin scaffolds	2010	Choi et al. [22]
Hybrid hydrogel scaffold composed of poloxamer 407 (PO) and the self-assembling oligopeptide EFK8	2010	Wang et al. [106]
Hybrid implant comprising a cellular layer (ASCs in alginate) encapsulating an acellular core (poly(ethylene) glycol-diacrylate)	2010	Moioli et al. [71]
Decellularized adipose tissue (DAT)	2010	Flynn [28]
3D aqueous-derived silk scaffolds	2009	Kang et al. [48]
Type I collagen scaffolds for in vivo adipogenesis	2009	Tsuji et al. [100]
Matrix based on small intestinal submucosa particles cultured with preadipocytes in a stirred bioreactor (spinner flask)	2008	Marra et al. [63]
Human placental decellularized matrix (PDM) and PDM combined with cross-linked hyaluronan (XLHA) scaffolds	2007	Flynn et al. [29]
Comparison of silk fibroin, collagen and poly-lactic acid (PLA)-based scaffolds	2007	Mauney et al. [64]

living substitutes since it mimics the natural extracellular matrix (ECM) environment surrounding ASCs. The next section will describe this approach in more detail.

21.2.2 The Self-Assembly Approach or Cell Sheet Technology in Tissue Engineering

We have recently reconstructed both connective and adipose tissues composed entirely of cultured human ASCs and their secreted endogenous ECM components. This was achieved by a methodology adapted from the self-assembly approach of tissue engineering [55, 104]. This strategy relies on the stimulation of mesenchymal cells with ascorbic acid (vitamin C) and serum during their culture, resulting in the formation of manipulatable cell sheets that can be assembled into thicker multilayer tissues. The self-assembly approach has been intensively developed by our research group LOEX from Université Laval in Québec City, Canada [6]. Smooth-muscle cells, dermal fibroblasts, and keratocytes (corneal stromal cells) have been used for the reconstruction of either tissue-engineered blood vessels [54], skin [59, 67], cornea [18], or urological structures [14].

My team uses human ASCs as a cell source for tissue reconstruction. These cells feature a high capacity for ECM production and organization, leading to the formation of dense reconstructed connective tissues [102, 104]. These can be used as stromal compartments onto which epithelial cell types such as keratinocytes can be seeded, resulting in reconstructed skin with features very similar to native human skin or to skin reconstructed using dermal fibroblasts, as will be described in the next section [99]. More importantly, we have been able to adapt our approach and to utilize ASCs not only as matrix-secreting cells, but also as progenitors to generate in vitro differentiated adipocytes. This was possible by including an adipogenic differentiation step concomitant with ECM deposition stimulated by ascorbic acid. Therefore, multilayered human adipose tissues have been produced in vitro. They feature adipocytes embedded into endogenous ECM and globally, their structural characteristics are reminiscent of native adult human tissue [104]. Importantly, these substitutes are functional, as defined by their capacity to initiate de novo synthesis and accumulation of triglycerides, to secrete a variety of adipokines (leptin, VEGF, angiopoietin-1, etc.) and to mediate beta-adrenergic stimulated lipolysis in vitro [104]. These reconstructed adipose tissues are stable and functional for many weeks in culture. They represent useful tools for in vitro assays detailing metabolic function or modulation of adipose tissue secretory function.

Despite the lack of synthetic or exogenous biomaterials, the abundant matrix secretion and deposition resulting from ascorbic acid supplementation facilitates handling of these tissues. They can be manipulated with forceps and sutured (Fig. 21.1A). Results from a pilot study evaluating the characteristics of adipose tissues reconstructed using the self-assembly approach after subcutaneous grafting onto the back muscle of athymic mice are promising (Fig. 21.1). Histological appearance of the grafts as seen on tissue cross-sections stained with Masson's trichrome reveals that the adipose substitutes comprise many adipocytes

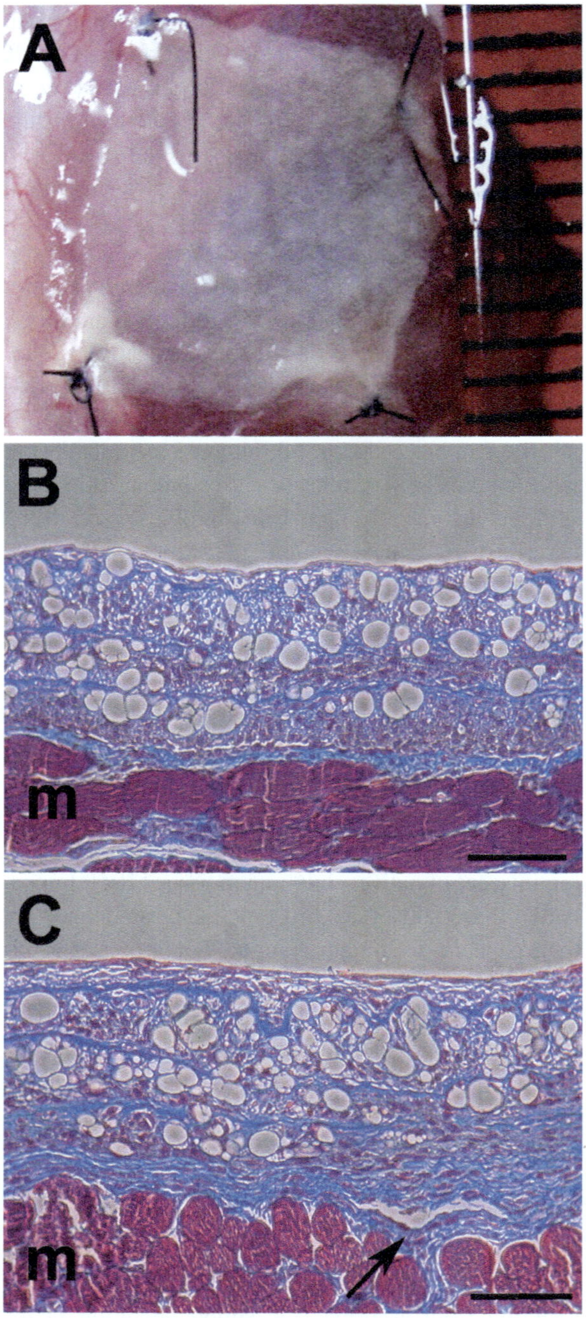

Fig. 21.1 Aspect of reconstructed adipose tissues after subcutaneous grafting onto athymic mice. (**A**) Macroscopic aspect of reconstructed adipose tissues 3 days after grafting (unit = mm). (**B**) Histological appearance of an adipose tissue graft excised with the underlying muscle 3 days after grafting. Many round adipocytes (*blank spaces*) are apparent. (**C**) Capillaries (*arrow*) can be detected on histological cross-sections of excised adipose tissues 7 days after grafting. Adipocytes are also well preserved. *m* murine muscle, *Bars* = 100 μm

(blank spaces) as well as an abundant ECM (in blue) (Fig. 21.1B, C). These substitutes performed well in vivo, as shown by the presence of capillaries detected as soon as 7 days after grafting (Fig. 21.1C, arrow). Short-term results highlighted the ability of these substitutes to maintain a high number of adipocytes within the graft over a 14-day period. Work is in progress to establish their behavior at longer time points following implantation using imaging techniques that will facilitate precise measurements and determination of volume retention.

21.3 ASCs for the Production of Human Skin Substitutes and Their Transplantation

21.3.1 Classical and New Cell Sources for Skin Reconstruction

Subcutaneous fat which is also named hypodermis represents the third and deepest layer of skin, a most needed tissue in reconstructive surgery. Skin tissue engineering is indeed a very active research area and many groups have contributed key developments to this field [7, 10, 11, 15, 39, 53, 62, 78, 97, 107]. First, cultured epidermal sheets were produced and revolutionized patient treatment. They allowed rapid reconstitution of the skin's barrier function [5, 19, 30]. However, regions grafted uniquely with this thin epidermal layer can display a lack of elasticity and reduced tensile strength, these properties being mostly associated with the underlying stroma. Since bilayered skin substitutes comprising both an epidermis and a dermis have been developed, the importance of the dermal compartment for tissue regeneration has become apparent. In general, the addition of autologous fibroblasts into these matrices was associated with improved wound healing of contracted scars and burns [23, 56]. Over the last decade, a wide variety of dermal matrices and engineering strategies have been developed to properly sustain keratinocyte growth and differentiation (reviewed in [8, 44, 61, 75, 90]). But while keratinocytes can be quickly amplified in vitro from a small biopsy (1–4 cm^2), the extent of skin loss may limit the availability of donor sites from which autologous dermal fibroblasts can be harvested and amplified in culture. This is one of the reasons why fibroblasts from different origins (skin, buccal, or vaginal mucosa) [65] as well as alternate sources of fibroblast-like mesenchymal cells have been evaluated in the past for skin regeneration, ranging from eschar tissue [103], bone marrow stromal cells [3, 42, 79], and adipose-derived cells or preadipocytes [3, 27, 45, 49, 103].

The concept of using adipose tissue as a source of mesenchymal cells for skin reconstruction is very attractive due to its abundancy, accessibility, and ease of extraction. It has been reported that human [27, 45, 99] and rodent [3] adipose-derived stromal cells can adequately support epidermal growth in vitro. One study reported that when type I collagen gels were populated with human ASCs instead of dermal fibroblasts, epidermal regeneration was achieved with slight differences. For example, a delayed deposition of type IV collagen was seen, as well as an increased alpha smooth muscle actin (αSMA) expression, leading to greater gel contraction [27]. It is not obvious from this small number of studies whether tangible benefits are in fact associated with the use of ASCs instead of dermal fibroblasts for skin tissue engineering. However, it is evident that these abundant cells have the potential to complement dermal fibroblast cultures. Side-by-side comparison of both cell types for each preferred tissue-engineering strategy will be important since differences in cell extraction, expansion, as well as the type of dermal matrices used can greatly impact on cell behavior. Data from our lab using ASCs for human skin reconstruction using the self-assembly approach are supportive of a future role for these cells in the development of new skin substitutes.

21.3.2 Skin Reconstruction with Human ASCs and the Self-Assembly Approach

We have used the self-assembly approach of tissue engineering to produce and compare skin substitutes reconstructed using human dermal fibroblasts or lipoaspirated ASCs. Human keratinocytes were seeded on three different types of stromal compartments consisting of cells and human ECM produced from either dermal fibroblasts (Fig. 21.2A, D, G), ASCs (Fig. 21.2B, E, H), or the same ASC population induced towards adipogenic differentiation in vitro (Fig. 21.2C, F, I) [99]. The bilayered skin substitutes thus formed were further cultured at the air–liquid interface to promote cell differentiation [57]. For these three types of reconstructed skin, the newly formed epidermis was reminiscent of normal human skin, already featuring a multilayered, stratified epidermis after 14 days of culture at the air–liquid interface (Fig. 21.2A–C). A *stratum corneum* layer had also formed, featuring terminally differentiated corneocytes essential to the skin's barrier function. For skin reconstructed using a stroma made of ASCs induced towards adipogenic differentiation, adipocytes can be seen on histological cross-sections of paraffin-embedded biopsies (Fig. 21.2C, arrows). They appear as blank spaces since adipocytes are filled with lipids that were extracted during processing and coloration.

We verified whether the type of stromal compartment influences the formation and differentiation of the epidermis. For this purpose, we assessed the expression of specific proteins commonly used as cell differentiation markers, namely keratin (K)14, K10, and transglutaminase, indicative of the basal, suprabasal, and granular epidermal layers, respectively [99]. Immunolabelings for these molecules as well as for filaggrin (Fig. 21.2D–F), another important enzyme expressed in the granular layer of native human skin, revealed an appropriate epidermal localization.

The epidermis for all three types of reconstructed skin also revealed that Ki67 expression was restricted to the basal layer, where proliferating cells are expected to be found. This, combined with K10 expression restricted to the suprabasal layers, suggest a tight recapitulation of the normal differentiation process [99]. Within the reconstituted epidermis, K19-expressing epithelial stem cells were also observed. They were present in their expected location in the basal layer of the substitutes where they could ensure the long-term renewal of the epithelium [58, 99]. These results highlight two important notions. First, it establishes that the self-assembly method can recreate the complex interactions between stromal cells, epithelial cells, and the diversified extracellular components composing the stroma, because when skin substitutes featuring a dermis made of bovine type I collagen gel were analyzed [32], the K19-expressing stem cells were not retained in the basal layer but migrated upwards to be later lost by shedding. Second, these results established a comparable efficiency for dermal fibroblast and

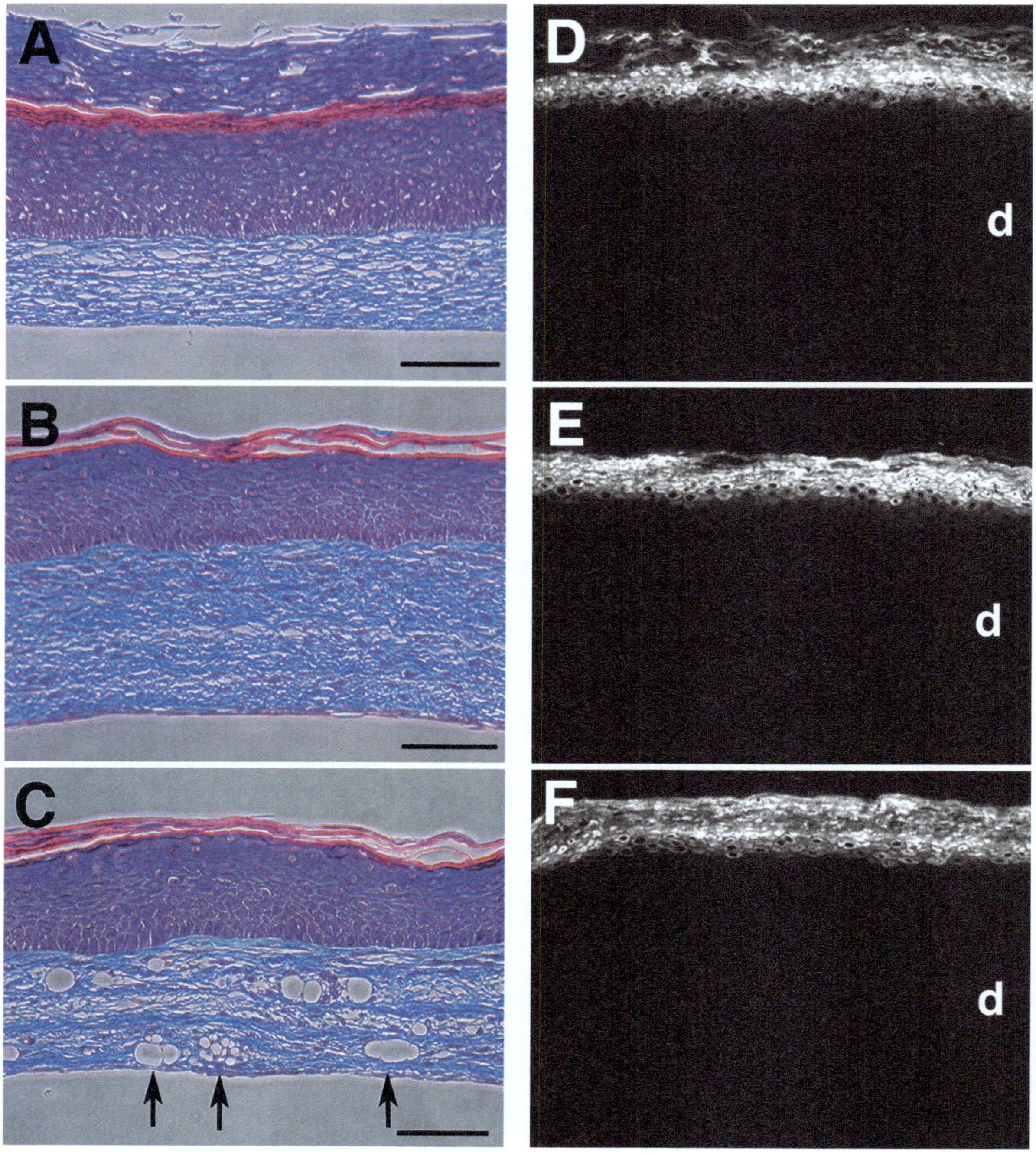

Fig. 21.2 Comparative analysis of human skin substitutes produced using different types of mesenchymal cells. (**A**–**C**) Histological appearance after Masson's trichrome staining, (**D**–**F**) Filaggrin expression and (**G**–**I**) type IV collagen detection on tissue cross-sections for the three types of substitutes. Skin was reconstructed using either (**A**, **D**, **G**) dermal fibroblasts, (**B**, **E**, **H**) ASCs or (**C**, **F**, **I**) the same ASC population induced towards adipogenic differentiation and featuring mature adipocytes within the stroma (*arrows*). For (**B**) and (**C**), keratinocytes as well as ASCs were from the same donor (age 33 year-old). (**A**–**C**) *Bars* = 100 μm, (**D**–**I**) 200×, *e* epidermis, *d* dermis

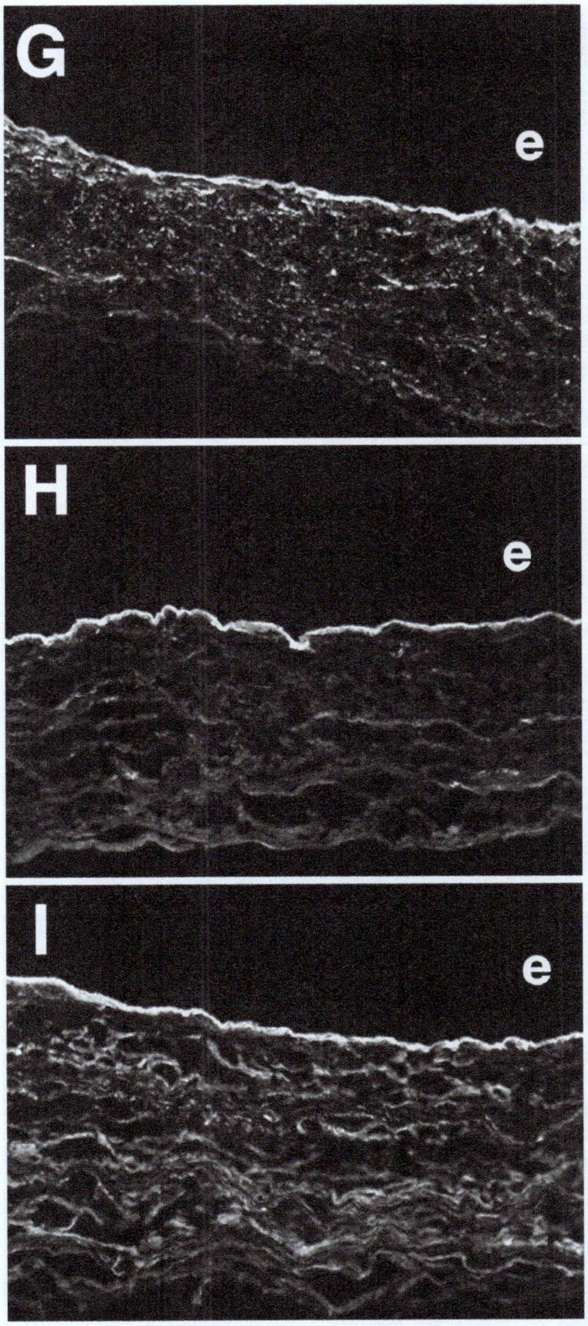

Fig. 21.2 (continued)

ASC-derived stromas in their capacity to retain epidermal basal and stem cells at their appropriate location.

This preservation of stem cell retention is likely due to the well-defined dermo-epidermal junction featured by all three types of skin substitutes, ensuring adequate anchorage of the epidermis to the stromal compartment. As seen for skin reconstructed using dermal fibroblasts, type IV collagen (Fig. 21.2G–I), laminin-5, as well as type VII collagen [99] are strongly expressed in substitutes produced using ASCs, as detected by specific immunolabelings. Analysis of the tissue ultrastructure by transmission electron microscopy confirmed the presence of a well-defined basement membrane comprising hemi-desmosomal structures [99]. These are the hallmarks confirming a strong cohesion between the epidermis and dermis. This, in combination with the dense expression of structural collagens in the stroma, indicates that the reconstructed skin possesses the structural elements necessary to provide excellent mechanical properties to the substitutes. This was observed not only for skin reconstructed using a stroma produced from dermal fibroblasts and noninduced ASCs, but also for the substitutes produced by seeding keratinocytes directly onto an adipocyte-containing stroma. It indicates that differentiated adipocytes did not negatively influence the structural organization of the skin substitutes.

Comparison between skin substitutes reconstructed in vitro using dermal fibroblasts or ASCs and then grafted on the back of athymic mice revealed a very nice morphology 21 days after grating ([99] and Fig. 21.3A, B). The epidermal compartment was usually thicker for substitutes reconstructed using ASCs (Fig. 21.3B vs. A). Cuboidal basal keratinocytes were apparent for both types of reconstructed skin as well as numerous suprabasal layers and the presence of a *stratum corneum* (Fig. 21.3A, B). A subset of grafted skin was followed up to 85 days after grafting (Fig. 21.3C–F). At that time, the thickness of the epidermal compartment was quite similar between skin substitutes produced using dermal fibroblasts (Fig. 21.3C) or ASCs (Fig. 21.3D). Blood vessels are visible in the stromal compartments (Fig. 21.3C, D, arrows), indicating tissue integration. The grafted skin looked healthy macroscopically and immunolabeling for type IV collagen revealed a continuous dermo-epidermal junction as well as the presence of tubular structures corresponding to capillaries (Fig. 21.3E, F), supporting a good survival, integration, and remodeling of the reconstructed skin.

These findings suggest that ASCs could complement or even be utilized in place of dermal fibroblasts for in vitro skin reconstruction using the self-assembly method. Our cell sheet technology allows the reconstruction of various kinds of stroma by combining individual adipose or connective sheets, produced using ASCs or dermal fibroblasts. A number of

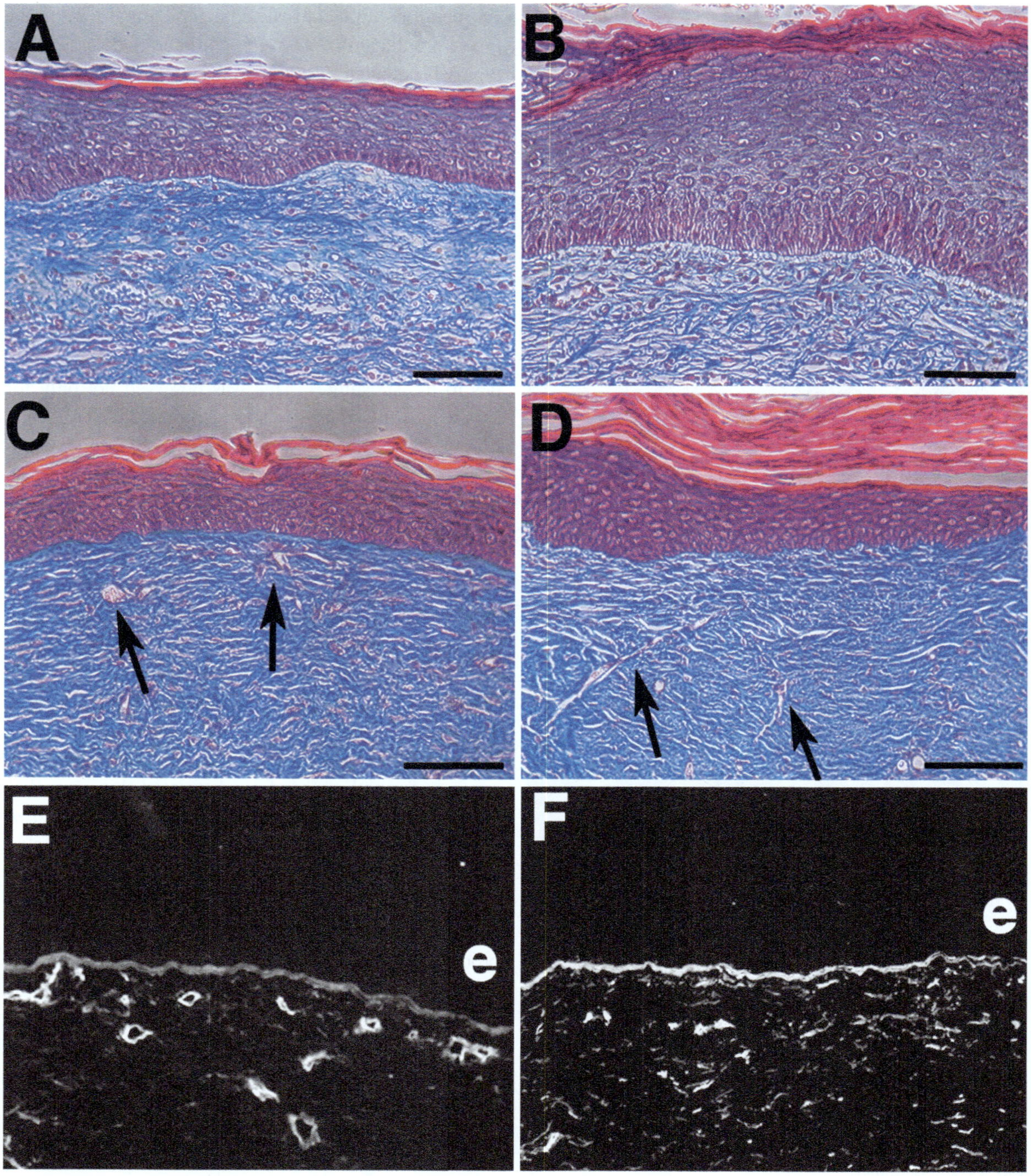

Fig. 21.3 Transplantation of reconstructed human skin. (**A**, **B**) Twenty-one days after being grafted onto athymic mice, skin substitutes produced using (**A**) dermal fibroblasts or (**B**) ASCs were harvested for analysis. A thick epidermal compartment was present, especially when ASCs were used for reconstruction. (**C–F**) The skin substitutes were also analyzed after 85 days in vivo. (**C**, **D**) Histological analysis revealed a compact epidermis firmly anchored to its underlying dermis enriched in capillaries (*arrows*). (**E**, **F**) Immunodetection of type IV collagen established its presence in basement membranes, both at the dermo-epidermal junction and surrounding capillaries. *Bars* = 100 μm, (**E**, **F**) 200×, *e* epidermis

combinations were tested by layering and mixing sheets produced from dermal fibroblasts, from uninduced ASCs (connective sheet) or adipogenic-induced adipose sheets. We described above the production of bilayered skin substitutes, containing or not functional adipocytes. In the next section, the production of a trilayered skin featuring a hypodermis in addition to the dermis and epidermis will be detailed.

21.3.3 Reconstruction of a More Complete Trilayered Skin

The potential of ASCs to differentiate into adipocytes was further exploited using our reconstruction method. An adipose layer was added underneath the bilayered skin previously described, in order to produce a more complete skin substitute [99]. The hypodermis, the skin's deepest layer, was therefore created both for bilayered skin comprising a dermis made from dermal fibroblasts, or a stroma made from ASCs.

Once again, our findings revealed that each type of skin substitute displayed a well-organized stratified squamous epithelia as visualized after 14 or 21 days of culture at the air–liquid interface, irrespective of the stromal compartment used. Figure 21.4A presents a histological view of a trilayered skin composed of keratinocytes that were seeded onto a multilayer stromal compartment formed by the superposition of three cell sheets: two dermal sheets produced from fibroblasts were layered onto an adipose sheet featuring round adipocytes (Fig. 21.4A). Appropriate expression of several markers used to assess keratinocyte proliferation and differentiation was established. Immunolabeling for the detection of K14 (Fig. 21.4B) and the enzyme transglutaminase (Fig. 21.4C) established their expression at the expected epidermal localization. The components of the dermo-epidermal junction such as laminin-1 [99] and collagen IV (Fig. 21.4D) were also strongly expressed. Interestingly, the more intense type IV collagen staining seen in the hypodermal compartment is in accordance with the formation of a basement membrane around adipocytes [41]. The presence of adipocytes within the new skin substitute did not seem to abrogate nor promote epidermal proliferation or stratification differently than dermal fibroblasts, or from the ASCs from which they originated. The impact of keratinocytes on the underlying adipocytes will certainly reveal interesting relationships between these two cell types.

Very few teams have investigated and successfully reconstructed adipocyte-containing skin substitutes [95, 96]. The first model described was elaborated from freshly isolated rat adipocytes that were embedded into a bovine type I collagen gel on top of which keratinocytes were cultured [96]. While keratinocyte proliferation and differentiation was supported, fat cells displayed an inhibitory effect on dermal fibroblasts and basement membrane formation was not reported. Our model differs by using in vitro differentiated human adipocytes. Also, in contrast to collagen gels in which exogenous collagen is added, the collagen is produced and organized by the cells themselves in the self-assembly approach. Such a self-assembled stromal compartment adequately supported the formation of a well-organized epidermis and basement membrane, but also provides remarkable mechanical properties to the substitute, before and after grafting to animals [81, 99].

A different strategy has recently been described in order to generate skin comprising an adipose layer. The authors favored a direct in vivo approach: they produced and cultured their substitutes for 1 week in vitro before assessing the outcome and adipogenic differentiation after 6 weeks in vivo [72]. Initially, they seeded ASCs on one side of a collagen sponge and then seeded dermal fibroblasts and keratinocytes on the opposite side. After 1 week of culture, the sponges were implanted on the backs of severe combined immunodeficient mice (*scid*), in presence or not of basic FGF. Six weeks after implantation, the specimens were harvested and the presence of bFGF seemed to favor the formation of adipose tissue as well as capillary development.

The mutual influences taking place between keratinocytes and adipose cells have not been extensively investigated yet since adipose tissue used to be considered mostly as a passive lipid-storage site. This view has changed and it is now recognized that adipose tissue is in fact an important paracrine organ. Many of the adipokines secreted by adipocytes have the potential to impact on the proliferation or differentiation of keratinocytes. For example, in conventional monolayer culture, leptin has been shown to promote keratinocyte proliferation and migration [92]. The effects of adipocytes on the proliferation and differentiation of rat skin-derived organoid hair follicle cells have been reported in the past using collagen gel matrices to sustain the coculture [69]. In this system, adipocytes seemed to inhibit the outgrowth of keratinocytes from the isolated organoid hair follicles when they were

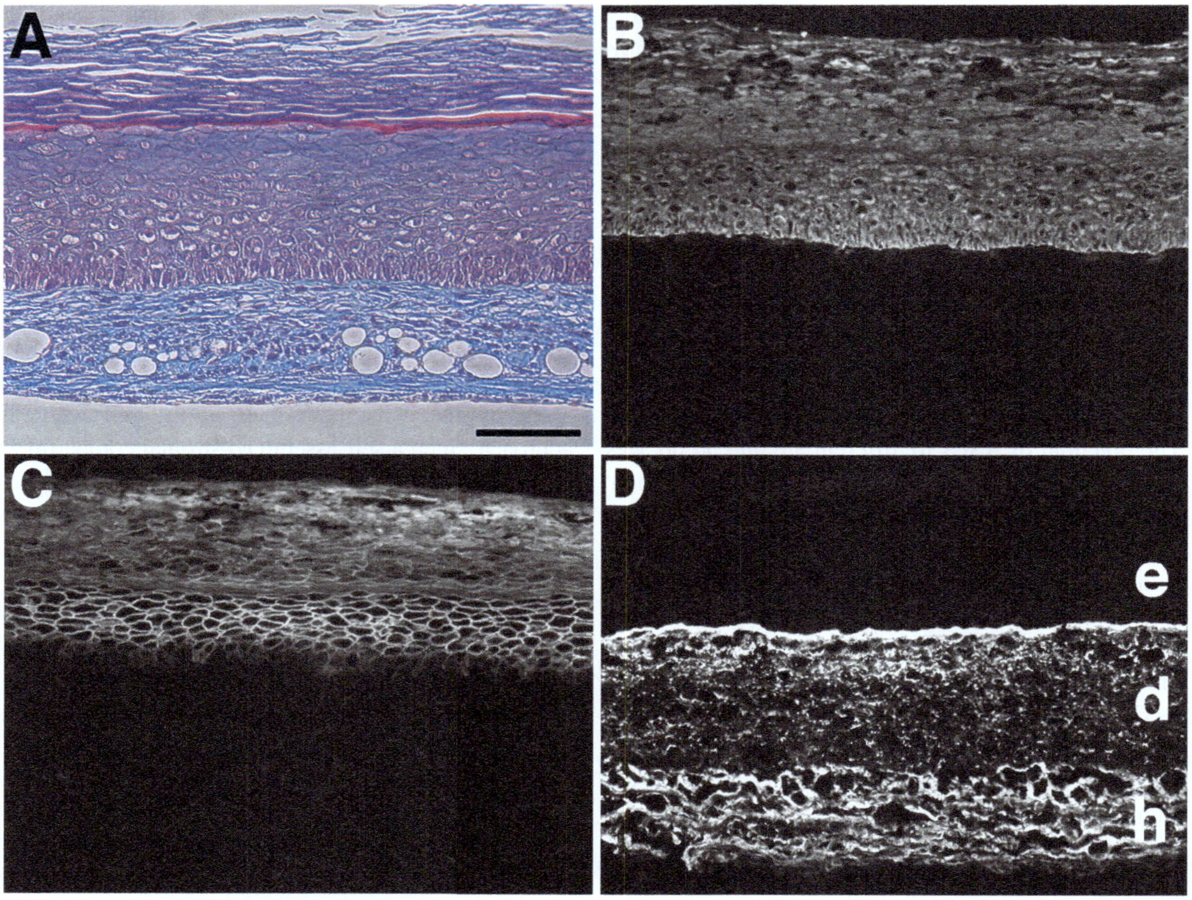

Fig. 21.4 Creation of a new human skin substitute comprising an epidermis, a dermis, and a hypodermis in vitro. (**A**) Histological appearance of a trilayer skin substitute engineered using keratinocytes, dermal fibroblasts, as well as ASCs that were induced to differentiate into adipocytes for the production of the hypodermal compartment. (**B**) K14 expression as well as (**C**) transglutaminase detection confirmed adequate epidermal regeneration. (**D**) Immunolabeling for type IV collagen revealed its strong expression at the dermo-epidermal junction as well as in the hypodermis. *Bar* = 100 μm, (**B–D**) 200×, *e* epidermis, *d* dermis, *h* hypodermis

placed in direct contact with each other. The authors noticed that epidermal and hair follicle differentiation seemed to be favored over epithelial proliferation, but only when adipocytes were closely apposed to the organoids and not when they were spatially separated.

Taken together, the development of these different types of new skin substitutes should help uncover the involvement of adipocytes in skin homeostasis and healing. The production of complete skin substitutes featuring in vitro differentiated adipocytes, either in direct contact with keratinocytes or recreating the hypodermal compartment located under the dermis therefore represents a clever approach to study in details and under controlled conditions the reciprocal influences between adipocytes and keratinocytes. In the future, these trilayered skin substitutes comprising a hypodermis could also be useful for the treatment of defects extending deep into the connective tissue. Their potential for regenerative medicine still remains to be determined.

21.4 ASCs as Vehicles for the Stimulation of Skin Wound Repair

While skin reconstruction using ASCs is still at its beginning, one cannot ignore the increasing variety of cell therapy strategies that are being investigated in an attempt to improve cutaneous wound healing. Different modes of delivery for ASCs have been described, ranging from cell suspensions injected intravenously or in the underlying muscle, to direct cell transplantation into wounds [20, 26]. Local application techniques include topical administration [50], the use of platelet-rich plasma gel (PRP) [13, 20], collagen matrix [73, 74], silk-chitosan scaffolds [1], collagen–elastin matrices [25], and naturally formed cellular aggregates [2]. The ability of ASCs to mediate repair of full-thickness wounds has also been tested using various models based on *db/db* mice [2, 74], mitomycin-C treated wounds in rats [73], or rat irradiated skin [26].

Of course, the delivery mode of these cells is important and can affect their capacity to mediate tissue repair. For local cell transplantation, a scaffold or matrix able to sustain ASCs proliferation and survival in the wounds is favorable. A study comparing ASCs delivered as 3D multicellular aggregates to cell suspensions directly applied into the wounds indicated that cell aggregates (spheres) promoted faster wound closure [2]. Those 3D spheres are enriched in ECM components produced naturally after ascorbic acid stimulation. The combination of cells, matrix, and endogenously produced growth factors secreted in higher amounts by the 3D spheres than by cells expanded in monolayer likely explain the increased healing.

Few studies have compared the regenerative potential of ASCs to dermal fibroblasts [25] or to bone-marrow derived mesenchymal cells [26]. However, most of the wound healing studies reported the benefits of using ASCs compared to an acellular matrix or to the delivery vehicle alone. The proposed mechanisms of action that would be responsible for this favorable wound repair by ASCs transplantation include enhanced reepithelialization, improved dermal regeneration, and increased vascularization of the wound. This is not surprising considering the important biological effects mediated by the factors secreted by ASCs that likely account for a large part of their healing properties (reviewed in [51]). More detailed characterization of ASCs and their secretome in comparison to other mesenchymal cell types will pinpoint their role and future involvement as therapies for in vivo skin repair and regeneration.

21.5 Conclusions

The discovery of adult multipotent stem cells within the stroma of adipose tissue opened the door to a wide range of possibilities in regenerative medicine. Growing autologous tissue from the patient's own stem cells in order to repair damaged tissue and restore tissue function is becoming a reality. A clear advantage of using ASCs in regenerative medicine and tissue engineering resides in the possibility to harvest a great amount of subcutaneous adipose tissue. This is usually performed by a lipoaspiration procedure associated with reduced morbidity at donor sites in comparison with the harvesting of dermis, for example. Our previous findings compared cell yields obtained for ASCs and dermal fibroblasts. At the time of extraction, 1 g of dermis contained more cells than 1 g of adipose tissue. However, the average proliferation of fibroblasts and ASCs was not significantly different when passaged once or twice under the same culture conditions [99]. This translates into the possibility of culturing and expanding more ASCs because a significant amount of adipose tissue can be harvested in the first place. This would reduce the period required for cell expansion in vitro before tissue reconstruction.

The skin substitutes presented here can be engineered in an autologous fashion from a single donor site such as a lipectomy biopsy. However, considering that in our hands, ASCs from lipoaspirated fat perform better than ASCs from excised fat [104], the preferred procedure would include the harvesting of a stamp-size piece of skin (1–4 cm^2) combined with a small volume of lipoaspirated subcutaneous fat (approximately 100 g). This type of autologous reconstruction would then take advantage of the properties of two important mesenchymal cell types. It would likely benefit patients requiring skin reconstruction by accelerating the production of complete skin substitutes.

In general, the quality of various tissues engineered in vitro depends on the quality of the cells as well as the method of production including the type of ECM components involved. Each type of engineering strategy should investigate the preservation of stem cells under their culture conditions because the long-term survival of the tissue/organ after transplantation depends on their presence for homeostasis maintenance. This is particularly true for organs in constant renewal such as skin (epidermis). In that respect, we have shown that the self-assembly method of tissue engineering recreates a 3D environment conducive to stem cell preservation [58, 59, 99].

New substitutes designed to provide even superior gain of function will certainly be developed in the future. For example, skin featuring a capillary network in vitro has been reconstructed using the self-assembly approach [34, 85] and could be used as skin flaps to promote vascularization. Further enhancement of skin substitutes by the addition of other cell types such as melanocytes [88, 89] or neuronal cells [36] has also been described and could help to reestablish skin's pigmentation and innervation, respectively. It will be interesting, considering their plasticity, to see if the use of ASCs for similar endeavors will benefit the development of new and improved tissues.

The substitutes described above are produced with the aim of being used as autologous substitutes for permanent replacement and regeneration of soft tissues. In addition, these tissue-engineered substitutes represent valuable tools for in vitro assays evaluating the toxicity or biological effects imparted by specific molecules [24]. The ability to produce tissue-engineered skin or adipose tissue in large numbers under controlled conditions will likely contribute to a better characterization of cell–cell and cell–matrix interactions that are so important to our understanding of tissue function. This chapter focused on the production of manipulatable adipose tissues and skin substitutes using ASCs as the primary mesenchymal cell source. However, ASC-based tissue engineering strategies are being developed and investigated for many other types of applications that also necessitate transplantation of solid tissues/organs. As mentioned previously, ASCs are studied for the reconstruction and regeneration of trachea [52], cornea [4], tendon [101], musculo-skeletal [40, 83], as well as urological tissues [46, 109]. It will therefore be very exciting to see the extent of ASC contributions in the field of regenerative medicine in the coming years.

Acknowledgments No competing financial interests exist. We are grateful to all members of the LOETA team of the LOEX. This work was supported by the Canadian Institutes for Health Research (CIHR) grant NRF-84368 and the Natural Sciences and Engineering Research Council of Canada (NSERC), grant NRF-326987. B.L. and M.P. are recipients of a studentship from the Fonds de la Recherche en Santé du Québec (FRSQ) and from the Natural Sciences and Engineering Research Council of Canada (NSERC), respectively. JF is a FRSQ scholar.

References

1. Altman AM, Yan Y, Matthias N et al (2009) IFATS collection: human adipose-derived stem cells seeded on a silk fibroin-chitosan scaffold enhance wound repair in a murine soft tissue injury model. Stem Cells 27:250–258
2. Amos PJ, Kapur SK, Stapor PC et al (2010) Human adipose-derived stromal cells accelerate diabetic wound healing: impact of cell formulation and delivery. Tissue Eng A 16:1595–1606
3. Aoki S, Toda S, Ando T et al (2004) Bone marrow stromal cells, preadipocytes, and dermal fibroblasts promote epidermal regeneration in their distinctive fashions. Mol Biol Cell 15:4647–4657
4. Arnalich-Montiel F, Pastor S, Blazquez-Martinez A et al (2008) Adipose-derived stem cells are a source for cell therapy of the corneal stroma. Stem Cells 26:570–579
5. Auger FA (1988) The role of cultured autologous human epithelium in large burn wound treatment. Transplantation/Implantation Today 5:21–24
6. Auger FA, Remy-Zolghadri M, Grenier G et al (2000) The self-assembly approach for organ reconstruction by tissue engineering. e-Biomed: J Tissue Eng Regen Med 1:75–86
7. Auger FA, Berthod F, Moulin V et al (2004) Tissue-engineered skin substitutes: from in vitro constructs to in vivo applications. Biotechnol Appl Biochem 39:263–275
8. Auxenfans C, Fradette J, Lequeux C et al (2009) Evolution of three dimensional skin equivalent models reconstructed in vitro by tissue engineering. Eur J Dermatol 19:107–113
9. Bauer-Kreisel P, Goepferich A, Blunk T (2010) Cell-delivery therapeutics for adipose tissue regeneration. Adv Drug Deliv Rev 62(7–8):798–813
10. Bell E, Ehrlich HP, Buttle DJ et al (1981) Living tissue formed in vitro and accepted as skin-equivalent tissue of full thickness. Science 211:1052–1054
11. Berthod F, Damour O (1997) In vitro reconstructed skin models for wound coverage in deep burns. Br J Dermatol 136:809–816
12. Billings E Jr, May JW Jr (1989) Historical review and present status of free fat graft autotransplantation in plastic and reconstructive surgery. Plast Reconstr Surg 83:368–381
13. Blanton MW, Hadad I, Johnstone BH et al (2009) Adipose stromal cells and platelet-rich plasma therapies synergistically increase revascularization during wound healing. Plast Reconstr Surg 123:56S–64S
14. Bouhout S, Perron E, Gauvin R et al (2010) In vitro reconstruction of an autologous, watertight, and resistant vesical equivalent. Tissue Eng A 16:1539–1548
15. Boyce ST, Kagan RJ, Greenhalgh DG et al (2006) Cultured skin substitutes reduce requirements for harvesting of skin autograft for closure of excised, full-thickness burns. J Trauma 60:821–829
16. Cai L, Johnstone BH, Cook TG et al (2007) Suppression of hepatocyte growth factor production impairs the ability of adipose-derived stem cells to promote ischemic tissue revascularization. Stem Cells 25:3234–3243
17. Caplan AI (1991) Mesenchymal stem cells. J Orthop Res 9:641–650
18. Carrier P, Deschambeault A, Audet C et al (2009) Impact of cell source on human cornea reconstructed by tissue engineering. Invest Ophthalmol Vis Sci 50:2645–2652

19. Carver N, Leigh IM (1991) Keratinocyte grafts and skin equivalents. Int J Dermatol 30:540–551
20. Cervelli V, De Angelis B, Lucarini L et al (2010) Tissue regeneration in loss of substance on the lower limbs through use of platelet-rich plasma, stem cells from adipose tissue, and hyaluronic acid. Adv Skin Wound Care 23:262–272
21. Choi JS, Yang HJ, Kim BS et al (2009) Human extracellular matrix (ECM) powders for injectable cell delivery and adipose tissue engineering. J Control Release 139:2–7
22. Choi JH, Gimble JM, Vunjak-Novakovic G et al (2010) Effects of hyperinsulinemia on lipolytic function of three-dimensional adipocyte/endothelial co-cultures. Tissue Eng Part C Methods 16(5):1157–1165
23. Coulomb B, Friteau L, Baruch J et al (1998) Advantage of the presence of living dermal fibroblasts within in vitro reconstructed skin for grafting in humans. Plast Reconstr Surg 101:1891–1903
24. Damour O, Augustin C, Black AF (1998) Applications of reconstructed skin models in pharmaco-toxicological trials. Med Biol Eng Comput 36:825–832
25. de Vries HJ, Middelkoop E, van Heemstra-Hoen M et al (1995) Stromal cells from subcutaneous adipose tissue seeded in a native collagen/elastin dermal substitute reduce wound contraction in full thickness skin defects. Lab Invest 73:532–540
26. Ebrahimian TG, Pouzoulet F, Squiban C et al (2009) Cell therapy based on adipose tissue-derived stromal cells promotes physiological and pathological wound healing. Arterioscler Thromb Vasc Biol 29:503–510
27. El-Ghalbzouri A, Van Den Bogaerdt AJ, Kempenaar J et al (2004) Human adipose tissue-derived cells delay re-epithelialization in comparison with skin fibroblasts in organotypic skin culture. Br J Dermatol 150:444–454
28. Flynn LE (2010) The use of decellularized adipose tissue to provide an inductive microenvironment for the adipogenic differentiation of human adipose-derived stem cells. Biomaterials 31:4715–4724
29. Flynn L, Prestwich GD, Semple JL et al (2007) Adipose tissue engineering with naturally derived scaffolds and adipose-derived stem cells. Biomaterials 28:3834–3842
30. Gallico GG III, O'Connor NE, Compton CC et al (1984) Permanent coverage of large burn wounds with autologous cultured human epithelium. N Engl J Med 311:448–451
31. Garcia-Olmo D, Herreros D, Pascual I et al (2009) Expanded adipose-derived stem cells for the treatment of complex perianal fistula: a phase II clinical trial. Dis Colon Rectum 52:79–86
32. Germain L, Michel M, Fradette J et al (1997) Skin stem cell identification and culture: a potential tool for rapid epidermal sheet production and grafting. In: Rouabhia M (ed) Skin substitute production by tissue engineering: clinical and fundamental applications. Landes Bioscience, Austin, TX
33. Germain L, Larouche D, Auger FA et al (2008) Human postnatal stem cells in organs produced by tissue engineering for clinical applications. In: Faraday AV, Dyer JT (eds) Progress in stem cell applications. Nova Science, New York
34. Gibot L, Galbraith T, Huot J et al (2010) A preexisting microvascular network benefits in vivo revascularization of a microvascularized tissue-engineered skin substitute. Tissue Eng A 16:3199–3206
35. Gimble JM, Katz AJ, Bunnell BA (2007) Adipose-derived stem cells for regenerative medicine. Circ Res 100:1249–1260
36. Gingras M, Bergeron J, Dery J et al (2003) In vitro development of a tissue-engineered model of peripheral nerve regeneration to study neurite growth. FASEB J 17:2124–2126
37. Gingras M, Champigny MF, Berthod F (2007) Differentiation of human adult skin-derived neuronal precursors into mature neurons. J Cell Physiol 210:498–506
38. Gomillion CT, Burg KJ (2006) Stem cells and adipose tissue engineering. Biomaterials 27:6052–6063
39. Green H, Kehinde O, Thomas J (1979) Growth of cultured human epidermal cells into a multiple epithelia suitable for grafting. Proc Natl Acad Sci USA 76:5665–5668
40. Guilak F, Estes BT, Diekman BO et al (2010) 2010 Nicolas Andry award: multipotent adult stem cells from adipose tissue for musculoskeletal tissue engineering. Clin Orthop Relat Res 468:2530–2540
41. Haraida S, Nerlich AG, Wiest I et al (1996) Distribution of basement membrane components in normal adipose tissue and in benign and malignant tumors of lipomatous origin. Mod Pathol 9:137–144
42. He L, Nan X, Wang Y et al (2007) Full-thickness tissue engineered skin constructed with autogenic bone marrow mesenchymal stem cells. Sci China C Life Sci 50:429–437
43. Hong SJ, Traktuev DO, March KL (2010) Therapeutic potential of adipose-derived stem cells in vascular growth and tissue repair. Curr Opin Organ Transplant 15:86–91
44. Huang S, Fu X (2010) Naturally derived materials-based cell and drug delivery systems in skin regeneration. J Control Release 142:149–159
45. Huh CH, Kim SY, Cho HJ et al (2007) Effects of mesenchymal stem cells in the reconstruction of skin equivalents. J Dermatol Sci 46:217–220
46. Jack GS, Zhang R, Lee M et al (2009) Urinary bladder smooth muscle engineered from adipose stem cells and a three dimensional synthetic composite. Biomaterials 30:3259–3270
47. Jiang Y, Jahagirdar BN, Reinhardt RL et al (2002) Pluripotency of mesenchymal stem cells derived from adult marrow. Nature 418:41–49
48. Kang JH, Gimble JM, Kaplan DL (2009) In vitro 3D model for human vascularized adipose tissue. Tissue Eng A 15:2227–2236
49. Keck M, Haluza D, Lumenta DB et al (2010) Construction of a multi-layer skin substitute: simultaneous cultivation of keratinocytes and preadipocytes on a dermal template. Burns 37(4):626–630
50. Kim WS, Park BS, Sung JH et al (2007) Wound healing effect of adipose-derived stem cells: a critical role of secretory factors on human dermal fibroblasts. J Dermatol Sci 48:15–24
51. Kim WS, Park BS, Sung JH (2009) The wound-healing and antioxidant effects of adipose-derived stem cells. Expert Opin Biol Ther 9:879–887
52. Kobayashi K, Suzuki T, Nomoto Y et al (2010) A tissue-engineered trachea derived from a framed collagen scaffold, gingival fibroblasts and adipose-derived stem cells. Biomaterials 31:4855–4863
53. Kumagai N, Nishina H, Tanabe H et al (1988) Clinical application of autologous cultured epithelia for the treatment of burn wounds and burn scars. Plast Reconstr Surg 82:99–110
54. L'Heureux N, Pâquet S, Labbé R et al (1998) A completely biological tissue-engineered blood vessel. FASEB J 12: 47–56

55. Labbé B, Marceau Fortier G, Fradette J (2011) Cell sheet technology for tissue engineering: the self-assembly approach using adipose-derived stromal cells. In: Gimble JM, Bunnell BA (eds) Adipose-derived stem cells: methods and protocols. Humana Press, Totowa
56. Lamme EN, Van Leeuwen RT, Brandsma K et al (2000) Higher numbers of autologous fibroblasts in an artificial dermal substitute improve tissue regeneration and modulate scar tissue formation. J Pathol 190:595–603
57. Larouche D, Paquet C, Fradette J et al (2009) Regeneration of skin and cornea by tissue engineering. Methods Mol Biol 482:233–256
58. Larouche D, Lavoie A, Paquet C et al (2010) Identification of epithelial stem cells in vivo and in vitro using keratin 19 and BrdU. Methods Mol Biol 585:383–400
59. Larouche D, Cuffley K, Paquet C et al (2011) Tissue-engineered skin preserving the potential of epithelial cells to differentiate into hair after grafting. Tissue Eng A 17(5–6):819–830
60. Lee EY, Xia Y, Kim WS et al (2009) Hypoxia-enhanced wound-healing function of adipose-derived stem cells: increase in stem cell proliferation and up-regulation of VEGF and bFGF. Wound Repair Regen 17:540–547
61. Mansbridge JN (2009) Tissue-engineered skin substitutes in regenerative medicine. Curr Opin Biotechnol 20:563–567
62. Mansbridge J, Liu K, Patch R et al (1998) Three-dimensional fibroblast culture implant for the treatment of diabetic foot ulcers: metabolic activity and therapeutic range. Tissue Eng 4:403–414
63. Marra KG, Defail AJ, Clavijo-Alvarez JA et al (2008) FGF-2 enhances vascularization for adipose tissue engineering. Plast Reconstr Surg 121:1153–1164
64. Mauney JR, Nguyen T, Gillen K et al (2007) Engineering adipose-like tissue in vitro and in vivo utilizing human bone marrow and adipose-derived mesenchymal stem cells with silk fibroin 3D scaffolds. Biomaterials 28:5280–5290
65. Merne M, Syrjanen S (2003) The mesenchymal substrate influences the epithelial phenotype in a three-dimensional cell culture. Arch Dermatol Res 295:190–198
66. Mesimaki K, Lindroos B, Tornwall J et al (2009) Novel maxillary reconstruction with ectopic bone formation by GMP adipose stem cells. Int J Oral Maxillofac Surg 38:201–209
67. Michel M, L'Heureux N, Pouliot R et al (1999) Characterization of a new tissue-engineered human skin equivalent with hair. In Vitro Cell Dev Biol 35:318–326
68. Miranville A, Heeschen C, Sengenes C et al (2004) Improvement of postnatal neovascularization by human adipose tissue-derived stem cells. Circulation 110:349–355
69. Misago N, Toda S, Sugihara H et al (1998) Proliferation and differentiation of organoid hair follicle cells co-cultured with fat cells in collagen gel matrix culture. Br J Dermatol 139:40–48
70. Mizuno H (2009) Adipose-derived stem cells for tissue repair and regeneration: ten years of research and a literature review. J Nippon Med Sch 76:56–66
71. Moioli EK, Chen M, Yang R et al (2010) Hybrid adipogenic implants from adipose stem cells for soft tissue reconstruction in vivo. Tissue Eng A 16:3299–3307
72. Morimoto N, Takemoto S, Kawazoe T et al (2008) In vivo culturing of a bilayered dermal substitute with adipo-stromal cells. J Surg Res 146:246–253
73. Nambu M, Ishihara M, Nakamura S et al (2007) Enhanced healing of mitomycin C-treated wounds in rats using inbred adipose tissue-derived stromal cells within an atelocollagen matrix. Wound Repair Regen 15:505–510
74. Nambu M, Kishimoto S, Nakamura S et al (2009) Accelerated wound healing in healing-impaired db/db mice by autologous adipose tissue-derived stromal cells combined with atelocollagen matrix. Ann Plast Surg 62:317–321
75. Paquet C, Larouche D, Bisson F et al (2010) Tissue engineering of skin and cornea: development of new models for in vitro studies. Ann NY Acad Sci 1197:166–177
76. Patrick CW Jr (2001) Tissue engineering strategies for adipose tissue repair. Anat Rec 263:361–366
77. Patrick CW, Uthamanthil R, Beahm E et al (2008) Animal models for adipose tissue engineering. Tissue Eng B Rev 14:167–178
78. Pellegrini G, Ranno R, Stracuzzi G et al (1999) The control of epidermal stem cells (holoclones) in the treatment of massive full-thickness burns with autologous keratinocytes cultured on fibrin. Transplantation 68:868–879
79. Perng CK, Kao CL, Yang YP et al (2008) Culturing adult human bone marrow stem cells on gelatin scaffold with pNIPAAm as transplanted grafts for skin regeneration. J Biomed Mater Res A 84:622–630
80. Pittenger MF, Mackay AM, Beck SC et al (1999) Multilineage potential of adult human mesenchymal stem cells. Science 284:143–147
81. Pouliot R, Larouche D, Auger FA et al (2002) Reconstructed human skin produced in vitro and grafted on athymic mice. Transplantation 73:1751–1757
82. Prockop DJ (1997) Marrow stromal cells as stem cells for nonhematopoietic tissues. Science 276:71–74
83. Rada T, Reis RL, Gomes ME (2009) Adipose tissue-derived stem cells and their application in bone and cartilage tissue engineering. Tissue Eng B Rev 15:113–125
84. Rehman J, Traktuev D, Li J et al (2004) Secretion of angiogenic and antiapoptotic factors by human adipose stromal cells. Circulation 109:1292–1298
85. Rochon MH, Fradette J, Fortin V et al (2010) Normal human epithelial cells regulate the size and morphology of tissue-engineered capillaries. Tissue Eng A 16:1457–1468
86. Rohrich RJ, Sorokin ES, Brown SA (2004) In search of improved fat transfer viability: a quantitative analysis of the role of centrifugation and harvest site. Plast Reconstr Surg 113:391–395; discussion 396–397
87. Sarugaser R, Lickorish D, Baksh D et al (2005) Human umbilical cord perivascular (HUCPV) cells: a source of mesenchymal progenitors. Stem Cells 23:220–229
88. Scuderi N, Onesti MG, Bistoni G et al (2008) The clinical application of autologous bioengineered skin based on a hyaluronic acid scaffold. Biomaterials 29:1620–1629
89. Scuderi N, Anniboletti T, Carlesimo B et al (2009) Clinical application of autologous three-cellular cultured skin substitutes based on esterified hyaluronic acid scaffold: our experience. In Vivo 23:991–1003
90. Shevchenko RV, James SL, James SE (2010) A review of tissue-engineered skin bioconstructs available for skin reconstruction. J R Soc Interface 7:229–258
91. Sinna R, Delay E, Garson S et al (2006) Scientific bases of fat transfer. Critical review of the literature. Ann Chir Plast Esthét 51:223–230

92. Stallmeyer B, Kampfer H, Podda M et al (2001) A novel keratinocyte mitogen: regulation of leptin and its functional receptor in skin repair. J Invest Dermatol 117:98–105
93. Sterodimas A, De Faria J, Correa WE et al (2009) Tissue engineering in plastic surgery: an up-to-date review of the current literature. Ann Plast Surg 62:97–103
94. Stillaert FB, Di Bartolo C, Hunt JA et al (2008) Human clinical experience with adipose precursor cells seeded on hyaluronic acid-based spongy scaffolds. Biomaterials 29:3953–3959
95. Sugihara H, Toda S, Miyabara S et al (1991) Reconstruction of the skin in three-dimensional collagen gel matrix culture. In Vitro Cell Dev Biol 27A:142–146
96. Sugihara H, Toda S, Yonemitsu N et al (2001) Effects of fat cells on keratinocytes and fibroblasts in a reconstructed rat skin model using collagen gel matrix culture. Br J Dermatol 144:244–253
97. Teepe RG, Kreis RW, Koebrugge EJ et al (1990) The use of cultured autologous epidermis in the treatment of extensive burn wounds. J Trauma 30:269–275
98. Toma JG, McKenzie IA, Bagli D et al (2005) Isolation and characterization of multipotent skin-derived precursors from human skin. Stem Cells 23:727–737
99. Trottier V, Marceau-Fortier G, Germain L et al (2008) IFATS collection: using human adipose-derived stem/stromal cells for the production of new skin substitutes. Stem Cells 26:2713–2723
100. Tsuji W, Inamoto T, Yamashiro H et al (2009) Adipogenesis induced by human adipose tissue-derived stem cells. Tissue Eng A 15:83–93
101. Uysal AC, Mizuno H (2010) Tendon regeneration and repair with adipose derived stem cells. Curr Stem Cell Res Ther 5:161–167
102. Vallee M, Cote JF, Fradette J (2009) Adipose-tissue engineering: taking advantage of the properties of human adipose-derived stem/stromal cells. Pathol Biol (Paris) 57:309–317
103. van den Bogaerdt AJ, van Zuijlen PP, van Galen M et al (2002) The suitability of cells from different tissues for use in tissue-engineered skin substitutes. Arch Dermatol Res 294:135–142
104. Vermette M, Trottier V, Menard V et al (2007) Production of a new tissue-engineered adipose substitute from human adipose-derived stromal cells. Biomaterials 28:2850–2860
105. Wang HS, Hung SC, Peng ST et al (2004) Mesenchymal stem cells in the Wharton's jelly of the human umbilical cord. Stem Cells 22:1330–1337
106. Wang Y, Zhao L, Hantash BM (2010) Support of human adipose-derived mesenchymal stem cell multipotency by a poloxamer-octapeptide hybrid hydrogel. Biomaterials 31:5122–5130
107. Wood FM, Kolybaba ML, Allen P (2006) The use of cultured epithelial autograft in the treatment of major burn wounds: eleven years of clinical experience. Burns 32:538–544
108. Young DA, Ibrahim DO, Hu D et al (2011) Injectable hydrogel scaffold from decellularized human lipoaspirate. Acta Biomater 7(3):1040–1049
109. Zhu WD, Xu YM, Feng C et al (2010) Bladder reconstruction with adipose-derived stem cell-seeded bladder acellular matrix grafts improve morphology composition. World J Urol 28:493–498
110. Zuk PA, Zhu M, Mizuno H et al (2001) Multilineage cells from human adipose tissue: implications for cell-based therapies. Tissue Eng 7:211–228

Industrial Approaches to Adipose Stem Cells Engineering

22

Filippo Boriani, Robert Warr, and Raimondo Ascione

Contents

22.1 Introduction 245

22.2 Technologies of Fat Processing 246
22.2.1 Simple Techniques of Fat Processing 246
22.2.2 Future Directions with ATDSC Processing 248

22.3 Technologies of Fat and Adipose Stem Cells Delivery 248

22.4 Integrated Systems 248

22.5 Industrial Applications of Adipose-Derived Stem Cells 251

22.6 Conclusion 254

References 254

F. Boriani (✉) • R. Warr
Department of Plastic Surgery, Frenchay Hospital, Frenchay Park Road, Bristol, BS16 1LE, UK
e-mail: borianifilippo@gmail.com; rpwarr@googlemail.com

R. Ascione
University of Bristol, The Bristol Heart Institute, Bristol Royal Infirmary, Bristol, BS2 8HW, UK
e-mail: r.ascione@bristol.ac.uk

22.1 Introduction

Although stem cells engineering represents a promising workhorse of future health care, nowadays its value is far from being relevant from a commercial point of view. Few companies currently base their income purely on the utilization of stem cells for regenerative medicine or on the sale of equipment in the biomedical field of regenerative cells. Many industries dealing with adipose tissue-derived stem cells (ATDSC) have the main part of their profit deriving from other traditional sectors of biomedical products, such as pharmacological production, commercialization of laboratory consumables and medical instruments. Currently, the lack of compelling evidence makes cells processing and engineering one of the most volatile sectors in the financial market of health and biomedical production. In addition, very little national funding is devoted to research in ATDSC engineering, which makes developing slow and mainly supported by commercial funding. Although stem cells industry has recently dedicated many efforts to embryonic stem cells, there has been little advance in translating these results into the clinical area due to the persistence of ethical, immunogenic, and tumorigenic issues. Concomitantly, potential advantages have been associated with the use of adult stem cells derived from bone marrow, blood and more recently from adipose tissue. Fat tissue is a virtually unlimited and cheap source of regenerative cells. Adipose stem cells can be harvested and injected with or without in vitro processing in the same patient and therefore represent an autologous potential therapeutic strategy for many diseases, which overpasses the problem of allologous tissue reactions.

Y.-G. Illouz and A. Sterodimas (eds.), *Adipose Stem Cells and Regenerative Medicine*,
DOI: 10.1007/978-3-642-20012-0_22,

Pioneer attempts of fat harvest and autotransplantation for reconstructive plastic surgery are well described in the literature and date back to the nineteenth century [3, 12]. Similar attempts are also described in the field of cardiac surgery. For example, omental flap has been used for reconstructive surgery following sternal wound infection/dehiscence. Furthermore, cardio-omentopexy had been demonstrated as a surgical technique for myocardial revascularization for ischemic heart disease induced by autologous tissue [2, 14, 18], before the era of direct coronary artery bypass grafting. Industrial involvement into ATDSC harvesting technology and engineering relates to all phases: from harvesting to delivery through the several more and less complicated types of fat processing and stem cells extraction, banking, culture, differentiations and clinical applications. We will explore in this chapter the technology currently available for ATDSC processing and delivery and discuss potential future development of this relatively new discipline. Of note, only demonstrative examples are given from the industrial world, with apologies to those industries that are not cited.

22.2 Technologies of Fat Processing

Once the lipoaspirate, an otherwise disposable by-product of aesthetic surgery, has been obtained, ATDSC can be extracted with specific ways of fat processing, including enzymatic digestion, filtration and centrifugation of the stromal-vascular fraction, where adipose-derived stem cells reside. However, autologous fat transplantation has been carried out for at least one century with minimal or no pretransfer fat processing, until the 1990s, when centrifugation was diffused by the American dermatologist Sydney Coleman.

22.2.1 Simple Techniques of Fat Processing

Simple techniques of fat processing preceding centrifugation had included filtration, lavage, separation with a decanter and oil absorption with blotting paper. Sydney Coleman confirmed that adipose tissue could be transferred satisfactorily if a strict protocol for preparation and injection of fat was respected.

22.2.1.1 Centrifugation

Centrifugation was discovered to concentrate adipocytes, separating them from adipo-toxic substances, such as lipids, proteases, lipases, and blood cells. Centrifugal forces vary by two parameters: revolutions per minute (rpm) and radius of centrifugation, although usually only the first parameter of the two is mentioned, being the second considered a standard feature of centrifuges. A more accurate measurement unit for this scope is the gravity acceleration (g), which is 9.8 m/s^2. Centrifugation separates the harvested fat into three layers:

1. An upper part of oil containing tryglycerides and chylomicrons deriving from cell lysis
2. A middle layer containing the purified fat
3. A lower layer with blood residues, serum and infiltration fluid

The recommendation for the choice of the centrifuge is a machine with a central rotor and sleeves that can be steam sterilized, in order to reduce the risk of contamination. Coleman designed several cannulas for harvest and several cannulas for injection, which differ significantly. Coleman's cannulas along with lipotransfer-specific equipment, all designed by Coleman himself, are commercialized by Byron Medical Inc, Tucson, Arizona. The rate of centrifugation proposed by the pioneer of fat processing is 3,000 rpm for 3 min [4]. This corresponds to 1,207 or 1,811 g, depending on either a 12- or an 18-cm radius of gyration, respectively. On the ground of possible cellular damage because of the centrifugal forces, many studies have been done investigating the effects of these forces on survival of adipocytes and adipose-derived stem cells [20, 24].

A recent study has shown that cell survival percentages are significantly lower when centrifugation rates of 1,200 g are used for longer than 5 min and 3,000 g for longer than 1 min [13].

The current recommendation is 1,200 g for 5 min as an optimal centrifugal force among all the tested forces for achieving valid results in fat transplantation, both in the short and long term.

22.2.1.2 Frozen Storage

Another technology of fat processing, which was introduced in 2001 by Shoshani and coworkers [21], is frozen storage for repeated procedures. Human adipose tissue extracted with lipoaspiration was preserved in a domestic refrigerator for 2 weeks at –18 °C. The fat

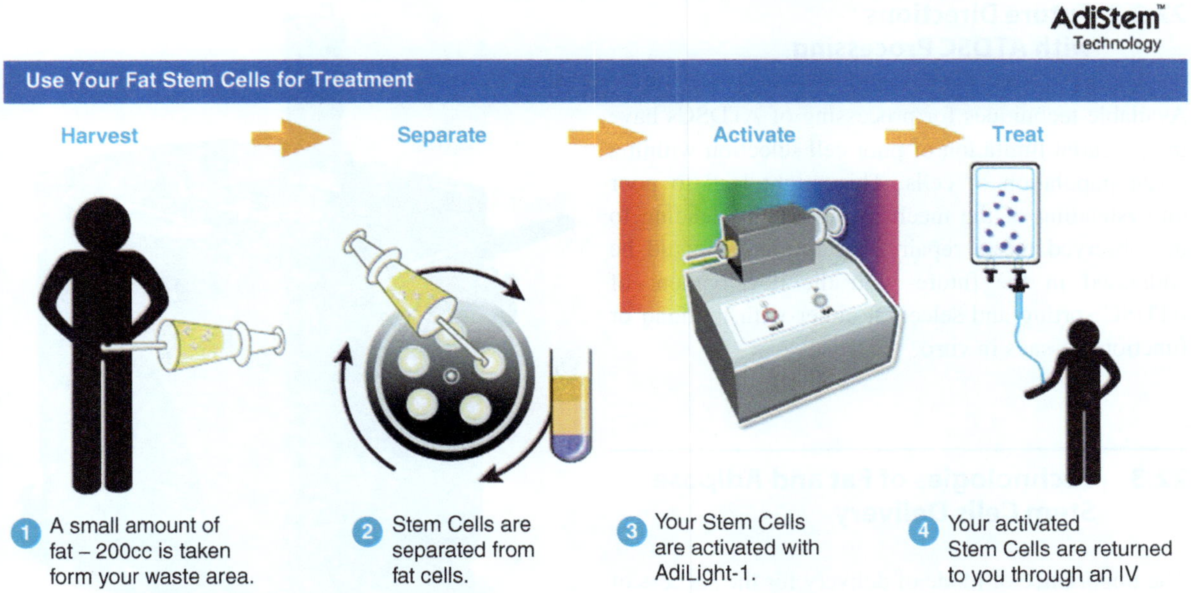

Fig. 22.1 Adistem™ technology for intravenous treatment

was then thawed and injected into nude mice. Fat-injected mice with freshly harvested fat and with defrosted fat were compared and no significant differences were found between the two groups, in terms of fat graft weight and volume.

For the specific purpose of harvesting stem cells from the lipoaspirate, the fat is divided into two parts. The first portion is washed thoroughly with sterile phosphate-buffered saline (PBS) to remove red blood cells and contaminating debris and then treated with 0.075% collagenase (type I, commercialized, among the various companies, also by Sigma-Aldrich, St. Louis, Missouri) in PBS at 37 °C for 30 min with gentle agitation. The enzyme is then inactivated with an equal volume of 10% bovine serum and the infranatant is centrifugated at 1,200 g for 5 min. The cellular pellet is then resuspended in 10% fetal bovine serum and passed through a 100-micrometer filter, in order to remove debris [23]. The second part of the lipoaspirate is centrifugated for 5 min at 1,200 g. The stromal vascular fraction obtained with the above mentioned fat processing can be expanded in culture or used directly as a source of stem cells for autotransfer. If adipose stem cells need to be expanded in culture, they are plated on gelatin-coated dishes (30,000 cells/cm^2) and cultured in an atmosphere of 5% carbon dioxide and humid air at 37 °C. The culture medium consists of 0.5 mM isobutyl-methylxanthine, 1 μM dexamethasone, 10 μM insulin, 200 μM indomethacin and 1% antibiotic/antimycotic. After 1 week, attached cells are trypsinized and cultured in the same medium, which has to be replaced every 3rd day.

22.2.1.3 Photomodulation

A recently commercialized modality of adipose stem cell processing, proposed by ISCI (International Stem Cell Institute, USA) and by Adistem Ltd (biotechnology company born in Hong Kong) is the Adilight1® a LED device which, according to the company, activates adipose stem cells by means of photomodulation (Fig. 22.1). Along with this product, Adistem also produces kits for extraction of stem cells from different volumes of lipoaspirated fat, for both clinical or research and laboratory utilization. The rationale for the use of this photoactivation would be the fact that large numbers of adipose stem cells lie dormant within the adipose tissue and therefore they need activation to become fully functional. Adistem also commercializes AdiStem PRP Kit®, a system that allows the extraction of growth factors (GFs) from the patient's platelets. These platelet-derived regenerative factors are then utilized to activate the adipose-derived cells harvested from the same patient. Adistem has also started a different type of fat processing for stem cells harvesting. This consists of the use of lecithin-liposomal factors instead of enzymes, which should make the extraction quicker.

22.2.2 Future Directions with ATDSC Processing

Available techniques for processing of ATDSCs have the potential limitation of poor cell selection within a given population of cells. This might lead to poor understanding of the mechanistic insights leading to the observed tissue repair. This limitation could be addressed in the future with the development of ATDSC sorting and selection either with immuno or functional assays in vitro.

22.3 Technologies of Fat and Adipose Stem Cells Delivery

The importance of mode of delivery for the success of stem cell therapy is becoming very obvious as new evidence become available. This concept applies also to ATDSC.

In the 1980s, Illouz [8–11] and Fournier [5, 6] invented a simple approach to fat transfer, after syringe harvesting, called by Fournier "microlipoinjection" [5]. This is based on injection of small drops of fat through 5–10 mL syringes, while withdrawing the needle. Some years later Coleman introduced a personally designed equipment for fat injection, consisting of small syringes with blunt cannulas, either curved or straight, much smaller than the cannulas utilized for aspiration. Cytori's Cellbrush™ is a recent innovative device intended for use on autologous fat delivery. This allows the surgeon, with a brush of thumb, to exactly control the amount of fat droplets injected into the area, with no need to look at the syringe at any time. In case of cell-enriched lipotransfer, consisting of fresh or cultured adipose stem cells, the cells are gently mixed with the processed lipoaspirate and after the 15 min needed for cell adherence to the centrifuged fat, the cell-enriched lipograft can be transplanted [23]. An injection syringe system delivers the cell-supplemented fat with an 18-gauge needle. Diffuse spreading of the complex lipograft is obtained by creating multiple planes in several directions, always from the depth to the more superficial aspects. The fat graft should be injected as the cannula is withdrawn and care should be taken in order not to inject excessive fat. An alternative modality of delivering the freshly cultured or isolated stromal vascular fraction is by resuspending it in saline and then diffusely infiltrating the lipografted area [23].

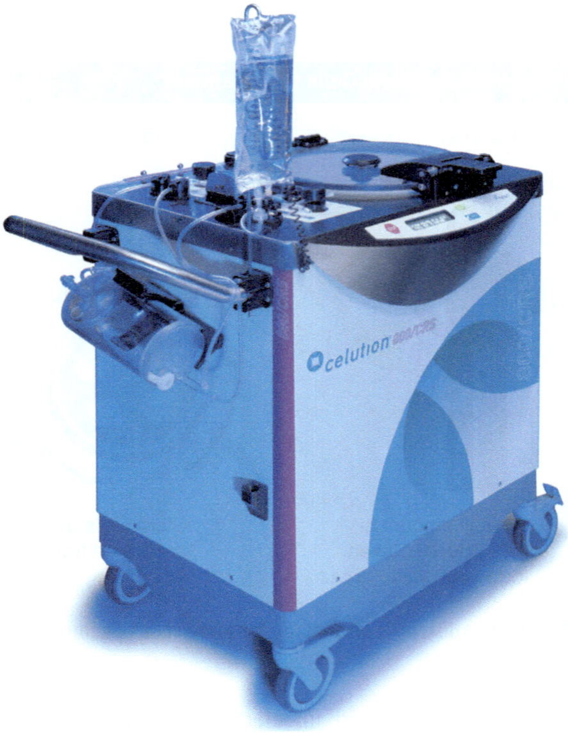

Fig. 22.2 Cytori's Celution®

22.4 Integrated Systems

The diffusion and establishment of fat transfer as an effective cosmetic and reconstructive technique, as well as the results in the medical research, validating the utility of this procedure, have stimulated the industrial world to commercialize sophisticated products for fat transfer. The most recent devices tend to integrate all or some of the three phases of harvesting, processing and injecting, in addition to the preliminary stage of infiltration, in a single or few complex systems. This has relevantly diminished the efforts and the labour of the surgeon during the procedure, although the preparation of the equipment may sometimes be arduous with these new complex devices. In addition, the cost of the machinery and of the consumables may be significant and biotechnology technicians may be required in the operation room. The most advanced integrated systems offered by the market allow a sophisticated fat-processing consisting of isolation and concentration of adipose-derived stem cells. In addition,

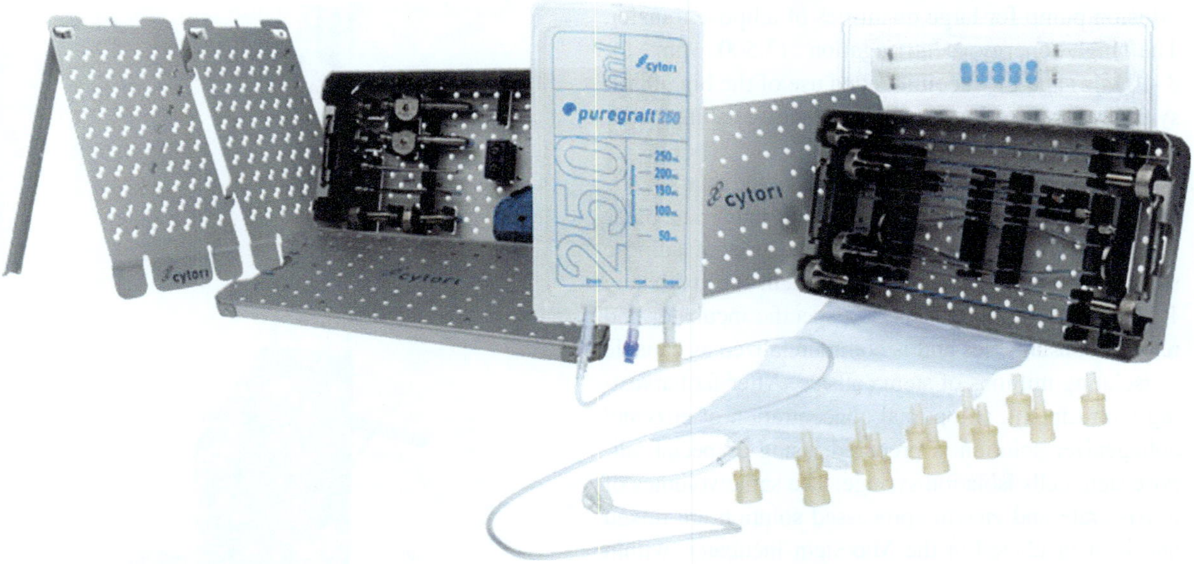

Fig. 22.3 Cytori's Puregraft™

these evolved integrated systems incorporate in a single or few equipments the stages of infiltration, harvest and injection of processed lipoaspirate along with adipose stem cells, what has been called cell-assisted lipotransfer. An example of complete and sophisticated integrated system for cell-enriched lipotransfer is Celution® 800/CRS System, the device commercialized by Cytori (Fig. 22.2). The Celution® System is a bedside completely automated device that isolates autologous adipose tissue-derived regenerative cells within about an hour, beside the patient. It consists of a closed system with disposables, allowing the clinician to produce a high yield cell suspension of stem and regenerative cells at the bedside for immediate use and this rapid processing can be directly included within the same procedure. The process starts with the harvest of adipose tissue by liposuction. During the fully automated isolation by the Celution® System, the adipose tissue is enzymatically digested and concentrated into a cell suspension. In the following stage, all lipid-laden adipocytes and debris are removed from the therapeutic ADRCs, which will contain a mixture of adult mesenchymal-like stem cells, endothelial progenitor cells and other adipose tissue stromal cells. At this stage the cell isolate is ready for delivery to the patient.

For extraction, processing and delivery of purified adipose tissue, without isolation of the regenerative component, Cytori has produced its Puregraft™ 250 (Fig. 22.3). It consists of an outer bag with a combination of inner membranes allowing gentle washing, removal of debris and tumescent fluid, and overall purification of the graft.

Tissue Genesis is also a company dealing with adipose-derived stem cells, offering its Tissue Genesis Cell Isolation System, including a harvest set, a cell-processing unit which works at the bedside and a delivery set.

Another example of integrated system for autologous fat transfer is the Lipokit® with Lee's 50 mL syringes, produced by the Korean Medi-Khan. With Lipokit several phases are offered within a single machine, called fat-processing unit (FPU): tumescent injection, liposuction, centrifugal harvest of pure adipose tissue with exclusion of oils and impurities and fat transfer under a closed-loop environment with nearly no exposure to contamination sources or surrounding air during the whole surgical procedure. For both the harvest and the transfer, the surgeon can use the same Lee's 50 mL autologous fat transfer syringes, which are equipped with a weight-mesh piston. This device can squeeze fat aspirates separating impurities and free oils from pure adipocytes during Lipokit centrifugation. The surgeon obtains low or high air pressure and vacuum power using foot pedal switches. Low or high air pressures are for infiltration and fat transfer and vacuum power for liposuction. For fat injection, the surgeon can either use syringes, even the 50 mL Lee's syringes themselves, or utilize the infusion pump for large quantities of adipose transfer. The high energy centrifugation (3,500 rpm for 8–10 min) and the recommended use of the big 50 mL syringes for injection are the two distinguishing peculiarities of the Lipokit system.

Medi-Khan has also launched on the market the Maxstem® device, which allows, with many analogies with the Cytori's Celution, the extraction of adipose stem cells in real time at patient's bed, in approximately 1 h and 20 min. It consists of a specific incubator and relative consumables and is commercialized as capable of isolating millions of stem cells/cc. After fat harvesting with Lipokit, an optimal concentration of enzyme/collagenizer solution is prepared, using a special adipose stem cells isolation syringe. The same volumes of lipoaspirate and enzyme-processed solution are mixed and then incubated in the MaxStem incubator, within rotatory motion and slope container for some certain minutes. The next stage is centrifugation at an optimized g force for 3–4 min. The suspension is then washed and centrifugated three more times, at desired g force and finally adipose regenerative cells are obtained at the bottom part of an adipose stem cells isolation syringe.

As a further possibility for stem cells delivery, Medi-Khan also commercializes a device for micronized autologous fat gel, an injectable fat filler. This represents an interesting autologous alternative to expensive artificial synthetic fillers. Injection is performed with a 26–30 gauge needle and is useful for the fine and elaborate injection areas.

For a gentle extraction of adipocytes, preadipocytes, and stem cells, the German Humanmed has produced Body-Jet (Fig. 22.4). Body-Jet Liposuction is a device delivering pulses of saline to gently dislodge fat for easier aspiration [25], with less vascular and nerve damage. This product is born from technological development of previous systems for waterjet-assisted treatment in various applications (e.g. helix hydro-jet). The aspirated fat is filtered, washed and collected inside a canister (LipoCollector/FillerCollector), from which it can be retrieved relatively easily with 50 mL syringes. With the syringe connector, smaller syringes are prepared for injection, with no need for centrifugation. The company recommends Bodyjet as a valid device for fat harvest and injection in breast augmentation and defines this surgical application BEAULI (Breast Augmentation Lipotransfer). According to Humanmed AG, this waterjet approach eliminates the necessity for the brute movement of an aspiration cannula which is necessary with traditional liposuction. Compared to traditional lipoaspiration, smaller volumes of tumescent fluids are injected into the body. With Body-Jet, the amount of fluid maintained at any given time within the body is therefore probably reduced, which enables the surgeon to inject less anaesthesia and perform a better anatomical assessment. According to publications, Body-Jet Liposuction is often performed in less time than traditional liposuction and patients experience less discomfort, swelling and bruising and shorter recovery due to the gentle water spray technology [1, 15, 16]. A recent study by C. Herold and coworkers [7] describes the results of fat harvesting by water-assisted liposuction (WAL) and subsequent autologous fat transfer. By means of MRI scan, autologous breast lipotransfer was evaluated in terms of breast volume,

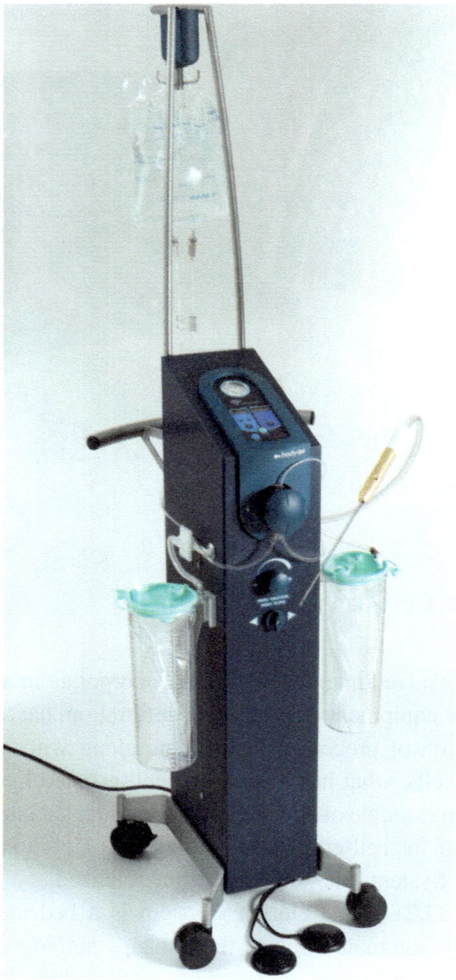

Fig. 22.4 Humanmed's Bodyjet

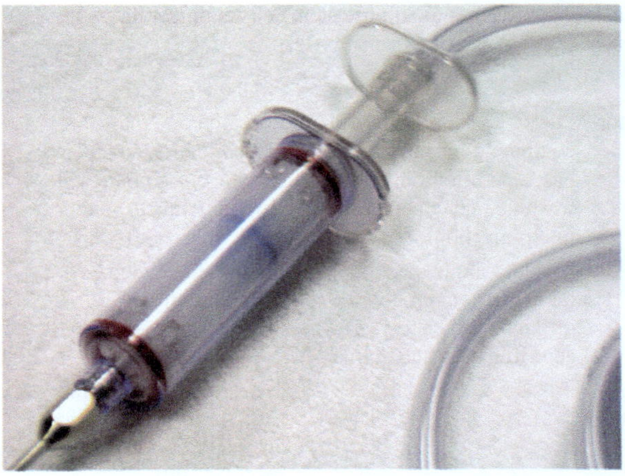

Fig. 22.5 Lipivage by Genesis Biosystems

preoperatively and postoperatively. A mean volume augmentation of 147 ± 18 mL was performed and $72 \pm 11\%$ of the transplanted adipose tissue persisted in the area of injection after 6 months. According to preliminary results of a study which is being conducted with WAL, there is no reduction in the amount of adipose stem cells in the lipoaspirate.

Another valid industrial device for relatively small-medium amounts of fat transfer is the LipiVage™ (Fig. 22.5), produced by the American company Genesis Biosystems. This innovative tool is a special syringe connected to an aspirator. Fat cells collect in the harvester, a sterile filtration chamber inside the device. Once inside, the lipoaspirate is gently lavaged and concentrated by the low vacuum and filter. Unwanted fluids pass through the lipoaspirate and filter, and are discarded into an attached waste canister. The resulting processed lipoaspirate consists of concentrated fat tissue, immediately ready for autologous transfer. As initially the Lipivage device was created for lipotransfer into the face and hands, a bigger system called High volume Lipivage has been created, for aspiration, processing and injection of larger quantities of fat tissue for body contouring.

22.5 Industrial Applications of Adipose-Derived Stem Cells

Over the last years there has been a growing interest from the industry on the use of ATDSC for tissue repair. A comparative summary of the industrial approaches to adipose stem cells engineering is reported in Table 22.1. The initiating industrial entity dealing with extraction, processing, delivery and research on adipose adult stem cells has been Cytori Therapeutics, especially since obtaining the CE Mark for its product Celution® in 2006. The Korean Medi-Khan is also investing relevantly in technology for autologous adipose tissue engineering. The two enterprises, along with Tissue Genesis, are currently leading in the field of adipose stem cells harvesting, banking, processing, engineering and experimenting. Medi-Khan is also producing human scaffolds through the process of frozen drying, for future adipose tissue medical engineering. For clinical use, the adipose stem cells isolation process can be done in any certified laboratory under Good Manufacturing Practice (GMP) conditions. The described commercial devices can perform the isolation of stem cells in a closed automated system in the operation room are still being evaluated by the FDA. The FDA has rules pertaining to the marketing of cells that have been more than "minimally manipulated," such as those digested with enzymes. Cytori's PureGraft System for pure fat grafting has just been FDA approved for use in the USA in 2010.

Clinical trials involving adipose regenerative cells selected with the technologies described above have been being conducted in Asia and Europe in the last few years with encouraging results. Lipoaspirate has been shown to be an optimal source for real-time cell therapy utilizing a patient's own cells, as initially described in April 2001 by a research group from the University of California at Los Angeles. Cytori was founded in 1997 as Macropore and in 2005 the name changed to Cytori which derives from "cyto" for "cell" and "ori" for "originate." Initially, Macropore produced and commercialized bioresorbable plates and screws. Between 2000 and 2002, Macropore management started considering alternative opportunities in regenerative medicine and discovered a common vision with a young company, StemSource, which had been founded by Dr Marc Hedrick. Dr Hedrick is currently Cytori's president and the new mission of and the focus of the company is based on his research work with adipose tissue and stem cells. In 2002, Macropore sold the rights for full commercialization of the original craniofacial and other products, which allowed it to acquire StemSource and fund development of the Celution System and its clinical application in terms of stem cells therapies. In 2005, Cytori signed a joint venture with Olympus Corporation worth up to \$55

Table 22.1 Comparison of products and research areas in the industrial approaches to adipose stem cells engineering

Company	Country	Products for ADSC isolation	Products for fat grafting	Research on ADSC/other products
Cytori	USA, but mainly operates in Europe and Japan	Celution® closed system for extraction of Vascular Stromal Fraction (VSF) in 1 h	PureGraft™ 250/Celbrush™: equipment for fat grafting, based on filtration, fat washing and controlled microinjection	Research on breast reconstruction, acute and chronic heart failure. Many other preclinical studies
Medikan	Korea	Maxstem® : bedside closed system for extraction of VSF in 1 h and 20 min	Lipokit: single machine for fat extraction, processing through high energy centrifugation and delivery	Filler-Geller, adipose-derived scaffolds
Tissue genesis	USA	Cell isolation system: bedside closed system for extraction of VSF in 1 h		Research on peripheral vascular disease, pelvic disorders. ADSC-coated vascular graft
Adistem/ISCI	Hong Kong, Mexico, USA	Use of lecithin-liposomal mixture for extraction. Adilight-1®: LED for ADSC photoactivation. Kits for stem cells isolation from lipoaspirate.		Parkinson's, rheumatoid arthritis, Alzheimer's, myocardial ischemia, lupus, multiple sclerosis, stroke, diabetes type 1 and type 2, COPD, spinal cord injuries, and autoimmune diseases
Humanmed	Germany		Body-Jet: single machine for waterjet-based fat extraction, processing through lavage and delivery	Study on preservation of ADSC in the waterjet lipoaspirate
Genesis Biosystems	USA		Lipivage™: vacuum system for extraction, filtration, lavage and reinjection of fat	
Cellerix	Spain			Inflammatory bowel disease/ Ontaril®: ADSC-based drug
RNL BIO	Korea			Adipose stem cell therapy for osteoarthritis and autoimmune disease
Medistem	USA			Multiple sclerosis, weight loss
Invitrogen	USA			Laboratory products for extraction, culturing, and differentiation of adipose stem cells

million in order to develop the next generation Celution System. Cytori started selling the Celution System in 2008 in cosmetic and reconstructive surgery institutions and practices. During 2008 and early 2009, the industry worked closely with some Celution utilizers in order to optimize the instrumentation, delivery tools and Celution System yield and processing time, so that it would be more suitable for the plastic surgery market. The product has received a CE Mark, for processing and re-injection of a patient's own cells during the same surgical procedure. FDA approval is in process. According to studies conducted by Cytori and other independent researchers [17, 26–28], the transplant weight of fat persisting after 6 months over the skull of nude mice normalized to a fat-only group is 250% with a fresh cells-enriched lipograft and 150% with cultured-enriched lipograft. The company also sells a system for adipose cell banking, which allows the preservation of fat tissue harvested in a single procedure, for multiple delivery sessions. Cytori, among the few other companies extracting adipose stem and regenerative cells, has a wide range of clinical and academic centres in Asia and Europe where studies on adipose stem cells are conducted. Cytori is now working with physicians to secure reimbursement in Europe for breast reconstruction following partial mastectomy and the RESTORE-2 study supports this effort. Preliminary results of this clinical research are encouraging, according to the investigators. More recently, the use of ATDSC for cardiovascular repair is gaining increasing interest particularly following the conduction of clinical studies by researchers of the Tulane University in collaboration with Cytori. These two studies (Apollo and Precise) focus on patients with myocardial infarction and assess the safety and efficacy of ATDSC processed with Cytori technology. In May 2010, based on the outcomes of the Apollo heart attack clinical trial, Cytori announced that adipose stem cells may improve heart vascularity and function when injected into a myocardial infarction and that they now plan a European Multicenter Pivotal Trial. It is reported that the trial showed a 47% reduction in the infarct size, as compared with the control group, along with an improvement in perfusion within the left ventricle. If confirmed this would be a major finding as infarct size is reported as predictor of poor clinical outcome. Similarly, data from the Precise trial on 27 patients with chronic heart attack suggest that the use of ATDSC is associated with a reduction in the extent of infarct size and a significant improvement in maximum oxygen consumption and patients' aerobic capacity. The research by Cytori and Tulane is the first to prove the possibility of using autologous, uncultured cells as a therapy for ischaemic cardiac disease. With a lower degree of advancement, Cytori is also conducting studies with adipose stem cells in spinal disc repair and wound healing. Independent investigators are being funded by several academic hospitals in Japan for studies on radiation injury, facial wasting, skin graft, cardiomyopathy, urinary incontinence, renal failure, liver disease, Crohn's disease, and calvarial repair. In many other areas, preclinical studies are currently being conducted and Cytori is obtaining revenue from the sale of devices and consumables used in these researches.

This research therefore suggests ATDSC as a further potential source of stem cell for myocardial repair and further studies are warranted to confirm these preliminary findings. Cytori has also announced the conduction of other clinical studies in the areas of wound healing. Concomitantly, independent investigators are being funded by several academic hospitals to assess the impact of ATDSC on other clinical conditions including radiation injury, facial wasting, skin graft, cardiomyopathy, urinary incontinence, renal failure, liver disease, Crohn's disease, and calvarial repair.

AdiStem is also contributing to research in adipose stem cells engineering in the fields of type II diabetes, breast reconstruction postlumpectomy, healing of diabetic ulcers and hair regrowth in baldness. Future research areas on which this industry is investigating include osteoarthritis, emphysema, stroke, heart failure and early stage Parkinson's disease. Adistem does not use any enzymes (such as collagenase) to digest the fat (or lipoaspirate), but rather a lecithin liposomal mixture (based on phosphatidyl choline) to emulsify the lipid containing cells (adipocytes) and then remove all non-lipid containing cells (stromal/stem cells). According to the company, this process allows for a quicker and more efficient extraction of the stromal cells. Furthermore, Adistem technology also includes the use of low level light frequencies to turn dormant stem cells into their active phase (S0 to S1). It is worthy noting that in collaboration with Adistem, the International Stem Cell Institute (ISCI) is currently planning clinical studies with lipoaspirate-derived regenerative cells for many indications, such as Parkinson's disease, rheumatoid arthritis, Alzheimer's disease, myocardial ischemia, systemic lupus erythematosus, multiple sclerosis, stroke, type 1 and type 2 diabetes mellitus, chronic obstructive pulmonary disease and spinal cord injuries.

Ontaril® (Cx401), produced by Cellerix (Madrid, Spain) is a medication obtained from adipose stem cells. It represents the first drug based on adipose regenerative cells, to be studied in a clinical trial. Preliminary results of ongoing prospective studies on Ontaril have shown ability to induce healing and regulate the immune system as well as anti-inflammatory properties. Ontaril is an injectable suspension containing regenerative cells. These are isolated from the patient's lipoaspirate and subsequently expanded. Ontaril is being evaluated in several clinical trials for two indications:

- Treatment of complex perianal fistulas with cryptoglandular origin (patients without inflammatory bowel disease).
- Treatment of complex perianal fistulas complicating Crohn's disease.

Traditional therapies for complex fistulas in Crohn's disease have low success rates and are expensive.

Surgery is often indicated for perianal fistulas without inflammatory disease, but the risk of faecal incontinence is high. Ontaril has been termed an Orphan Drug by the European Medicines Agency.

Invitrogen (USA) is a supplier of products for laboratory-based research and has recently commercialized STEMPRO®, defined as a Human Adipose-Derived Stem Cell Kit. This consists of adipose stem cells derived from a donor of human lipoaspirate and ready for culture and expansion with a specific medium.

Medistem Inc. has recently announced [19] encouraging preliminary results on three patients with multiple sclerosis who were treated with their own adipose-derived stem cells. This treatment was based on the rationale that the stromal vascular fraction of adipose tissue exhibits immunomodulation properties, in addition to regenerative activities. Another possible application of adipose stem cells which has been proposed by Medistem is obesity, based on the clinical findings that various stem cell populations, in particular obtained from liposuction mononuclear cells, are able to induce weight loss in subsets of patients.

RNL BIO CO. LTD is a Korean company in the area of embryonic and adult stem cell research. Many are the diseases target of studies and clinical trials by RNL, such as myocardial infarction, osteoarthritis, spinal cord injury, limb ischaemia, as well as cosmetic procedures. This company also offers a service of banking for adipose stem cells.

22.6 Conclusion

In the current global industrial scenario, adipose-derived stem cells remain a promising biotechnological weapon in plastic reconstructive and aesthetic surgery, while they are being tested as a further source of stem cells for tissue repair in other clinical areas [22] including cardiovascular.

Although this cell type and related technology is emerging, it does not have at the present a dominant position in the developing market due to limited interest over the years by scientists, biomedical industry and non-commercial funding bodies. Recently, this trend appears changing and the renewed interest in ATDSC might lead to the development of refined technologies. Fat regenerative cells have potential practical advantages when compared with other sources of adult stem cells. Their abundance, ease of harvest and processing, coupled with easy culturing in vitro might gradually make this cell type replace other more common sources of adult stem cells including bone marrow. This has led to a major interest by industry, triggering a large body of experimental and clinical work. Cytori and its Celution technology are leading the effort in developing the infrastructure for the advancement and utilization of adipose-derived therapies.

References

1. Araco A, Gravante MD, Araco F, Delogu D, Cervelli V (2007) Comparison of power water-assisted and traditional liposuction: a prospective randomized trial of postoperative pain. Aesthet Plast Surg 31:259–265
2. Beck CS, Leighninger DS (1954) Operations for coronary artery disease. J Am Med Assoc 156:1226–1233
3. Billings EJ Jr, May JW Jr (1989) Historical review and present status of free fat grafting autotransplantation in plastic and reconstructive surgery. Plast Reconstr Surg 83:368–381
4. Coleman SR (2002) Hand rejuvenation with structural fat grafting. Plast Reconstr Surg 110(7):1731–1744, discussion 1745–7
5. Fournier PF (1985) Microlipoextraction et microlipoinjection. Rev Chir Esth langue Franc 10:36–40
6. Fournier PF (1990) Facial recontouring with fat grafting. Dermatol Clin 8:523–537
7. Herold K, Ueberreiter F, Cromme MN, Busche PM (2010) Vogt: MRT-volumetrie der mamma zur kontrolle der fettresorptionsrate nach autologem lipotransfer. Handchir Mikrochir Plast Chir 42:129–134
8. Illouz YG (1985) De l'utilization de la graisse aspire' pour combler les defects cutane's. Rev Chir Esth Langue Franc 10:13
9. Illouz YG (1986) The fat cell graft: a new technique to fill depressions. Plast Reconstr Surg 78:122
10. Illouz YG (1988) Present results of fat injection. Aesthet Plast Surg 12:175–181
11. Illouz YG (1989) Fat injection: hope or reality? In: Illouz YG, De Villers YT (eds) Body sculpturing by lipoplasty. Churchill Livingstone, Edinburgh, pp 389–395
12. Illouz YG (1996) History and current concepts of lipoplasty. Clin Plast Surg 23:721–730
13. Kim IH, Yang JD, Lee DG et al (2009) Evaluation of centrifugation technique and effect of epinephrine on fat cell viability in autologous fat injection. Aesthet Surg J 29: 35–39
14. Knock FE (1958) Cardioomentopexy and implantation of multiple omental loops for revascularization of the heart. Surg Forum 9:230–232
15. Krahl D, Stutz JJ (2009) Water-jet assisted liposuction for patients with lipoedema: histologic and immunohistologic analysis of the aspirates of 30 lipoedema patients. Aesthet Plast Surg 33:153–162
16. Man D, Meyer H (2007) Water jet-assisted lipoplasty. Aesthet Surg J 27(3):342–346

17. Moseley TA, Zhu M, Hedrick MH (2006) Adipose-derived stem and progenitor cells as fillers in plastic and reconstructive surgery. Plast Reconstr Surg 118(Supp l):121S–128S
18. O'Shaughnessy L (1937) Surgical treatment of cardiac ischemia. Lancet 232:185–194
19. Riordan NH, Ichim TE, Min WP et al (2009) Non-expanded adipose stromal vascular fraction cell therapy for multiple sclerosis. J Transl Med 24:7–29
20. Rohrich RJ, Morales DE, Krueger JE et al (2000) Comparative lipoplasty analysis of in vivo-treated adipose tissue. Plast Reconstr Surg 105:2152–2158
21. Shoshani O, Ullmann Y, Shupak A et al (2001) The role of frozen storage in preserving adipose tissue obtained by suction-assisted lipectomy for repeated fat injection procedures. Dermatol Surg 27:645–647
22. Sterodimas A, de Faria J, Correa WE et al (2009) Tissue engineering in plastic surgery: an up-to-date review of the current literature. Ann Plast Surg 62:97–103
23. Sterodimas A, de Faria J, Nicaretta B et al (2009) Cell assisted lipotransfer. Aesthet Surg J 30:78–82
24. Sterodimas A, de Faria J, Nicaretta B et al (In press) Fat transplantation versus adipose derived stem cells enriched lipograft: a study. Aesthet Surg J
25. Taufig AZ (2006) Water-jet assisted liposuction. In: Liposuction – principles and practice. Springer, New York, pp 326–330
26. Yoshimura K, Sato K, Aoi N et al (2008) Cell-assisted lipotransfer for facial lipoatrophy: efficacy of clinical use of adipose-derived stem cells. Dermatol Surg 34:1178–1185
27. Yoshimura K, Sato K, Aoi N et al (2008) Cell-assisted lipotransfer for cosmetic breast augmentation: supportive use of adipose-derived stem/stromal cells. Aesthet Plast Surg 32:48–55, discussion 56–57
28. Zhu M, Zhengyu Z, Chen Y et al (2009) Supplementation of fat grafts with adipose-derived regenerative cells (ADRCs) improves long-term graft retention. Ann Plast Surg 63: 670–675
29. Zuk PA, Zhu M, Ashijan P et al (2002) Human adipose tissue is a source of multipotent stem cells. Mol Biol Cell 13:4279–4295

Future Research in Adipose Stem Cell Engineering

23

Jeanne Adiwinata Pawitan

Contents

23.1	**Introduction**	257
23.2	**Adipose Tissue-Derived Mesenchymal Stem Cells**	258
23.3	**Prospective Studies on AT-MSC**	259
23.3.1	Identity of AT-MSCs	259
23.3.2	Potentials of AT-MSCs	260
23.3.3	Preserving the Stemness of AT-MSCs	261
23.3.4	Hierarchy of AT-MSCs	262
23.4	**Formulas for Effective and Efficient Isolation, Expansion, and Differentiation**	262
23.4.1	Culture Condition	262
23.4.2	Culture Vessel	262
23.4.3	Medium	263
23.5	**Methods in Adipose Stem Cell Engineering**	264
23.5.1	Scaffold	264
23.5.2	Mechanical Stimulation	266
23.5.3	Means of Timely Delivery of Bioactive Factors and Cells	266
23.6	**Future Animal and Clinical Studies in Adipose Stem Cell Engineering**	267
23.6.1	Approach	267
23.6.2	Type and Structure of Tissue/Organ	267
23.6.3	Natural Development	268
23.6.4	Type of Cell	268
23.6.5	Scaffold	268
23.6.6	Safety, Integration, and Efficacy	269
23.7	**Conclusion**	270
References		270

J.A. Pawitan
Department of Histology, Faculty of Medicine, University of Indonesia, Jl. Salemba 6, Jakarta 10430, Indonesia
e-mail: jeanneadiwip@fk.ui.ac.id; jeanneadiwip@gmail.com

23.1 Introduction

Adipose stem cells have a bright prospect in regenerative medicine especially for tissue engineering. The term tissue engineering actually refers to regeneration of tissue/organ using three components, i.e., cell, scaffold and growth factors, though the three components are not always simultaneously used. Therefore, in a broader sense, tissue engineering may use only cells, scaffold, or growth factors [25]. In this chapter, the cell for adipose stem cell engineering refers to adipose tissue-derived stem cells. Compared to bone marrow stem cells, adipose tissue-derived stem cells have many advantages as adipose tissue is easier to get in larger volumes and at lower risks [35], and the most important for patient, the procedure to get adipose tissue is less painful compared to bone marrow puncture.

Recently, researches in tissue/organ engineering are accumulating. For adipose stem cell engineering, studies on various adipose tissue-derived stem cells have been conducted [9, 28, 35, 39, 46], and various biomaterials [11, 45] and methods for tissue/organ engineering using various kinds of cells have been studied [6, 7, 12, 18, 19, 26, 27, 43, 49].

However, problems to good manufacturing practice arose, as researches in adipose tissue-derived stem cells used various protocols or modifications of the protocols that lead to nonstandardized products. To use adipose tissue-derived stem cells for regenerative medicine, it is very important to follow good manufacturing practices in processing the adipose tissue to yield the desired stem cells in sufficient amount without using any animal-derived materials in the process [45].

Y.-G. Illouz and A. Sterodimas (eds.), *Adipose Stem Cells and Regenerative Medicine*,
DOI: 10.1007/978-3-642-20012-0_23, © Springer-Verlag Berlin Heidelberg 2011

Adipose tissue-derived stem cells are actually composed of myriads of stem cells that originated from a mixed population of activated pericytes at various stages of differentiation [10, 35]. Studies showed that adipose tissue-derived stem cells were not homogenous [54]. Moreover, some attempts to propagate certain adipose tissue-derived stem cells by serial passages met some problems as most culturing method lead to spontaneous differentiation [28, 35, 39, 46] and senescence [9].

Therefore this chapter highlights the advances in adipose tissue-derived stem cell researches, and focuses on prospective researches that are needed to overcome the hurdles in adipose stem cell engineering.

Various studies isolated adipose tissue-derived stem cells from the stromal vascular fraction (SFV) or alternatively called processed lipoaspirate (PLA) and the stem cells that grew after culture were composed of various kinds of stem cells that can be identified from their surface markers. These stem cells are supposed to come from the perivascular niches, and therefore the term vascular stem cells is actually more appropriate compared to adipose stem cells for adipose tissue-derived stem cells [35]. In identifying the stem cells, different studies used different set of markers and therefore got different results, even when using the same markers (Table 23.1). Moreover, some of these cells may express two or three markers [3, 10, 35]. These various kinds of adipose tissue-derived stem cells that attach on plastic when they are grown in culture are collectively called as adipose tissue-derived mesenchymal stem cells (AT-MSCs) [35].

Table 23.1 Adipose tissue stem cell identification using various markers [3, 35, 39]

Surface markers	Percentage of surface marker bearing stem cell
CD13 [3, 39]	5.6±3.9 [3], SVF: 37.0±0.2 [39], P0: 79.5±9.79 [39]
CD14 [3]	10.9±9.6 [3]
CD15 [3]	2.0±1.7 [3]
CD29 [39]	SVF: 47.7±13.3 [39], P0: 71.1±30.3 [39]
CD31 (endothelial marker) [3, 35, 39]	1.8±1.5 [3], NA [35], SVF: 21.8±10.8 [39], P0: 24.4±17.4 [39]
CD34 (hematopoietic and endothelial marker) [3, 35, 39]	6.9±3.0 [3] , 22.5 [35], SVF: 60.0±11.5 [39], P0: 59.2±25.4 [39]
CD44 [39]	SVF: 63.8±14.5 [39], P0: 84.1±8.2 [39],
CD45 [3]	9.0±6.9 [3]
CD73 [3, 39]	1.6±1.5 [3] , SVF: 25.0±6.2 [39], P0: 74.7±10.2 [39]
CD90 [3, 39]	29.2±20.8 [3] , SVF: 54.8±10.9 [39], P0: 76.6±9.6 [39]
CD105 [3, 39]	24.1±22.1 [3] , SVF: 4.9±3.5 [39], P0: 42.6±17.7 [39]
CD133 [3]	1.2±1.5 [3]
CD140b (pericyte marker) [35]	NA [35]
SSEA1 (stem cell marker) [35]	25.7 [35]
OCT4 (embryonic stem cell marker) [35]	1.94 [35]
Wnt5a (niche marker) [35]	NA [35]
Telomerase (stem cell marker) [35]	1.03 [35]
α-SMA (smooth muscle actin) [35]	NA [35]
STRO-1(stem cell marker) [35]	14.90 [35]

Astori et al, 2007[3] Studied crude SVF, Lin et al, 2008 [35] studied freshly isolated SVF that were attached to plastic, Mitchel et al, 2006 [39] studied SVF and P0

SVF stromal vascular fraction, *P0* primary culture, *NA* not applicable

23.2 Adipose Tissue-Derived Mesenchymal Stem Cells

Adipose tissue-derived mesenchymal stem cells showed similar properties to the previously characterized bone marrow-derived mesenchymal stem cells (BM-MSCs) [35]. The AT-MSCs bear similar CD markers to BM-MSCs, and in vitro can be stimulated to differentiate into cells of the mesodermal lineage such as osteogenic, chondrogenic, adipogenic and myogenic (skeletal muscle) cells, and even into neuronal protein expressing neuron-like cells [66]. Another studies showed that AT-MSCs under proper condition differentiated into endothelium and showed angiogenic capacity [47], and into functional cardiomyocyte-like cells that expressed specific cardiac markers and showed appropriate physiological response to adrenergic and cholinergic agonist [46]. In addition to their differentiation potential, AT-MSCs have a potential to prevent graft versus host disease (GvHD) [63].

Minimal criteria for BM-MSCs and other tissue MSCs according to the International Society for Cellular Therapy (ISCT) are adherence to plastic under standard culture condition, expression of specific surface markers, i.e., CD73, CD90, and CD105, and lack the expression of CD34, CD45, CD11b or CD14, CD19 or CD79a, and HLA class II, and differentiation capacity into osteoblasts, chondroblasts, and adipocytes under standard differentiating conditions [16]. However, some studies showed that some cells in the primary culture and early passages of AT-MSCs expressed CD34, a fact that opposed the minimal criteria, though the CD34 cells decreased upon further passages [35, 39], indicating differentiation into other kind of stem cells or into maturation. Further, in suitable culture condition, CD34+ and CD34– AT-MSCs can differentiate into one another [35, 47]. Several studies showed that serial passages of AT-MSCs lead to spontaneous differentiation [35, 39, 46] and senescence that begin within early passages, which is indicated by loss of proliferation capacity and change in morphology [28].

Further, all of the surface markers proposed by ISCT are also found on fibroblast that also lacks the expression of CD34, CD45, and CD14 [22, 61], and fibroblast is claimed to be able to differentiate into osteoblastic, chondrogenic, and adipogenic cells [22], and even into hepatocyte-like cell [37]. The same as MSCs, fibroblast also has immunosuppression property [22]. However, another study showed that fibroblasts have no osteogenic and adipogenic differentiation potential [61].

Considering the various kinds of stem cells that are labeled as AT-MSCs, questions about their identity arise when they are to be used in adipose stem cell engineering such as:

- Are AT-MSCs real stem cells, each with multipotential capacities, or are the various differentiation potentials due to accumulated potentials of the various AT-MSCs as a whole? Is fibroblast one of the AT-MSCs?
- If the multipotential capacities are accumulated potentials of the various AT-MSCs, does one type of AT-MSC have only one or more than one differentiation capacity? How are the differentiation potentials of the various AT-MSCs?
- In using AT-MSCs for adipose stem cell engineering, which should be chosen, whole AT-MSCs or enriched fraction of the variable individual stem cells?
- Which fraction of the variable AT-MSCs has the potential to prevent GvHD?
- What is the underlying mechanism of the potential to prevent GvHD?
- Which pathways are switch on or off, when the AT-MSCs differentiate into a certain kind of cells? And thus which substances are involved in the differentiation?
- As they differentiate into other cells upon passage, are the CD34-bearing AT-MSCs less differentiated compared to the other AT-MSCs?

23.3 Prospective Studies on AT-MSC

Considering the questions above, future researches should address the identity of the various stem cells that are present in AT-MSCs, the potentials of these various stem cells and the involved factors including factors that may play a role in the stemness of these cells, efforts to preserve stemness, and the hierarchy of these various stem cells. All these knowledge is invaluable to decide which of the various AT-MSCs and which treatment designs are to be used in certain adipose stem cell engineering.

23.3.1 Identity of AT-MSCs

To date, little is known about the identity of the various kinds of stem cells that are present in the so-called AT-MSCs. Our own experience showed that culture of adipose tissue-derived stromal vascular fraction yielded plastic adherent cells with various morphologies (Figs. 23.1 and 23.2). Therefore vast studies should be conducted to get more information, and to identify the suitable markers. The first step is to get enriched fractions of all kinds of these stem cells, either using flow cytometry equipment or antibody-linked magnetic beads, or both. A study isolated CD34 stem cells from the stromal vascular fraction of adipose tissue, which is the source of AT-MSCs, and showed that CD34 might be co-expressed with CD45 and CD133 [3]. In case there is co-expression, the fraction involved should be further fractionated. A problem may arise as fractionation may yield a minimal amount of cell in the fractions, but this can be overcome by new developed methods that can analyze a single cell [58].

Further, to reveal their identity, each fraction should be analyzed in many aspects, such as the gene expression, protein expression, or miRNA profile. Gene expression profile can be analyzed using either PCR

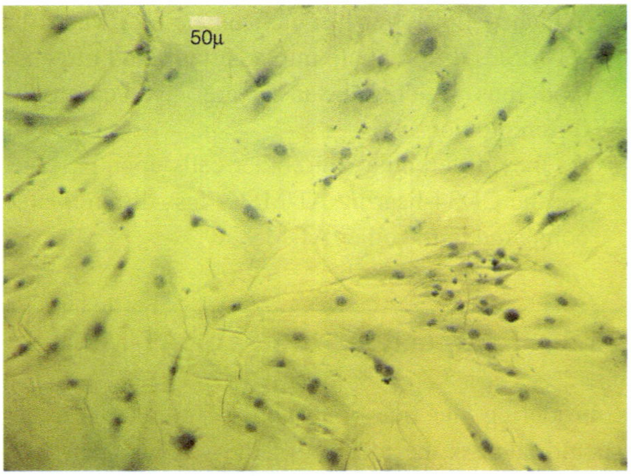

Fig. 23.1 Plastic adherent cells from culture of adipose tissue-derived stromal vascular fraction using DMEM-F12-based medium, passage 3 (hematoxylin and eosin stained, and photographed using Nikon Diaphot inverted microscope and Nikon E 8400 digital camera, magnification 200×)

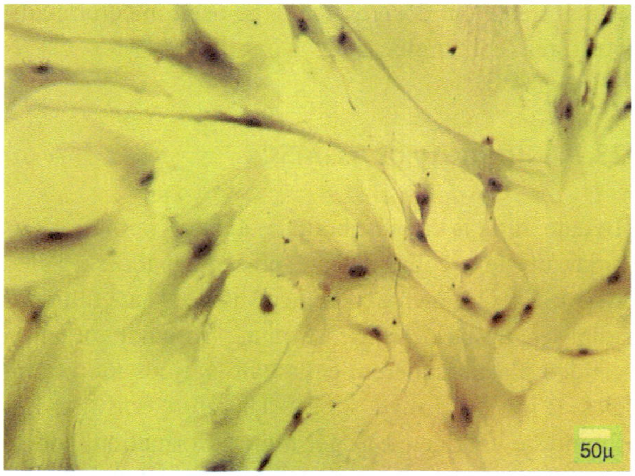

Fig. 23.2 Plastic adherent cells from culture of adipose tissue-derived stromal vascular fraction using DMEM-based medium with the addition of leukemia inhibitory factor, passage 3 (hematoxylin and eosin stained, and photographed using Nikon Diaphot inverted microscope and Nikon E 8400 digital camera, magnification 200×)

array or microarray; protein expression can be analyzed using ELIZA array or proteomic analysis either using the supernatant to reveal the cell secretions or using the cell lysate; and miRNA profile that reflects posttranslational regulation can be analyzed using microRNA PCR array or microarray [2, 50–52]. Further, epigenetic regulation of gene expression may be analyzed by analyzing methylation of the genome. The results of the analyses above may reveal the genes that are on and off in each fraction, which allow us to deduce the function, and may reveal the cell secretions that may be useful in certain co-culturing.

23.3.2 Potentials of AT-MSCs

In a certain adipose stem cell engineering choosing the appropriate kind of stem cell might be important. There are possibilities that the various differentiation potentials of AT-MSCs are due to accumulation of the potentials of myriads of stem cells that are present in the AT-MSCs; a certain fraction may have only one or two kinds of potential, while another fraction may have a broader potential. This supposition is supported by a study that showed a successful induction of AT-MSC using smooth muscle cell induction medium to make a proportion of the AT-MSCs to express smooth muscle cell-specific α actin (ASMA) and end-stage smooth muscle differentiation marker, the myoglobin heavy chain (MHC). The MHC is only present in cells with contractile ability. As seen in fluorescent staining, not all, but only part of the induced AT-MSCs expressed ASMA and MHC [48]. Therefore, studies to reveal the differentiation potentials of each fraction are highly required as a base in choosing the appropriate fraction to be used in certain tissue/organ engineering.

Another concern is to reveal whether the differentiation potential of a fraction needs the presence of another fraction, or the presence of AT-MSCs as a whole. A study in the Cold Spring Harbor Laboratory (CHSL) on bone marrow showed that nestin-expressing BM-MSCs, which outnumbered the hematopoietic stem cells (HSCs) by 10:1, were either found in clusters around the HSCs or in direct contact with HSCs to form a supporting niche for the HSCs. The MSCs are supposed to provide molecular signaling that might control HSCs maintenance, homing, and function in the bone marrow [14].

Studies showed that a proportion of AT-MSCs may spontaneously differentiate [28, 35, 39, 46], and therefore lost their stemness. Another study on BM-MSCs showed that long-term culture and serial passages reduced differentiation potential and caused morphological changes that are typical to cellular aging due to telomere shortening. Senescence is supposed to begin from the moment of in vitro culturing [9]. A study on bone marrow showed that a population of very small embryonic stem cell-like cells that expressed stage-specific embryonic antigen (SSEA), Oct4 and Nanog was present in addition to hematopoietic stem cells and BM-MSCs [31]. Oct4 and Nanog are transcription factors and together with SSEA are early developmental markers that are the characteristics for embryonic stem cells, and indeed these cells were shown to be pluripotent as are embryonic stem cells [31]. Further, another study on adipose tissue revealed that a small proportion of AT-MSCs expressed Oct4 and SSEA-1 [35].

It is very interesting to know whether Oct4- and SSEA-1-bearing AT-MSCs also have the property of embryonic stem cells as their counterpart in bone marrow, the very small embryonic stem cell-like cells, and whether they are development remnants of embryonic stem cells from the inner cell mass of an embryo, and how they remain undifferentiated, while others are differentiated. In addition, questions arise whether the various differentiation potentials of AT-MSCs are attributed to this population, and further, whether they spontaneously differentiate and enter senescence in culture.

To answer these questions, explorative studies are needed. Further, the various analyses as above are also needed to reveal what happen when the various AT-MSCs differentiate, either spontaneously or induced, and when they enter senescence. The results of those analyses may reveal novel mechanisms and pathways that are involved in differentiation, aging, stemness, and pluripotency in addition to that have been published before. Knowledge about the mechanisms and pathways that govern differentiation, aging, stemness, and pluripotency may be invaluable to allow us to engineer the fate of the cell to be used in adipose stem cell engineering.

One of the various potentials of AT-MSCs that is very important in adipose stem cell engineering is the immunosuppressive and anti-inflammatory potential [63]. Studies on BM-MSCs and AT-MSCs suggested that immunosuppressive and anti-inflammatory properties are due to the ability of MSCs to suppress lymphocyte's alloreactivity due to a shift from pro-inflammatory (TNF-α, IFN-γ, and IL-12) to anti-inflammatory (IL-4, IL-10) cytokine production at the site of injury, and secretion of other soluble factors (IL-6, vascular endothelial growth factor, nitric oxide, and IFN-γ-induced indoleamine 2,3-dioxygenase) by MSCs in the presence of T cell, which suppress B and T cell as well as NK cell proliferation, suppress differentiation and maturation of the antigen-presenting dendritic cells, and suppress monocyte differentiation into antigen-presenting cells [33, 44, 63].

Further, a proportion of MSCs express MHC class I molecules, which prevent NK cell-mediated lysis and promote tolerance. Moreover, most MSCs do not express MHC class II molecules on their cell surface, though intracellular MHC class II is present. Both MHCs expression can be induced by IFN-γ, and inflammation degree is supposed to play a role in MHC class II expression. However, when MSCs have differentiated, MHC class II can no longer be induced. Whether re-expression of MHC class II occurs, when the differentiated allogeneic MSCs are transplanted into an incompatible host is unclear. In human, low number of allogeneic MSCs can stimulate immune response by increasing IgG and IFN-γ production, while excess MSCs suppress immune response. Therefore, studies on the regulation of MHC class II in AT-MSCs need to be conducted, if allogeneic AT-MSCs are to be used [44].

The question is whether immunosuppressive and anti-inflammatory potentials are due to AT-MSCs as a whole, or only due to a certain kind of AT-MSCs, as MHC class I molecules are only present in a proportion of MSCs. The same question may be raised on the immunostimulatory potential of allogeneic AT-MSCs. Therefore, studies to reveal the AT-MSCs that are responsible for immunosuppression and anti-inflammation as well as those responsible for immune rejection and their complete mechanisms are very important. If only a fraction is responsible for immunosuppression or immunostimulation property, the fraction with desirable property can be enriched, and that with undesirable property can be depleted.

23.3.3 Preserving the Stemness of AT-MSCs

In adipose stem cell engineering different approaches can be used, as using unexpanded whole AT-MSCs (in the form of freshly isolated SVF/PLA), expanded AT-MSCs (cultured SVF/PLA), fractionated AT-MSCs or AT-MSC-derived differentiated cells (induction of SVF/PLA, cultured SVF/PLA, or fractionated AT-MSCs to a certain kind of cells, such as chondrocytes for cartilage engineering, osteocytes for bone engineering, etc.).

When undifferentiated cells are to be used, whether fractionated, expanded, or unexpanded, the cells should be able to retain their stemness, especially when they are not directly used after processing; thus they should be kept in culture or cryopreserved. Therefore studies on preserving the stemness of the various AT-MSCs in culture and after cryopreservation are very important, and once again, knowledge of the mechanisms and pathways that are involved in stemness is invaluable in designing the studies.

23.3.4 Hierarchy of AT-MSCs

An intriguing question is whether the various AT-MSCs follow a certain hierarchy of differentiation [59]. For instance, whether CD34-bearing AT-MSCs have the potential to differentiate into cells of hematopoietic lineage and endothelium as their counterpart the CD34-bearing BM-MSCs, and whether they may transdifferentiate into other lineages. Another interesting issue is whether CD133-bearing cells are upstream in the hierarchy compared to CD34-bearing cells, and then at the downstream are the non-CD34-bearing cells. In vivo, AT-MSCs may follow a certain differentiation hierarchy. However, studies showed that in vitro, AT-MSCs may transdifferentiate even into neuron-like cells when properly induced [66]. Knowledge of in vivo and in vitro differentiation hierarchy (if any) may be valuable when undifferentiated cells are to be used, i.e., by choosing the AT-MSCS at the nearest upstream hierarchy to the cell needed, so that they will readily differentiate into the desired type of cells.

23.4 Formulas for Effective and Efficient Isolation, Expansion, and Differentiation

Adipose tissue is readily available and can be obtained by liposuction or excision. In general, after thoroughly washing and mincing of excised samples, the method to isolate AT-MSCs composed of enzymatic digestion of extra-cellular matrix by collagenase, and stem cell retrieval from the infranatant of the digest. However, after that, there are various minor differences in the methods, such as the speed in centrifugation, with or without the use of lysis buffer and/or different mesh filters. Further, there are differences in the culture medium and conditions for primary culture, expansion, and differentiation, which yield variable results [3, 39, 47, 66].

Successful stem cell expansion and differentiation depends on the type of stem cell, culture condition, culture vessel and medium [45]. Therefore, studies to find standardized formulas for effective and efficient isolation, expansion, and differentiation of the various AT-MSCs regarding culture condition, culture vessel and medium are very important.

23.4.1 Culture Condition

Culture condition includes pH, oxygen and carbon dioxide percentage, and temperature. Most cultures are done at pH 7.4, 20% oxygen, 5% carbon dioxide and 37°C. However, different cell type or different culture aim may require different condition. A study showed that pH 7.1 favored the expansion of megakaryocyte progenitors [64]. Oxygen demand may vary for different cell lineages [45], and 40% oxygen retained embryonic stem cell state in mouse embryonic stem cell as shown by Oct-4 expression and alkaline phosphatase activity [32]. Further, lower temperature may reduce oxidative damage and maintain viability and population doubling time in MSCs, as extended expansion in usual condition lead to reduced viability due to oxidative damage [45]. Another important issue is the sterility of culture. Therefore handling and the materials used in culture should ensure that the culture is not contaminated, either by microorganism or other unimportant materials.

The knowledge from the studies above can be considered in designing prospective studies to establish an optimized standard culture condition for the isolation, expansion, and differentiation of every AT-MSCs to produce the required number of long-term viable desired cells for successful adipose stem cell engineering.

23.4.2 Culture Vessel

The cells in the body interact with each other, especially with adjacent cells and extracellular matrix. Adherence to the extracellular matrix is mediated by transmembrane proteins called integrins. Integrins may act as cell surface receptors that upon binding with certain extracellular matrix protein will activate certain intracellular signaling pathway, which control certain gene expressions, cytoskeletal organization, cell morphology, and in vitro differentiation in the involved stem cell. In murine embryonic stem cell culture, coating the culture vessel with certain extracellular matrix proteins such as laminin and fibronectin cannot stop spontaneous differentiation into primitive ectoderm, even in the presence of leukemia inhibiting factor (LIF). However, coating the culture vessel surface with type I and IV collagen, gelatin or poly-D-lysine can force the cells to remain undifferentiated. For MSCs, coated culture vessel facilitate cell adherence, and best adhering function is achieved by coating with fibronectin, followed by collagen I, collagen IV, vitronectin, and laminin-1 [45], but vitronectin and collagen I perform better for osteogenic differentiation [53]. Recently, a number of commercial matrices can be used for culture vessel coating, such as Matrigel and Cartrigel, but their animal origins make them unsuitable for human adipose stem cell engineering [11].

Therefore extracellular matrix proteins, whether they are secreted by the cell itself or provided in the culture system, either in the medium or by coating the culture vessel surface may play a role in cell adhesion, proliferation, and differentiation. Thus, developing various human-derived substitutes for matrices and extracellular matrix proteins, and designing future studies on coated culture vessels that are needed by each AT-MSC are of great importance to improve the ability to keep the stem cells in undifferentiated state, or to increase differentiation efficiency toward the desired cells when required.

23.4.3 Medium

Most complete medium formula for culture contains serum and additives. Fetal bovine serum (FBS) seems to be essential for cell culture, but some batches are not safe when the cultured cells are to be used in regenerative medicine. Animal-derived serum may contain xenoproteins that may cause rejection, and may contain transmissible infectious agents. Contamination with certain xenoproteins that contain N-glycolylneuraminic acid (Neu5Gc) may be reduced by transferring the culture into human serum-containing medium. However, certain xenoproteins may be internalized by MSCs, and sialic acid-containing xenoprotein may be incorporated to the cell surface of the cultured cells [8]. Therefore it is of great importance to develop substitute to FBS, though good manufacturing practice-compliant FBS batches are available. Therefore studies to compare the efficiency and efficacy of various available FBS substitutes, as well as development of new substitutes should be conducted. In addition, studies on the possible use of additives to increase growth, direct differentiation, and prevent senescence are of great importance.

23.4.3.1 Human Serum, Plasma, and Platelet-Derived Factors as FBS Substitute

Several studies substituted FBS with human serum, plasma, or platelet-derived factors. The platelet-derived factors can be obtained from concentrated buffy coat-derived platelet-rich plasma that contains at least 10^9 platelets/mL. The platelets are either fragmented by repeated freeze–thaw cycles or activated by thrombin to release the platelet's alpha granules content. The platelet-derived factors consist of platelet-derived growth factors (PDGFs), basic fibroblast growth factor (bFGF), epidermal growth factor (EGF), transforming growth factor β1(TGF-β1), and insulin-like growth factor (IGF) [8, 29]. Though in some studies the use of platelet-derived growth factors as FBS substitute showed promising results, their effects on various stem cell cultures were variable depending on the type of stem cells, platelet growth factor content, mode of preparation, methods of growth factor release, and leukocyte contamination. Further, comparison of pooled human serum, blood type AB human serum and various protocols to release the platelet-derived factors showed that both AB human serum and thrombin-activated platelet release of platelet-derived factors were superior for the adhesion and expansion of AT-MSCs compared to FBS. Studies on BM-MSCs showed that allogeneic AB human serum was not superior to FBS and reduced growth due to senescence was observed. Thus necessitates the use of autologous human substitute [8].

Therefore, efficacy and efficiency studies in using serum or platelet-rich plasma either autologous or allogeneic from blood type AB individuals for the isolation, expansion, and differentiation of the various AT-MSCs need to be conducted.

23.4.3.2 Studies to Develop New FBS Substitutes

Studies showed that whole growth factors from platelets were able to replace FBS for AT-MSC as well as BM-MSC cultures [8, 29, 55]. A study on BM-MSCs showed that a combination of TGF-β, PDGF, and β FGF was required to grow BM-MSCs in a serum-free medium, and could maintain the growth for up to five passages [41]. Therefore, it is interesting to test whether individual growth factors or in combinations may also work for the various AT-MSC isolation, expansion, and differentiation. Further, studies on the various growth factor effects on the various AT-MSCs should also include the activated pathways to reveal the intermediate substances that are produced by the cultured cells. These substances in turn may be tested again for their efficacy and efficiency as additives in various AT-MSCs isolation, propagation, and differentiation.

23.4.3.3 Studies on Potential Additives

Various additives were used to enhance MSC proliferation and/or differentiation. A study used dexamethasone on human cord blood-derived MSC [30], while another study used lithium chloride on BM-MSCs, which stimulated Wnt signaling to increase proliferation. Low levels of Wnt signaling lead to increased proliferation, while high Wnt levels lead to osteogenic differentiation [15]. Recently, a dipeptide, the Glutamax is available as a more stable alternative to L-glutamime, and was claimed to support cell growth better than l-glutamine [57]. Further, various additives were used in the differentiation of AT-MSCs into chondrocytes, osteocytes, lipocytes, etc. [46, 47, 66].

A study showed that senescence in vascular endothelial cells were mediated by integrin β4 that was supposed to suppress the expression of Ca^{2+}-independent phosphatidylcholine-specific phospholipase C (PC-PLC), increase the level of p53, and reactive oxygen species (ROS) [36]. Moreover, ROS from cellular origin are involved in redox signaling, cell damage and senescence. Addition of selenite (100 nM) in BM-MSC culture increased superoxide dismutase 1 (SOD 1) expression and reduced ROS accumulation [17].

Therefore, search for potential additives in the various AT-MSC cultures should take advantage of the knowledge of previous studies on other cell cultures. Testing various substances with antioxidant property may find new additives that are suitable for certain AT-MSC culture. As different AT-MSC may require different additives for isolation, expansion and differentiation, testing should be done on the whole as well as the various AT-MSC fractions. Further, studies on various signaling biomolecules that are involved in various signaling pathways that are active in embryonic stem cells and in early embryogenesis may reveal their potential as additives to maintain self-renewal and confer pluripotency to AT-MSC, as well as to direct differentiation into various desirable lineages.

23.5 Methods in Adipose Stem Cell Engineering

Many studies on tissue/organ engineering have been published. A study reported successful urinary bladder engineering using urothelial and muscle cells that were grown in culture, which were seeded on a collagen, or a composite of collagen and polyglycolic acid scaffold. Urinary bladder is a simple organ that consists of three layers, and the study engineered the mucosa and muscular layer, while the third layer, the serosa was constructed using an omental wrap [4].

A study even has succeeded to engineer a complex organ, the kidney. The study showed that embryo-derived cloned renal cells that were seeded onto a biodegradable scaffold ex vivo and implanted into the subcutaneous space succeeded to form functional renal tissue, in terms of excreting urine-like fluid. Histological examination of the renal tissue showed extensive vascularization and glomeruli- and tubule-like structures, with clear continuity between the glomeruli and tubules [24].

To be successful, tissue/organ engineering should involve timely regulated proliferation and differentiation on an appropriate surface of three-dimensional scaffold [45]. To achieve these goals, appropriate scaffold and means to supply nutrients and deliver substances, as well as means to provide mechanical stimuli when appropriate to direct timely proliferation and differentiation are highly required. Therefore, further studies in these fields should be conducted.

23.5.1 Scaffold

Preparing three-dimensional scaffolds for tissue/organ engineering should consider many factors, i.e., the material, material porosity and pore size, mechanical property, stability and degradability, bioactivity, hydrophobicity, biocompatibility, toxicity, and immunogenicity [45]. Therefore, this section addressed those factors, and prospective studies in the use and development of various scaffolds.

23.5.1.1 Material

To date, various scaffolds can be obtained from either natural biomaterials that lack mechanical strength or modified synthetic biomaterials that are mechanically strong with desired bioactive properties to facilitate cellular growth, differentiation, and tissue/organogenesis. Natural biomaterials that can be used to produce scaffolds are either proteins such as collagen, fibrin and silk, or polysaccharides such as agarose, alginate, hyaluronan, and chitosan, while synthetic biomaterials are polymer-, peptide-, or ceramic-based biomaterials [62]. These scaffolds can be engineered into a certain desirable shape with suitable micro-architecture (desirable porosity and pore size). Various studies developed and used various scaffolds such as porous collagen, cellulose porous microspheres, alginate hydrogels, porous gelatin microspheres (CultiSpher G), porous biomatrix (Cellfoam), tantalum-coated porous biomaterial (TCPB), polyester nonwoven fabric porous disk carriers (Fibra-cel), nylon filtration screen, nonwoven polyethylene terephthalate (PET), and porous polyvinyl format (PVF) [45].

23.5.1.2 Porosity and Pore Size

As cell growth needs nutrients, the scaffold should contain a network of interconnected pores and channels to facilitate penetration of nutrients and metabolites and to allow penetration of cells and the formation of cellular associations. Pore size has an impact on cellular adhesion, viability, formation, and distribution of extracellular matrix by a certain cell type. Further, it is supposed that pore size can be used to prevent or to support the growth of certain cells. For bone tissue engineering, the recommended pore size is between 100 and 500 μm, and for fibroblast, larger than 500 μm [45].

23.5.1.3 Mechanical Property

The various tissues in our body have different mechanical property, for instance, loose connective tissue is soft, cartilage is viscoelastic and bone is hard and rigid. Therefore, in tissue engineering, scaffold choice should be appropriate with the natural property of the tissue to be engineered. For instance, polyglycolic acid (PGA) yarns that are woven into an orthotropic porous textile have similar mechanical properties to a native articular cartilage [40]. Therefore, it is a suitable scaffold material for articular cartilage engineering. Moreover, a certain material's mechanical property, such as stiffness of agarose gels can be set by regulating the cross-linking densities, to get various degrees of stiffness, and indeed, human MSCs cultured on agarose gels differentiated into neuronal, muscle, or bone lineages when the stiffness approximated that of brain, muscle, and bone, respectively [11]. Further, various uses of scaffold need various mechanical properties; for instance, when a scaffold needs to be sutured to fix it in a certain location, it needs high tearing strength; when it is used to support a tubular system and should be able to avoid stricture, so it needs a property to endure compression [25].

23.5.1.4 Bioactivity

Bioactivity of scaffold material is essential for proper tissue/organ engineering. Many synthetic polymers with hydrophobicity and low surface energy lack cell recognition signals. Thus are not suitable for tissue/organ engineering [65]. Bioinert synthetic polymers can be made biofunctional by combining them with certain bioactive fibers, extracellular matrix (glycosaminoglycans, proteoglycans, and glycoproteins), proteins, peptides or soluble bioactive molecules such as cytokines, chemokines, growth factors or hormones to provide beneficial effect on cell adhesion, growth, and differentiation [11, 13, 38, 45]. In the future, development of various bioactive scaffolds that are suitable for various tissue/organ engineering is very prospective.

23.5.1.5 Stability and Degradability

When biodegradable scaffold is needed, various biomaterials are recently known as biodegradable, such as alginate hydrogels that have controllable biodegradability [45], or a combination of alginate with gelatin [5, 45]. However, certain biodegradable scaffold may induce inflammation due to the acidity of the degraded product as in the case of poly (L-lactic-co-glycolic acid) [45, 60]. Therefore, studies on the stability and degradation rate of the various biodegradable materials are very important in designing smart scaffold with delivery properties.

23.5.1.6 Biocompatibility

The material of a scaffold used in tissue/organ engineering should be biocompatible in term that it is nontoxic, does not cause cytotoxicity, or irritation to the surrounding tissue, has no/low immunogenicity, and can integrate well into the host tissue [11, 45]. However, when immunorejection problems arise, immunomodulatory molecules may be incorporated into the material [11].

As the site of the damaged tissue/organ to be repaired often troubled by inflammation and increased immunological surveillance, a good scaffold biomaterial should be able to protect the cells that are incorporated into the scaffold from host immune system [11].

23.5.1.7 Prospective Studies in the Use and Development of Scaffolds

Recently, various techniques are available to produce various scaffolds with various properties [45]. Therefore, in the future, development of various materials with various porosities, mechanical properties, and bioactivities, which are biocompatible for human, and studies on their effective and efficient application as scaffold for various tissue/organ engineering are of great importance.

In nature, the extracellular matrix has surface features such as pores, pits, grooves and ridges, and surface chemical properties, which are important for cell growth, movement, and orientation. Therefore, development of scaffolds for the various tissue/organ engineering is suggested to mimic the natural features of the extracellular matrix of the tissue/organ to be engineered, for instance, differentiation into osteoblastic phenotype was favored by 150-nm micro-grooved patterned surface coated by fibrinogen or calcium phosphate. Further, various micro- and nano-scale patterning techniques, such as various soft lithography, photolithography, electron-beam lithography, Langmuir–Blodgett lithography, hot embossing imprint lithography, nano-imprinting, three-dimensional (3D) printing, electrospinning, layer-by-layer microfluidic patterning, ion milling, reactive ion etching, and chemical or physical vapor deposition, and electrochemical deposition are recently available to create scaffolds with certain surface feature and coating [11].

Therefore, further studies using these micro- and nanoscale patterning technologies are needed to tailor various scaffolds with certain mechanical property, porosity, surface pattern, and coating that mimics the natural extracellular matrix and microenvironment of the various AT-MSCs, which allow them to be undifferentiated or differentiated into the desired type of cell. In addition, studies on the use of these scaffolds in various tissue/organ engineering need to be conducted to find the most effective and efficient technique for adipose stem cell engineering.

23.5.2 Mechanical Stimulation

Mechanical stimulation induces changes in the extracellular matrix that are transmitted to the cell surface via deformation at the integrin-binding sites and cause mechanobiological responses. These responses depend on the type of mechanical stimulus, cell type, and stimulation site. Various mechanical stimulations such as shear stress, stretching, cyclic strain, and physiological deformational loading were successfully applied to induce differentiation into various kinds of cell. Various devices and bioreactors to provide controlled mechanical stimulation have been used in various studies [45]. Therefore, studies on the level and type of mechanical stimulation needed by the various AT-MSCs to remain undifferentiated or to be differentiated and incorporated in a certain tissue or organ are of great importance.

23.5.3 Means of Timely Delivery of Bioactive Factors and Cells

Complex tissues and organs contain not only many kinds of cells that form a network and cooperate with each other, but also blood vessels and innervations. Therefore, complex tissue/organ engineering needs to take these facts and the natural development of the tissue/organ into account. Besides careful selection of stem cells or progenitors to be used, construction of a suitable smart scaffold, which is able to timely deliver the growth factors, signaling molecules, and other bioactive factors that play a crucial role in the development, vascularization and innervation of a certain tissue/organ is of great importance.

Scaffold material-based nanoparticle and microsphere technology enables regulated release of bioactive factors or cells that are encapsulated in scaffold material nanoparticles, by choosing scaffold material with a certain degradation rate. The various bioactive factor or cell containing nanoparticles can then be dispersed into the scaffold bulk. Smart biomaterials that may respond to temperature, pH, electrical signals or metabolites [11] may be used to control the timely release of the needed bioactive factors and various AT-MSCs or progenitors, which lead to timely regulated proliferation and differentiation. To provide vascularization and innervation, AT-MSC-derived endothelial progenitors, angiogenic factors, and nerve growth factors can be encapsulated in smart biomaterials that enable timely and subsequent release when required. Therefore future studies in developing various smart nanoparticles and microspheres that enable timely release of various bioactive factors and cells, followed by testing their functional use in various tissue/organ engineering will be very exciting.

Another means is by using growth factor or bioactive material encoding gene that is transferred into a certain type of cells either using a vector or plasmid [25]. Therefore future studies are highly needed to construct vectors and plasmid that can be controlled to be timely "on or off," thus to deliver the growth factors or bioactive materials subsequently, and only as long as they are needed.

23.6 Future Animal and Clinical Studies in Adipose Stem Cell Engineering

Designing future studies should take some considerations into account, such as the approach to be used, the type, structure, and natural development of the tissue/organ to be engineered, type of cell, scaffold requirement, integration into surrounding tissue, efficacy, and safety of the animal or clinical study.

23.6.1 Approach

Tissue/organ engineering in a broader sense may use various approaches. The very simple mode is by injecting cells into peripheral circulation, and expecting the cells to home to the tissues/organs to be repaired [6], or delivering the cells into the blood vessel near the site of injury/damage using smart catheters [56] or exactly to the site of injury/damage by injection or surgery [1]. Another approach is by implanting/injecting a scaffold that is seeded by cells and/or bioactive materials and the tissue/organ will develop in vivo [19, 21, 25] or by tissue/organ engineering ex vivo and then implant the engineered tissue/organ into the animal/patient [25].

In designing future studies, the approach plays a role in determining whether animal studies should be done. In an animal study, the used animal should mimic the clinical situation of the tissue/organ to be engineered as close as possible [25]. When the approach has been proven safe and effective in animals, or do not pose a great risk to the patient, clinical studies can begin.

For complex tissue/organs, ex vivo tissue/organ engineering may be suitable, as nutrient supply as well as growth and bioactive factors can be provided as required and suitable mechanical stimulation can be given, while for in vivo tissue/organ engineering, implanting scaffold construct needs good vascular supply on the site of implantation, and smart scaffold material to timely deliver the required growth and bioactive factors.

23.6.2 Type and Structure of Tissue/Organ

The type, structure, site, and surrounding condition of the damaged tissue/organ to be repaired by tissue/organ engineering play a role in the selection of the approach. Organs of low complexity such as heart or liver, which consist of one type of parenchymal cells and are only partly damaged do not need tissue/organ engineering, and may benefit from a simple approach by injecting a certain stem cells or progenitors into peripheral blood or the nearest supplying arteries, like have been done by many studies in regenerative medicine. However, questions arise concerning whether the stem cells really differentiate into the desired cells, or only provide the needed bioactive and growth factors and thus suitable microenvironment for resident stem cells to differentiate into the desired cells, or just develop new blood vessels and thus provide nutrient and oxygen supply to repair the damage. Further, for this type of approach, knowledge of homing mechanisms of the involved cells is an advantage. Therefore, study on homing mechanism of various kinds of cells into various organs is an exciting area to explore.

However, when a very complicated organ with various functional structures and cells such as the kidney is to be engineered, organ engineering should consider careful cell type selection, use of smart scaffold that can timely deliver various needed bioactive factors, careful design of the scaffold to mimic the natural organ architecture and mechanical property, and that allow blood vessels to grow to form the glomerulus and kidney vasculature, and later enable innervations to grow in the scaffold when the organ is implanted, so that the engineered organ can integrate and function well in the host.

23.6.3 Natural Development

Natural development of the tissue/organ to be engineered should be taken into account to get a success in an animal/clinical study. Therefore, knowledge of tissue/organ development cues in embryo such as the type of cells involved, cell–cell interactions, extracellular matrix, growth factors and bioactive factors, the "on and off" signaling pathways, and microenvironment that play a role in every stage of development of the tissue/organ is very important.

In designing future animal/clinical studies, subsequent timely delivery of growth and bioactive factors as in natural bio-environment, which may cause the "on or off" condition of certain signaling pathways should be carefully managed. This condition may lead to proliferation, differentiation, and extracellular matrix and bioactive factors secretion that may mimic the microenvironment as in natural development and finally benefit the development of the tissue/organ to be engineered.

23.6.4 Type of Cell

In adipose stem cell engineering, the selection of appropriate cell type from the various AT-MSCs, whether the AT-MSCs as a whole, certain enriched fraction, partly differentiated to progenitors, or fully differentiated into a mature type of cell should be decided. Studies on the identity and potentials of the various AT-MSCs may help in the decision. In the future, testing the various AT-MSCs in various tissue/organ engineering in animal studies is very attractive as a prerequisite for clinical studies.

Adipose tissue contains only 5,000 stem cells per gram [35], while the need for tissue/organ engineering is millions of stem cells, for instance, an animal study on adult male Sprague–Dawley rats with brain infarct due to acute ischemic stroke used 2.0×10^6 AT-MSCs [34], and a clinical trial on patients with complex perianal fistulas used 20–40 million AT-MSCs [20]. Therefore, the isolated stem cells should be expanded in vitro to get sufficient cells.

Various approaches can be used to expand the stem cells whether differentiated or undifferentiated, such as using conventional culture flask and medium, to large-scale expansion using various scaffolds [11] and bioreactors [8, 45]. For the purpose of expansion, various studies on the expansion of the various AT-MSCs, whether differentiated or undifferentiated, using various approaches are highly important to get a standardized protocol for each kind of the various AT-MSCs or their induced differentiation into various lineages without senescence. In addition, when undifferentiated AT-MSCs are needed, the protocol should ensure to prevent spontaneous differentiation and senescence, or determine the highest passage possible with acceptable spontaneous differentiation and senescence.

23.6.5 Scaffold

When a scaffold is needed, proper selection of scaffold material to tailor smart scaffold that mimics the natural structure and extracellular matrix property of the tissue/organ to be engineered is of great importance. Therefore, studies using various smart scaffold materials to construct natural macroscopic and microscopic architecture that is suitable for certain tissue/organ engineering are very prospective. In constructing the smart scaffold, it is also interesting to combine several smart biomaterials to achieve the properties and architecture of the natural extracellular matrix and timely delivery of various factors that happen in natural development of the tissue/organ to be engineered.

The overall macroscopic shape of a scaffold may interfere with nutrient supply. A thin scaffold of less than 100 μm may get enough nutrients from diffusion of the surrounding environment, but thicker scaffold of more than 100 μm needs to develop a capillary network inside the scaffold, and therefore should contain large interconnected pores. Vascular endothelial growth factor may be used to promote vascularization. Further, some scaffold materials such as porous polyvinyl alcohol or polyacrylate that contains traces of methacrylic acid can induce vascularization without the use of growth factors [45]. Studies on vasculogenesis in embryo revealed the role of various growth and bioactive factors [23]. Therefore, in the future, studies on developing capillary network and blood vessels in a large scaffold using various smart materials of various properties that can timely deliver the needed growth factors, and AT-MSC-derived endothelial and smooth muscle progenitors is of great importance for the survival of the construct.

23.6.6 Safety, Integration, and Efficacy

In tissue/organ engineering, animal studies and clinical trials using BM-MSCs and other kinds of cells are accumulating, and some good results in safety and efficacy were reported [25]. Therefore, it is of great importance that animal studies and clinical trials in adipose tissue engineering investigate the integration of the engineered tissue/organ, and short- and long-term safety and efficacy of the whole procedure.

23.6.6.1 Safety

Safety issues address the safety of the procedure including the absence of rejection problems. In designing a study, various issues that have been mentioned in previous sections should be taken into account. However, when the scaffold material turns out to be suitable for various issues, but immunogenicity, and/or allogeneic AT-MSCs are used, studies to overcome host immune response are very important. When nonimmunogenic scaffold is used, the cells may be incorporated in nanoparticles of the scaffold material. However, when the suitable scaffold material chosen is immunogenic, several pretreatments to protect the engineered tissue/organ are available, such as coating with polyethylene glycol, alginate, polysulfone, or heparin. A drawback to some of these approaches is a decrease in nutrient and oxygen diffusion [42]. Therefore, studies to develop methods of coating to evade host immune surveillance without decrease in nutrient and oxygen supply are of great importance.

Another approach is to protect the scaffold's or cell's immunogenic surface by attaching regulators of complement activation (RCA) such as: factor H, N-terminal short consensus repeats 1–5 of factor H that is alternatively called as decay accelerating factor, APT070 (a modified fragment of human complement receptor 1), or CR2 (CD21) through a linker such as Pluronic® that will attach to surfaces while retaining the surface activity. All of these approaches protect the implant from host innate immune response, either by preventing cascade system activation or binding the regulators of innate immune response, such as the RCA. Studies to find other mode of native immunogenic evasions are accumulating such as using peptides, which are part of streptococcal M protein that bind C4b-binding protein (a member of the RCA). In this approach, the peptides will capture RCA from plasma and thus provide autologous complement activation control [42]. However, most of these approaches have not been translated into animal or clinical studies, and future researches need to be conducted to prove whether they can really help to evade host innate immune response in animal studies and clinical trials of adipose stem cell engineering.

Though AT-MSCs themselves have immunosuppression potentials, cautions should be taken when using allogeneic AT-MSCs. Therefore, in the future, animal studies need to be conducted to prove whether allogeneic AT-MSCs are really safe in adipose stem cell engineering.

Moreover, as stem cells have various potentials, and even application of stem cells on a certain site cannot guarantee that they will stay on site and do not wander to unexpected sites, long-term safety studies are needed to monitor the possible side effects.

23.6.6.2 Integration and Efficacy

Clinical trials in tissue/organ engineering using AT-MSCs are just beginning. Some are recruiting participants [6, 12, 19, 49], one is ongoing but does not recruit more participants [7], one is not yet recruiting [18], two are suspended [26, 27], and one is completed [43]. A tissue engineering clinical trial on enterocutaneous fistulas due to Crohn's disease, using fibrin glue as scaffold material and expanded whole AT-MSCs or unexpanded cells of the stromal vascular fraction showed that tissue integration and functional efficacy was higher in the expanded cells receiving group [21]. Another clinical trial on complex perianal fistulas, using fibrin glue and AT-MSCs implants showed promising results, but there was a 17.6% recurrence rate after 1 year [20].

Therefore, in designing an animal study or a clinical trial, long-term efficacy should be considered, and long-term follow-up is needed to monitor recurrence. It is important to know whether the engineered tissue/organ will last for the whole life, or the treatment should be repeated after a certain time. Further, in vivo tissue/organ engineering or implanted ex vivo tissue/organ engineering product should be able to integrate with the surrounding tissue of the host to attain functional efficacy, and to survive.

Therefore, in the future, animal and clinical studies on adipose stem cell engineering should be based on successful animal/clinical studies on tissue/organ engineering using BM-MSCs or embryonic stem cell lines, and should include safety and efficacy studies.

23.7 Conclusion

Adipose stem cell engineering is very promising, but many studies are needed to reveal the identity of AT-MSCs, and the pathways that are involved in their differentiation, stemness, and senescence. That knowledge is useful in designing studies to overcome the problems concerning their expansion and differentiation. In addition, studies on various scaffolds and bioactive materials to tailor smart scaffold that are suitable for adipose stem cell engineering should be stress on their potential to engineer tissues/organs that can integrate with the surroundings and be functionally effective and safe.

References

1. Amado LC, Saliaris AP, Schuleri KH et al (2005) Cardiac repair with intramyocardial injection of allogeneic mesenchymal stem cells after myocardial infarction. Proc Natl Acad Sci USA 102:11474–11479
2. Arikawa E, Sun Y, Wang J et al (2008) Cross-platform comparison of SYBR® Green real-time PCR with TaqMan PCR, microarrays and other gene expression measurement technologies evaluated in the MicroArray Quality Control (MAQC) study. BMC Genomics 9:328, (12 pages) http://www.biomedcentral.com/1471-2164/9/328
3. Astori G, Vignati F, Bardelli S et al (2007) In vitro and multicolor phenotypic characterization of cell subpopulations identified in fresh human adipose tissue stromal vascular fraction and in the derived mesenchymal stem cells. J Transl Med 5:55, (10 pages) Available from: http://www.translational-medicine.com/content/5/1/55
4. Atala A, Bauer SB, Soker S et al (2006) Tissue-engineered autologous bladders for patients needing cystoplasty. Lancet 367:1241–1246
5. Balakrishnan B, Jayakrishnan A (2005) Self-cross-linking biopolymers as injectable in situ forming biodegradable scaffolds. Biomaterials 26:3941–3951
6. Barrenechea EA, Lucero FQ (2008) Safety and efficacy of autologous adipose-derived stem cell transplantation in patients with type 1 diabetes. http://clinicaltrials.gov/ct2/show/NCT00703599?term=adipose+tissue+and+stem+cell+and+clinical+trial&rank=8. Accessed 21 Sep 2010
7. Barrenechea EA, Lucero FQ (2008) Safety and efficacy of autologous adipose-derived stem cell transplantation in type 2 diabetics. http://clinicaltrials.gov/ct2/show/NCT00703612?term=adipose+tissue+and+stem+cell+and+clinical+trial&rank=9. Accessed 21 Sep 2010
8. Bieback K, Schallmoser K, Klüter H et al (2008) Clinical protocols for the isolation and expansion of mesenchymal stromal cells. Transfus Med Hemother 35:286–294
9. Bonab MM, Alimoghaddam K, Talebian F et al (2006) Aging of mesenchymal stem cell in vitro. BMC Cell Biol 7:14. doi:10.1186/1471-2121-7-14, http://www.biomedcentral.com/1471-2121/7/14
10. Caplan AI (2009) Why are MSCs therapeutic? new data: new insight. J Pathol 217:318–324
11. Chai C, Leong KW (2007) Biomaterials approach to expand and direct differentiation of Stem Cells. Mol Ther 15:467–480
12. Chem RC (2009) Autologous adipose-derived stem cell transplantation in patients with lipodystrophy. http://clinicaltrials.gov/ct2/show/NCT00715546?term=adipose+tissue+and+stem+cell+and+clinical+trial&rank=15. Accessed 21 Sep 2010
13. Chen G, Zhou P, Mei N et al (2004) Silk fibroin modified porous poly (epsiloncaprolactone) scaffold for human fibroblast culture in vitro. J Mater Sci Mater Med 15:671–677
14. CHSL news letter (2010) News and features. Researchers discover that one type of stem cell creates a niche for another type within bone marrow. http://www.cshl.edu/Article-Enikolopov/researchers-discover-that-one-type-of-stem-cell-creates-a-niche-for-another-type-within-bone-marrow. Accessed 2 Sep 2010
15. De Boer J, Wang HJ, Van Blitterswijk C (2004) Effects of Wnt signaling on proliferation and differentiation of human mesenchymal stem cells. Tissue Eng 10:393–401
16. Dominici M, Le Blanc K, Mueller I et al (2006) Minimal criteria for defining multipotent mesenchymal stromal cells. The International Society for Cellular Therapy position statement. Cytotherapy 8:315–317
17. Ebert R, Ulmer M, Zeck S et al (2006) Selenium supplementation restores the antioxidative capacity and prevents cell damage in bone marrow stromal cells in vitro. Stem Cells 24:1226–1235
18. Evans KFK (2008) The role of lipoaspirate injection in the treatment of diabetic lower extremity wounds and venous stasis ulcers. http://clinicaltrials.gov/ct2/show/NCT00815217?term=adipose+tissue+and+stem+cell+and+clinical+trial&rank=17. Accessed 21 Sep 2010
19. Garcia-Olmo D (2009) Long-term safety and efficacy of adipose-derived stem cells to treat complex perianal fistulas in patients participating in the FATT-1 randomized controlled trial. http://clinicaltrials.gov/ct2/show/NCT01020825?term=adipose+tissue+and+stem+cell+and+clinical+trial&rank=18. Accessed 21 Sep 2010
20. Garcia-Olmo D, Herreros D, Pascual I et al (2009) Expanded adipose-derived stem cells for the treatment of complex perianal fistula: a phase II clinical trial. Dis Colon Rectum 52:79–86
21. Garcia-Olmo D, Herreros D, Pascual M et al (2009) Treatment of enterocutaneous fistula in Crohn's Disease with adipose-derived stem cells: a comparison of protocols with and without cell expansion. Int J Colorectal Dis 24(1):27–30
22. Haniffa MA, Collin MP, Buckley CD et al (2009) Mesenchymal stem cells: the fibroblasts' new clothes? Haematologica 94:258–263

23. Hanjaya-Putra D, Gerecht S (2009) Vascular engineering using human embryonic stem cells. Biotechnol Prog 25:2–9
24. Hipp J, Atala A (2004) Tissue engineering, stem cells, cloning, and parthenogenesis: new paradigms for therapy. J Exp Clin Assist Reprod 1:3, (10 pages) http://www.jexpclinassistreprod.com/content/1/1/3
25. Ikada Y (2006) Challenges in tissue engineering. J R Soc Interface 3:589–601
26. Kaneko S (2010) Liver regeneration therapy by intrahepatic arterial administration of autologous adipose tissue derived stromal cells. http://clinicaltrials.gov/ct2/show/NCT01062750?term=adipose+tissue+and+stem+cell+and+clinical+trial&rank=6. Accessed 21 Sep 2010
27. Kaneko S (2010) Liver regeneration therapy using autologous adipose tissue derived stromal cells. http://clinicaltrials.gov/ct2/show/NCT00913289?term=adipose+tissue+and+stem+cell+and+clinical+trial&rank=5. Accessed 21 Sep 2010
28. Kern S, Eichler H, Stoeve J et al (2006) Comparative analysis of mesenchymal stem cells from bone marrow, umbilical cord blood, or adipose tissue. Stem Cells 24:1294–1301
29. Kilian O, Flesch I, Wenisch S et al (2004) Effects of platelet growth factors on human mesenchymal stem cells and human endothelial cells in vitro. Eur J Med Res 9:337–344
30. Kogler G, Sensken S, Wernet P (2006) Comparative generation and characterization of pluripotent unrestricted somatic stem cells with mesenchymal stem cells from human cord blood. Exp Hematol 34:1589–1595
31. Kucia M, Reca R, Campbell FR et al (2006) A population of very small embryonic-like (VSEL) CXCR4(+)SSEA-1(+) Oct-4(+) stem cells identified in adult bonemarrow. Leukemia 20:857–869
32. Kurosawa H, Kimura M, Noda T et al (2006) Effect of oxygen on in vitro differentiation of mouse embryonic stem cells. J Biosci Bioeng 101:26–30
33. Le Blanc K, Ringden O (2007) Immunomodulation by mesenchymal stem cells and clinical experience. J Intern Med 262:509–525
34. Leu S, Lin YC, Yuen CM et al (2010) Adipose-derived mesenchymal stem cells markedly attenuate brain infarct size and improve neurological function in rats. J Transl Med 8:63
35. Lin G, Garcia M, Ning H et al (2008) Defining stem and progenitor cells within adipose tissue. Stem Cells Dev 17:1053–1064
36. Liu X, Yin D, Zhang Y et al (2007) Vascular endothelial cell senescence mediated by integrin beta4 in vitro. FEBS Lett 581:5337–5342
37. Lysy PA, Smets F, Sibille C et al (2007) Human skin fibroblasts: From mesodermal to hepatocyte-like differentiation. Hepatology 46:1574–1585
38. Mann BK, West JL (2002) Cell adhesion peptides alter smooth muscle cell adhesion, proliferation, migration, and matrix protein synthesis on modified surfaces and in polymer scaffolds. J Biomed Mater Res 60:86–93
39. Mitchel JB, Mc Intosh K, Zvonic S et al (2006) Immunophenotype of human adipose-derived cells: temporal changes in stromal-associated and stem cell–associated markers. Stem Cells 24:376–385
40. Moutos FT (2006) A biomimetic 3-D woven composite scaffold that recreates the anisotropic, nonlinear and viscoelastic behavior of articular cartilage. Trans Orthop Res Soc 31:788
41. Ng F, Boucher S, Koh S et al (2008) PDGF, TGF-beta, and FGF signaling is important for differentiation and growth of mesenchymal stem cells (MSCs): transcriptional profiling can identify markers and signaling pathways important in differentiation of MSCs into adipogenic, chondrogenic, and osteogenic lineages. Blood 112(2):295–307
42. Nilsson B, Korsgren O, Lambris JD et al (2010) Can cells and biomaterials in therapeutic medicine be shielded from innate immune recognition? Trends Immunol 31:32–38
43. Oh GS (2009) Safety and efficacy of autologous cultured adipocytes in patient with depressed scar. http://clinicaltrials.gov/ct2/show/NCT00992147?term=adipose+tissue+and+stem+cell+and+clinical+trial&rank=10. Accessed 21 Sep 2010
44. Patel SA, Sherman L, Munoz J et al (2008) Immunological properties of mesenchymal stem cells and clinical implications. Arch Immunol Ther Exp 56:1–8
45. Placzek MR, Chung IM, Macedo HM et al (2009) Stem cell bioprocessing: fundamentals and principles. J R Soc Interface 6:209–232
46. Planat-Bénard V, Menard C, André M et al (2004) Spontaneous cardiomyocyte differentiation from adipose tissue stroma cells. Circ Res 94:223–229
47. Planat-Benard V, Silvestre JS, Cousin B et al (2004) Plasticity of human adipose lineage cells toward endothelial cells. Physiological and therapeutic perspectives. Circulation 109:656–663
48. Rodríguez LV, Alfonso Z, Zhang R et al (2006) Clonogenic multipotent stem cells in human adipose tissue differentiate into functional smooth muscle cells. PNAS 103:12167–12172
49. Ruiz-Salmeron RJ, de la Cuesta A (2010) Intra arterial infusion of autologous mesenchymal stem cells from adipose tissue in diabetic patients with chronic critical limb ischemia. http://clinicaltrials.gov/ct2/show/NCT01079403?term=adipose+tissue+and+stem+cell+and+clinical+trial&rank=2. Accessed 21 Sep 2010
50. SABioscience (2008) Human mesenchymal stem cell PCR array. http://www.sabiosciences.com Accessed 29 Sep 2008
51. SABiosciences (2008) Cell development and differentiation miRNA PCR array. http://www.sabiosciences.com/mirna_pcr_product/HTML/MAH-103A.html Accessed 21 Jan 2009
52. SABiosciences (ed) (2008) User manual. Multi-analyte profiler ELISArray™ kit multi-protein profiling ELISA kits. SABiosciences Co, Frederick
53. Salasznyk RM, Williams WA, Boskey A et al (2004) Adhesion to vitronectin and collagen I promotes osteogenic differentiation of human mesenchymal stem cells. J Biomed Biotechnol 2004:24–34
54. Savitz SI, Dinsmore JH, Wechsler LR et al (2004) Cell therapy for stroke. NeuroRx 1:406–414
55. Schallmoser K, Bartmann C, Rohde E et al (2007) Human platelet lysate can replace fetal bovine serum for clinical-scale expansion of functional mesenchymal stromal cells. Transfusion 47:1436–1446
56. Silva GV, Perin EC, Dohmann HFR et al (2004) Catheter–based transendocardial delivery of autologous bone-marrow-derived mononuclear cells in patients listed for heart transplantation. Tex Heart Inst J 31:214–219

57. Sotiropoulou PA, Perez SA, Salagianni M et al (2006) Characterization of the optimal culture conditions for clinical scale production of human mesenchymal stem cells. Stem Cells 24:462–471
58. Tang F, Hajkova P, Barton SC et al (2006) MicroRNA expression profiling of single whole embryonic stem cells. Nucleic Acids Res 34(2):e9
59. Tuan RS, Boland G, Tuli R (2003) Adult mesenchymal stem cells and cell-based tissue engineering. Arthritis Res Ther 5:32–45
60. Uebersax L, Hagenmuller H, Hofmann S et al (2006) Effect of scaffold design on bone morphology in vitro. Tissue Eng 12:3417–3429
61. Wagner W, Wein F, Seckinger A et al (2005) Comparative characteristics of mesenchymal stem cells from human bone marrow, adipose tissue, and umbilical cord blood. Exp Hematol 33:1402–1416
62. Willerth SM, Sakiyama-Elbert SE (2008) Combining stem cells and biomaterial scaffolds for constructing tissues and cell delivery (July 09, 2008).In: StemBook (ed) The stem cell research community. StemBook. doi:10.3824/stembook.1.1.1, http://www.stembook.org. Accessed 21 Sep 2010
63. Yañez R, Lamana ML, García-Castro J, et al (2006) Adipose tissue-derived mesenchymal stem cells (AD-MSCs) have immunosuppressive properties applicable for the in vivo control of the graft-versus-host disease (GvHD). Stem cells online doi: 10.1634/stemcells.2006-0228. 33pages. http://www.StemCells.com
64. Yang H, Miller WM, Papoutsakis ET (2002) Higher pH promotes megakaryocytic maturation and apoptosis. Stem Cells 20:320–328
65. Yang J, Shi G, Bei J et al (2002) Fabrication and surface modification of macroporous poly(L-lactic acid) and poly(L-lactic-coglycolic acid) (70/30) cell scaffolds for human skin fibroblast cell culture. J Biomed Mater Res 62:438–446
66. Zuk PA, Zhu M, Ashjian P et al (2002) Human adipose tissue is a source of multipotent stem cells. Mol Biol Cell 13:4279–4295

Conclusions and Future Directions

24

Yves-Gérard Illouz and Aris Sterodimas

Recent years have seen an almost unprecedented excitement about medical research. It was well caught by Aristotle 2500 years ago in the opening line of his *Metaphysics*, when he wrote that "all men by nature desire to know," and broadened by Francis Bacon in the sixteenth century to encompass the power of science to improve the human condition.

The early practice of medicine relied largely on palliative management of pain and distress. As science has been evolved, pharmaceuticals to change the body's physiology, vaccines to prevent communicable diseases, surgery to remove diseased parts, have largely remained the standard medical therapies.

Until very recently, most scientists and clinicians believed that damaged or diseased human tissue could only be replaced by donor transplants or with totally artificial parts. Tissue engineering promises a more advanced approach in which organs or tissues can be repaired, replaced, or regenerated for more targeted solutions. This approach also responds to clinical needs that cannot be met by organ donation alone. As with all research, our ability even to contemplate the possibilities offered by tissue engineering-derived therapies is a result of many years of research. This science of tissue engineering dates to the mid-1970s, and many papers have been published in the literature. The development of new research tools, new knowledge about pathways of cell differentiation are opening new ways in cell transplantation therapy for human diseases.

Liposuction was devised as a means of removing fat from the body for cosmetic purposes safely and using small incisions (Chaps. 3 and 4). The procedure has been improved over the years, making liposuction the most common performed aesthetic procedure worldwide. Autologous fat tissue has been considered as an ideal filler for soft tissue augmentation because it is biocompatible, versatile, natural-appearing, readily available, abundant, inexpensive, and can be harvested easily and repeatedly, with minimal trauma to the donor sites (Chap. 5). In the last 25 years, several different techniques of lipoinjection have been developed to correct various problems such as buttocks, trochanteric depressions, breast augmentation, scar depressions, thighs and legs, small wrinkles and depressions of the face, nasolabial fold, upper outer breast quadrant, liposuction sequel (Chap. 6). A large volume of adipose tissue can be easily harvested through liposuction. Mesenchymal adherent stem/progenitor cells, similar to those from bone marrow, can be extracted from liposuction aspirates with a high efficiency and thus adipose tissue is now regarded as a potential source of adult stem/progenitor cells for regenerative medicine. Many features of adipose stem/progenitor cells, such as physiological functions and localization, have been clarified in the past decade (Chap. 7). The stromal enriched lipograft has already been used in clinical practice and the results have proved superior to traditional autologous fat grafting (Chap. 16).

Y.-G. Illouz
Plastic Surgery Department, Saint Louis Hospital, Avenue Claude-Vellefaux, Paris, France
e-mail: yillouz2000@hotmail.com

A. Sterodimas (✉)
Plastic and Reconstructive Surgery Department, IASO General Hospital, 264 Mesogeion Avenue, Athens, Greece
e-mail: aris@sterodimas.com; www.sterodimas.com

Y.-G. Illouz and A. Sterodimas (eds.), *Adipose Stem Cells and Regenerative Medicine*,
DOI: 10.1007/978-3-642-20012-0_24, © Springer-Verlag Berlin Heidelberg 2011

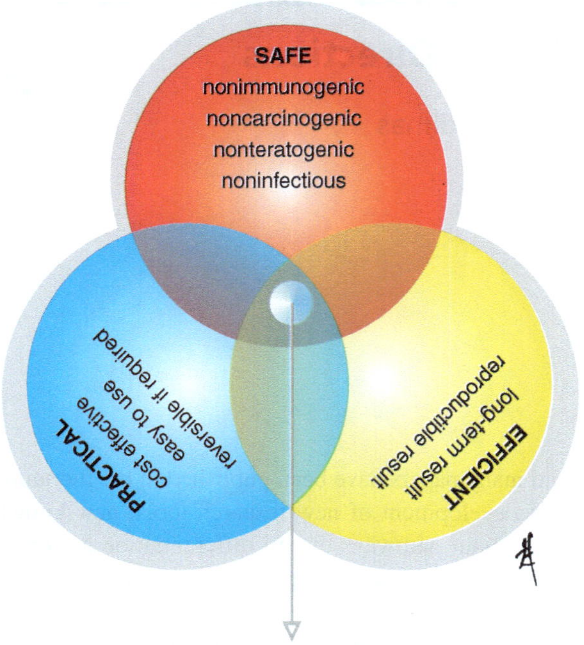

Fig. 24.1 Tissue engineering principles

Bioengineers, life scientists, reconstructive surgeons and physicians across all specialties are synergistically coupling expertise in areas such as cell culture technology, tissue transfer, cell differentiation, angiogenesis, computer modeling, and polymer chemistry to use adipose tissue as a base for tissue regeneration and augmentation. Tissue engineering principles are shown in Fig. 24.1. Adipose tissue is just like a new continent waiting for exploration. It is not a "lazy," only fat storing region; it is a dynamic hard working tissue (Chap. 1). It is like a treasure offering the widest treatment potential and also being the most abundant, fertile, easily accessible tissue for "cell therapists." With the support of basic and clinical studies, adipose tissue seems to be rising as a distinguishing tissue. Adipose-derived stem cells (ADSCs) are becoming the cells of choice for an increasing number of clinical trials and they promise to be in the next decade the preferential cell type used in cell therapies. There is a growing body of experimental evidence from both in vitro and in vivo studies demonstrating the multipotentiality of ADSCs from adipose tissue isolated from humans and other species. Employing stem cell-inclusive tissue engineering methods as a means to generate adipose tissue may offer patients a new reconstructive option using their own healthy cells and may offer a clinically feasible "just-in-time" solution (Chap. 9). The properties of an ideal biomaterial for tissue engineering are shown in Fig. 24.2.

Adipose tissue is a dynamic "player" in endocrine physiology and serves as a source of cytokine secretion (Chap. 10). The ability of ADSCs to secrete several biologic factors plus evidence at a basic science level lends way to ADSCs playing a major role in tissue engineering and organ regeneration. Therapeutic interventions of obesity might be achieved by: targeting vasculature through angiogenesis inhibitors leading to regression of preexisting adipose vasculature and shrinkage of fat mass; isolation and transplantation of a homogenous adipose stem cell population giving rise to mature brown cells; inducing pluripotent stem cells in vitro that could involve the overexpression of adipokines or the promotion of a beneficial brown cell fate (Chap. 13).

ADSCs contain angiogenic and vasculogenic cytokines with regenerative properties, and there has been considerable interest in testing the impact of ADSCs on acute disorders. Transient ischemic injury underlies the pathophysiology of myocardial infarction, stroke, renal ischemia, obstruction of the small intestine, and related conditions associated with trauma to the brain, spinal cord, and extremities. In the myocardium, several recent studies demonstrate the benefit of ADSCs to include direct contribution to neovasculogenesis and de novo cardiogenesis, in addition to indirect paracrine action for cardioprotection, neoangiogenesis, and neuron spindle formation. Furthermore, treatment with ADSCs has demonstrated a diminished propensity for arrhythmogenic risk without any evidence for uncontrolled growth, systemic complications, or tumorigenic change (Chap. 14). Engrafted adipose stem cells also improve the myocardial function with the capacity for contribution both via direct differentiation and indirect regenerative paracrine signaling in the host myocardial microenvironment. Although the differentiation of ADSCs to neuron did not achieve remarkable success, Schwann cell differentiation of ADSCs in vitro and in vivo and beneficial roles of ADSCs in the reduction of inflammation and apoptosis enhanced angiogenesis, and nerve regeneration promise future alternatives to naïve neural and glial cells for neural disease or disorder treatments (Chap. 15). Theoretically, acute diseases present situations where rapid intervention can most efficiently prevent the development of long-term

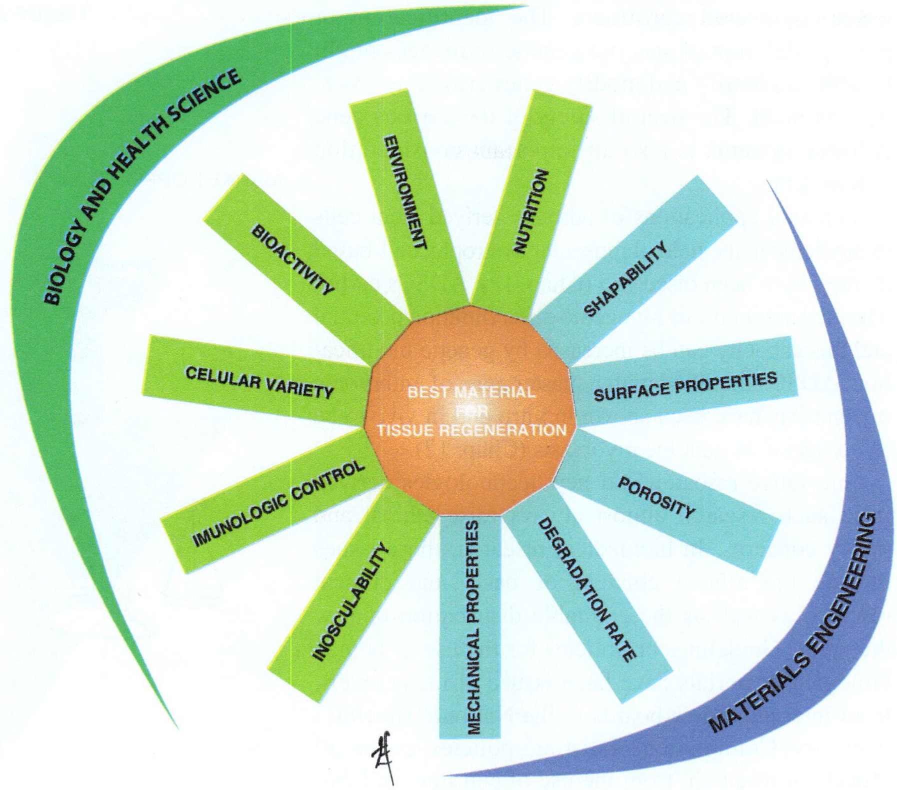

Fig. 24.2 The properties of an ideal biomaterial for tissue engineering

pathologies. In contrast, chronic diseases may be less amenable to cell-based therapies. Preclinical animal models suggest that ADSCs can be used to treat type 1 diabetes mellitus, chronic heart failure, and neurodegenerative diseases due to inborn errors of metabolism. Since these chronic diseases account for the majority of healthcare costs in most national economies, there is interest in using ADSCs for their treatment. There is particular interest in the use of ADSCs and adipose transplantation therapy for lipodystrophy and associated metabolic disorders such as diabetes. After preclinical animal model data supporting the safety and efficacy of these approaches, it has been possible to test ADSCs in important human conditions. A positive outcome in a single disease alone could have substantial economic and public health benefits.

The use of adipose stem cells for tissue-engineering applications brings numerous concerns to address, including standardization of methods for tissue preparation, cell isolation, cell culture and nutrient support, and vascularization (Chaps. 2 and 8). The methods for harvesting the tissue may have an effect on the ability of the cells to proliferate and differentiate during in vitro culture. ADSCs differentiate into epithelial-like cells when transplanted into kidneys and their epithelial-like and smooth muscle differentiations in the bladder have been reported. Treatment of stress urinary incontinence by injecting ADSCs into the urethra has been demonstrated in animal models and human patients. Improvement of erectile function in animal models of different types of erectile dysfunction has been also reported (Chap. 18). Tissue engineering strategies can combine cell delivery with appropriate scaffold biomaterials that can mimic the unique physiological and biomechanical properties of ear, nose, and throat. The potential role of mesenchymal stem cells derived from adipose tissue as a source of stem cells for oral tissues regeneration has been also reported (Chaps. 19 and 20). ADSCs with the proper triggering can be converted to osteoblasts, chondrocytes, tenocytes, and myoblasts (Chap. 17). Advances in tissue engineering for the production of solid tissues/organs using ADSCs as building blocks have been reported (Chap. 21).

Genetic engineering and the use of adult stem cells may hold the key to future development of tissue-engineered constructs. The identification or perhaps deletion of specific genetic sequences might be able to identify and modify genes critical to tissue development. The overall safety of the various gene delivery systems is also an important consideration (Chap. 11).

Potential applications of adipose-derived stem cells in medicine in the field of muscular dystrophy cell-based therapy have been discussed (Chap. 12). ADSCs harbor a limited autonomous myogenic differentiation potential and this capacity can be increased by genetic modification. ADSCs modified with a master gene of embryonic myogenesis have been shown in vitro and in vivo to be as myogenic as genuine myoblasts (Chap. 12).

Innovative research and new technologies derived from such research almost always raise ethical and policy concerns. In biomedical research, these issues include the ethical conduct of basic and clinical research as well as the equitable distribution of new therapies. Guidelines or policies for the use of human biological materials have been issued at many levels, from internal review boards to the National Bioethics Advisory Commission. Existing policies cover all aspects of research, from the use of cell lines in laboratories, to human subject protections, that will surface in the consideration of stem cell research. It is essential that there be a public that is educated and informed about the ethical and policy issues raised by stem cell research and its applications. Informed public discussion of these issues should be based on an understanding of the science associated with stem cell research, and it should involve a broad cross-section of society. The policy issues raised by stem cell research are not unique, but this research has received a significant amount of public attention and there is much to gain by open reflection on the implications of this sensitive area of research (Fig. 24.3).

Over the last decade, the potential advantages of using adult adipose tissue-derived stem cells have become clear to the industrial sector (Chap. 22). Further research is necessary to investigate on safety, efficacy, and ideal applications of adipose-derived stem cells. This might lead to major investment and funding from the industrial sector with interest in adipose stem cells processing and engineering for tissue repair. The application of ADSCs in the field of regenerative medicine has made progressive advances including recent clinical applications. The clinical

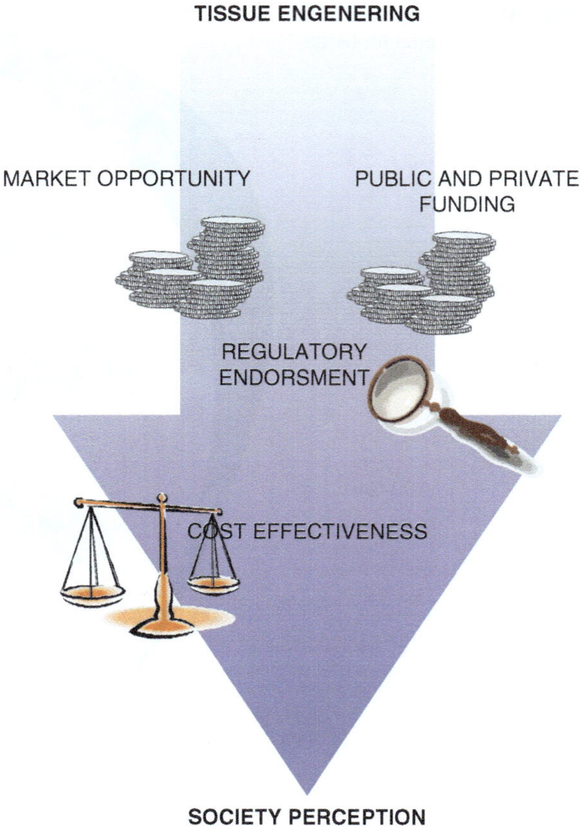

Fig. 24.3 Adipose tissue engineering and public perception

investigators who have published their findings have always addressed the important regulatory questions before advancing from animal models to clinical setting. International standardization of ADSCs would significantly advance clinical translation. The next steps will include documenting the reproducibility of the current preclinical and clinical findings, and controlled testing of the safety and efficacy of ADSCs in a range of human conditions (Chap. 23). Although the current animal applied strategies for adipose tissue engineering have advanced in the last years, complete understanding of the mechanisms of interactions among adipose stem cells, growth factors, and biomaterials in tissue engineering is needed in order to advance the end goal of developing "off-the-shelf" tissue engineering products (Chap. 16).

The enthusiasm over what unquestionably represents a markedly innovative technique with huge therapeutic potential must be balanced though against stringent standards of scientific and clinical investigation.

Index

A
Abdominoplasty, 22, 23, 25–27
Adipose derived stem cells, 142, 147–149, 156–158, 160, 166–176, 181–189, 198–218, 222, 224, 225, 229–240, 274, 276, 357–370
Adiposeness, 134
Adipose stem cell research, 114–115
Adipose stem cells, 9–10, 103–108, 111–118, 123
Adipose tissue, 1–10, 13–19, 22, 31–33, 37, 39, 48, 52, 67–80
Adipose tissue organ, 133–134
Angiogenesis, 135–137
Angiogenic factors, 104–106
Autologous fat graft, 57–63

B
Biomaterial, 93–100
Biomolecules, 17, 103–108
Bladder, 199–200
Bone marrow derived stem cells, 142, 149
Bone regeneration, 182–184
Brain ischemic stroke, 159–160
Breast reduction, 36

C
Cell-based myocardial regeneration, 143–145
Cell biology, 3–5
Centrifugation, 246, 249, 250, 252
Chondrogenesis, 184–186

D
Dental pulp, 222–226
Dentistry, 221–227

F
Facial contour, 60, 61
Fat grafting, 45–51, 53
Fat processing, 246–249
Fat survival, 48
Frozen storage, 246–247
Future animal studies, 269
Future clinical studies, 267–270

G
Gene therapy, 111–118
Genetic modification, 130
Gluteal augmentation, 60, 61
Gynecomastia, 36

H
Heart disease, 142, 148, 149

K
Kidney, 198–199

L
Ligament regeneration, 186–188
Liposuction, 19–28, 67–80

M
Mesenchymal stem cells, 83, 86–89, 91, 92
Methods of liposuction, 38
Micro fat grafting, 67, 76–79
Muscle regeneration, 188

N
Neural induction, 157–159

O
Obesity, 133–138

P
Penis, 198, 201
Periodontal ligament (PDL), 221, 222, 225–226
Pharmacological applications, 121–131
Photmodulation, 247
Plastic and reconstructive surgery, 165–177

R
Regenerative medicine, 103, 104, 111, 113, 114, 166, 229, 238–240, 257, 263, 267, 273, 276

S
Scaffold, 93–100
Solid organ transplantation, 229–240
Spinal cord injury, 158–160

Y.-G. Illouz and A. Sterodimas (eds.), *Adipose Stem Cells and Regenerative Medicine*,
DOI: 10.1007/978-3-642-20012-0,

Stem cell bio-banking, 84–85
Stem cell culture, 16–17
Stem cells, 67–80, 133–138
Storage of adipose-derived stem cells, 83–92
Stromal enriched lipograft (SEL), 63, 167–170, 176
Stromal vascular fraction, 193–195
Syringe liposculpture, 46–47, 51, 54

T
Testes, 198, 201–202
Tissue engineering, 5, 9–10, 15, 93–95, 99–100, 229–240, 273–276
Tissue regeneration, 79, 80

U
Urology, 198–218

GPSR Compliance

The European Union's (EU) General Product Safety Regulation (GPSR) is a set of rules that requires consumer products to be safe and our obligations to ensure this.

If you have any concerns about our products, you can contact us on ProductSafety@springernature.com

In case Publisher is established outside the EU, the EU authorized representative is:

Springer Nature Customer Service Center GmbH
Europaplatz 3
69115 Heidelberg, Germany

Batch number: 10731293

Printed by Printforce, the Netherlands